JPL Publication 96-9

GaAs Solar Cell Radiation Handbook

B. E. Anspaugh

July 1, 1996

National Aeronautics and
Space Administration

Jet Propulsion Laboratory
California Institute of Technology
Pasadena, California

The research described in this publication was carried out by the Jet Propulsion Laboratory, California Institute of Technology, and was sponsored by the U.S. Air Force Wright Patterson Laboratory and the National Aeronautics and Space Administration

TABLE OF CONTENTS

History of GaAs Solar Cell Development 1-1
    1.1   Advantages of Gallium Arsenide 1-1
    1.2   Early Work 1954 - 1964 1-3
    1.3   Hiatus 1964 - 1972 1-8
    1.4   Rebirth 1972 - 1-10
    1.5   Development of LPE Growth Techniques 1-11
    1.6   Development of OMCVD Growth Techniques 1-15
    1.7   GaAs Solar Cells in Space 1-18
           References for Chapter 1 1-21

Photovoltaic Equations 2-1
    2.1   Basic Solar Cell Equation 2-1
    2.2   Diffusion Current 2-4
    2.3   Generation-Recombination Current 2-7
    2.4   Temperature Dependence of I_{sc} 2-11
    2.5   Spectral Response for p/n Solar Cells 2-12
    2.6   Spectral Response for p/n Solar Cells with a $Ga_{1-x}Al_xAs$ Window 2-16
           References for Chapter 2 2-20

Instrumentation Techniques for Measuring GaAs Solar Cells 3-1
    3.1   Light Sources and Solar Simulators 3-1
    3.2   Current-Voltage Characteristics 3-9
    3.3   Special Considerations for GaAs Solar Cells 3-11
    3.4   Spectral Response Measurements 3-13
    3.5   Diffusion-Length Measurements 3-15
    3.6   Irradiation Methods 3-16
           References for Chapter 3 3-18

General Radiation Effects 4-1
    4.1   The Theory of Radiation Damage 4-1
          4.1.1   Displacement Damage 4-2
          4.1.2   Atomic Displacements 4-5
          4.1.3   Primary Displacement Cross Sections 4-7
          4.1.4   Secondary Displacements 4-9
          4.1.5   Ionization 4-12
           References for Chapter 4 4-18

Radiation Effects in Solar Cells 5-1
    5.1   Dependence on Lifetime and Diffusion Length 5-1
    5.2   The Concept of Damage Equivalence 5-5
    5.3   Calculation of Damage Coefficients for Space Radiation 5-7
    5.4   Electron Space Radiation Effects 5-10
    5.5   Proton Space Radiation Effects 5-12
    5.6   Miscellaneous Radiation Effects 5-21
          5.6.1   Low Energy Proton Irradiation of Partially Covered Solar Cells 5-21

5.6.2   Radiation Rate Effects 5-23

5.6.3   Annealing of Irradiated GaAs Cells 5-25

References for Chapter 5 5-29

Electrical Performance of GaAs Solar Cells 6-1

6.1   1 MeV Electron Radiation Data 6-1

6.2   GaAs Solar Cell Behavior with Temperature and Solar Intensity 6-34

Spacecraft Flight Data for GaAs Solar Arrays 7-1

7.1   NTS-II 7-1

7.2   LIPS-II 7-1

7.3   LIPS-III 7-1

7.4   High Efficiency Solar Panel (HESP) on the Combined Release and Radiation
Effects Satellite (CRRES) 7-2

7.5   EURECA 7-4

7.6   STRV-1 A and B 7-5

7.7   UoSAT-5 7-5

7.8   PASP-PLUS Experiment on the Advanced Photovoltaic and Electronic
Experiments (APEX) Spacecraft 7-6

7.9   Advanced Solar Cell Orbit Test (ASCOT) Flight Experiment 7-6

References for Chapter 7 7-12

The Space Radiation Environment 8-1

8.1   Geomagnetically Trapped Radiation 8-1

8.1.1   Trapped Protons 8-5

8.1.2   Trapped Electrons 8-6

8.2   Trapped Radiation at Other Planets 8-8

8.3   Solar Flare Protons 8-9

References for Chapter 8 8-18

GaAs Solar Array Degradation Calculations 9-1

9.1   General Procedure 9-1

9.2   Rear-Incidence Radiation 9-3

9.3   Rough Degradation Calculations 9-3

9.4   Computer-Calculated Equivalent Fluence 9-5

9.5   Example of Calculation for Rear-Incidence Radiation on a Thin Solar Panel . . . 9-7

References for Chapter 9 9-63

APPENDIX A-1

LIST OF FIGURES

Figure 1.1.   Optical Absorption Coefficients for GaAs and Silicon   1-1

Figure 1.2.   Carrier Concentrations in GaAs and Silicon with AM0 Spectrum Incident   1-2

Figure 1.3.   Maximum Efficiency η_{max} as a Function of Energy Gap E_G for Various
              Temperatures (© 1973 IEEE, used with permission)   1-4

Figure 1.4.   GaAs Solar Cell Efficiency vs. Surface Recombination Velocity (from Ellis
              and Moss)   1-9

Figure 1.5.   Energy Band Diagram of an $Al_xGa_{1-x}As/GaAs$ Solar Cell   1-11

Figure 1.6.   Design of Hughes Research Lab's AlGaAs/GaAs Solar Cell   1-12

Figure 1.7.   GaAs Solar Cell Structure Developed by ASEC   1-16

Figure 1.8.   I-V Curves for a GaAs/Ge Solar Cell with an Active GaAs/Ge Junction as
              Measured Under Different Simulators (© 1990 IEEE, used with permission) . .   1-17

Figure 2.1.   Equivalent Circuit of an Illuminated Solar Cell   2-1

Figure 2.2.   Current-Voltage Curves of a Solar Cell under Illumination and in the Dark . . .   2-3

Figure 2.3.   Energy Band Diagram in Equilibrium of a $p-Ga_{1-x}Al_xAs/p-GaAs/n-GaAs$
              Device   (© 1973 IEEE, used with permission)   2-17

Figure 3.1.   Comparison of the NASA SP 8005 and the Willson Composite AM0 Solar
              Spectra   3-3

Figure 3.2.   Normalized Spectral Irradiance of Spectrolab X25 Mark II Solar Simulator . . .   3-4

Figure 3.3.   Normalized Spectral Irradiance of Spectrolab XT-10 Solar Simulator   3-5

Figure 3.4.   Normalized Spectral Irradiance of ASEC Tungsten/Xenon Solar Simulator   3-6

Figure 3.5.   Normalized Spectral Irradiance of ELH Tungsten-Halogen Lamp   3-7

Figure 3.6.   Normalized Spectral Irradiance of the JPL LAPSS with and without Schott
              UV Filter   3-8

Figure 3.7.   Comparison of the Irradiance of an X-25 Solar Simulator with the Spectral
              Response of a GaAs Solar Cell   3-13

Figure 4.1.   Zinc-blende structure of GaAs (from Pons and Bourgoin)   4-6

Figure 4.2.   Stopping Power and Range Curves for Electrons in GaAs
              Reference: S.M. Seltzer, EPSTAR Program   4-15

Figure 4.3.   Stopping Power and Range Curves for Protons in GaAs
              Reference: J.F. Ziegler, TRIM Program   4-16

Figure 5.1.   Normalized Degradation of I_{sc} for GaAs/Ge Solar Cells vs. Proton and
              Electron Fluence   5-6

Figure 5.2.   Relative Damage Coefficients for Electron Irradiation of GaAs/Ge Solar Cells   5-11

Figure 5.3.   Geometries Used in Omnidirectional Proton Damage Coefficient Calculation .   5-12

Figure 5.4.   Experimental I_{sc} and V_{oc}, P_{max} Relative Damage Coefficients for GaAs/Ge
              Solar Cells and Normally Incident Proton Irradiation, Normalized to 10 MeV .   5-15

Figure 5.5. I_{sc} Relative Damage Coefficients for Omnidirectional Space Proton Irradiation
              of Shielded GaAs/Ge Solar Cells   5-16

Figure 5.6. V_{oc} or P_{max} Relative Damage Coefficients for Omnidirectional Space Proton
              Irradiation of Shielded GaAs/Ge Solar Cells   5-17

Figure 5.7.   GaAs/Ge Solar Cell and Coverglass Geometry Used in Low Energy Proton
              Gap Irradiations   5-22

Figure 5.8.   GaAs/Ge Solar Cell Performance after Irradiation with 500 keV Protons to a
              Fluence of 1×10^{13} p/cm^2   5-23

Figure 5.9. P_{max} Annealing of Electron and Proton Irradiated Hughes LPE GaAs Solar
              Cells   5-27

Figure 6.1. I_{sc}/cm^2 vs. 1 MeV Electron Fluence for ASEC GaAs/Ge Solar Cells 6-4
Figure 6.2. V_{oc} vs. 1 MeV Electron Fluence for ASEC GaAs/Ge Solar Cells 6-5
Figure 6.3. P_{max}/cm^2 vs. 1 MeV Electron Fluence for ASEC GaAs/Ge Solar Cells 6-6
Figure 6.4. I_{mp}/cm^2 vs. 1 MeV Electron Fluence for ASEC GaAs/Ge Solar Cells 6-7
Figure 6.5. V_{mp} vs. 1 MeV Electron Fluence for ASEC GaAs/Ge Solar Cells 6-8
Figure 6.6. Normalized I_{sc} vs. 1 MeV Electron Fluence for ASEC GaAs/Ge Solar Cells . . . 6-9
Figure 6.7. Normalized V_{oc} vs. 1 MeV Electron Fluence for ASEC GaAs/Ge Solar Cells . 6-10
Figure 6.8. Normalized P_{max} vs. 1 MeV Electron Fluence for ASEC GaAs/Ge Solar Cells 6-11
Figure 6.9. Normalized I_{mp} vs. 1 MeV Electron Fluence for ASEC GaAs/Ge Solar Cells . 6-12
Figure 6.10 Normalized V_{mp} vs. 1 MeV Electron Fluence for ASEC GaAs/Ge Solar Cells . 6-13
Figure 6.11. I_{sc}/cm^2 vs. 1 MeV Electron Fluence for Spectrolab GaAs/Ge Solar Cells 6-14
Figure 6.12. V_{oc} vs. 1 MeV Electron Fluence for Spectrolab GaAs/Ge Solar Cells 6-15
Figure 6.13. P_{max}/cm^2 vs. 1 MeV Electron Fluence for Spectrolab GaAs/Ge Solar Cells . . . 6-16
Figure 6.14. I_{mp}/cm^2 vs. 1 MeV Electron Fluence for Spectrolab GaAs/Ge Solar Cells . . . 6-17
Figure 6.15. V_{mp} vs. 1 MeV Electron Fluence for Spectrolab GaAs/Ge Solar Cells 6-18
Figure 6.16. Normalized I_{sc} vs. 1 MeV Electron Fluence for Spectrolab GaAs/Ge Solar
Cells 6-19
Figure 6.17. Normalized V_{oc} vs. 1 MeV Electron Fluence for Spectrolab GaAs/Ge Solar
Cells 6-20
Figure 6.18. Normalized P_{max} vs. 1 MeV Electron Fluence for Spectrolab GaAs/Ge Solar
Cells 6-21
Figure 6.19. Normalized I_{mp} vs. 1 MeV Electron Fluence for Spectrolab GaAs/Ge Solar
Cells 6-22
Figure 6.20 Normalized V_{mp} vs. 1 MeV Electron Fluence for Spectrolab GaAs/Ge Solar
Cells 6-23
Figure 6.21. I_{sc}/cm^2 vs. 1 MeV Electron Fluence for Hughes LPE and ASEC OMCVD
GaAs/Ge Solar Cells 6-24
Figure 6.22. V_{oc} vs. 1 MeV Electron Fluence for Hughes LPE and ASEC OMCVD
GaAs/Ge Solar Cells 6-25
Figure 6.23. P_{max}/cm^2 vs. 1 MeV Electron Fluence for Hughes LPE and ASEC OMCVD
GaAs/Ge Solar Cells 6-26
Figure 6.24. I_{mp}/cm^2 vs. 1 MeV Electron Fluence for Hughes LPE and ASEC OMCVD
GaAs/Ge Solar Cells 6-27
Figure 6.25. V_{mp} vs. 1 MeV Electron Fluence for Hughes LPE and ASEC OMCVD
GaAs/Ge Solar Cells 6-28
Figure 6.26. Normalized I_{sc} vs. 1 MeV Electron Fluence for Hughes LPE and ASEC
OMCVD GaAs/Ge Solar Cells 6-29
Figure 6.27. Normalized V_{oc} vs. 1 MeV Electron Fluence for Hughes LPE and ASEC
OMCVD GaAs/Ge Solar Cells 6-30
Figure 6.28. Normalized P_{max} vs. 1 MeV Electron Fluence for Hughes LPE and ASEC
OMCVD GaAs/Ge Solar Cells 6-31
Figure 6.29. Normalized I_{mp} vs. 1 MeV Electron Fluence for Hughes LPE and ASEC
OMCVD GaAs/Ge Solar Cells 6-32
Figure 6.30 Normalized V_{mp} vs. 1 MeV Electron Fluence for Hughes LPE and ASEC
OMCVD GaAs/Ge Solar Cells 6-33
Figure 6.31. Average I_{sc}/cm^2 as a Function of Temperature, Pre-Irradiation 6-39
Figure 6.32. Average V_{oc} as a Function of Temperature, Pre-Irradiation 6-40
Figure 6.33. Average P_{max}/cm^2 as a Function of Temperature, Pre-Irradiation 6-41

Figure 6.34.   Average I_{sc}/cm^2 as a Function of Intensity, Pre-Irradiation 6-42
Figure 6.35.   Average V_{oc} as a Function of Intensity, Pre-Irradiation 6-43
Figure 6.36.   Average P_{max}/cm^2 as a Function of Intensity, Pre-Irradiation 6-44
Figure 6.37.   Average I_{sc}/cm^2 as a Function of Temperature, after 1 x 10^{14} e/cm^2 6-49
Figure 6.38.   Average V_{oc} as a Function of Temperature, after 1 x 10^{14} e/cm^2 6-50
Figure 6.39.   Average P_{max}/cm^2 as a Function of Temperature, after 1 x 10^{14} e/cm^2 6-51
Figure 6.40.   Average I_{sc}/cm^2 as a Function of Intensity, after 1 x 10^{14} e/cm^2 6-52
Figure 6.41.   Average V_{oc} as a Function of Intensity, after 1 x 10^{14} e/cm^2 6-53
Figure 6.42.   Average P_{max}/cm^2 as a Function of Intensity, after 1 x 10^{14} e/cm^2 6-54
Figure 6.43.   Average I_{sc}/cm^2 as a Function of Temperature, after 1.1 x 10^{15} e/cm^2 6-59
Figure 6.44.   Average V_{oc} as a Function of Temperature, after 1.1 x 10^{15} e/cm^2 6-60
Figure 6.45.   Average P_{max}/cm^2 as a Function of Temperature, after 1.1 x 10^{15} e/cm^2 6-61
Figure 6.46.   Average I_{sc}/cm^2 as a Function of Intensity, after 1.1 x 10^{15} e/cm^2 6-62
Figure 6.47.   Average V_{oc} as a Function of Intensity, after 1.1 x 10^{15} e/cm^2 6-63
Figure 6.48.   Average P_{max}/cm^2 as a Function of Intensity, after 1.1 x 10^{15} e/cm^2 6-64
Figure 7.1. I_{sc} vs. Time in Orbit for the Spectrolab 200-μm GaAs/Ge Solar Cells with
               500-μm Coverglasses on STRV-1B 7-8
Figure 7.2. I_{sc} vs. Time in Orbit for the ASEC 90-μm GaAs/Ge Solar Cells with 500-μm
               Coverglasses on STRV-1B 7-8
Figure 7.3.   Normalized P_{max} vs. Time in Orbit for the 90-μm GaAs/Ge Solar Cells with
               100-μm-CMX Coverglasses on PASP-Plus Experiment 4 7-9
Figure 7.4.   Normalized P_{max} vs. Time in Orbit for the 90-μm GaAs/Ge Solar Cells with
               100-μm-CMX Coverglasses on PASP-Plus Experiment 6 7-9
Figure 7.5.   Normalized P_{max} vs. Time in Orbit for the 178-μm GaAs/Ge Solar Cells with
               150-μm-CMX Coverglasses on PASP-Plus Experiment 11 7-10
Figure 7.6.   Normalized P_{max} vs. Equivalent 1 MeV Electron Fluence for GaAs/Ge Solar
               Cells on ASCOT and on PASP-Plus Module 4 7-10
Figure 7.7.   Normalized V_{oc} vs. Equivalent 1 MeV Electron Fluence for GaAs/Ge Solar
               Cells on ASCOT and on PASP-Plus Module 4 7-11
Figure 7.8.   Normalized I_{sc} vs. Equivalent 1 MeV Electron Fluence for GaAs/Ge Solar
               Cells on ASCOT, PASP-Plus Module 4, and on STRV-1B 7-11
Figure 8.1.   Motion of a Positive Charged Particle in a Magnetic Field of Two Intensities
               Giving Rise to a Gradient 8-3
Figure 8.2.   Forces on a Charged Particle in a Converging Magnetic Field. $F_\parallel$ Acts Along
               the Magnetic Field and is Responsible for "Mirroring" 8-3
Figure 8.3.   Fluence Probability Curves for Protons of Energy Greater than 1 MeV for
               Various Exposure Times 8-12
Figure 8.4.   Fluence Probability Curves for Protons of Energy Greater than 4 MeV for
               Various Exposure Times 8-12
Figure 8.5.   Fluence Probability Curves for Protons of Energy Greater than 10 MeV for
               Various Exposure Times 8-13
Figure 8.6.   Fluence Probability Curves for Protons of Energy Greater than 30 MeV for
               Various Exposure Times 8-13
Figure 8.7.   Fluence Probability Curves for Protons of Energy Greater than 60 MeV for
               Various Exposure Times 8-14
Figure 8.8.   Propagation of Solar Flare Protons from Sun to Earth 8-16
Figure A.1.   Optical Absorption Coefficients for GaAs and Si A-5

Figure A.2. Optical Absorption Coefficients for $Ga_{1-x}Al_xAs$ as a Function of Aluminum
Concentration, x A-6
Figure A.3. Band Gap Energy vs. Aluminum Concentration, x, in $Ga_{1-x}Al_xAs$ A-7
Figure A.4. Bandgap Energy of GaAs, Si, and Ge as a Function of Temperature A-8
Figure A.5. Drift Mobility of Electrons and Holes in GaAs at 300 K vs. Impurity
Concentration A-9

LIST OF TABLES

Table 3-1. Solar Spectral Irradiance [Ref. 3.6] 3-2

Table 4-1. Displacement Rates for Incident Electrons 4-11

Table 4-2. Displacement Rates for Incident Protons 4-11

Table 5-1. Electron Damage Coefficients for GaAs Solar Cells 5-18

Table 5-2. Proton Damage Coefficients for GaAs Solar Cells -- I_{sc} 5-19

Table 5-3. Proton Damage Coefficients for GaAs Solar Cells -- P_{max}, V_{oc} 5-20

Table 6-1. Least-Square Fits to the Normalized Electrical Parameters of ASEC GaAs/Ge Cells Irradiated with 1 MeV Electrons (see Eq. 6-1) 6-3

Table 6-2. Least-Square Fits to the Normalized Electrical Parameters of Spectrolab GaAs/Ge Cells Irradiated with 1 MeV Electrons (see Eq. 6-1) 6-3

Table 6-3. I_{sc}/cm^2 vs. Temperature and Intensity, Pre-Irradiation 6-35

Table 6-4. V_{oc} vs. Temperature and Intensity, Pre-Irradiation 6-35

Table 6-5. I_{mp}/cm^2 vs. Temperature and Intensity, Pre-Irradiation 6-36

Table 6-6. V_{mp} vs. Temperature and Intensity, Pre-Irradiation 6-36

Table 6-7. P_{max}/cm^2 vs. Temperature and Intensity, Pre-Irradiation 6-37

Table 6-8. Fill Factor vs. Temperature and Intensity, Pre-Irradiation 6-37

Table 6-9. Efficiency vs. Temperature and Intensity, Pre-Irradiation 6-38

Table 6-10. I_{sc}/cm^2 vs. Temperature and Intensity, after 1×10^{14} e/cm^2 6-45

Table 6-11. V_{oc} vs. Temperature and Intensity, after 1×10^{14} e/cm^2 6-45

Table 6-12. I_{mp}/cm^2 vs. Temperature and Intensity, after 1×10^{14} e/cm^2 6-46

Table 6-13. V_{mp} vs. Temperature and Intensity, after 1×10^{14} e/cm^2 6-46

Table 6-14. P_{max}/cm^2 vs. Temperature and Intensity, after 1×10^{14} e/cm^2 6-47

Table 6-15. Fill Factor vs. Temperature and Intensity, after 1×10^{14} e/cm^2 6-47

Table 6-16. Efficiency vs. Temperature and Intensity, after 1×10^{14} e/cm^2 6-48

Table 6-17. I_{sc}/cm^2 vs. Temperature and Intensity, after 1.1×10^{15} e/cm^2 6-55

Table 6-18. V_{oc} vs. Temperature and Intensity, after 1.1×10^{15} e/cm^2 6-55

Table 6-19. I_{mp}/cm^2 vs. Temperature and Intensity, after 1.1×10^{15} e/cm^2 6-56

Table 6-20. V_{mp} vs. Temperature and Intensity, after 1.1×10^{15} e/cm^2 6-56

Table 6-21. P_{max}/cm^2 vs. Temperature and Intensity, after 1.1×10^{15} e/cm^2 6-57

Table 6-22. Fill Factor vs. Temperature and Intensity, after 1.1×10^{15} e/cm^2 6-57

Table 6-23. Efficiency vs. Temperature and Intensity, after 1.1×10^{15} e/cm^2 6-58

Table 8-1. Characteristic Orbital Parameters for Trapped 1 MeV Electrons and 1 MeV Protons at 2000 km Altitude Near the Equator [8.4] 8-4

Table 8-2. Solar Cycle Maxima [8.31] 8-9

Table 8-3. Cumulative Fluence-Energy Spectra for the Major Flares during Solar Cycle 22 8-15

Table 9-1. Conversion Factors for Protons to Electrons for GaAs Solar Cells (Converting from 10 MeV Protons to 1 MeV Electrons) 9-2

Table 9-2. Rough Degradation Calculation. 9-4

Table 9-3. Catalog of Materials and Thicknesses for an Example Substrate 9-7

Table 9-4. Annual Equivalent 1 MeV Electron Fluence for GaAs Solar Cells from Trapped Electrons During Solar Max, 0° Inclination (Infinite Backshielding) 9-9

Table 9-5. Annual Equivalent 1 MeV Electron Fluence for GaAs Solar Cells from Trapped Electrons During Solar Min, 0° Inclination (Infinite Backshielding) 9-10

ix

Table 9-6.   Annual Equivalent 1 MeV Electron Fluence for GaAs Solar Cells
               from Trapped Protons During Solar Max and Min, 0° Inclination
               (Infinite Backshielding)   9-11

Table 9-7.   Annual Equivalent 1 MeV Electron Fluence for GaAs Solar Cells
               from Trapped Protons During Solar Max and Min, 0° Inclination
               (Infinite Backshielding)   9-12

Table 9-8.   Annual Equivalent 1 MeV Electron Fluence for GaAs Solar Cells
               from Trapped Protons During Solar Max and Min, 0° Inclination
               (Infinite Backshielding)   9-13

Table 9-9.   Annual Equivalent 1 MeV Electron Fluence for GaAs Solar Cells
               from Trapped Electrons During Solar Max, 10° Inclination (Infinite
               Backshielding)   9-14

Table 9-10.   Annual Equivalent 1 MeV Electron Fluence for GaAs Solar Cells
               from Trapped Electrons During Solar Min, 10° Inclination (Infinite
               Backshielding)   9-15

Table 9-11.   Annual Equivalent 1 MeV Electron Fluence for GaAs Solar Cells
               from Trapped Protons During Solar Max and Min, 10° Inclination
               (Infinite Backshielding)   9-16

Table 9-12.   Annual Equivalent 1 MeV Electron Fluence for GaAs Solar Cells
               from Trapped Protons During Solar Max and Min, 10° Inclination
               (Infinite Backshielding)   9-17

Table 9-13.   Annual Equivalent 1 MeV Electron Fluence for GaAs Solar Cells
               from Trapped Protons During Solar Max and Min, 10° Inclination
               (Infinite Backshielding)   9-18

Table 9-14.   Annual Equivalent 1 MeV Electron Fluence for GaAs Solar Cells
               from Trapped Electrons During Solar Max, 20° Inclination (Infinite
               Backshielding)   9-19

Table 9-15.   Annual Equivalent 1 MeV Electron Fluence for GaAs Solar Cells
               from Trapped Electrons During Solar Min, 20° Inclination (Infinite
               Backshielding)   9-20

Table 9-16.   Annual Equivalent 1 MeV Electron Fluence for GaAs Solar Cells
               from Trapped Protons During Solar Max and Min, 20° Inclination
               (Infinite Backshielding)   9-21

Table 9-17.   Annual Equivalent 1 MeV Electron Fluence for GaAs Solar Cells
               from Trapped Protons During Solar Max and Min, 20° Inclination
               (Infinite Backshielding)   9-22

Table 9-18.   Annual Equivalent 1 MeV Electron Fluence for GaAs Solar Cells
               from Trapped Protons During Solar Max and Min, 20° Inclination
               (Infinite Backshielding)   9-23

Table 9-19.   Annual Equivalent 1 MeV Electron Fluence for GaAs Solar Cells
               from Trapped Electrons During Solar Max, 30° Inclination (Infinite
               Backshielding)   9-24

Table 9-20.   Annual Equivalent 1 MeV Electron Fluence for GaAs Solar Cells
               from Trapped Electrons During Solar Min, 30° Inclination (Infinite
               Backshielding)   9-25

Table 9-21.   Annual Equivalent 1 MeV Electron Fluence for GaAs Solar Cells
               from Trapped Protons During Solar Max and Min, 30° Inclination
               (Infinite Backshielding)   9-26

Table 9-22.   Annual Equivalent 1 MeV Electron Fluence for GaAs Solar Cells
               from Trapped Protons During Solar Max and Min, 30° Inclination
               (Infinite Backshielding)   9-27
Table 9-23.   Annual Equivalent 1 MeV Electron Fluence for GaAs Solar Cells
               from Trapped Protons During Solar Max and Min, 30° Inclination
               (Infinite Backshielding)   9-28
Table 9-24.   Annual Equivalent 1 MeV Electron Fluence for GaAs Solar Cells
               from Trapped Electrons During Solar Max, 40° Inclination (Infinite
               Backshielding)   9-29
Table 9-25.   Annual Equivalent 1 MeV Electron Fluence for GaAs Solar Cells
               from Trapped Electrons During Solar Min, 40° Inclination (Infinite
               Backshielding)   9-30
Table 9-26.   Annual Equivalent 1 MeV Electron Fluence for GaAs Solar Cells
               from Trapped Protons During Solar Max and Min, 40° Inclination
               (Infinite Backshielding)   9-31
Table 9-27.   Annual Equivalent 1 MeV Electron Fluence for GaAs Solar Cells
               from Trapped Protons During Solar Max and Min, 40° Inclination
               (Infinite Backshielding)   9-32
Table 9-28.   Annual Equivalent 1 MeV Electron Fluence for GaAs Solar Cells
               from Trapped Protons During Solar Max and Min, 40° Inclination
               (Infinite Backshielding)   9-33
Table 9-29.   Annual Equivalent 1 MeV Electron Fluence for GaAs Solar Cells
               from Trapped Electrons During Solar Max, 50° Inclination (Infinite
               Backshielding)   9-34
Table 9-30.   Annual Equivalent 1 MeV Electron Fluence for GaAs Solar Cells
               from Trapped Electrons During Solar Min, 50° Inclination (Infinite
               Backshielding)   9-35
Table 9-31.   Annual Equivalent 1 MeV Electron Fluence for GaAs Solar Cells
               from Trapped Protons During Solar Max and Min, 50° Inclination
               (Infinite Backshielding)   9-36
Table 9-32.   Annual Equivalent 1 MeV Electron Fluence for GaAs Solar Cells
               from Trapped Protons During Solar Max and Min, 50° Inclination
               (Infinite Backshielding)   9-37
Table 9-33.   Annual Equivalent 1 MeV Electron Fluence for GaAs Solar Cells
               from Trapped Protons During Solar Max and Min, 50° Inclination
               (Infinite Backshielding)   9-38
Table 9-34.   Annual Equivalent 1 MeV Electron Fluence for GaAs Solar Cells
               from Trapped Electrons During Solar Max, 60° Inclination (Infinite
               Backshielding)   9-39
Table 9-35.   Annual Equivalent 1 MeV Electron Fluence for GaAs Solar Cells
               from Trapped Electrons During Solar Min, 60° Inclination (Infinite
               Backshielding)   9-40
Table 9-36.   Annual Equivalent 1 MeV Electron Fluence for GaAs Solar Cells
               from Trapped Protons During Solar Max and Min, 60° Inclination
               (Infinite Backshielding)   9-41
Table 9-37.   Annual Equivalent 1 MeV Electron Fluence for GaAs Solar Cells
               from Trapped Protons During Solar Max and Min, 60° Inclination
               (Infinite Backshielding)   9-42

Table 9-38.   Annual Equivalent 1 MeV Electron Fluence for GaAs Solar Cells
               from Trapped Protons During Solar Max and Min, 60° Inclination
               (Infinite Backshielding)   9-43

Table 9-39.   Annual Equivalent 1 MeV Electron Fluence for GaAs Solar Cells
               from Trapped Electrons During Solar Max, 70° Inclination (Infinite
               Backshielding)   9-44

Table 9-40.   Annual Equivalent 1 MeV Electron Fluence for GaAs Solar Cells
               from Trapped Electrons During Solar Min, 70° Inclination (Infinite
               Backshielding)   9-45

Table 9-41.   Annual Equivalent 1 MeV Electron Fluence for GaAs Solar Cells
               from Trapped Protons During Solar Max and Min, 70° Inclination
               (Infinite Backshielding)   9-46

Table 9-42.   Annual Equivalent 1 MeV Electron Fluence for GaAs Solar Cells
               from Trapped Protons During Solar Max and Min, 70° Inclination
               (Infinite Backshielding)   9-47

Table 9-43.   Annual Equivalent 1 MeV Electron Fluence for GaAs Solar Cells
               from Trapped Protons During Solar Max and Min, 70° Inclination
               (Infinite Backshielding)   9-48

Table 9-44.   Annual Equivalent 1 MeV Electron Fluence for GaAs Solar Cells
               from Trapped Electrons During Solar Max, 80° Inclination (Infinite
               Backshielding)   9-49

Table 9-45.   Annual Equivalent 1 MeV Electron Fluence for GaAs Solar Cells
               from Trapped Electrons During Solar Min, 80° Inclination (Infinite
               Backshielding)   9-50

Table 9-46.   Annual Equivalent 1 MeV Electron Fluence for GaAs Solar Cells
               from Trapped Protons During Solar Max and Min, 80° Inclination
               (Infinite Backshielding)   9-51

Table 9-47.   Annual Equivalent 1 MeV Electron Fluence for GaAs Solar Cells
               from Trapped Protons During Solar Max and Min, 80° Inclination
               (Infinite Backshielding)   9-52

Table 9-48.   Annual Equivalent 1 MeV Electron Fluence for GaAs Solar Cells
               from Trapped Protons During Solar Max and Min, 80° Inclination
               (Infinite Backshielding)   9-53

Table 9-49.   Annual Equivalent 1 MeV Electron Fluence for GaAs Solar Cells
               from Trapped Electrons During Solar Max, 90° Inclination (Infinite
               Backshielding)   9-54

Table 9-50.   Annual Equivalent 1 MeV Electron Fluence for GaAs Solar Cells
               from Trapped Electrons During Solar Min, 90° Inclination (Infinite
               Backshielding)   9-55

Table 9-51.   Annual Equivalent 1 MeV Electron Fluence for GaAs Solar Cells
               from Trapped Protons During Solar Max and Min, 90° Inclination
               (Infinite Backshielding)   9-56

Table 9-52.   Annual Equivalent 1 MeV Electron Fluence for GaAs Solar Cells
               from Trapped Protons During Solar Max and Min, 90° Inclination
               (Infinite Backshielding)   9-57

Table 9-53.   Annual Equivalent 1 MeV Electron Fluence for GaAs Solar Cells
               from Trapped Protons During Solar Max and Min, 90° Inclination
               (Infinite Backshielding)   9-58

Table 9-54.   Annual Equivalent 1 MeV Electron Fluence for GaAs Solar Cells
                 from Trapped Electrons During Solar Min, 0° Inclination at
                 Synchronous Altitude vs. Longitude (Infinite Backshielding)   9-59

Table 9-55.   Annual Equivalent 1 MeV Electron Fluence for GaAs Solar Cells
                 from Trapped Protons During Solar Min, 0° Inclination at
                 Synchronous Altitude vs. Longitude (Infinite Backshielding)   9-60

Table 9-56.   Annual Equivalent 1 MeV Electron Fluence for GaAs Solar Cells
                 from Trapped Protons During Solar Min, 0° Inclination at
                 Synchronous Altitude vs. Longitude (Infinite Backshielding)   9-61

Table 9-57.   Annual Equivalent 1 MeV Electron Fluence for GaAs Solar Cells
                 from Trapped Protons During Solar Min, 0° Inclination at
                 Synchronous Altitude vs. Longitude (Infinite Backshielding)   9-62

Table A-1.   Some Useful Physical Constants A-1

Table A-2.   Properties of GaAs, Si, and Ge at 300 K A-2

Table A-3.   Properties of GaAs and $Ga_{1-x}Al_xAs$ at 300 K A-4

Table A-4.   Indices of Refraction of Some Common Solar Cell Materials at 300 K A-10

Table A-5.   Physical and Optical Properties of Coverglass Materials A-11

Table A-6.   Ranges, Stopping Powers, and Straggling of Protons Incident on GaAs A-12

Table A-7.   Ranges, Stopping Powers, and Straggling of Electrons Incident on GaAs A-14

Table A-8.   Ranges, Stopping Powers, and Straggling of Protons Incident on CMG Glass .   A-16

Table A-9.   Ranges, Stopping Powers, and Straggling of Electrons Incident on CMG Glass   A-18

Table A-10.   Equivalent Fluence Program for GaAs Solar Cells A-20

xiii

ACKNOWLEDGEMENTS

The author would like to express his appreciation to all those who contributed directly or indirectly to the production of this book. Two major contributors are Jack Carter, Jr. and Herb Tada, both formerly of TRW, who were the principal authors of the first editions of the Solar Cell Radiation Handbook. I had many interesting, and I hope, mutually beneficial interactions with several individuals who contributed to bringing GaAs solar cells to a state of production readiness: Sanjiv Kamath and Bob Loo of Hughes Research Laboratory; Peter Iles, Milton Yeh, Frank Ho, and Scott Khemthong at the Applied Solar Energy Corp. (ASEC); and Terry Cavicchi at Spectrolab. These companies were also very generous in supplying us GaAs solar cells for measurement and evaluation. The early JPL work on GaAs solar cells was generously supported by NASA Headquarters and by Joe Wise, Pat Rahilly, and Terry Trumble at Wright Patterson Air Force Base. George Datum of Lockheed Missiles and Space Co. (now Lockheed Martin) is worthy of special mention for continued ongoing support of the work at JPL. Tetsuo Miyahira and Robert Weiss of JPL performed most of the electron irradiations and took thousands of I-V curves, which are distilled into the data presented in Chapter 6. Gil Downing, retired from JPL, and Alan Rice of Caltech were of enormous help in performing the proton irradiations of hundreds of GaAs solar cells. Jim Hix, Diane Engler, and Jose Miranda of the JPL Design Section wrote and operated the computer programs that tabulate and plot the electrical performance of solar cells as a function of temperature and incident intensity. Richard C. Willson of JPL and Carol Riorden of National Renewable Energy Laboratory (NREL) contributed the solar irradiance spectral data. I relied heavily on assistance from Henry Garrett, Guy Spitale, and Rene Aguero of JPL, from Herb Sauer of the National Oceanic and Atmospheric Administration (NOAA), and from Jim Vette (retired) and E. G. Stassinopoulos of Goddard Space Flight Center in preparing the chapter on the radiation environment. Dean Marvin of the Aerospace Corporation was more than generous in sharing flight data on GaAs solar cells from the PASP-Plus, STRV-1B and ASCOT flight experiments. Much of the historical material reported in Chapter 1 was adopted from a report written by P.A. Crossley, G.T. Noel, and M. Wolf, of RCA. And finally, special thanks goes to Barbara Anspaugh, Johanne St. Henri, and Susan Foster for many hours spent editing the manuscript, and to our cat, Jezebel, who walked across the keyboard so many times during the writing, that surely a part of the book is hers.

Chapter 1

History of GaAs Solar Cell Development

1.1   Advantages of Gallium Arsenide

Gallium Arsenide (GaAs) has been of interest as a photovoltaic material for many years. This interest arises primarily for three reasons. First, the bandgap of 1.42 eV at 300 K is very nearly ideal for a photovoltaic device operating in our solar spectrum. Second, GaAs solar cells should be capable of operating at higher temperatures than silicon (Si) cells. Third, GaAs solar cells are expected to be very radiation resistant.

GaAs has a direct bandgap which gives the material a high optical absorption coefficient, α. Figure 1.1 shows the absorption coefficients of GaAs and Si as a function of wavelength [1.1,1.2]. The α for GaAs rises very steeply at the band edge (λ = .88 μm) to values greater than 10^4 cm^{-1}, whereas the absorption coefficient for Si rises much more gradually because it is an indirect bandgap material. Figure 1.2 displays the carrier concentration produced in GaAs and Si with the Air Mass Zero[*] (AM0) spectrum incident. As this figure shows, 99% of the AM0 solar photons will be absorbed in GaAs within 2.6 μm of the front surface, whereas for Si the AM0 photons must penetrate to a depth of 25 μm for 99% absorption. A significant number of AM0 photons penetrate

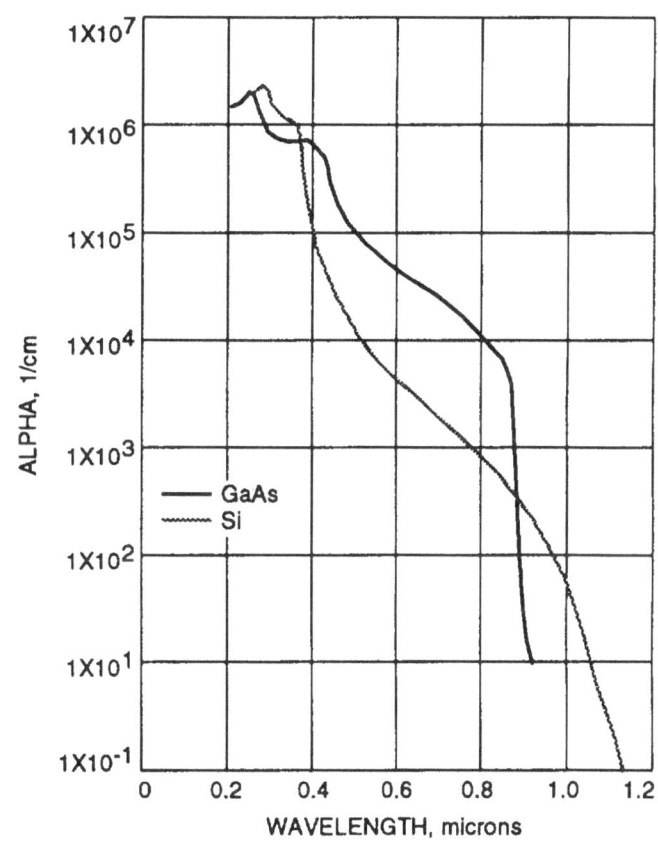

Figure 1.1.   Optical Absorption Coefficients for GaAs and Silicon

---

[*]Air Mass Zero is a term used to refer to the solar spectrum in both intensity and spectral content at a distance of 1 AU from the sun.

to depths as great as 200 μm in Si, which is why Si solar cells must be made much thicker than GaAs cells.

Since GaAs solar cells can be very thin, they may be expected to be radiation resistant. In Si solar cells, the major effect of particulate radiation in degrading the performance is the production of defects in the lattice. These defects then decrease the minority carrier diffusion length. Si cells require large minority carrier diffusion lengths so that minority carriers produced deep in the cell have a reasonably high probability of

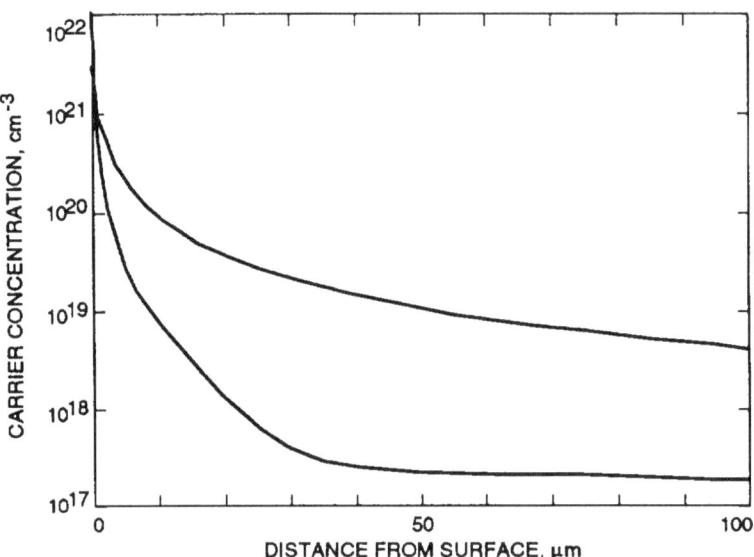

Figure 1.2.   Carrier Concentrations in GaAs and Silicon with AM0 Spectrum Incident

diffusing to the junction. The thin GaAs solar cells do not require such long diffusion lengths, hence the cells should be able to endure a significant amount of diffusion length degradation before cell performance is affected. Since the initial minority carrier lifetime is low, a much larger particle fluence is required to significantly reduce the existing recombination rate [1.3]. In addition, the probability of radiation interaction with thin layers is small, so less degradation of those critically sensitive layers in the region of the junction is to be expected.

Modern GaAs solar cells have indeed lived up to these expectations. Good quality GaAs solar cells are being routinely produced which are as thin as 8 μm, although they do need to be fabricated on thicker carrier substrates for mechanical strength. They have also proven to be very radiation resistant to the electron and proton radiation encountered in the space environment, with one exception: they are susceptible to damage by protons having energies near 300 keV because protons of this energy penetrate to the junction region. Protons tend to produce most damage near the end of their path, so protons with energies near 300 keV produce a lot of damage in the active area of the cell. The small volume of damage is significant in relation to the volume for light absorption and carrier generation. But most proton energies found in the space environment do not stop in the relatively small photo generation volume and will produce less damage to the cell. Most of these protons will be producing their heavy end-of-track damage much deeper in the cell than in the critically sensitive photo generation region. Contrasting this behavior with how Si solar cells are affected by radiation, we note that protons of nearly all energies will cause regions of damage somewhere in the large photo generation volume, so all proton energies will be more effective in damaging cell performance than would be the case for GaAs cells. But low energy protons can be stopped by a modest amount of shielding, so coverglasses with thicknesses of 25 μm or greater are effective in protecting GaAs cells from the most damaging protons.

The expected superior performance at high temperature arises because the high bandgap implies a higher open circuit voltage (V_{oc}) for GaAs solar cells (≈ 1 volt for GaAs as compared to ≈ 0.6 volt for Si). The most sensitive solar cell electrical parameter to temperature is V_{oc}, and like Si cells, this V_{oc} decreases with temperature at about 2 mV/°C. Percentage-wise, this decrease translates into a lesser decrease in power output as a function of temperature for the cell with the higher V_{oc}.

1.2   Early Work 1954 - 1964

Interest in the use of GaAs as a material for making solar cells began around 1954 when a survey paper on "Semiconducting Intermetallic Compounds" was published by Welker [1.4, 1.5]. Welker published curves showing the output of a GaAs "photocell" as a function of illumination intensity, but no details of the cell construction were included. The construction of this cell was no doubt based on Welker's studies of III-V compounds published in 1953 [1.6], which showed that certain of these compounds were promising semiconductor materials. Welker's pioneering work was followed by numerous other studies on the properties of III-V compounds. These studies were collected and reported by Madelung [1.7] in 1964. Madelung's report also included an extensive bibliography.

In 1955, Gremmelmaier reported the characteristics of two GaAs solar cells which had measured efficiencies of 1% and 4% while illuminated with "sea-level sunlight." [1.8] The cells were made of polycrystalline GaAs by growing thin p-layers (0.01mm) onto n-type substrates. These cells had promising photovoltaic properties and led Gremmelmaier to predict much higher efficiencies for cells made from crystalline materials.

In 1955, an RCA group, under funding from the U.S. Army Signal Corps, published a theoretical study on the suitability of various semiconductor materials for solar cell use [1.9, 1.10]. This study

1-3

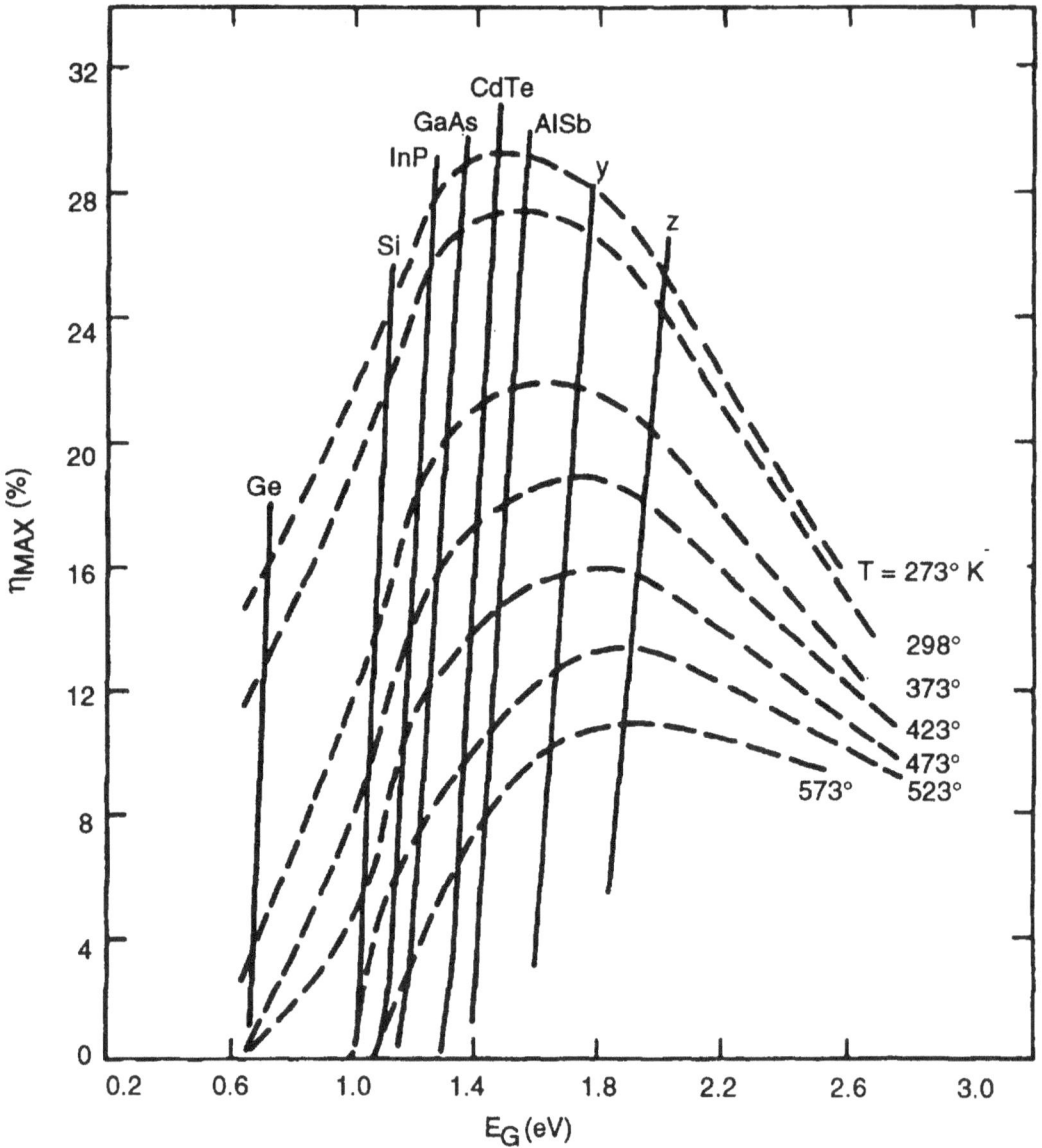

Figure 1.3.   Maximum Efficiency η_{max} as a Function of Energy Gap E_G for Various Temperatures (© 1973 IEEE, used with permission)

showed that a number of III-V compounds (GaAs, indium phosphide (InP), aluminum antimonide (AlSb)) and one II-VI compound, cadmium telluride (CdTe), should be capable of producing photovoltaic cells with higher efficiencies than cells made from Si. Figure 1.3 from reference [1.10] summarizes the results of their calculations. As the figure shows, they found that the efficiency is a strong function of both temperature and bandgap. The general shape of these curves results from the fact that as the bandgap increases, V_{oc} increases, but I_{sc} decreases because less of the solar spectrum can be absorbed. The two effects combine to produce a maximum in P_{max} for materials with bandgaps near 1.5 eV for devices operating near room temperature. They concluded that GaAs should be the most promising III-V compound for future solar cell development. The RCA group followed up their theoretical work with construction of a working GaAs solar cell. This cell, made by diffusing cadmium (Cd) into n-type GaAs, produced measured efficiencies as high as 6.5% [1.11], although the cell area was small (≈ 1 mm^2). There was difficulty in obtaining material of sufficiently low impurity concentration and adequately large single crystal samples. Measurements and analysis of these cells pointed out both good and bad features. On the positive side, the open circuit voltages, V_{oc}, were in the vicinity of 0.9 V as expected, and the series resistance, R_s, was reasonably low, indicating that they had a workable contact system. On the negative side, the reverse saturation current, I_o, was about 10^5 times larger than predicted by theory, and the short circuit current, I_{sc} was about a factor of 10 lower than expected. They believed the low I_{sc} was not caused by carrier recombination at the surface but was due to low minority carrier lifetimes in the material, and that the low lifetimes were due to defects in the GaAs crystal.

A report published by Nasledov et al. in 1959 revealed that the Soviets were also making GaAs solar cells [1.12]. Their cells were made on small n-type polycrystalline wafers by diffusion of Cd using a sealed ampoule system. The contacts on these cells were made by pressing tin into the diffused layer, and the resulting relatively large series resistance no doubt contributed to the low measured efficiencies of $\approx 2.8\%$ on 1.8 mm^2 cells. The Soviets did improve their contact system in later cells, but made them with very deep junctions ($\approx 10\mu$m), which caused a very poor spectral response in the blue region and detracted from their efficiency.

The work at RCA continued with the objective of improving GaAs material and cell technology. A major goal was to develop cells which could operate at higher temperatures than Si cells. The RCA group worked on several diffusion techniques, including the use of a two-temperature-zone sealed ampoule, use of zinc (Zn) and Cd as diffusants, open-tube diffusion, resistance-heated furnaces, radiant-heated furnaces, etc. Along with these variations, they experimented with various combinations of diffusion process parameters, including diffusion time; temperature; dopant material and concentration; and post-diffusion cooldown rates. Attempts were made to improve the minority carrier lifetimes (estimated to be $\approx 10^{-10}$ sec) by post-diffusion treatments, such as annealing and low-temperature diffusion of copper (Cu) or nickel (Ni) into the cell.

The most effective way of improving the efficiencies was found to be the reduction of the junction depth by a combination of shallow diffusion followed by surface etching. Even crystal growing techniques were developed so that suitably large and pure starting crystalline wafers could be available for cell fabrication. Crystal dopants of both tin (Sn) and germanim (Ge) were used to produce an n-type base starting material. Techniques were developed for vacuum deposition of silver contacts, followed by solder coating of the contacts using solder preforms, thus avoiding a solder dipping step that had been found to severely degrade cell performance. Quarter-wave antireflection coatings of silicon monoxide (SiO) were developed and applied by vacuum evaporation. All this work resulted in a steady improvement in cell size and efficiency, culminating in peak production line efficiencies for 2 x 2 cm^2 cells of 8.5% in 1962. However, maximum cell efficiencies of 13 to 14% were observed [1.13]. Although most of these cells were p/n, a few n/p cells were fabricated. The n/p cells were considered promising because the electron mobilities in GaAs are much higher than hole mobilities, but the cell efficiencies turned out to be lower than in the p/n cells. Epitaxial growth of GaAs on GaAs was also investigated during this time period, but the cells made by this technique had relatively low efficiencies [1.5].

GaAs cell work continued at RCA under Air Force funding through the 1964 time period. During this time a pilot line was established with the objective of optimizing production processes. New crystal-growing techniques were pursued, notably the development of a gradient-freeze technique in which the ampoule containing GaAs remained in a fixed place in a multizoned furnace. The furnace was equipped with power supplies which were programmed to move a temperature profile along the ampoule. This technique produced crystals of a more uniform quality, which is more suitable for a production line environment. Junctions were produced by diffusion in an open tube using Zn as the diffusant in a carrier gas of hydrogen. The procedure used produced junctions ≈ 1.5 μm deep, but the front surface was subsequently etched away leaving a junction depth of 0.5 μm. Vacuum-evaporated nickel-silver (Ni-Ag) contacts were added to the cells, followed by vacuum evaporation of antireflection coatings of SiO. The pilot line yielded cells of $\approx 9.5\%$ average efficiency. These cells were extensively tested in both proton and electron radiation. It was found that under 5.6 MeV electron bombardment or under high energy proton exposure to 1.8 to 9.5 MeV protons, these cells were superior in radiation resistance to n/p Si cells. Wysocki [1.14, 1.15] found that the critical fluence (the fluence required to reduce cell efficiency by 25%) for the GaAs cells was a factor of 10 higher than for Si cells, the main degradation factor being a rapid loss in I_{sc}. But the performance of the two cell types was comparable when exposed to 0.8 MeV electrons. The GaAs cells were worse than Si cells when the exposure was to low energy protons (0.1 to 0.4 MeV). This was explained by noting that the entire GaAs cell operating region is near the junction, and the ranges of these low energy protons are such that they come to rest in the vicinity of the junction where they produce most of their damage. The higher energy particles tend to pass on through the cells without producing such damage spikes, so they have much less effect on the performance of thin solar cells. However, Si cells have much lower light absorption coefficients and necessarily rely on the deeper regions of the cells for production and collection of carriers. Hence they are damaged more by penetrating radiation. This explanation was supported by the experimental spectral response measurements, which showed that the Si cells mainly lost red response after irradiation, since the carriers generated deep in the Si by the longer wavelength photons had ever-increasing difficulty in reaching the junction by diffusion. The spectral response of GaAs cells, on the other hand, showed a loss in the blue, since the longer wavelengths only penetrated to regions very near the junction.

It is interesting to note that the RCA workers began to realize that the low minority lifetimes seen in GaAs were related to the direct bandgap nature of the material, and that lifetimes longer than $\approx 10^8$ seconds could not be expected in GaAs with carrier concentrations of the order of 10^{17} cm^{-3}. In a paper comparing the status of GaAs with other materials, Loferski [1.16] postulated that direct recombination processes were probably occurring in GaAs and limiting the cell output.

The temperature dependence of GaAs cell efficiency was of great interest at that time. GaAs solar cells had been predicted to lose efficiency at a lower rate than Si cells as operating temperatures were raised [1.17]. Measurements on cells from the pilot line confirmed this prediction. It was found that the efficiency fall-off was between 0.02 to 0.03% per °C for GaAs cells, whereas for Si cells the fall-off was measured to be 0.035 to 0.045% per °C. The GaAs cells were also shown to be superior to Si cells at light levels up to 800 mW/cm^2.

A parallel effort in developing GaAs solar cells at Texas Instruments (TI) was reported in 1960 [1.18]. The TI workers produced p/n cells using techniques very similar to those used at RCA. The TI cells were produced by diffusing Zn into the 1 x 2 cm substrates to make p/n cells. They used electroless nickel to make the contacts, and applied antireflection coatings of vacuum-deposited, quarter-wave SiO. Efficiencies achieved were $\approx 7\%$, but it does not appear that TI continued with their GaAs solar cell effort.

A very interesting development was taking place at the Philco Applied Research Laboratory [1.19] where Maxwell and coworkers observed that germanium (Ge) substrates might work well for the deposition of GaAs solar cells since there is a very small lattice mismatch between the two materials and they have very nearly equal coefficients of thermal expansion. The Philco group successfully deposited good quality GaAs layers on Ge substrates. The growth method was to bubble pure hydrogen gas through $AsCl_3$ then pass the gas over metallic gallium which was held in the hot end of a two-zone furnace. The resulting gaseous mixture was then passed over Ge blanks at the cool end of the furnace and GaAs was deposited on the Ge. They found that the best quality GaAs layers resulted when grown on Ge substrates oriented intermediately between the polar (111) and the nonpolar (100) faces. The dark I-V diode characteristics measured for the grown material was very promising. Junctions were grown on the 1 cm x 1 cm substrates by diffusion of Zn to a junction depth of 1-2 μm. Back contacts were made using tin and antimony (Sn-Sb) and the front contacts were made by the evaporation of gold and silver (Au-Ag). Their cells were plagued by low shunt resistances. The shunt resistances could be eliminated either by etching away nearly all the top layer at the expense of I_{sc}, which usually became vanishingly small, or by cutting the cells into smaller pieces until the shunted area was removed. An I-V curve of one of their small area cells (0.05 cm^2), measured under illumination of a 100 mW/cm^2 tungsten simulator, gave an efficiency of $\approx 2.5\%$ with a fill factor of ≈ 0.68. In view of the fact that the simulator they used was a very poor choice for measuring GaAs solar cells this growth technique was seen to be promising, but apparently the group was not successful in curing their major shunting problem.

1.3 Hiatus 1964 - 1972

Further development of GaAs solar cells essentially ceased in the U.S. between 1964 and 1972. The GaAs cells of that era could not compete with conventional Si cells for normal space mission requirements. Cell costs were more than an order of magnitude higher than for Si cells, partly because of the higher cost of starting material and partly because GaAs is inherently a much more difficult material to work with than Si [1.5]. The efficiencies of these GaAs cells were disappointingly low in comparison to the theoretical predictions (maximum measured efficiency of 13% compared to the theoretical efficiency of $\approx 26\%$ shown in Figure 1.3).

The direct bandgap and high optical absorption coefficient of GaAs dictates that the cells should be very thin and that junction depths need to be correspondingly shallow. With so much of the electrical activity taking place near the front surface, the properties of the front surface are expected to play an important role in the performance of these cells. The mathematical factor which delineates the ability of the front surface to act as a sink for carriers is known as the surface recombination velocity. It was recognized by the early GaAs workers that GaAs would probably have large recombination velocities, and that this might be preventing the cells from approaching the maximum theoretical efficiencies. The thin emitter layers would cause such cells to have high series resistances. Other factors were postulated to be limiting factors as well, such as the very short minority carrier diffusion length, but the diffusion lengths did not have to be nearly as large as in silicon because the distances the carriers had

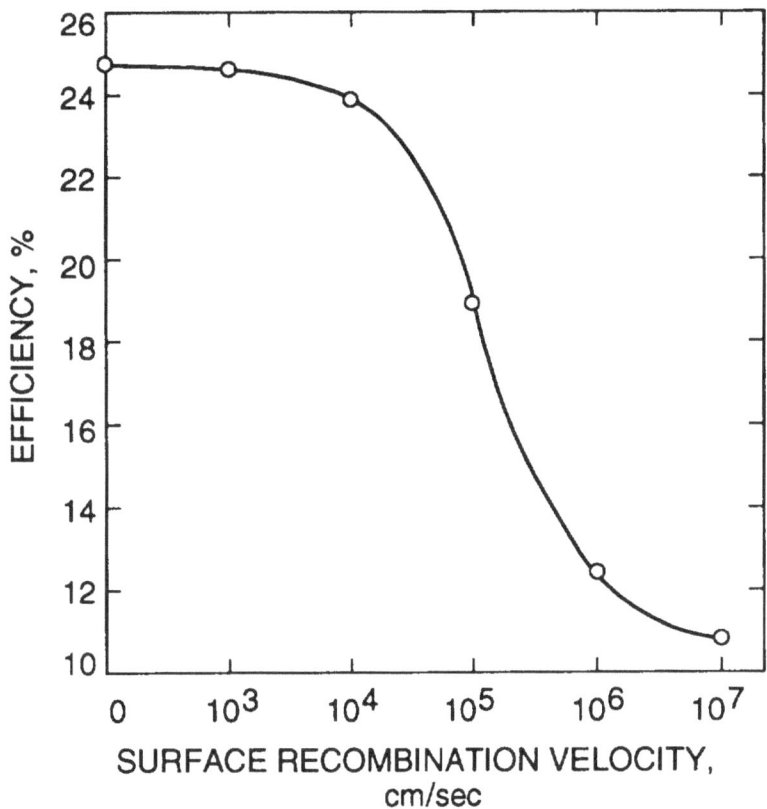

Figure 1.4.   GaAs Solar Cell Efficiency vs. Surface Recombination Velocity (from Ellis and Moss)

to travel were much smaller. Also poor quality GaAs would probably cause a significant amount of carrier recombination in the junction region. In 1972 Ellis and Moss [1.20] showed theoretically that surface recombination was indeed the most probable cause for the poor efficiencies obtained for GaAs cells in practice. Figure 1.4, extracted from Figure 4 of reference [1.20] shows the effect that surface recombination velocity has on GaAs solar cell efficiency. Although this calculation was made for n/p cells, the conclusions are equally valid for p/n cells. Since values of surface recombination velocity in the vicinity of 10^6 cm/sec are commonly found in GaAs, this could readily account for the low efficiencies obtained in the early GaAs cells. Ellis and Moss showed that high surface recombination velocities primarily degraded I_{sc}, but had very little affect on V_{oc}. To cure this problem, they proposed the introduction of a concentration gradient in the emitter dopant. Such a gradient can produce an electric field of the polarity required to nudge the minority carriers toward the junction. These authors also

1-9

calculated that the role of minority carrier diffusion length in the substrate material was important, and that better substrates would be required to exploit the full potential of GaAs solar cells.

1.4   Rebirth 1972 -

Up until 1970, all reported GaAs solar cells had been homojunction cells, where the same material (GaAs) was used for both base and emitter. In 1970 a Russian group [1.21] reported a heterojunction solar cell (heterojunctions are junctions formed between two semiconductors with different energy gaps [1.22]) consisting of a p-type emitter of $Ga_{1-x}Al_xAs$ grown on an n-type base of GaAs by liquid phase epitaxy (LPE). These authors reported AM0 efficiencies of their cells in the 10 to 11% range. Two papers by Woodall and Hovel [1.23, 1.24] followed shortly thereafter, describing a heteroface cell consisting of a p-type $Ga_{1-x}Al_xAs$ layer on a p/n GaAs cell. (A heteroface cell is a p/n homojunction cell to which a semiconductor having a larger energy gap has been added [1.22].) Woodall and Hovel wrote: "The $Ga_{1-x}Al_xAs$ heterojunction system is probably unique among heterojunctions in that very few interface states are expected to exist at the boundary between the two materials because of their extremely close lattice match, while at the same time a growth technique is available that is capable of producing high-quality single-crystal material of controlled doping level. The use of a $Ga_{1-x}Al_xAs$ layer on a GaAs substrate is therefore very attractive for solar-cell applications since the layer can be made thick and heavily doped to reduce the series resistance, will be transparent to most wavelengths which are absorbed efficiently by the GaAs, and should greatly reduce the recombination velocity at the GaAs "surface" (i.e., the interface between the two materials)" [1.23]. They also noted that longer diffusion lengths should be possible in the p-region because lower doping densities are required there when a highly doped $Ga_{1-x}Al_xAs$ window is present.

The energy band structure of their cells is shown in Figure 1.5 [1.24]. The wide-gap front structure will pass all photons with energies less than E_{g1}, but those photons having energies less than E_{g2} will be absorbed in the underlying GaAs device and produce hole-electron pairs there. $Ga_{1-x}Al_xAs$ does have a wide, indirect bandgap which is suitable for such a front structure. Hovel and Woodall [1.24] found that the $Ga_{1-x}Al_xAs$ bandgap increased as x increased, reporting an E_{g1} of 1.69 eV for x = 0.23 up to an E_{g1} of 2.094 eV for x = 0.86. At an x-value of ≈ 0.85 the GaAlAs layer has a highly absorbing direct bandgap of 2.6 eV (λ = 0.477 μm) and an indirect bandgap of 2.1 eV (λ = 0.59 μm), and transmits most of the light from the solar spectrum into the GaAs cell. These bandgap energies increase with x, so that it is advantageous to use a high x-value and to make the GaAlAs layer as thin as possible to allow maximum light transmission (see Figure A.3 in the Appendix). Figure 1.5 also illustrates still another attribute of the heteroface structure. The energy discontinuity ΔE_c, between the p-Ga$_{1-x}$Al$_x$As and the p-GaAs, prevents photogenerated electrons in the p-GaAs layer from entering the window layer and being lost. This discontinuity arises from the difference in electron affinities (electron affinity is the energy difference between the bottom of the conduction band to the vacuum level

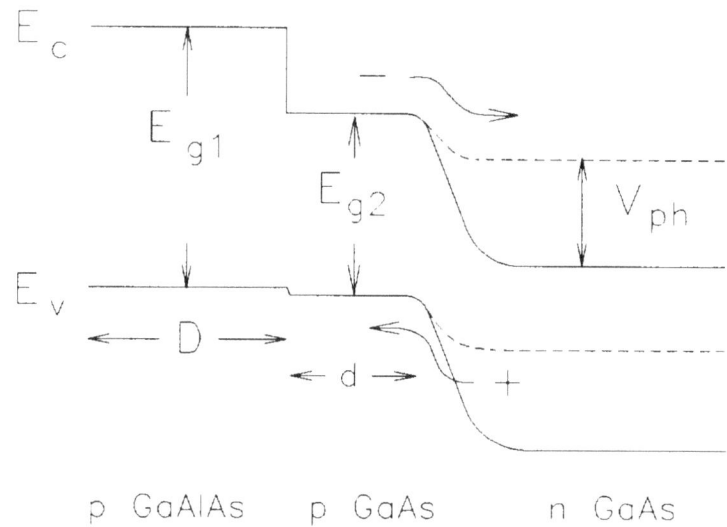

Figure 1.5. Energy Band Diagram of an Al$_x$Ga$_{1-x}$As/GaAs Solar Cell

[1.22]) in the two materials. Hovel and Woodall grew several cells using LPE. They started with n-type GaAs wafers with Si or Sn as dopants. The p-type layers were grown by placing the substrates in a melt of Ga, Al, GaAs chunks, and Zn. During the growth of the Zn-doped Ga$_{1-x}$Al$_x$As layer, Zn diffused into the n-type substrate and simultaneously formed a p-type layer in the GaAs region also. Cell efficiencies between 11.7 and 12.8% were achieved with an AM0 light source. Their cells were small, ≈ 0.2 cm^2, and they used much greater window thicknesses (2 to 20 μm) and junction depths (0.6 to 1.0 μm) than have been found to be optimum today. But the authors demonstrated the principle of an effective heteroface GaAs solar cell, and this work sparked a renaissance in GaAs solar cell research.

1.5   Development of LPE Growth Techniques

This work was followed by a substantial effort by Kamath, Loo, and Ewan at Hughes Research Lab (HRL), beginning in 1974 under internal research and development (IR&D) funding, supplemented by funding by the Air Force Aero Propulsion Lab (AFAPL) [1.25]. They used LPE techniques to make their cells, but used beryllium (Be) as a dopant instead of Zn. They used a very large volume chamber of GaAs solution and developed techniques for inserting graphite substrate holders into this "infinite melt" solution for growing the epitaxial layers. The infinite-melt growth system improved the day-to-day consistency and uniformity in their cells. The use of large quantities of GaAs became affordable because of the increased demand for this material in LEDs and microwave diodes. The cost of GaAs dropped

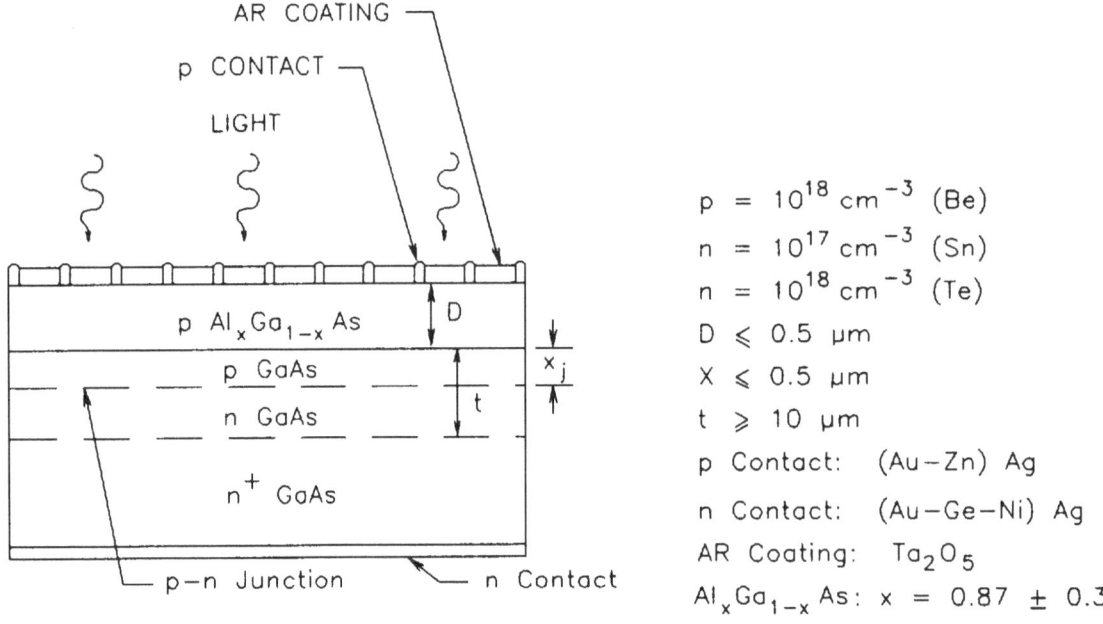

Figure 1.6.   Design of Hughes Research Lab's AlGaAs/GaAs Solar Cell

from $\approx$ \$5/g in 1959 to \$0.50/g in 1976. The HRL group addressed the problem of low diffusion lengths in the available GaAs substrates by growing a high quality n-type buffer layer on the starting n-type GaAs wafers. They found that a such a buffer layer of ≈ 10 μm thickness resulted in considerable improvement in cell efficiencies. A cross-section of the Hughes cell is shown in Figure 1.6. Considerable effort was expended in optimizing the growth parameters, dopant concentrations, window compositions, window thicknesses, junction thicknesses (since radiation damage was a major consideration), contact design, and antireflection coatings.

Many results of this optimization are apparent from Figure 1.6. It is advantageous to use a high value for x in the $Ga_{1-x}Al_xAs$ structure to produce a high bandgap, hence they used an x of 0.87. Since some light absorption does occur in this layer, this effect is minimized by making the layer very thin. The radiation testing of these cells showed that the junction thickness, x_j, should also be very thin. Electron radiation experiments showed that cells made with junction depths of 0.3 μm lost 18% of their maximum power, P_{max}, after exposure to 1 x 10^{15} e/cm^2 of 1 MeV electrons, whereas cells with junction depths of 0.5 μm lost 14% after this exposure and cells with junction depths of 1.0 μm lost 58% [1.26]. So the junction depths were made as shallow as practical, yet deep enough to allow the application of a contact with enough rigor to permit soldering or welding. The antireflection coatings of tantalum pentoxide (Ta_2O_5) and the Ag coated contacts were adapted from Si technology used for space cells.

The HRL group produced a cell with an AM0 efficiency of $\approx 13\%$ late in 1975, and by early 1977 the efficiency had increased to 17.5% [1.27] on a 2 x 2 cm solar cell. By 1978 these cells were produced in some quantity, with average efficiencies of 16% routinely achieved (yield $>$ 50%) and a maximum efficiency of 18%. The cells were used in a space-qualification testing program which included electron and proton radiation testing, post-irradiation annealing, thermal cycling, humidity testing, ultrasonic welding to the contacts, and contact pull strength tests. The HRL cells were found to be superior to Si cells in all respects except weight, fragility, and cost.

Work progressed in many other laboratories as well. At IBM, Hovel and Woodall developed the theory for computing spectral response and I_{sc} of the $Ga_{1-x}Al_xAs/GaAs$ heteroface cells and compared their theoretical results with experimental data [1.28]. They also attacked the problem of n-type substrates with poor hole diffusion lengths by leaching out the lifetime killing defects in a Ga or Ga-Al melt. They also made cells with junction depths greater than 1 μm so that almost all the light would be absorbed in the high quality grown p-layer, thus minimizing the role of the substrate [1.29]. These same workers explored the effect of junction depth on the cell's performance, making cells with junction depths between 0.05 μm and 0.3 μm, but having no windows. A maximum efficiency of $\approx 11\%$ was measured for the cells with x_j of 0.05 μm [1.30]. At AEG Telefunken in Germany, Huber and Bogus [1.31] produced GaAs cells with aluminum arsenide (AlAs) windows in an attempt to minimize absorption losses in the window, but only achieved efficiencies of $\approx 13.5\%$. Johnston and Callahan at Bell Telephone Laboratories also made cells with AlAs windows by using vapor phase epitaxy, where their window layer also served as the emitter. Their cells achieved efficiencies of $\approx 18.5\%$ as measured in sunlight [1.32]. James and Moon at Varian pointed out that a thick $Ga_{1-x}Al_xAs$ layer would be advantageous for concentrator cells because it would reduce the series resistance. They built such cells and measured an efficiency of 17.5% (active area) in sunlight concentrated 284 times [1.33]. The effect of making very thin $Ga_{1-x}Al_xAs$ windows was examined by Hovel and Woodall [1.30], who made windows as thin as 0.2 μm on cells which achieved efficiencies of 18.5%. The same effect was explored by Sahai et al. at Rockwell [1.34], who used LPE to make $Ga_{1-x}Al_xAs/GaAs$ cells with windows as thin as 0.05 μm and incorporated dual layer AR coatings to achieve measured AM0 efficiencies of $\approx 17\%$.

On still another front, a number of papers began to appear on the design of concentrator cells made from GaAs [1.35, 1.36, 1.37]. These authors explained that higher efficiencies are achievable with GaAs cells under higher concentration ratios (with their necessarily higher operating temperatures) because (1) I_{sc} increases as intensity increases and also at higher operating temperatures because the bandgap decreases at higher temperatures, (2) V_{oc} increases because of the higher I_{sc}, yet V_{oc} decreases with temperature at a slower rate than it does for Si, and (3) the fill factor increases because the generation-recombination current in the depletion region becomes a relatively less important part of the forward bias current. In addition, some radiation damage in GaAs cells is expected to anneal at temperatures near the projected operating temperatures of concentrator cells, so GaAs cells were found to be very attractive candidates for concentrator applications.

Kamath et al. developed graphite cassettes capable of holding at least 40 large GaAs wafers for scaling up the LPE processing to production quantities. The problems of assembling GaAs solar cells into panels were investigated by Spectrolab as they manufactured small panels to be flown on the LIPS-II and San Marco satellites. The results of this effort were reported in 1984 [1.38]. At about the same time, Spectrolab began transferring the LPE technology from Hughes Research Lab to their facility, and in late 1985 established a production capability of 20,000 cells per year. This facility and the process steps used in production are described in reference [1.39]. Cells measuring 2 x 4 cm cells and as thin as 125 μm were produced on this LPE production line. The production cells had average efficiencies of 18% with a maximum efficiency of $\approx 20\%$. The cells were thinned from a starting thickness of 250 μm by an etch in a late stage of processing. A contact system was developed wherein the gridlines were deposited directly on the p-GaAs surface, but the contact pads were deposited on top of the AlGaAs window layer. This robust contact system allowed weldable contacts with more than adequate pull strengths to be made to the cells [1.40, 1.41].

Theoretical work continued in modeling the electrical characteristics of the $Ga_{1-x}Al_xAs/GaAs$ cells. Building on the work of Ellis and Moss [1.20] and Hovel and Woodall [1.28], the effect of a graded bandgap was calculated by Hutchby et al. [1.42, 1.43] at NASA Langley and by Sutherland and Hauser [1.44] at North Carolina State University. In England, Debney [1.45] also performed a theoretical analysis of various GaAs structures, including the graded bandgap design, with regard to radiation hardness. Kamath and coworkers also developed a computer model for predicting and optimizing the performance of their cells [1.46].

1.6   Development of OMCVD Growth Techniques

In 1982, the Air Force, who had been the major sponsor of the LPE GaAs work at HRL, adopted the philosophy that metallic-oxide-chemical-vapor-deposition (MOCVD) processing of GaAs solar cells would be more amenable to large-quantity cell production, and they began to support work at Applied Solar Energy Corp. (ASEC) to develop GaAs solar cells using this technique. This program, called the Manufacturing Technology for GaAs Solar Cells or (MANTECH) program, had goals focussing on the ultimate production of large quantities (5000 cells per week) of large (>2 x 4 cm^2), uniformly consistent cells, with AM0 efficiencies over 16% and high radiation resistance. ASEC, with the help of supplemental funding from a commercial solar panel builder, was very successful in this program. By demanding a large quantity of GaAs substrates, ASEC was able to assist substrate suppliers in decreasing the price by half for substrates and at the same time improving the quality and shape (near rectangular) of the substrates.

The cell design developed for this program is shown in Figure 1.7 [1.47]. The design is very similar to the cell design developed at HRL, but there are important differences. An example is the deposition of front contacts to the p-GaAs layer, which avoids a possible loss of integrity if there is subsequent moisture corrosion of the AlGaAs layer. As can be seen from the plots of light absorption in $Ga_{1-x}Al_xAs$ (Figure A.2 in the Appendix), it is desirable to have an x-value as high as possible for maximum window effect; but if x is too high the AlGaAs layer becomes hygroscopic and degrades severely in humidity. A good balance between these competing requirements was found for x-values of ≈ 0.85. The dual antireflection (AR) coating using TiO_2 and Al_2O_3 was designed considering the AlGaAs layer itself to be a part of the AR system since it too was $\approx 1/4$ wavelength in thickness. A comprehensive review of the role of the $Ga_{1-x}Al_xAs$ layer was published by the ASEC workers [1.48]. 1000 cells were produced on a pre-production line by 1984. Cells from this line were used to establish production yields and carry out space qualification tests. These tests included application of coverglasses, tests of solderability and weldability to the contacts, contact adhesion, humidity tests, and radiation tests. The cells met all the required specifications with the exception of the radiation specification, probably because the p-n junction was slightly deeper (0.6 μm) than planned. Based on the knowledge gained from the 1000-cell run, the cell design was changed and an additional 4000-cell production run was completed, yielding average efficiencies of over 16.5%. These cells, with junction depths decreased to $< 0.5 \mu$m, succeeded in meeting the radiation specification, and ASEC concluded that they had a process capable of manufacturing space-ready cells in large quantities [1.49].

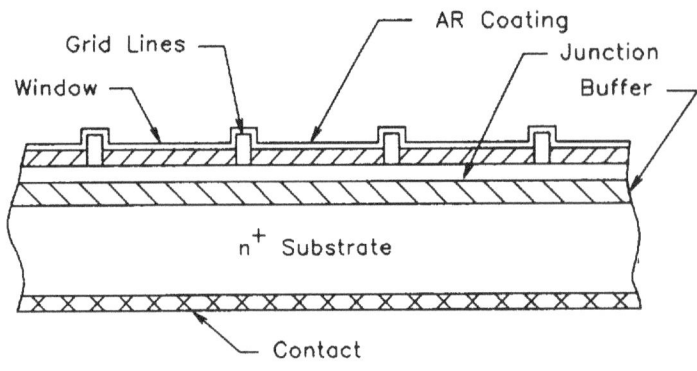

ELEMENT	THICKNESS (μm)	COMPOSITION	DOPANT CONCENTRATION $\times 10^{18}$ cm^{-3}
Grid (p−Contact)	3.4	Ag−Au Zn Au	−
AR Coating	0.1	Ti O_x/Al_2O_5	−
Window (p+)	0.1	Al_x Ga_{1-x}As (Zn)	2 to 4
Junction (p)	0.45	GaAs (Zn)	2
Buffer	< 10	GaAs (Se)	0.2 to 0.5
Substrate (n+)	355	GaAs (Si Dopant)	1 to 4
n Contact	3.4	Ag−Au Ge Ni Au	−

Figure 1.7.   GaAs Solar Cell Structure Developed by ASEC

Although the AlGaAs/GaAs heteroface cells had leapfrogged the efficiencies of Si cells, they were heavy. Since GaAs is quite fragile, the cells had to be made on fairly thick substrates, typically about 300 μm thick. Since the density of GaAs is about twice that of Si, the weight of these cells was equivalent to the weight of a Si cell about 600 μm thick. Since Si cells were routinely made in thicknesses of 200 μm, and sometimes as thin as 100 μm, the power-to-weight ratio of GaAs cells did not compare favorably with Si cells. This problem was attacked by several workers who grew GaAs cells on Ge substrates, or on Si substrates with thin intervening Ge layers [1.50 - 1.53].

In order to pursue the ideas of using an underlying substrate to grow GaAs solar cells, the Air Force funded ASEC to develop rugged, thin GaAs cells in 1986. ASEC chose to use Ge as a substrate because Ge and GaAs have very close lattice constants and thermal-expansion coefficients. Also Ge is a much tougher material and costs about half as much as GaAs [1.54]. A major part of the early work was to passivate the Ge/GaAs junction. This has the disadvantage of not using the possible additional output from the bottom cell. The advantage of an inactive second junction is that it allows a looser specification in the Ge substrate properties, which is in turn reflected in lower Ge costs [1.55]. It also removes the requirement of having to match the current collected from the GaAs junction with the current collected from the GaAs/Ge junction. The consequence of having mismatched junctions is a kink in the

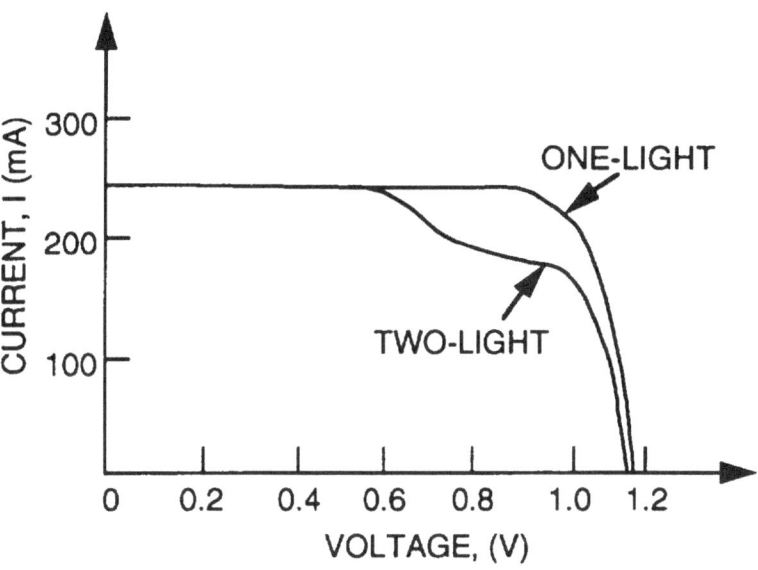

Figure 1.8.   I-V Curves for a GaAs/Ge Solar Cell with an Active GaAs/Ge Junction as Measured Under Different Simulators (© 1990 IEEE, used with permission)

I-V curve near P_{max}, which occurs when one junction becomes current saturated and goes into reverse bias. The resulting I-V curve is the superposition of one I-V curve in forward bias with an I-V curve in reverse bias, and the result is the kink. Such a cell may exhibit a normal I-V curve when measured under a solar simulator with excessive output in the red end of the spectrum, but when measured under true AM0, or with a simulator with a closer spectral match in the red, the kink will appear. An example of the output of a developmental cell with an active Ge/GaAs junction as measured with a red-rich (one light) simulator and with a red-neutral (two-light) simulator is shown in Figure 1.8 [1.56]. ASEC grew a thin quantum barrier between the Ge layer and the GaAs layer to passivate this junction. The barrier region incorporated an energy-band gradient layer by growing a thin AlGaAs layer immediately under the base layer which acts as a minority-carrier reflector. By January of 1988, ASEC produced a 2 x 4 cm GaAs/Ge cell with an AM0 efficiency of 20.5%, albeit on a relatively thick 200 μm substrate [1.57]. The company then turned its attention to making thinner cells with large areas. Cells measuring 4 x 4 cm and 85 μm thick appeared by 1990 with efficiencies routinely in the 17.5% range. ASEC has continued development of this cell design so that today, cells averaging over 19% efficiency are available from their production line. As of this writing, cells are available from ASEC as large as 6 x 6 cm^2 and as thin as 75 μm with efficiencies over 20% [1.58], including cells with all contacts on the rear surface (coplanar back contacts either as a wraparound or a wrapthrough contact) with slightly reduced efficiencies [1.60], and they have delivered over 4,000,000 cm^2 of high-efficiency GaAs/GaAs or GaAs/Ge cells. [1.61]. A thorough review of the implementation of this technology on a large number of space solar arrays by a major aerospace panel manufacturer was given by Datum [1.62].

Spectrolab also began production of GaAs/Ge cells using an MOCVD reactor. Under U.S. Air Force funding, they explored the use of this cell type with the GaAs/Ge junction both active and inactive. With the second junction active, they computed that cell efficiency should improve by $\approx 4\%$, but to achieve this the cell would have to have low recombination velocities at both the rear surface and the emitter of the Ge layer, a thin Ge emitter, good optical coupling of the infrared wavelengths into the Ge cell, long diffusion lengths in the Ge cell, and a highly reflecting back contact on the Ge cell. Dual-junction cells with efficiencies of $\approx 19\%$ were reported [1.59]. Spectrolab grew an additional $p+$ GaAs cap on the AlGaAs window in the areas where the front contacts of titanium-palladium-silver (Ti-Pd-Ag) were deposited, thereby preventing metal diffusion from the contacts into the thin junction during interconnect attachment. Spectrolab also produced cells with inactive second junctions, achieving average efficiencies of over 18%. Their cells ranged in size from 2 x 2 cm up to as large as 6.5 x 6.5 cm, and some cells were fabricated with a wrapthrough contact system. GaAs/Ge cells measuring 2 x 4 cm were used to assemble solar panels for the UOSAT-F satellite. The panels incorporated welded interconnects on cells using the GaAs cap under the contacts described above, and panel efficiencies between 17.9% and 18.6% were measured [1.63].

1.7   GaAs Solar Cells in Space

Other solar cell manufacturers around the world have produced GaAs cells. Japan launched an engineering test satellite (ETS-V) in 1987 and two communication satellites in 1988 into synchronous orbit with panels made with Japanese GaAs solar cells [1.64]. GaAs cells made by CISE in Italy and by EEV in England were used to manufacture experimental panels launched into low-earth orbit in 1992 [1.65]. Cells made by CISE, manufactured by an LPE process using GaAs wafers, were also used on the solar panels for ARSENE, a French satellite [1.66]. In England, the Defence Research Agency (DRA) built experimental satellites (STRV-1 A and B) carrying GaAs solar cells from EEV and Epi Materials Ltd. in England; Telefunken Systemtechnik in Germany; and CISE in Italy. Power for this satellite was generated by panels bearing GaAs solar cells made by ASEC and Spectrolab [1.67].

In the 1990s, GaAs cells have achieved an ever-increasing acceptance by spacecraft manufacturers. Array designers have become more sophisticated and are willing to evaluate more advanced cell designs for their specific mission needs. It has been recognized that the trade-off studies between various solar cell designs need to go beyond the cell level to the panel level or even to the system level, and for some missions the advantage goes to GaAs. One of the advantages is the higher efficiency remaining at end-of-life (EOL). For spacecraft flying at low-earth orbit (LEO), where aerodynamic drag can be a problem, it is advantageous to use panels as small as possible, so GaAs panels are an attractive choice. Reduced drag and lower moment of inertia of the smaller panels also translate into a reduced requirement for station-keeping fuel, needed to offset orbital decay and to maintain satellite orientation. The reduced weight of the panels can also reduce the weight and complexity of the panel deployment mechanisms. Even if the GaAs cells are heavier than Si cells, the net result is often that the total system weight may be less. This type of analysis has also shown in some cases that even though the cost of GaAs cells may be several times that of Si cells, the overall system cost using GaAs cells may be lower.

In the 1990s, NASA is operating under the philosophy of making smaller, better, and cheaper satellites. These smaller satellites often have body-mounted solar panels, with limited areas available for the panels. In such cases the higher efficiency of GaAs/Ge cells may be the only option possible to meet the area and power requirements for the mission. Another interesting advantage for GaAs cells occurs for missions where several satellites are carried into orbit in a single vehicle. The higher-efficiency GaAs cells require less stowage space and, in one example, enabled six satellites with GaAs/Ge arrays to be injected into orbit as compared to only four satellites if Si cell arrays had been used [1.68].

Several satellites are either under design or have been launched with GaAs solar panels. Among these are small satellites such as Miniature Seeker Technology Integration (MSTI), Deep Space Probe Scientific Experiment (DSPSE), and the University of Surrey Satellite (UoSAT). The choice of GaAs for these satellites was driven by power considerations. The IRIDIUM satellites under development by Motorola for a telephone system are examples of commercial satellites where the choice of GaAs was driven by stowage area considerations. NASA missions, where the choice was cost driven, include Earth Orbiting Satellite (EOS), Tropical Rainfall Measuring Mission (TRMM), Fast Auroral Snapshot (FAST), and Submillimeter Wave Astronomy Satellite (SWAS). Air Force missions which have chosen GaAs arrays because of power considerations include Follow-on Early Warning System (FEWS), and Brilliant Eyes (BE).

It is clear that GaAs solar cell technology is mature and widely accepted. The recognition of the advantage of using high-efficiency solar cells has led to increased interest in cascade cells. The experience in producing GaAs cells using MOCVD growth techniques has shown that only small additional growth times may be needed to add a top cell and a tunnel diode to a bottom cell like GaAs. The higher efficiency of these cells, even though more costly at the cell level, may well lead to additional reductions in overall system weights and costs.

References for Chapter 1

1.1   D.E. Aspnes, S.M. Kelso, R.A. Logan, and R. Bhat, "Optical Properties of Al$_x$Ga$_{1-x}$As," *Journal of Applied Physics*, 60 (2), 754, July 15, 1986.

       H.C. Casey, Jr., D.D. Sell, and K.W. Wecht, "Concentration Dependence of the Absorption Coefficient for n- and p-type GaAs between 1.3 and 1.6 eV," *Journal of Applied Physics*, 46, 250, 1975.

1.2   M.A. Green, Editor, *High Efficiency Silicon Solar Cells*, Trans Tech Publications, Switzerland, 1987.

1.3   W.P. Rahilly and B. Anspaugh, "Status of AlGaAs/Ga Heteroface Solar Cell Technology," *Proceedings of the 16th IEEE Photovoltaic Specialists Conference*, 21, 1982.

1.4   H. Welker, "Semiconducting Intermetallic Compounds," *Physica*, 20, 893, 1954.

1.5   P.A. Crossley, G.T. Noel, and M. Wolf, "Review and Evaluation of Post Solar Cell Development Efforts," RCA Astro-Electronics Division Report to NASA, Contract NASW-1427, 1967.

1.6   H. Welker, "Über Neue Halbleitende Verbindungen," *Zeitschrift fur Naturforschung*, 7a, 744, 1952; 8a, 248, 1953.

1.7   O. Madelung, (Translated by D. Meyerhofer), *Physics of III-V Compounds*, John Wiley & Sons, Inc., New York, 1954.

1.8   R. Gremmelmaier, "GaAs-Photoelement," *Zeitschrift fur Naturforschung*, 10a, 501, 1955.

1.9   J.J. Loferski, P. Rappaport, and Linder, "Investigation of Materials for Photovoltaic Solar Energy Converters," RCA Contract #DA-36-039-SC-64643, Interim Reports and Final Report to U.S. Army Signal Corps, 1957.

1.10   J.J. Loferski, "Principles of Photovoltaic Solar Energy Conversion," *Proceedings of the 10th IEEE Photovoltaic Specialists Conference*, 1, 1973.

1.11   D.A. Jenny, J.J. Loferski, and P. Rappaport, "Photovoltaic Effect in GaAs p-n Junctions and Solar Energy Conversion," *Physical Review*, 101, 1208, 1956.

1.12   D.N. Nasledov and B.V. Tsarenkov, "Spectral Characteristics of GaAs Photocells," *Fiz. Tver. Tela*, 1, 1467, 1959; translation *Soviet Physics-Solid State*, 1, 1346, 1960.

1.13   A.R. Gobat, M.F. Lamorte, and G.W. McIver, "Characteristics of High-Conversion-Efficiency Gallium Arsenide Solar Cells," *IRE Transactions on Military Electronics*, ME-6, 20, 1962.

1.14   J.J. Wysocki, P. Rappaport, E. Davison, and J.J. Loferski, "Low-Energy Proton Bombardment of GaAs and Si Solar Cells," *IEEE Transactions on Electron Devices*, ED-13, 420, 1966.

1.15   J.J. Wysocki, "Radiation Properties of GaAs and Si Solar Cells," *Journal of Applied Physics*, 34, 2915, 1963.

1.16   J.J. Loferski, "Recent Research on Photovoltaic Solar Energy Converters," *Proceedings of the IEEE*, 51, 667, 1963.

1.17   J.J. Wysocki and P. Rappaport, "Effect of Temperature on Photovoltaic Solar Energy Conversion," *Journal of Applied Physics*, 31, No. 3, 571, 1960.

1.18   E.G. Bylander, A.J. Hodge, and J.A. Roberts, "Characteristics of a High Solar Conversion Efficiency Gallium Arsenide p-n Junction," *Journal of the Optical Society of America*, 50, 983, 1960.

1.19   K.H. Maxwell, L.C. Bobb, H. Holloway, and E. Zimmerman, "Solar Cells from Epitaxial Gallium Arsenide on Germanium," *Proceedings of the 5th Photovoltaic Specialists Conference*, B-2-1, 1965.

1.20   B. Ellis and T.S. Moss, "Calculated Efficiencies of Practical GaAs and Si Solar Cells Including the Effect of Built-in Electric Fields," *Solid State Electronics*, 13, 1, 1970.

1.21   Zh. I. Alferov, V.M. Andreev, M.B. Kagan, I.I. Protasov, and V.G. *Trofim, Fiz Tekh. Poluprov.*, 4, 2378, 1970. English Translation: *Soviet Physics-Semiconductors*, 4, 2047, 1971.

1.22   S.M. Sze, *Physics of Semiconductor Devices*, Second Edition, John Wiley & Sons, New York, 1981.

1.23   J.M. Woodall and H.J. Hovel, "High-efficiency $Ga_{1-x}Al_xAs$-GaAs Solar Cells," *Applied Physics Letters*, 21, No. 8, 379, 1972.

1.24   H.J. Hovel and J.M. Woodall, "$Ga_{1-x}Al_xAs$ P-P-N Heterojunction Solar Cells," *Journal of the Electrochemical Society*, 120, 1246, 1973.

1.25   J. Ewan, G.S. Kamath, and R.C. Knechtli, "Large Area GaAlAs/GaAs Solar Cell Development," *Proceedings of the 11th IEEE Photovoltaic Specialists Conference*, 409, 1975.

1.26   S. Kamath, et al. (unnamed), "GaAs Solar Cells," Interim Report to AFAPL Contract Nos. F33615-77-C-3150 and F33615-79-C-2039, January 1981.

1.27   R.C. Knechtli, S. Kamath, and R. Loo, "GaAs Solar Cell Development," *Report of the Solar Cell High Efficiency and Radiation Damage Conference, NASA Conference Publication 2020*, 149, 1977.

1.28   H.J. Hovel and J.M. Woodall, "Theoretical and Experimental Evaluations of $Ga_{1-x}Al_xAs$ - GaAs Solar Cells," *Proceedings of the 10th IEEE Photovoltaic Specialists Conference*, 25, 1973.

1.29   H.J. Hovel and J.M. Woodall, "Diffusion Length Improvements in GaAs Associated with Zn Diffusion During $Ga_{1-x}Al_xAs$ Growth," *Proceedings of the 11th IEEE Photovoltaic Specialists Conference*, 409, 1975.

1.30   H.J. Hovel and J.M. Woodall, "Improved GaAs Solar Cells with Very Thin Junctions," *Proceedings of the 12th IEEE Photovoltaic Specialists Conference*, 945, 1976.

1.31   D. Huber and K. Bogus, "GaAs Solar Cells with AlAs Windows," *Proceedings of the 10th IEEE Photovoltaic Specialists Conference*, 100, 1973.

1.32   W.D. Johnston, Jr. and W.M. Callahan, "Vapor-Phase-Epitaxial Growth, Processing and Performance of AlAs-GaAs Heterojunction Solar Cells," *Proceedings of the 12th IEEE Photovoltaic Specialists Conference*, 934, 1976.

1.33   L.W. James and R.L. Moon, "GaAs Concentrator Solar Cell," *Applied Physics Letters*, 26, No. 8, 467, 1975.

1.34   R. Sahai, D.D. Edwall, E. Cory, and J.S. Harris, "High Efficiency Thin Window $Ga_{1-x}Al_xAs$/GaAs Solar Cells," *Proceedings of the 12th IEEE Photovoltaic Specialists Conference*, 989, 1976.

1.35   R.Y. Loo, R.C. Knechtli, and G.S. Kamath, "GaAs Solar Cells for Concentrator Systems in Space," *Proceedings of the Space Photovoltaic Research and Technology Conference, NASA Conference Publication 2256*, 123, 1982.

1.36   N. Kaminar, P. Borden, M. Grounner, P. Gregory, R. LaRue, and P. Cheeseman, "A Passive Cooled 1000x GaAs Module with Secondary Optics," *Proceedings of the 16th IEEE Photovoltaic Specialists Conference*, 675, 1982.

1.37   J.G. Werthen, H.C. Hamaker, G.F. Virshup, C.R. Lewis, and C.W. Ford, "High-Efficiency AlGaAs-GaAs Cassegrainian Concentrator Cells," *Proceedings of the Space Photovoltaic Research and Technology Conference, NASA Conference Publication 2408*, 61, 1985.

1.38   D. Zemmrich, N. Mardesich, B. MacFarlane, and R. Loo, "Gallium Arsenide: Solar Panel Assembly Technology," *Proceedings of the 17th IEEE Photovoltaic Specialists Conference*, 315, 1984.

1.39   N. Mardesich, M. Gillanders, and B. Cavicchi, "Production and Characterization of High Efficiency LPE AlGaAs Space Solar Cells," *Proceedings of the 18th IEEE Photovoltaic Specialists Conference*, 105, 1985.

1.40   B.T. Cavicchi, H.G. Dill, and D.K. Zemmrich, "Novel Front Contact Design for Improved GaAs Solar Cell Performance," *Proceedings of the 19th IEEE Photovoltaic Specialists Conference*, 67, 1987.

1.41   D.R. Lillington, M.S. Gillanders, G.F.J. Garlick, B.T. Cavicchi, G.S. Glenn, and S. Tobin, "Gallium Arsenide Welded Panel Technology for Advanced Spaceflight Applications," *Proceedings of the Space Photovoltaic Research and Technology Conference, NASA Conference Publication No. 3030*, 1988.

1.42   J.A. Hutchby and R.L. Fudurich, "Theoretical Analysis of $Ga_{1-x}Al_xAs$-GaAs Graded Band Gap Solar Cells," *Journal of Applied Physics*, 47, 3140, 1976.

1.43   J.A. Hutchby, "High Efficiency Graded Band Gap $Ga_{1-x}Al_xAs$-GaAs Solar Cell," *Applied Physics Letters*, 26, 457, 1975.

1.44   J.E. Sutherland and J.R. Hauser, "Computer Analysis of Heterojunction and Graded Bandgap Solar Cells," *Proceedings of the 12th IEEE Photovoltaic Specialists Conference*, 939, 1976.

1.45   B.T. Debney, "A Theoretical Evaluation and Optimization of the Radiation Resistance of Gallium Arsenide Solar-Cell Structures," *Journal of Applied Physics*, 50, No. 11, 7210, 1979.

1.46   S. Kamath, R. Knechtli, R. Loo, J. Ewan, G. Wolff, and G. Vendura, "High Efficiency GaAs Solar Cell Final Report," Final Report to the Air Force Aeropropulsion Laboratory, Tech. Report No. AFAPL-TR-78-96, 1978.

1.47   Anon, "Manufacturing Technology for GaAs Solar Cells," Applied Solar Energy Corp. Report to U.S. Air Force on Phase III of A.F. Contract F-33615-81-C-5150, 1984.

1.48   P.A. Iles, K.I. Chang, D. Leung, and Y.C.M. Yeh, "The Role of the AlGaAs Window Layer in GaAs Heteroface Solar Cells," *Proceedings of the 18th IEEE Photovoltaic Specialists Conference*, 304, 1985.

1.49   P.A. Iles et al., "Manufacturing Technology for GaAs Solar Cells," Applied Solar Energy Corporation Final Review to U.S. Air Force on A.F. Contract F33615-81-C-5150, 1986.

1.50   B-Y. Tsaur, J.C.C. Fan, G.W. Turner, F.M. Davis, and R.P. Gale, "Efficient GaAs/Ge/Si Solar Cells," *Proceedings of the 16th IEEE Photovoltaic Specialists Conference*, 1143, 1982.

1.51   S.M. Vernon, M.B. Spitzer, S.P. Tobin, and R.G. Wolfson, "Heteroepitaxial (Al)GaAs Structures on Ge and Si for Advanced High-Efficiency Solar Cells," *Proceedings of the 17th IEEE Photovoltaic Specialists Conference*, 434, 1984.

1.52   B-Y. Tsaur, J.C.C. Fan, G.W. Turner, and B.D. King, "GaAs/Ge/Si Solar Cells," *Proceedings of the 17th IEEE Photovoltaic Specialists Conference*, 440, 1984.

1.53   M. Kato, K. Mitsui, K. Mizuguchi, N. Hayafuji, S. Ochi, Y. Yukimoto, T. Murotani, and K. Fujikawa, "MOCVD AlGaAs/GaAs Solar Cells on GaAs, Ge and Ge/Si Substrate," *Proceedings of the 18th IEEE Photovoltaic Specialists Conference*, 14, 1985.

1.54   K.I. Chang, Y.C.M. Yeh, P.A. Iles, and R.K. Morris, "Heterostructure Solar Cells," *Proceedings of the Space Photovoltaic Research and Technology Conference, NASA Conference Publication No. 2475, 1986.*

1.55   P.A. Iles, F. Ho, and Y.C.M. Yeh, "Status of GaAs/Ge Solar Cells," *Proceedings of the Space Photovoltaic Research and Technology Conference, NASA Conference Publication No. 3107, 1989.*

1.56   P.A. Iles, Y.M. Yeh, F.H. Ho, C.L. Chu, and C. Cheng, "High-Efficiency (>20% AM0) GaAs Solar Cells Grown on Inactive-Ge Substrates," *IEEE Electron Device Letters*, 11, No. 4, 140, 1990.

1.57   P.A. Iles et al., "Rugged Thin GaAs Solar Cell Development," Program Review of Air Force Contract F33615-84-C-2403, January 1988.

1.58   C. Chu, P. Iles, H. Yoo, B. Reed, and J. Krogen, "Recent Technology Advances in Large Area, Lightweight GaAs/Ge Solar Cells," *Proceedings of the 22nd IEEE Photovoltaic Specialists Conference, 1512, 1991.*

1.59   D.R. Lillington, D.D. Krut, B.T. Cavicchi, E. Ralph, and M. Chung, "Progress Toward the Development of Dual Junction GaAs/Ge Solar Cells," *Proceedings of the Space Photovoltaic Research and Technology Conference, NASA Conference Publication No. 3107, 1989.*

1.60   H. Yoo, J. Krogen, C. Chu, P. Iles, and K.M. Bilger, "Development of Coplanar Back Contact for Large Area, Thin, GaAs/Ge Solar Cells," *Proceedings of the 22nd IEEE Photovoltaic Specialists Conference, 1463, 1991.*

1.61   C. Chu and P.A. Iles, "Contact Processing Methods for Production III-V Solar Cells," *Proceedings of the 23rd IEEE Photovoltaic Specialists Conference, 728, 1993.*

1.62   G.C. Datum and S.A. Billets, "Gallium Arsenide Solar Arrays - A Mature Technology," *Proceedings of the 22nd IEEE Photovoltaic Specialists Conference, 1422, 1991.*

1.63   B. Smith, M. Gillanders, P. Vijayakumar, D. Lillington, H. Yang, and R. Rolph, "Production Status of GaAs/Ge Solar Cells and Panels," *Proceedings of the Space Photovoltaic Research and Technology Conference,* NASA Conference Publication No. 3121, 1991.

1.64   S. Matsuda, Y. Yamamoto, and M. Uesugi, "NASDA Activities in Space Solar Power System Research, Development and Applications," *Proceedings of the Space Photovoltaic Research and Technology Conference,* NASA Conference Publication No. 3210, 1992.

1.65   C. Flores, R. Campesato, F. Paletta, G.L. Timo, E. Rossi, L. Brambilla, A. Caon, R. Contini, and F. Svelto, "The Flight Data of the GaAs Solar Cell Experiment on Eureca," *Proceedings of the 23rd IEEE Photovoltaic Specialists Conference, 1369, 1993.*

1.66   L. Brambilla, A. Caon, R. Contini, G. D'Accolti, E. Rossi, G. Verzeni, C. Flores, F. Paletta, E. Rapp, and F. Viola, "GaAs Photovoltaic Technology Application: Arsene Solar Array," *Proceedings of the 23rd IEEE Photovoltaic Specialists Conference*, 1453, 1993.

1.67   C. Goodbody and N. Monekosso, "The STRV-1 A & B Solar Cell Experiments," *Proceedings of the 23rd IEEE Photovoltaic Specialists Conference*, 1459, 1993.

1.68   P. Iles, Private Communication, January 1995.

Chapter 2
Photovoltaic Equations

In this chapter, we will present some of the equations that may prove to be useful for anyone working with GaAs solar cells. We will discuss the solar cell equation and those factors that are the most important in affecting solar cell performance. Since solar cell performance is a strong function of temperature, and this occurs primarily because of its affect on open circuit voltage, V_{oc}, we will examine the terms in the solar cell equation that are responsible for this behavior. The short-circuit current, I_{sc}, is also affected by temperature, and this dependence will be covered also. A great deal of this discussion is based on the books by Sze [2.1] and Hovel [2.2].

2.1   Basic Solar Cell Equation

Figure 2.1 shows an equivalent circuit of a solar cell. It may be thought of as a current generator in parallel with loss mechanisms. The light generator generates a current, I_L. In this model there are two diodes in parallel with the light generator. The first diode represents a bias-dependent dark

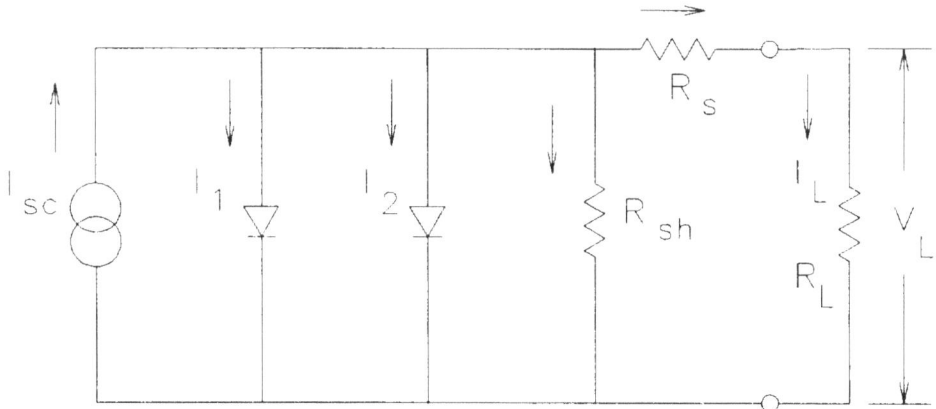

Figure 2.1.   Equivalent Circuit of an Illuminated Solar Cell

current, which is assumed to be due to the diffusion of minority carriers into the junction from its neighboring n- and p-type layers. The second diode represents a model for losses due to carrier generation and recombination through defect centers located in the space-charge region. A third loss mechanism in parallel with the light generator is a shunt resistance, R_{sh}. Finally, there is a series resistance, R_s, arising from the resistivity of the semiconductor material and from the ohmic contacts attached to the cell. The solar cell equation incorporating these loss mechanisms is expressed as follows:

$$I = I_L - I_1 - I_2 - I_{sh} \qquad or$$

$$I = I_L - I_{01}\left[\exp\left(\frac{(V \pm I R_s)}{kT}\right) - 1\right] - I_{02}\left[\exp\left(\frac{(V \pm I R_s)}{2kT}\right) - 1\right] - \frac{V \pm I R_s}{R_{sh}} \qquad (2\text{-}1)$$

Historically, this is the equation that has been used to describe typical solar cell behavior. In this book, the Boltzmann constant, k, will be expressed in units of eV/K when it is clear that a voltage is to be calculated (eg V/kT or E_g/kT); otherwise, k will be expressed in units of Joule/K. T is the absolute temperature, and at T = 300 K, kT = 0.026 volts. Most of the solar cell's temperature dependence arises from the I_{01} and I_{02} terms and will be discussed in subsequent sections. The light-generated current, I_L, is the current density J_L multiplied by the cell's illuminated area (excluding front contact area), but the I_{01} and I_{02} terms require their respective current densities to be multiplied by the total cell area. In most of today's large-area cells, where the cell areas are at least 4 cm^2 and the top contact is only about 5% of the total area, this distinction is usually not important. Typical values for I_{01} of 4 x 10^{-19} A/cm^2 and 9 x 10^{11} A/cm^2 for I_{02} have been reported by Gillanders et al. for LPE cells made by Spectrolab [2.3]. The generation-recombination term is derived from a simplified theory that probably overestimates the value for I_2. Section 2.3 discusses an alternate derivation for I_2.

Figure 2.2 shows the ideal dark and illuminated I-V curves for a typical solar cell, with the illuminated curve occurring in the fourth quadrant, signifying that the cell is delivering power to a load. In the laboratory, this load is often a power supply capable of acting as a current sink. The current directions shown in Figure 2.1 represent the illuminated configuration, and we use the + sign in the V ± IR$_s$ terms of eq. (2-1). If the cell is placed in the dark, the I_L term disappears, the power supply must supply current to the cell, and the current through R$_s$ changes direction, so the − sign is appropriate in the V ± IR$_s$ terms.

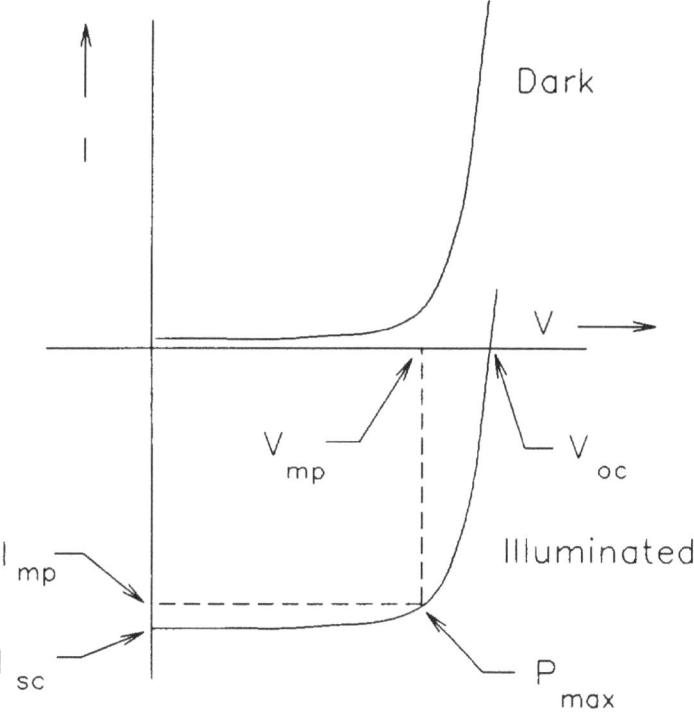

Figure 2.2.   Current-Voltage Curves of a Solar Cell under Illumination and in the Dark

It is often useful to combine the two diodes in the model expressed by eq. (2-1) into a single diode, as follows:

$$I = I_L - I_0 \left[\exp\left(\frac{(V \pm IR_s)}{AkT} \right) - 1 \right] - \frac{V \pm IR_s}{R_{sh}} \qquad (2\text{-}2)$$

where we have used the term I_0 to be a combination of I_{01} and I_{02} and have introduced the A factor or "ideality factor." The A factor will vary between 1 and 2, depending on whether the diffusion current or the generation-recombination current is dominant. In most solar cells, the R_s term is small ($R_s < 0.1$ ohm) and the R_{sh} term is large ($R_{sh} > 1$ x 10^4 ohms). Under these assumptions the term on the right goes away, as does the term involving R_s, resulting in:

$$I = I_L - I_0 \left[\exp\left(\frac{V}{AkT} \right) - 1 \right] \qquad (2\text{-}2a)$$

which is the basic solar cell equation most often used in practice. Again assuming small R_s and large R_{sh}, eq. (2-2) may be solved for the open circuit voltage, V_{oc} (where $I = 0$):

$$V_{oc} = AkT \ln\left(\frac{I_L}{I_0} + 1 \right) \approx AkT \ln\left(\frac{I_L}{I_0} \right) \qquad (2\text{-}3)$$

where we have taken the light-generated current, I_L to be equal to the short-circuit current, I_{sc}.

2.2   Diffusion Current

From eq. (2-3) it is apparent that V_{oc} is directly proportional to temperature and to $\ln I_L$. Experimentally, we know that V_{oc} decreases with temperature, so it must follow that the temperature dependence must be dominated by the behavior of I_0 (either via I_{01}, or I_{02}, or both). In this section we will look at the temperature dependence of I_{01}. The first loss mechanism, I_1 in eq. (2-1), represents a dark current that arises from the diffusion of carriers into the depletion layer from the neighboring p- and n-type layers. The term J_{01} in the equation for J_1 is known as the reverse-saturation current density (because at large negative biases, the current density J_1 is equal to J_{01}), and is given by:

$$J_{01} = \frac{qD_p p_{n0}}{L_p} + \frac{qD_n n_{n0}}{L_n} \qquad (2\text{-}4)$$

This equation is an expression of the ideal diode law, first derived by Shockley [2.4], in which q is the electronic charge, D_n and D_p are carrier diffusion coefficients for electrons and holes, L_n and L_p are minority-carrier diffusion lengths for electrons and holes, p_{n0} is the equilibrium concentration of holes in an n-type semiconductor, and n_{p0} is the equilibrium concentration of electrons in a p-type semiconductor. Equation (2-4) may be rewritten as:

$$J_{01} = q n_i^2 \left[\frac{1}{N_D} \left(\frac{D_p}{\tau_p} \right)^{1/2} + \frac{1}{N_A} \left(\frac{D_n}{\tau_n} \right)^{1/2} \right] \qquad (2\text{-}5)$$

by using $L_p = \sqrt{D_p \tau_p}$ and $L_n = \sqrt{D_n \tau_n}$. τ_p and τ_n are the minority-carrier lifetimes, and we use the law of mass action, $np = n_i^2$, to find equilibrium values of $n_{p0} = n_i^2/p_{p0} \approx n_i^2/N_A$ by assuming that the equilibrium concentration of holes is $\approx$ the number of acceptors, N_A, in p-type material. Similarly, $p_{n0} = n_i^2/n_{n0} \approx n_i^2/N_D$ for n-type material. n_i is the intrinsic-carrier concentration and N_A and N_D are the acceptor and donor concentrations in p- and n-type material, respectively.

In practice, one of the terms in eqs. (2-4) and (2-5) can usually be neglected, i.e. in an abrupt p^+n junction, where $p_{n0} \gg n_{p0}$, the second term is much smaller than the first. The temperature dependence of the reverse-saturation current may be examined with the help of eq. (2-5). Considering only the first term, eq. (2-5) reduces to:

$$ J_{01} \propto q D_p \frac{p_{n0}}{L_p} \propto q \sqrt{\frac{D_p}{\tau_p}} \frac{n_i^2}{N_D} \tag{2-6} $$

If we now follow Sze [2.1] and assume that D_p/τ_p is proportional to T^γ where γ is a constant, then we may rewrite (2-6) as:

$$ J_{01} = \left[T^3 \exp\left(\frac{-E_g}{kT} \right) \right] T^{\frac{\gamma}{2}} = T^{(3 + \frac{\gamma}{2})} \exp\left(-\frac{E_g}{kT} \right) \tag{2-7} $$

In writing this equation, we have used:

$$ n_i^2 = np = N_C N_V \exp\left(\frac{-E_g}{kT} \right) $$

$$ \text{with} \quad N_C = 2 \left(\frac{2\pi m_{de} kT}{h^2} \right)^{3/2} M_c $$

$$ \text{and} \quad N_V = 2 \left(\frac{2\pi m_{dh} kT}{h^2} \right)^{3/2} \tag{2-7a} $$

$$ \text{yielding} \quad n_i = 4.9 \times 10^{15} \left(\frac{m_{de} m_{dh}}{m_o^2} \right)^{3/4} M_c^{1/2} T^{3/2} \exp\left(\frac{-E_g}{2kT} \right) $$

where

      n = occupied number of carriers in the conduction band
      p = hole density near the top of the valence band
      N_c = effective densities of states in the conduction band
      N_v = effective densities of states in the valence band
      m_{de} = density-of-state effective mass for electrons

$m_{de} = (m_1^* m_2^* m_3^*)^{1/3}$

m_1^*, etc. are the effective masses along the principal axes of the ellipsoidal energy surface. For GaAs this surface is spherically symmetrical, so $m_{de} = m^*$

m_{dh} = density-of-state effective mass of the valence band = $(m_{lh}^{3/2} + m_{hh}^{3/2})^{2/3}$

m_{lh} and m_{hh} are the light- and heavy-hole masses

m_o = the free-electron mass

M_c = the number of equivalent minima in the conduction band

E_g = bandgap energy

The room-temperature intrinsic-carrier concentration for GaAs is $\approx 2 \times 10^6$ cm^{-3} in comparison to the value for Si of $\approx 1.5 \times 10^{10}$ cm^{-3}, the difference arising primarily from the difference in bandgap energies.

The bandgap energy itself is a function of temperature and is given by reference [2.5]:

$$E_g(T) = E_g(0) - \frac{\alpha T^2}{(T + \beta)} \qquad (2\text{-}8)$$

The values of $E_g(0)$, α, and β are given in Figure A.4 in the Appendix, along with a plot of E_g vs. T.

In the temperature dependence expressed by eq. (2-7), the exponential term dominates, so J_{01} will increase exponentially with temperature. When eq. (2-7) is substituted in eq. (2-3) for V_{oc}, it is apparent that the exponential term in eq. (2-7) not only determines the temperature dependence of V_{oc}, but the value for V_{oc} itself is largely governed by this term. Since I_L increases with temperature, it will also have a slight effect on V_{oc}'s temperature dependence, as will be discussed in Section 2.4.

In the case of GaAs cells with a $Ga_{1-x}Al_xAs$ window, Hovel and Woodall [2.6] have computed a modification to eqs. (2-4) and (2-5), as follows:

$$J_{01} = \frac{q n_i^2}{N_a} \frac{D_g}{L_g} \left[\frac{\sinh \dfrac{d}{Lg} + \dfrac{S_g L_g}{D_g} \cosh \dfrac{d}{L_g}}{\dfrac{S_g L_g}{D_g} \sinh \dfrac{d}{L_g} + \cosh \dfrac{d}{L_g}} \right]$$

$$+ \frac{q n_i^2}{N_d} \frac{D_p}{L_p} \coth \left(\frac{(H-D)-(d+W)}{L_p} \right)$$

where the first term is the current density from the p-GaAs region and the second term is the contribution from the base. H is the width of the cell, D is the width of the window layer, and W is the depletion width. The contribution from the $Ga_{1-x}Al_xAs$ layer has been assumed to be negligible. The g subscripts refer to the properties of the minority carriers in the p-GaAs region, which has a thickness d; for example S_g is the recombination velocity at the GaAlAs - GaAs interface. The p subscripts refer to minority carriers in the n-GaAs region. The second term ≈ 1 in the usual configuration, where the n-GaAs portion is at least twice the diffusion length, but the first term lowers J_{01} by roughly a factor of 10, when reasonable values are substituted in. However, the correction terms do not appear to modify the arguments given above about the temperature dependence of J_{01}.

2.3 Generation-Recombination Current

Under certain conditions the generation-recombination current, expressed as I_{02} in eq. (2-1), may be the dominant solar-cell loss mechanism. In this case, the temperature dependence of V_{oc} will be somewhat different and will be developed in this section. I_{02} is an internal current flow due to the generation and recombination of carriers that occur in the depletion layer. When the thermal equilibrium of a physical system is disturbed (i.e., $pn \neq n_i^2$), the system tries to return itself to equilibrium and this so-called generation-recombination current is a manifestation of that process. In 1957, Sah, Noyce, and Shockley developed a theory describing this generation-recombination current [2.7]. They made the simplifying assumptions that the mobilities, lifetimes, and doping levels on both sides of the junction were equal, and that the carrier recombination was due to the presence of a single recombination center at an energy level, E_t, near the intrinsic Fermi level. According to this theory the recombination rate, U (in units of cm^{-3}/sec) is [2.8]:

$$U = \frac{p\,n - n_i^2}{(n + n_1)\,\tau_{p0} + (p + p_1)\,\tau_{n0}} \tag{2-9}$$

where τ_{p0} and τ_{n0} are the hole and electron lifetimes in heavily doped n- and p-type material, and n_1 and p_1 are the free-carrier densities that would occur if the Fermi level coincided with the trap level:

$$n_1 = N_C \exp\left(\frac{E_t - E_C}{kT}\right)$$

$$p_1 = N_V \exp\left(\frac{E_V - E_t}{kT}\right) \tag{2-10}$$

The recombination current in the depletion region can be calculated by integrating the recombination rate over the depletion width:

$$J_{rg} = q \int_{x_1}^{x_2} U \cdot dx \tag{2-11}$$

Under forward bias, minority carriers are injected from each side toward the center of the depletion width, and recombine at recombination centers if a significant concentration of opposite carriers also exists there. Recombination is the dominant process with forward bias, and the generation of carriers in the depletion region is negligible in comparison. The recombination rate reaches a maximum at the center of the depletion width and is given by:

$$\textit{Ideal Case:} \quad J_r = \frac{q\,n_i\,W}{\tau_0} \cdot \frac{\exp\left(\dfrac{V}{2\,k\,T}\right)}{\dfrac{(V_{bi} - V)}{k\,T}} \cdot \frac{\pi}{2} \tag{2-12}$$

for the case of equal mobilities, lifetimes, and doping levels on both sides of the junction, and where W is the depletion layer width, and τ_0 (assumed $= \tau_{p0} = \tau_{n0}$) is the lifetime in the depletion region. V_{bi} is the built-in voltage of the junction, and V is the forward-bias voltage.

Under reverse bias, the injection of carriers into the depletion region is greatly diminished and the generation current becomes dominant:

$$\textit{Ideal Case:} \quad J_g = \frac{q\,n_i\,W}{2\,\tau_0} \qquad (2\text{-}13)$$

In the more general case of the Sah-Noyce-Shockley (S-N-S) theory, the lifetimes on the two sides of the junction and the energy position of the recombination centers were allowed to vary, but the doping levels were kept the same. Under these assumptions the recombination current density under forward bias is given by [2.8]:

$$\textit{Recombination Current (S-N-S):} \quad J_r = \frac{q\,n_i\,W}{\sqrt{\tau_{p0}\,\tau_{n0}}} \cdot \frac{2\sinh\left(\dfrac{V}{2\,kT}\right)}{\dfrac{V_{bi}-V}{kT}} \cdot \frac{\pi}{2} \qquad (2\text{-}14)$$

where the forward-bias voltage V is assumed to be restricted to the range V > 4kT, but remains at least 10kT less than V_{bi}. Here an average lifetime in the depletion region is computed from the lifetimes on either side of the junction: $\tau_0 = (\tau_{p0}\,\tau_{n0})^{1/2}$. Under reverse bias, the generation current is given by:

$$\textit{Generation Current (S-N-S):} \quad J_g = \frac{q\,n_i\,W}{2\sqrt{\tau_{p0}\,\tau_{n0}}}\left[\cosh\left(\frac{E_t - E_i}{kT} + \frac{1}{2}\ln\frac{\tau_{p0}}{\tau_{n0}}\right)\right]^{-1} \qquad (2\text{-}15)$$

Choo [2.9] has extended the theory to include the case where the doping levels and lifetimes are different on either side of the junction. Hovel has examined this extended theory [2.8] and concluded that the modification to eqs. (2-14) and (2-15) for generation and recombination currents is sufficiently small that they are accurate within the limitations of the theory. Eqs. (2-14) and (2-15) should therefore be used for I_2 in eq. (2-1), depending on whether forward or reverse bias is applied. Note that I_2 is proportional to n_i, whereas I_1 is proportional to n_i^2. Therefore I_2 is expected to vary with temperature through the weaker exponential term, $\exp(-E_g/2kT)$, in contrast to the I_1 dependence of $\exp(-E_g/kT)$. We also note that the recombination-generation currents are dependent on the depletion-layer width, W.

As a matter of historical interest we note that J_{02} is often written as:

$$J_{02} = \frac{qW}{2} \sigma v_{th} N_t n_i \tag{2-16}$$

In the derivation of this equation it was assumed that there was a single trap in the middle of the gap with a density N_t. The lifetime in the depletion region, τ, is related to the trap density through:

$$\tau_p = \frac{1}{\sigma_p v_{th} N_t} \quad and \quad \tau_n = \frac{1}{\sigma_n v_{th} N_t} \tag{2-17}$$

where σ_n and σ_p are the electron- and hole-capture cross sections, W is the width of the depletion region, and v_{th} is the thermal carrier velocity $= (3kT/m^*)^{1/2}$ with m^* the carrier effective mass. According to Hovel [2.8], eq. (2-16) overestimates the magnitude of I_2 by a factor of between 5 and 10, and is therefore a less accurate formulation than eq. (2-14). However in either formulation, I_2 has a $1/\tau$ dependence, therefore a linear dependence on the trap density, N_t. Since one of the effects of ionizing radiation is to increase the trap density, we would expect to see I_2 increase after solar cells have been exposed to a radiation environment.

The depletion width W, in eqs. (2-12 to 2-15), is related to the doping density, the temperature, and the bias voltage. For an abrupt junction, the depletion width can be computed by using the following expression (commonly known as the C-V formula) giving capacitance per unit area vs. applied voltage:

$$\frac{1}{C} = \frac{W}{\varepsilon_s} = \left(\frac{2 (V_{bi} \pm V - 2kT)}{q \varepsilon_s N_B} \right)^{1/2} \tag{2-18}$$

In this formulation ϵ_s is the semiconductor permittivity (permittivity of free space multiplied by the dielectric constant [see Tables A-1 and A-2 in the Appendix]), V_{bi} is a built-in voltage arising because of the charge distribution that exists on either side of a p-n junction, V is the applied bias voltage (use the + sign for reverse bias, the − sign for forward bias), and N_B is the doping density on the side of the junction that is most lightly doped (e.g. $N_B = N_A$ if $N_D \gg N_A$). Sze [2.1] gives values for V_{bi} between 1.2 and 1.4 volts for GaAs, increasing as the doping density increases from $\approx 10^{14}$ to 10^{17} cm^{-3}. If the depletion layer has been formed by a linear gradation in the doping density, the width depends on applied voltage as:

$$\frac{1}{C} = \frac{W}{e_s} = \left(\frac{12\,(V_{bi} \pm V)}{q\,e_s\,a} \right)^{1/3} \tag{2-19}$$

where a is the impurity gradient in units of cm^{-4}.

2.4   Temperature Dependence of I_{sc}

A qualitative argument for the slow variation of I_{sc} with temperature can be given [2.10]. When the junction is suitably located and the diffusion lengths are sufficiently great, the short-circuit current can be approximated by:

$$\tag{2-20} \quad I_{sc} \approx q\,g_0\,(L_p + L_n)$$

where g_0 is the generation rate of electron-hole pairs per unit volume. Sze [2.1] gives the temperature dependence of mobility, μ, as $T^{-1.0}$ in n-type and $T^{-2.1}$ for p-type GaAs. Using $L = (D\tau)^{1/2}$, and the Einstein relation $D = (kT/q)\mu$, we find that the temperature dependence of L is at most $T^{-1/2}$. Hall, Shockley, and Read [2.11, 2.12] have shown that the temperature dependence of the hole lifetime in the n-type region is given by the following equation, when it is assumed that recombination is determined by the existence of a single-level recombination center:

$$\tau_p = \tau_{p0} \left[1 + \exp\left(\frac{E_T - E_F}{kT} \right) \right] \tag{2-21}$$

where τ_{p0} is the hole lifetime in material in which all the traps are full, E_T is the energy level of the trap, and E_F is the Fermi energy level. The electron lifetime in p-type material is:

$$\tau_n = \tau_{n0} + \tau_{p0} \exp\left(\frac{E_T + E_F - 2E_i}{kT} \right) \tag{2-22}$$

where E_i is the intrinsic energy level. The Fermi level for an n-type semiconductor is near the conduction band and the exponential term in eq. (2-21) is very small. The Fermi level decreases with increasing temperature until it reaches the center of the bandgap. As doping levels increase, the rate of Fermi level decrease becomes smaller, so the exponential term remains small until high temperatures are reached and E_F becomes nearly equal to E_T. In a p-type semiconductor, the Fermi level is near the valence band, and it rises toward mid-bandgap with temperature in an analogous fashion. Thus the lifetime is expected to be relatively constant in temperature regions of practical solar cell applications, and the diffusion-length dependence is primarily determined by the temperature dependence of the mobility.

Most of the dependence of I_{sc} on temperature occurs because the bandgap decreases with increasing temperature (see Figure A.4 in the Appendix). As the bandgap decreases, photons of longer wavelength will be able to create electron-hole pairs, and more of the solar energy spectrum may be utilized, causing an increase in I_{sc}.

2.5   Spectral Response for p/n Solar Cells

The absorption of electromagnetic radiation, referred to as the optical injection of carriers, is fundamental to the operation of a solar cell. In the absorption process, a photon is absorbed and an electron-hole pair is created if the photon has energy greater than the bandgap energy. Optical energy at wavelength λ is continuously absorbed as it penetrates into the material, so that a light beam with incident intensity F_0 has intensity F after penetration to a depth x:

$$F = F_0 \exp[-\alpha(\lambda)x] \tag{2-23}$$

where $\alpha(\lambda)$ is the optical absorption in cm^{-1}, and F_0 is the number of incident photons per cm^2 per sec per unit bandwidth. At a depth x, the generation rate is dF/dx, or:

$$G(\lambda) = \alpha(\lambda) F_0 [1-R(\lambda)] \exp(-\alpha(\lambda) x) \tag{2-24}$$

where $G(\lambda)$ is the carrier generation rate at depth x, and $R(\lambda)$ is the reflection loss from the front surface. The generation rate is equal to the loss rate, consisting of loss due to carrier recombination and due to current flow. Considering electrons generated in unit volume of p-type material under low injection conditions (light-produced carrier density $\ll n_{p0}$), which almost always holds for solar cell operating conditions, the recombination rate is proportional to the number of excess electrons and inversely proportional to the lifetime of the carriers. The current flow arises because of carrier diffusion, which is a random thermal movement of particles under the influence of a concentration gradient. There will be a particle flux in a direction opposite to and proportional to the concentration gradient, with the diffusion constant D_n being the proportionality constant, e.g. for electron current density:

$$J_n = q D_n \frac{d(n_p - n_{p0})}{dx} \qquad \text{in p-type cells} \qquad (2\text{-}25)$$

and for holes:

$$J_p = -q D_p \frac{d(p_n - p_{n0})}{dx} \qquad \text{in n-type cells} \qquad (2\text{-}26)$$

where we have used the terminology (n_p - n_{p0}) to be the excess carrier concentration, i.e. that induced by the light. We assume here that there are no electric fields outside of the depletion layer that might be induced by concentration gradients, etc. If there were, the current density equations would have an additional term, e.g. for electrons = $q\mu_n n_p \mathscr{E}$. The net particle loss out of the unit volume will be $-dJ_n/dx$. Bringing this all together for electrons in p-type semiconductors and writing J for the current densities we have:

$$G(\lambda) = \frac{n_p - n_{p0}}{\tau_p} - \frac{1}{q} \frac{dJ_n}{dx} \qquad (2\text{-}27)$$

and for holes in n-type material:

$$G(\lambda) = \frac{p_n - p_{n0}}{\tau_p} + \frac{1}{q} \frac{dJ_p}{dx} \qquad (2\text{-}28)$$

In an n/p solar cell with uniform doping on each side of an abrupt-step junction, eqs. (2-24), (2-26), and (2-28) may be combined as a one-dimensional description of the top n-type layer:

$$D_p \frac{d^2(p_n - p_{n0})}{dx^2} + \alpha F_0 (1 - R) \exp(-\alpha x) - \frac{(p_n - p_{n0})}{\tau_p} = 0 \qquad (2\text{-}29)$$

The general solution to this equation is:

$$(p_n - p_{n0}) = A \cosh\left(\frac{x}{L_p}\right) + B \sinh\left(\frac{x}{L_p}\right) - \frac{\alpha F_0 (1 - R) \tau_p}{(\alpha^2 L_p^2 - 1)} \exp(-\alpha x) \qquad (2\text{-}30)$$

We impose two boundary conditions for evaluating the constants A and B. When carriers reach the front surface they will recombine, and this is expressed as:

$$D_p \frac{d(p_n - p_{n0})}{dx} = S_p(p_n - p_{n0}) \qquad at \ x = 0 \tag{2-31}$$

where S_p is a measure of the enthusiasm with which the carriers recombine at the surface. S_p, called the surface recombination velocity, has units of cm/sec. At the junction, the minority carriers are swept up by the strong electric field in the depletion region, so the second boundary condition is stated as:

$$p_n - p_{n0} = 0 \qquad at \ x = x_j \tag{2-32}$$

where x_j is the junction width. Using these boundary conditions, the excess hole density is found to be:

$$(p_n - p_{n0}) = \left[\frac{\alpha F_0 (1 - R) \tau_p}{(\alpha^2 L_p^2 - 1)} \right] x$$

$$\left[\frac{\left(\frac{S_p L_p}{D_p} + \alpha L_p \right) \sinh \left(\frac{x_j - x}{L_p} \right) + \exp(-\alpha x_j) \left(\frac{S_p L_p}{D_p} \sinh \frac{x}{L_p} + \cosh \frac{x}{L_p} \right)}{\frac{S_p L_p}{D_p} \sinh \frac{x_j}{L_p} + \cosh \frac{x_j}{L_p}} - \exp(-\alpha x) \right] \tag{2-33}$$

Differentiating eq. (2-33) with respect to x and using eq. (2-26), the hole current density at the junction edge is found to be:

$$J_p = \left[\frac{q F_0 (1 - R) \alpha L_p}{(\alpha^2 L_p^2 - 1)} \right] x$$

$$\left[\frac{\left(\frac{S_p L_p}{D_p} + \alpha L_p \right) - \exp(-\alpha x_j) \left(\frac{S_p L_p}{D_p} \cosh \frac{x_j}{L_p} + \sinh \frac{x_j}{L_p} \right)}{\frac{S_p L_p}{D_p} \sinh \frac{x_j}{L_p} + \cosh \frac{x_j}{L_p}} - \alpha L_p \exp(-\alpha x_j) \right] \tag{2-34}$$

To find the electron current collected from the base, eqs. (2-24), (2-25), and (2-27) are used with the boundary conditions:

$$n - n_{po} = 0 \qquad\qquad at\ x = x_j + W \qquad\qquad (2\text{-}35)$$

$$D_n \frac{d(n_p - n_{po})}{dx} = -S_n (n_p - n_{po}) \qquad\qquad at\ x = H \qquad\qquad (2\text{-}36)$$

where W is the width of the depletion region and H is the total width of the cell. Eq. (2-35) is analogous to eq. (2-32) and states that the excess electron density vanishes at the edge of the depletion region, and eq. (2-36) states that recombination takes place at the rear surface of the cell. Usually there is an ohmic contact there and the recombination velocity is considered to be infinite.

Using these boundary conditions, the electron distribution in a uniformly doped p-type base is given by:

$$(n - n_{po}) = \left[\frac{\alpha F_0 (1-R) \tau_n}{(\alpha^2 L_n^2 - 1)} \exp[-\alpha(x_j + W)] \right] x$$

$$\left[\cosh\left(\frac{x - x_j - W}{L_n}\right) - \exp[-\alpha(x - x_j - W)] - \frac{\frac{S_n L_n}{D_n}\left[\cosh\left(\frac{H'}{L_n}\right) - \exp(-\alpha H')\right] + \sinh\left(\frac{H'}{L_n}\right) + \alpha L_n \exp(-\alpha H')}{\frac{S_n L_n}{D_n}\sinh\left(\frac{H'}{L_n}\right) + \cosh\left(\frac{H'}{L_n}\right)} \right] x$$

$$\sinh\left(\frac{x - x_j - W}{L_n}\right)$$

$$(2\text{-}37)$$

where $H' = H - (x_j + W)$ is the total base thickness.

The electron current at the junction edge is:

$$J_n = \left[\frac{qF_0(1-R)\,\alpha L_n}{(\alpha^2 L_n^2 - 1)} \exp[-\alpha(x_j + W)] \right] x$$

$$\left[\alpha L_n - \frac{\dfrac{S_n L_n}{D_n}\left(\cosh\dfrac{H'}{L_n} - \exp(-\alpha H')\right) + \sinh\dfrac{H'}{L_n} + \alpha L_n \exp(-\alpha H')}{\dfrac{S_n L_n}{D_n}\sinh\dfrac{H'}{L_n} + \cosh\dfrac{H'}{L_n}} \right] \qquad (2\text{-}38)$$

Some photocurrent is generated within the depletion region. Assuming that the field in this region is high enough so that all the photogenerated carriers produced in this region are accelerated out and therefore collected, the photocurrent generated in this region is given by:

$$J_{dr} = qF_0(1 - R)\exp(-\alpha x_j)[1 - \exp(-\alpha W)] \qquad (2\text{-}39)$$

The total short-circuit photocurrent at a given wavelength is given by the sum of eqs. (2-34), (2-38), and (2-39), and the spectral response is equal to this sum divided by $qF_0(1 - R)$.

2.6   Spectral Response for p/n Solar Cells with a $Ga_{1-x}Al_xAs$ Window

The above equations are appropriate for either Si or GaAs p/n solar cells. The equations may be cast into the appropriate form for n/p cells by simply interchanging the p's and n's in the above equations. However, they do not apply to GaAs cells with a $Ga_{1-x}Al_xAs$ window. Hovel and Woodall [2-6] have worked out the equations pertaining to p/n GaAs cells with a window, which include collection from the $Ga_{1-x}Al_xAs$ window. Their formulation is based on a solar cell with the energy band diagram shown in Figure 2.3.

The minority-carrier continuity equation for the photogenerated carriers in the $Ga_{1-x}Al_xAs$ layer is:

$$D_a \frac{d^2(n_p - n_{p0})}{dx^2} + \beta F_0 \exp(-\beta x) - \frac{(n_p - n_{p0})}{\tau_a} = 0 \qquad (2\text{-}40)$$

In this and the following formulation, the a subscripts will pertain to the $Ga_{1-x}Al_xAs$ layer, and D will denote its thickness. S_a is the surface recombination velocity, τ_a is the minority-carrier lifetime, D_a is the diffusion coefficient, L_a is the diffusion length, and β is the optical absorption coefficient.

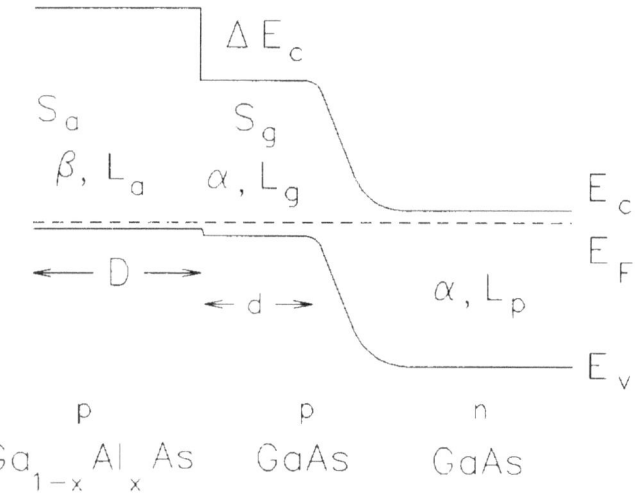

Figure 2.3.   Energy Band Diagram in Equilibrium of a p-Ga$_{1-x}$Al$_x$As/p-GaAs/n-GaAs Device
(© 1973 IEEE, used with permission)

The current collected from the Ga$_{1-x}$Al$_x$As layer at the p-GaAs interface is given by:

$$J_D = \left[\frac{q F_0 \beta L_a}{(\beta^2 L_a^2 - 1)} \right] x$$

$$\left[\frac{\beta L_a + S_a \dfrac{\tau_a}{L_a} \left(1 - \exp(-\beta D) \cosh \dfrac{D}{L_a} \right) - \exp(-\beta D) \sinh \dfrac{D}{L_a}}{S_a \dfrac{\tau_a}{L_a} \sinh \dfrac{D}{L_a} + \cosh \dfrac{D}{L_a}} - \beta L_a \exp(-\beta D) \right] \quad (2\text{-}41)$$

The boundary condition at the surface is:

$$D_a \frac{d(n_p - n_{p0})}{dx} = S_a(n_p - n_{p0}) \qquad at \ x = 0 \qquad (2\text{-}42)$$

and at the interface the boundary condition is:

$$n_p - n_{p0} = 0 \qquad at \ x = D^- \qquad (2\text{-}43)$$

The collection from the p-layer is dependent on the photons which pass through the Ga$_{1-x}$Al$_x$As layer with energy greater than the GaAs bandgap. The continuity equation for electrons produced in the p-GaAs layer is:

$$D_g \frac{d^2(n_p - n_{p0})}{dx^2} + \alpha F_0 \exp(-\beta D) \exp[-\alpha(x - D)] - \frac{(n_p - n_{p0})}{\tau_g} = 0 \qquad (2\text{-}44)$$

Here, the g subscripts are used for the p-GaAs layer, and d is used to denote its thickness. S_g is the surface recombination velocity, τ_g is the minority-carrier lifetime, D_g is the diffusion coefficient, L_g is the diffusion length, and α is the optical absorption coefficient. The boundary condition at the interface on the GaAs side is:

$$D_g \frac{d(n_p - n_{p0})}{dx} = S_g(n_p - n_{p0}) - \frac{J_D}{q} \qquad (2\text{-}45)$$

i.e. the current in the p-GaAs at the interface is equal to the component due to recombination at the interface, minus the current injected from the alloy layer. The difference in the two boundary conditions (2-45) and (2-43) is the result of the large (0.7 eV) energy barrier, ΔE_c, at the interface. The boundary condition at the junction edge is:

$$n_p - n_{p0} = 0 \qquad at \ x = (D + d) \qquad (2\text{-}46)$$

The photocurrent at the junction edge arising from both the $Ga_{1-x}Al_xAs$ and GaAs layers is:

$$J_{(D+d)} = \left[\frac{q F_0 \exp(-\beta D)\, \alpha L_g}{(\alpha^2 L_g^2 - 1)} \right] x$$

$$\left[\frac{\alpha L_g + S_g \frac{\tau_g}{L_g}\left(1 - \exp(-\alpha d)\cosh\frac{d}{L_g}\right) - \exp(-\alpha d)\sinh\frac{d}{L_g}}{S_g \frac{\tau_g}{L_g}\sinh\frac{d}{L_g} + \cosh\frac{d}{L_g}} + \right. \qquad (2\text{-}47)$$

$$\left[\frac{J_D}{S_g \frac{\tau_g}{L_g}\sinh\frac{d}{L_g} + \cosh\frac{d}{L_g}} \right]$$

The last term in (2-47) describes how the current collected from the $Ga_{1-x}Al_xAs$ layer is attenuated by the surface recombination term S_g, and the minority-carrier lifetime in the p-layer, τ_g.

The photocurrent collected from the depletion region, again assuming all photogenerated carriers produced there are collected, is:

$$J_W = q F_0 \exp(-\beta D)\ (1 - \exp(-\alpha W)\ \exp(-\alpha d) \qquad (2\text{-}48)$$

The photocurrent collected from the base is:

$$J_{(D+d+W)} = \left[\frac{q F_0 (1-R)\ \alpha L_p}{(\alpha^2 L_p^2 - 1)}\ \exp[-\beta D]\ \exp[-\alpha (d+W)] \right] x$$

$$\left[\alpha L_p - \frac{\dfrac{S_p L_p}{D_p}\left(\cosh\dfrac{H'}{L_p} - \exp(-\alpha H')\right) + \sinh\dfrac{H'}{L_p} + \alpha L_p \exp(-\alpha H')}{\dfrac{S_p L_p}{D_p} \sinh\dfrac{H'}{L_p} + \cosh\dfrac{H'}{L_p}} \right] \qquad (2\text{-}49)$$

which reduces to the following equation in the limit: $H' >> L_p$

$$J_{(D+d+W)} = \frac{q F_0 \exp(-\beta D)\ \alpha L_p}{(\alpha L_p + 1)}\ \exp{-\alpha (d+W)} \qquad (2\text{-}50)$$

The total short-circuit current density is given by the sum of eqs. (2-47), (2-48), and (2-49). The external spectral response, defined to be the short-circuit current density collected (mA/cm²), divided by the light intensity (mW/cm²) incident on the external surface of the cell, is computed by:

$$SR(\lambda) = \frac{J_{(D+d)}(\lambda) + J_{(D+d+W)}(\lambda) + J_W(\lambda)}{F_0(\lambda)} \qquad (2\text{-}51)$$

2-19

References for Chapter 2

2.1   S.M. Sze, *Physics of Semiconductor Devices*, Second Edition, John Wiley & Sons, New York, 1981.

2.2   H.J. Hovel, *Solar Cells*, in R.K. Willardson and A.C. Beer, Eds., *Semiconductors and Semimetals*, Volume 11, Academic Press, New York, 1975.

2.3   M. Gillanders, B. Cavicchi, D. Lillington, and N. Mardesich, "Pilot Production Experience of LPE GaAs Solar Cells," *Proceedings of the 19th IEEE Photovoltaic Specialists Conference*, 289, 1987.

2.4   W. Shockley, "The Theory of p-n Junctions in Semiconductors and p-n Junction Transistors," Bell Syst. Technical Journal, 28, 435, 1949; *Electrons and Holes in Semiconductors*, D. Van Nostrand, Princeton, New Jersey, 1950.

2.5   C.D. Thurmond, "The Standard Thermodynamic Functions for the Formation of Electrons and Holes in Ge, Si, GaAs, and GaP," *Journal of the Electrochemical Society*, 122, No. 8, 1133, 1975.

2.6   H.J. Hovel and J.M. Woodall, "Theoretical and Experimental Evaluations of $Ga_{1-x}Al_xAs$ - GaAs Solar Cells," *Proceedings of the 10th IEEE Photovoltaic Specialists Conference*, 25, 1973.

2.7   C.T. Sah, R.N. Noyce, and W. Shockley, "Carrier Generation and Recombination in P-N Junctions and P-N Junction Characteristics," *Proceedings of the IRE*, 45, 1228, 1957.

2.8   H.J. Hovel, "The Effect of Depletion Region Recombination Currents on the Efficiencies of Si and GaAs Solar Cells," *Proceedings of the 10th IEEE Photovoltaic Specialists Conference*, 34, 1973.

2.9   S.C. Choo, "Carrier Generation-Recombination in the Space-Charge Region of an Asymmetrical p-n Junction," *Solid-State Electronics*, 11, 1069, 1968.

2.10   J.J. Wysocki and P. Rappaport, "Effect of Temperature on Photovoltaic Solar Energy Conversion," *Journal of Applied Physics*, 31, 571, 1960.

2.11   W. Shockley and W.T. Read, "Statistics of the Recombination of Holes and Electrons," *Physical Review*, 87, 835, 1952.

2.12   R.N. Hall, "Electron-Hole Recombination in Germanium," *Physical Review*, 87, 387, 1952.

Chapter 3

Instrumentation Techniques for Measuring GaAs Solar Cells

In this section, the commonly used instrumentation techniques for assessing radiation effects in solar cells will be discussed, with emphasis on those special procedures peculiar to the measurement of GaAs solar cells. The most commonly used measurement in the analysis of radiation effects in solar cells is the current-voltage characteristic under illumination. The major concern is the interaction of photons in the semiconductor in order to produce hole-electron pairs. Since this interaction is strongly dependent on optical wavelength, it is extremely critical that the light sources used for illumination match the solar (or other) spectrum in which the cell is designed to be used.

3.1   Light Sources and Solar Simulators

The spectral irradiance of the sun at a distance of 1 astronomical unit (AU) is of primary importance in solar cell analysis for solar cell applications in outer space. This irradiance, just outside the Earth's atmosphere, is known as air mass zero (AM0), because it penetrates zero air mass at the point of measurement. The astronomical unit is defined to be the average Earth-Sun distance of 1.49597890×10^8 km. The values of solar spectral irradiance proposed by Johnson [3.1] were used extensively until about 1970. Johnson's results indicated that the solar constant (also referred to as the total irradiance, or the luminosity) was 139.5 mW/cm^2, and also that the solar spectrum closely approximated that of a 6000 K black body. Thekaekara et al. reviewed several high-altitude spectral measurements of the Sun in 1971 [3.2] and published a solar irradiance spectrum, which became commonly known as the Thekaekara spectrum, or more formally as the NASA SP 8005 spectrum. Integration of the Thekaekara spectrum resulted in a total irradiance of 135.3 $\pm$ 2.1 mW/cm^2. Labs and Neckel reported values for the solar irradiance in 1968 [3.3], with revisions in 1981 [3.4] and 1984 [3.5]. They concluded that the solar constant lies between 136.8 and 137.7 mW/cm^2. In 1985, R.C. Willson published a composite solar spectral irradiance at 1 AU [3.6], which was based on the best experimental observations over the world at the time. This composite was based on the results of Donnelly and Pope [3.7] for wavelengths between 0.2 and 0.3 μm; Arvesen, Griffin, and Pearson [3.8] for wavelengths between 0.3 and 0.4 μm and between 1.3 and 2.5 μm; and Labs and Neckel [3.3] for wavelengths between 0.4 and 1.3 μm and between 2.5 and 3.0 μm. Willson's composite spectrum is tabulated in Table 3-1. The total irradiance used for calculating the numbers in the column headed "percentage of total" was 136.8 mW/cm^2. In reference [3.6], Willson also tabulated the values measured for the total

Table 3-1.   Solar Spectral Irradiance [Ref. 3.6] (used with permission of the author)

Lambda (Microns)	Flux mW cm^2-micron	Integral 0 to Lambda (mW/cm^2)	Percent of Total	Lambda (Microns)	Flux mW cm^2-micron	Integral 0 to Lambda (mW/cm^2)	Percent of Total
0.205	1.10	0.0055	00.004	0.705	142.0	64.8260	47.422
0.215	4.83	0.0352	00.026	0.715	138.0	66.2260	48.446
0.225	6.65	0.0926	00.068	0.725	136.0	67.5960	49.448
0.235	6.73	0.1595	00.117	0.735	132.0	68.9360	50.429
0.245	6.01	0.2232	00.163	0.745	128.0	70.2360	51.380
0.255	9.98	0.3031	00.222	0.755	126.0	71.5060	52.309
0.265	29.20	0.4990	00.365	0.765	124.0	72.7560	53.223
0.275	21.20	0.7510	00.549	0.775	121.0	73.9810	54.119
0.285	18.30	0.9485	00.694	0.785	118.0	75.1760	54.993
0.295	58.50	1.3325	00.975	0.795	116.0	76.3460	55.849
0.305	53.70	1.8935	01.385	0.805	114.0	77.4960	56.691
0.315	72.90	2.5265	01.848	0.815	110.0	78.6160	57.510
0.325	87.80	3.3300	02.436	0.825	108.0	79.7060	58.307
0.335	102.00	4.2790	03.130	0.835	105.0	80.7710	59.086
0.345	99.70	5.2875	03.868	0.845	101.0	81.8010	59.840
0.355	102.0	6.2960	04.606	0.855	98.6	82.7990	60.570
0.365	115.0	7.3810	05.399	0.865	96.8	83.7760	61.285
0.375	113.0	8.5210	06.233	0.875	94.7	84.7335	61.985
0.385	106.0	9.6160	07.034	0.885	92.4	85.6690	62.669
0.395	121.0	10.7510	07.865	0.895	92.0	86.5910	63.344
0.405	163.0	12.1710	08.903	0.905	89.8	87.5000	64.009
0.415	170.0	13.8360	10.121	0.915	87.4	88.3860	64.657
0.425	166.0	15.5160	11.350	0.925	85.7	89.2515	65.290
0.435	167.0	17.1810	12.568	0.935	84.1	90.1005	65.911
0.445	193.0	18.9810	13.885	0.945	82.3	90.9325	66.520
0.455	201.0	20.9510	15.326	0.955	80.6	91.7470	67.116
0.465	199.0	22.9510	16.789	0.965	78.9	92.5445	67.699
0.475	199.0	24.9410	18.245	0.975	77.3	93.3255	68.270
0.485	189.0	26.8810	19.664	0.985	75.6	94.0900	68.830
0.495	196.0	28.8060	21.072	0.995	73.9	94.8375	69.376
0.505	190.0	30.7360	22.484	1.05	66.1	98.6875	72.193
0.515	183.0	32.6010	23.849	1.15	54.0	104.6925	76.586
0.525	186.0	34.4460	25.198	1.25	44.7	109.6275	80.196
0.535	192.0	36.3360	26.581	1.35	38.4	113.7825	83.235
0.545	186.0	38.2260	27.963	1.45	32.3	117.3175	85.821
0.555	184.0	40.0760	29.317	1.55	27.5	120.3075	88.008
0.565	183.0	41.9110	30.659	1.65	23.7	122.8675	89.881
0.575	183.0	43.7410	31.998	1.75	19.2	125.0125	91.450
0.585	181.0	45.5610	33.329	1.85	15.2	126.7325	92.708
0.595	176.0	47.3460	34.635	1.95	13.1	128.1475	93.744
0.605	174.0	49.0960	35.915	2.05	10.66	129.3355	94.613
0.615	171.0	50.8210	37.177	2.15	8.44	130.2905	95.311
0.625	166.0	52.5060	38.410	2.25	7.22	131.0735	95.884
0.635	164.0	54.1560	39.617	2.35	6.04	131.7365	96.369
0.645	160.0	55.7760	40.802	2.45	5.30	132.3035	96.784
0.655	152.0	57.3360	41.943	2.55	4.83	132.8100	97.154
0.665	156.0	58.8760	43.069	2.65	4.19	133.2610	97.484
0.675	152.0	60.4160	44.196	2.75	3.65	133.6530	97.771
0.685	149.0	61.9210	45.297	2.85	3.20	133.9955	98.022
0.695	145.0	63.3910	46.372	2.95	2.81	134.2960	98.241

solar irradiance on several flights involving balloons, rockets, and spacecraft. The average value reported for these measurements lies between 136.7 and 137.0 mW/cm². In 1985, Wehrli of the World Radiation Center also published a very detailed composite AM0 spectrum [3.9]. Wehrli's spectrum covers the wavelength range between 0.1995 μm and 10.075 μm, in small-wavelength steps (920 steps). This spectrum is also based on the Neckel and Labs observations in the wavelength region between 330 and 869 nm, and appears to be in substantial agreement with Willson's spectrum. The Wehrli spectrum integrates to a total solar irradiance of 136.7 mW/cm².

A plot showing both the Thekaekara spectrum and Willson's spectrum is presented in Figure 3.1. GaAs solar cells respond to wavelengths between 0.3 and 0.9 μm, and Si solar cells respond to wavelengths between 0.3 and 1.2 μm, so it is apparent that these irradiance spectra deviate markedly in the wavelength regions that are significant for solar cell applications.

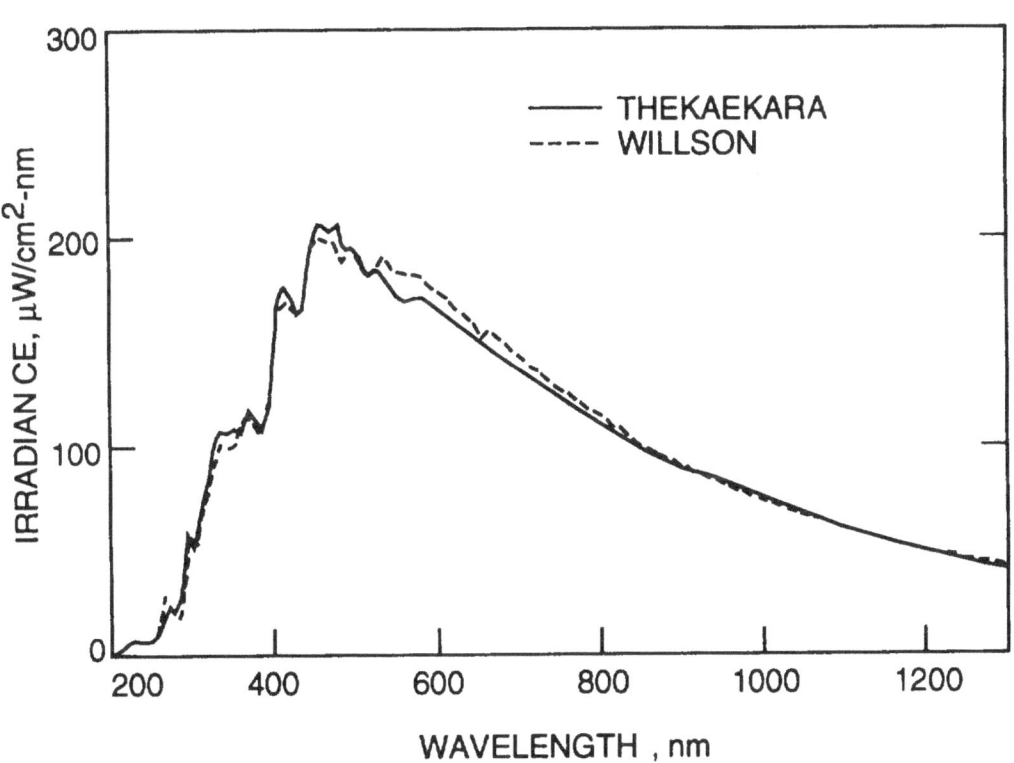

Figure 3.1.   Comparison of the NASA SP 8005 and the Willson Composite AM0 Solar Spectra

There is evidence that the solar intensity varies over the course of a solar cycle. Willson and Hudson reported the results of the total solar irradiance over a complete solar cycle [3.10] as measured by their Active Cavity Radiometer Irradiance Monitor (ACRIM I), which flew on the Solar Maximum Mission satellite during 1980 and 1989. They found that the total irradiance varied between approximately 136.7 mW/cm² at the time of minimum sunspot activity and 136.9 mW/cm² at solar maximum. A more recent publication [3.11] which includes data from the ACRIM II radiometer, which was launched aboard the Upper Atmosphere Research Satellite (UARS) in 1991, supports the adoption of a value of 136.8 mW/cm² for the solar constant.

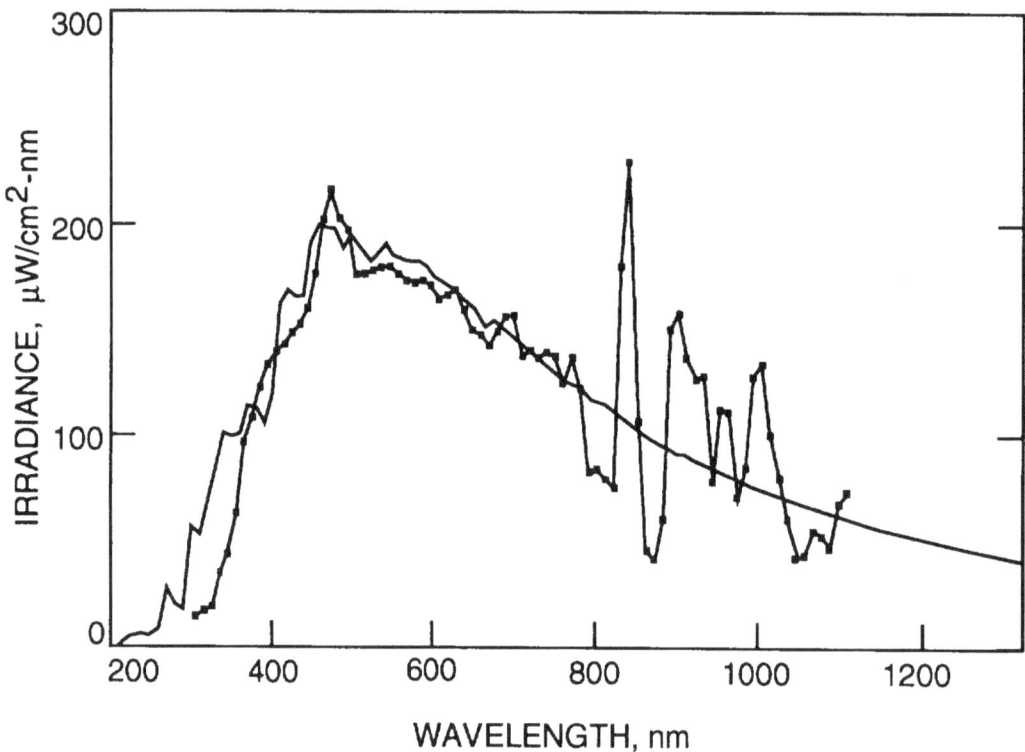

Figure 3.2.   Normalized Spectral Irradiance of Spectrolab X25 Mark II Solar Simulator

The most common solar simulation technique in use today is the use of xenon arc lamps with filters to remove undesired line spectra in the near-infrared region. Such simulators produce a spectrum that has a very good overall match to the solar spectrum, but they have some intense line spectra in the wavelength region above 750 μm. The spectral irradiance of two popular simulators, the Spectrolab X-25 and the Spectrolab XT10 simulators, are shown in Figures 3.2 and 3.3, along with plots of the AM0 spectrum. In these and the following figures showing normalized irradiances, the normalization was

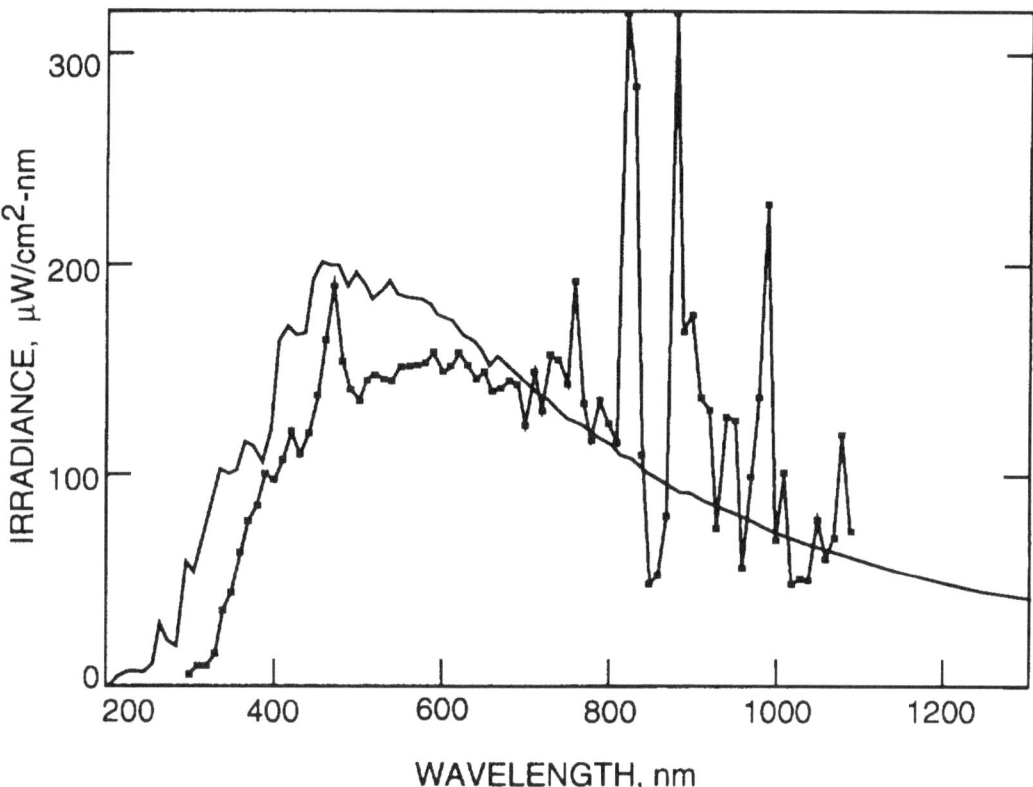

Figure 3.3.   Normalized Spectral Irradiance of Spectrolab XT-10 Solar Simulator

performed by adjusting the entire measured simulator irradiance by a constant multiplicative factor so that the power in the wavelength region between 0.3 μm and 1.090 μm was equal to the power of the AM0 spectrum in this same wavelength region. A simulation technique that is receiving a great deal of interest and developmental effort is the use of a dual light source consisting of a filtered xenon arc lamp, as in the simulators described above, with an additional tungsten lamp. Appropriate bandpass filters are interposed in each light beam so that the xenon lamp produces the short wavelengths and the tungsten lamp generates the long wavelengths, thereby eliminating most of the line spectra. Dual-light-source simulators are expected to become more popular as the development of multijunction heterostructure cells reachès maturity. In a dual-junction cell for example, both the upper and the lower cell must be illuminated by a spectrum that very closely matches its particular region of response, otherwise one or the other of the cells will act as a load for the other cell and distort the measurement. The spectral irradiance of a combination xenon-tungsten source simulator used by the Applied Solar Energy Corporation is shown in Figure 3.4.

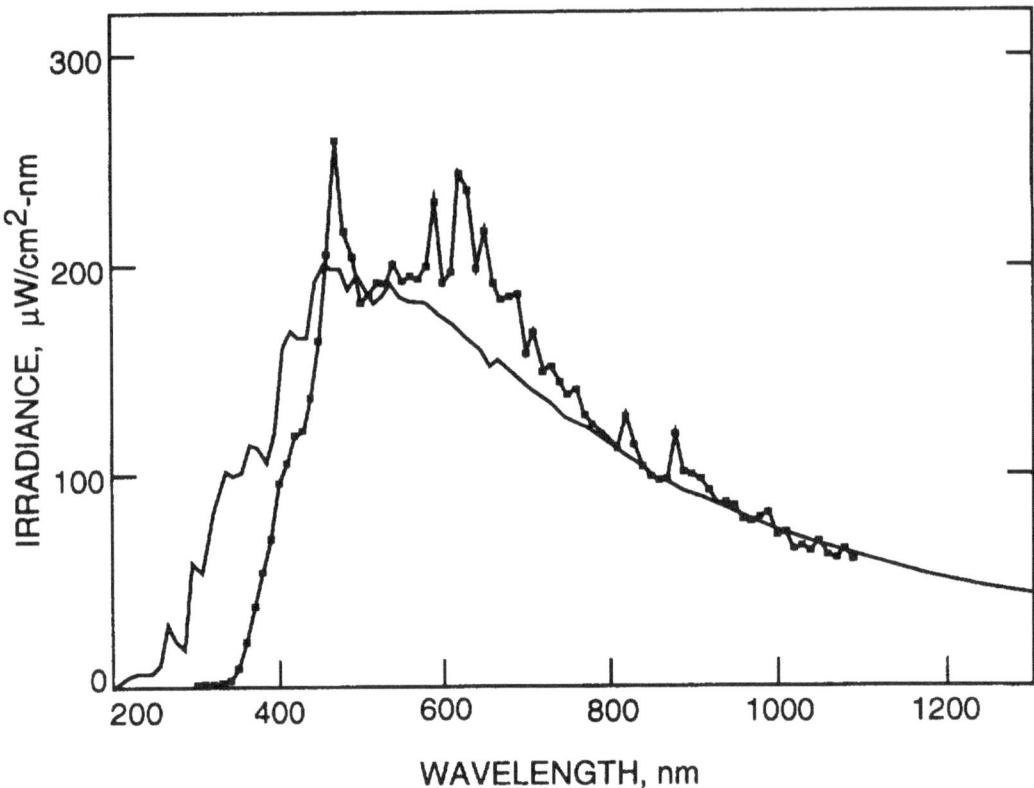

Figure 3.4.   Normalized Spectral Irradiance of ASEC Tungsten/Xenon Solar Simulator

Tungsten lamps have been used as light sources, both with and without some type of filtering, but the spectral output of such sources is a very poor match to the solar spectrum. The peak output of tungsten sources occurs in the red or near-infrared region, depending on the temperature of the tungsten filament. Since this is the wavelength region of the solar cell response that is most changed by radiation, the use of such sources is likely to show a much more severe cell degradation than would be shown in the real environment. The spectral irradiance of a General Electric Model ELH lamp, which is a tungsten-halogen projector bulb/reflector assembly operating at a high temperature, is shown in Figure 3.5. This lamp incorporates a dichroic reflector which lets most of the longer-wavelength light pass out the rear surface of the bulb. It is clear that use of such a lamp as a simulator for examining solar cell parameters could result in large measurement errors.

An important development in the field of solar simulation is the use of pulsed xenon arc lamps for solar cell and solar array testing [3.12, 3.13, 3.14]. When xenon arc lamps are operated in a pulsed mode at high current densities, the high intensity peaks in the 0.8 to 0.9 μm region are greatly attenuated, as shown in the spectral irradiance curves shown in Figure 3.6. The pulsed-xenon simulators produce

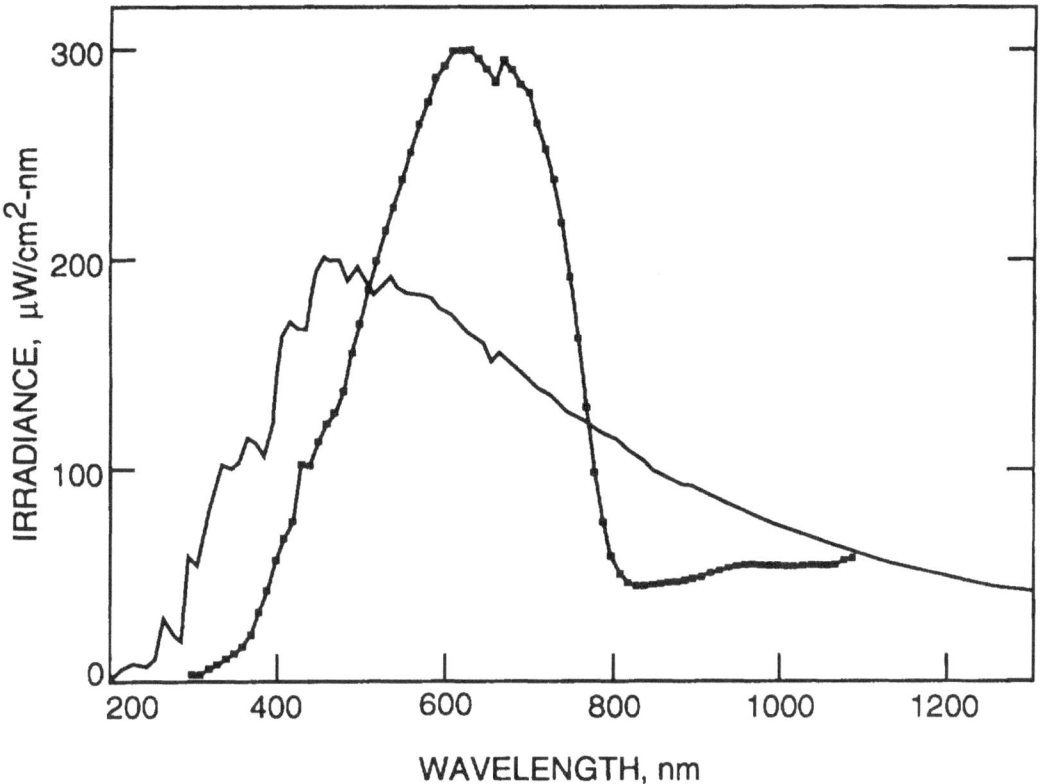

Figure 3.5.   Normalized Spectral Irradiance of ELH Tungsten-Halogen Lamp

an overabundance of energy in the ultraviolet (UV) region, so a much better simulation of the AM0 spectrum is obtained by using a UV absorption filter. Figure 3.6 shows the output of a pulsed simulator operated without a filter in comparison with the spectrum produced using a 1-mm-thick Schott" GG-395 filter [3.15]. This particular filter attenuates UV wavelengths below 0.395 μm, and the result is an irradiance spectrum that closely matches the AM0 spectrum at wavelengths greater than 0.36 μm. The onset of the spectral response for both Si and GaAs solar cells occurs at ≈ 0.36 μm, so this filter is quite suitable for solar cell measurements. Other filters are available from Schott with cutons at lower wavelengths. These types of simulators are commonly known as Large Area Pulsed Solar Simulators (LAPSS) because they have the capability of uniformly illuminating a large area (up to 5 m in diameter) at source-to-target distances of ≈ 11 meters, with the full AM0 intensity. These systems produce a pulse of light that lasts for ≈ 2 milliseconds. Solar cell or panel data can be accumulated during a time period of ≈ 1 millisecond in the central portion of the pulse. The measurement therefore requires the use of an electronic load which can sweep through the entire load voltage range during this time, together with a fast data collection system that allows the simultaneous reading of the cell voltage, cell current, and standard cell output (usually current) during the load sweep. In addition to the advantage of a close

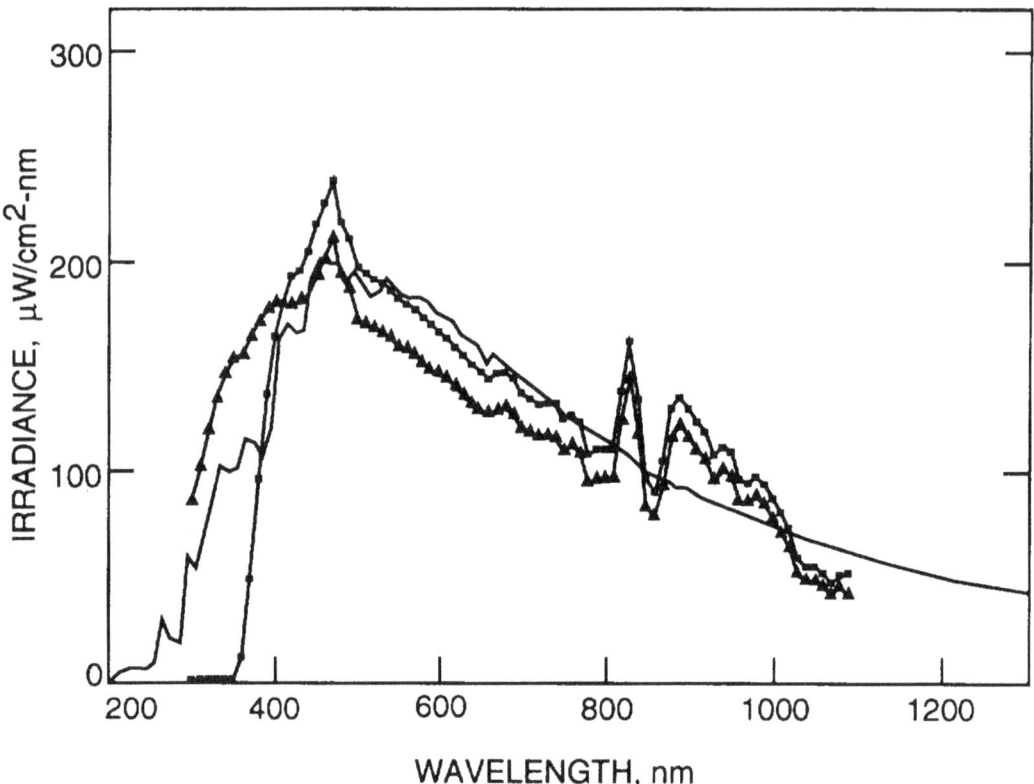

Figure 3.6.   Normalized Spectral Irradiance of the JPL LAPSS with and without Schott UV Filter

spectral match to the Sun, these systems have the very desirable attribute of not heating up the test specimen during measurement. This is a tremendous advantage for the measurement of solar panels which have a very difficult temperature control problem under continuous AM0 illumination. A disadvantage of the LAPSS simulator systems arises during the testing of Si solar cells made with back surface fields (BSF). These cells have a very high capacitance that varies with bias voltage. This capacitance is high enough so that it does not allow the cell output to follow the quickly changing load voltage. The accurate measurement of BSF cells and panels requires breaking the load sweep into several smaller load steps with a separate light pulse for each load range [3.16]. This has not been a problem with GaAs solar cells because of the rapid decay of excess carriers.

The setting of solar simulator intensities to match the AM0 intensity is an extremely important aspect of solar cell measurement. This is usually accomplished by the use of a solar cell which has been calibrated to be used as an intensity standard. The short circuit current, I_{sc}, of a solar cell is directly proportional to the light intensity falling on the cell, so these intensity standards are usually loaded to operate in the short-circuit mode with fixed resistors, the resistance value depending on the cell type and size. Primary standards for this purpose have been calibrated aboard a variety of flight vehicles, including the Shuttle, high-flying aircraft [3.17, 3.18, 3.19], rockets [3.20, 3.21], and high-altitude balloons. Secondary standards are produced by simultaneously measuring a primary standard along with the secondary standard in a solar simulator beam. For this calibration procedure to be accurate, the secondary standard must have a spectral response that exactly matches that of the primary standard. Another method of producing intensity standards requires the very accurate measurement of the cell's spectral response and the very accurate measurement of the irradiance of an illumination source. These data, along with a measurement of the short circuit current of the cell in the illumination source can be used to calculate the output of the cell in AM0 as follows:

$$I_L(AM0) = I_L(simulator) \cdot \frac{\int SR(\lambda) E(\lambda)_{space} \, d\lambda}{\int SR(\lambda) E(\lambda)_{sim.} \, d\lambda} \tag{3-1}$$

where SR is the measured spectral response of the cell under calibration. To produce calibration standards of the accuracy required by most users today requires very difficult, accurate and painstaking measurements of the quantities noted above. The primary standard cells commonly in use are generated by a NASA/JPL program wherein solar cells are flown on high-altitude balloons. The cells are mounted on a solar tracker, which is in turn mounted on top of the balloon, and the calibration data is telemetered to a ground station. Measurements are taken while the balloon is at or above 120,000 ft. (36,600 m) and within 1 hour of solar noon [3.22]. In 1984, a group of calibration standard cells were flown on the Shuttle, then recovered and flown on a balloon. The two sets of measurements agreed to within 1% and were considered to be a validation of the accuracy of the balloon flight calibration method [3.23]. When the effects of atmospheric absorption are properly accounted for, the results of the calibrations performed aboard the NASA Lewis high-altitude aircraft are in substantial agreement with the balloon flight data [3.24].

3.2   Current-Voltage Characteristics

The measurement of current-voltage (I-V) characteristics is the primary means of evaluating a solar cell. The evaluation is made by applying a load resistance across the illuminated cell, varying the load resistance from zero to infinity, and measuring the resultant current into the load and the voltage across the cell. The measurement is simple in principle, but attention to several practical details is necessary to insure accurate results. The first requirement is a suitable fixture to hold the cell during the measurement. The fixture must incorporate electrical contact probes which will make proper contact to the front cell contacts, and especially in the case of GaAs cells, not induce any damage to the underlying shallow junction. Electrical and thermal contact with the rear cell surface is usually assured by applying a vacuum to the cell through small holes suitably located in the fixture. That part of the fixture in contact with the rear surface of the solar cell is made of an electrically (and thermally) conducting metal block such as copper or brass.

The fixture must also provide some means of holding the cell at a fixed, controllable temperature. The vacuum hold-down feature is normally sufficient to clamp the cell thermally to the fixture. Control of the fixture temperature may be done by flowing water through passages in the fixture (it's best if the water passages do not intersect the vacuum passages). The water source in this case would be a temperature-controlled water bath. An alternate temperature-control method is the use of a thermoelectric (TE) module installed between the fixture and its mounting surface. The TE module is powered by a small power supply. The TE module has the advantage that it can either heat or cool the fixture by reversing the polarity of the supply voltage, which is easily accomplished by using an inexpensive temperature controller. The TE system can have a lower heat capacity than the water-cooling system so that temperature changes may be made quickly. By carefully monitoring both the cell temperature and the fixture temperature with the system under AM0 illumination, it has been found that cell temperatures usually run between $\approx 0.6°C$ and $1°C$ warmer than the fixture, due to the thermal impedance of the solar cell. Two standard cell measurement temperatures have been adopted. One, $28°C$, was probably originally selected because it also happens to be 300 K, a nice round number. Many laboratories use a measurement temperature of $25°C$. This temperature is favored by solar cell manufacturers because cell output is higher. Most measurements quoted in this publication were made at $28°C$.

To insure that the voltages measured are true voltages at the cell itself, a dual set of probes is used for the top-surface probes. One set, which may be one or more probes depending on cell size and configuration, is designed to carry cell current to the load. The external circuit connected to the current probes may incorporate insertion elements such as low-resistance, precision resistors for measuring cell current. The other set of probes carries no current and is used to measure cell voltage. Since connecting the cell to a variable resistive load cannot yield a true short-circuit current reading, a power supply is often used as a load. Variation of the power supply voltage then changes cell load. The power supply must be capable of sinking current at zero voltage. The power supply may be either a bipolar supply or two supplies connected in series opposition to achieve short-circuit current measurement. In the recent past, the cell voltage and current probes were connected to an X-Y recorder, and the power supply voltage was changed manually. The resulting I-V curve produced on the recorder was then used to establish the important electrical parameters. It is much more common today to use a computer to perform these chores. The computer is used to vary the power supply through a suitable voltage range, measure both current and voltage at selected load points, plot the data, and extract the desired electrical parameters from the digital data. Since the cell current when loaded near short-circuit condition is proportional to the illumination intensity, the current output will follow the short-term fluctuations of the arc-lamp-based simulators, and the current plot is likely to be very uneven in this region. It may be necessary to take an average of as many as 30 current readings at each load point in this region. Maximum power, P_{max}, is simply found by computing the power produced at each load point and selecting the maximum product. The series resistance of a solar cell may also be determined from I-V characteristics at two or more different illumination levels [3.25, 3.26].

3.3   Special Considerations for GaAs Solar Cells

There are several special considerations that must be given to the measurement of GaAs cell I-V characteristics. The probes on the fixture should be designed to apply a pressure of no more than 180 pounds/in², otherwise the shallow junctions may be damaged [3.27]. One way of successfully applying the requisite pressure is to build the probes out of phosphor bronze, so that the contact area is a known, rectangular dimension. The use of a weight on a lever arm may then be used to apply the required force. If the phosphor bronze contact assembly is made to be removable, then it is easy to replace the probes with a geometry that will match any practical front-contact configuration.

When measuring the I-V characteristic of GaAs cells with a power supply acting as the load, it is extremely important to be sure that at the time the probes make or break contact with the cell there is zero voltage on the probes. Successful systems have assured this condition by programming the power supply to deliver zero voltage or by shorting out the cell probes with a relay contact before cell insertion or removal. The reason for these cautionary measures is that GaAs cells are easily damaged by transients from some electronic loads. It is characteristic of voltage- or current-controlled power supplies to attempt to maintain a programmed output at its terminals. If such power supplies are presenting a nonzero voltage to the fixture when contact with the cell is made or broken, the power supply may force a very high voltage to the contact with possibly disastrous results. In addition to using the methods described above to program zero voltage, a bypass diode may be connected in parallel with the GaAs cell under test, so that transient voltages of the wrong polarity will be shunted by the diode.

Since GaAs cells are sensitive to reverse bias conditions, screening measurements of cells intended for use in radiation testing or in general applications may be specified. Solar cell manufacturers will generally screen their cells before delivery. One screening procedure is to test each cell at reverse currents of $\approx 30\%$ of I_{sc}. Cells which will be used on solar panels that might be shaded during spacecraft operation may be subjected to reverse currents as high as I_{sc}. Therefore, a second screening of such cells using reverse currents of $\approx 110\%$ to 167% times I_{sc} is a reasonable test. It has been shown that such screening measurements need not be performed at low temperatures, but screening tests performed at temperatures somewhat above room temperature may be necessary to flag cells likely to fail under reverse bias [3.28]. It is important to ramp up the reverse currents somewhat gradually in performing the screening tests, rather than immediately applying the full output power of a current-regulated supply to these cells. Experience has shown that (1) GaAs cells made on Ge substrates have been less susceptible to reverse current breakdown than GaAs cells made on GaAs substrates, and (2) cells which have passed reverse bias screening tests can withstand repeated exposure to reverse bias, thereby reducing the chance of operational degradation [3.29].

The measurement of GaAs solar cells as a function of temperature, using conventional solar simulators, has some special problems. Figure 3.7 shows the spectral response of a GaAs/Ge solar cell measured at a temperature of 28°C, superimposed on the spectral irradiance output curve of an X-25 solar simulator. The dashed lines represent the estimated spectral response of the cell at temperatures of +100°C and -100°C. This increase in response toward longer wavelengths at higher temperature is due to the decrease of bandgap with temperature (see Figure A.4 in the Appendix). At the higher temperature, the spectral response becomes sensitive to one of the large output lines in the xenon spectrum and causes a larger increase in I_{sc} with temperature than would occur in natural sunlight. As the cell temperature is lowered, the spectral response moves toward a valley in the xenon spectrum, resulting in a faster-than-normal decrease in I_{sc}. The net effect is that the I_{sc} temperature coefficients measured in this temperature range with a xenon-arc-based solar simulator are likely to be somewhat high.

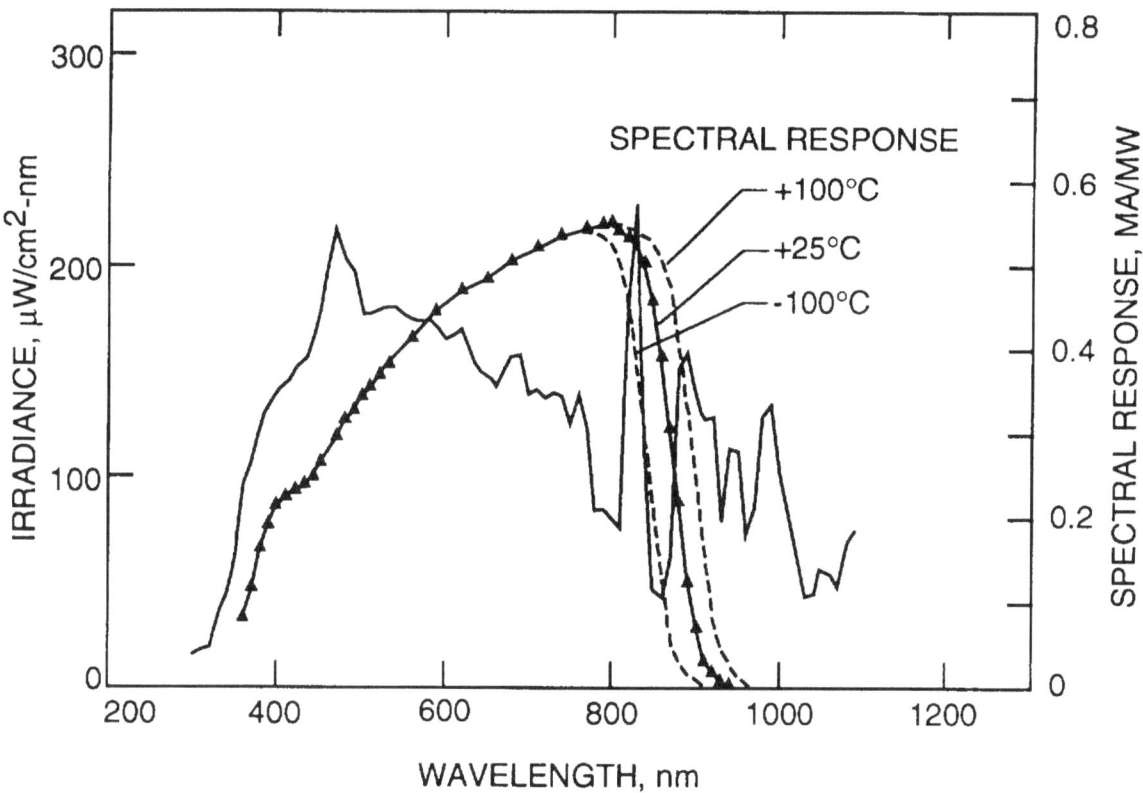

Figure 3.7.   Comparison of the Irradiance of an X-25 Solar Simulator with the Spectral Response of a GaAs Solar Cell

3.4   Spectral Response Measurements

Spectral response measurements are very useful for evaluating changes in solar cells due to radiation effects.   The spectral response (amps/watt) is a measure of the short-circuit current density generated by the cell under a range of monochromatic illuminations at known power densities.   The spectral response is often reported in terms of relative units when absolute values of the incident light intensities are not accurately known.   Various schemes have been used to measure the spectral response of solar cells.   High-resolution-monochromators are used when extreme accuracy is desired.   Narrow bandpass filters can be used as a source of nearly monochromatic light also.   When a monochromator is used, there are two methods to normalize the solar cell output to the light intensity.   Tungsten-halogen light sources operating at high intensity (e.g. projector bulbs) are usually used in monochromators, and the entrance slit width can be varied to control the optical power density illuminating the cell under test.   In some systems, the entrance slit width can be automatically controlled to maintain a constant optical power density on the solar cell.   An alternate approach is to maintain a constant slit width and allow the optical power density on the cell to vary with wavelength. This variation is then factored into the spectral response calculation. Either prism or grating monochromators may be used for measuring spectral response. A grating monochromator requires the insertion of filters to block higher order wavelengths and several filters may be required over appropriate wavelength regions. One advantage of using a grating monocromator is that large wavelength ranges can be covered with suitable gratings. A prism monochromator does not require the high-order filtering, so its use is simpler. One disadvantage with this type of monochromator is that at long wavelengths, the change in transmitted wavelength becomes very sensitive to rotation of the prism and makes long wavelength calibration extremely critical.

One disadvantage of the spectral response measurement techniques described above is that the solar cell response is determined at very low minority-carrier injection levels due to the low light intensities incident on the cell. Solar cells irradiated with neutrons and protons have response characteristics which are dependent upon the concentration of injected minority carriers. In such cases the cell must be illuminated with a light source similar in intensity and spectral content to the intended space environment during the spectral response evaluation. This can be achieved by chopping the monochromatic light and measuring the test cell output with a lock-in amplifier tuned to the chopper frequency. A DC bias light may then be used to illuminate the solar cell to achieve the required injection level, but the lock-in amplifier will not respond to the DC cell current.

Some solar cells made today consist of two cells stacked one atop the other, with the top cell responding to short wavelengths and passing the long wavelengths through to the bottom cell. The bandgap of each cell is chosen to be optimum for its particular wavelength range. When the spectral response of such a cell is measured without a bias light, during the short wavelength measurement the bottom cell becomes a load for the top cell and, conversely, during long wavelength measurement the top cell becomes a load for the bottom cell. In both cases the spectral response measurement will be inaccurate. The use of a bias light is necessary during the measurement of stacked cells, with the insertion of suitable filters to stimulate the bottom cell with long-wavelength DC-bias light during measurement of the top cell, and a short-wavelength bandpass filter to illuminate the top cell during measurement of the bottom cell. If a triple bandgap cell is under measurement, the use of three filters is required [3.30].

3.5   Diffusion-Length Measurements

In Si solar cells, the current collection is dominated by the behavior of the minority carriers in the base layer. The primary action of radiation on Si cells is the degradation of diffusion lengths in the base layer, which is manifested by decreases in the response to long wavelengths; however the effect of radiation on the emitter diffusion lengths is all but imperceptible. In such cells, where the dominant action is only on one side of the cell, the measurement of diffusion lengths is relatively straightforward and is discussed in some detail in references [3.31 and 3.32]. In GaAs solar cells, however, radiation affects the diffusion lengths in both the emitter and base regions and if an assessment of radiation effects on diffusion lengths is to be made, both L_p and L_n need to be measured.

One technique that has proved successful in measuring the diffusion lengths in GaAs solar cells has been to make a measurement of the spectral response of the cell. The equations developed for the spectral response of the cell in Chapter 2 are used with least square fitting techniques to find the diffusion lengths that best fit the experimental data. Use of this technique requires some knowledge about the construction of the cell. For instance eq. (2-47) gives the current collected at the p/n junction from the p-layers of $Ga_{1-x}Al_xAs$ and GaAs. To find a value of L_g that fits the measured spectral response, one has to know the window thickness, D and the junction depth, d. Knowledge of the doping density will give a value for the mobility, μ (see Figure A.5 in the Appendix) which can be used with the Einstein relation $D = (kT/q)\mu$, to find the diffusion constant. The use of appropriate absorption coefficients, β, for the $Ga_{1-x}Al_xAs$ (which requires a knowledge of x) and α for GaAs, along with an assumption for the surface recombination velocity at the $Ga_{1-x}Al_xAs$/GaAs interface, permits an estimate to be made for L_a. In actual practice the effect of the GaAlAs layer is ignored in the fitting procedure except for its effect on attenuating the light incident on the GaAs surface and for the optical role it plays in the reflectance of the incident light beam [3.33]. An example of the use of this technique is given in reference [3.34].

Another method commonly used to measure the minority carrier diffusion lengths in GaAs devices is the electron-beam-induced current (EBIC) technique. In this method, the sample is placed in a scanning electron microscope (SEM) with the diode turned sideways so that the electron beam can be traversed along the edge of the cell from the p-side into the n-side. During such a traverse, the current collected from the p/n junction is measured. The subsequent plot of collected current vs. distance from the junction is fit to the equation:

$$J = J_0 \exp\left(-\frac{X}{L}\right) \qquad\qquad (3\text{-}2)$$

For an example of this type of measurement on a GaAs diode see reference [3.35].

3.6   Irradiation Methods

The evaluation of solar cell radiation effects requires a wide range of specialized equipment and instrumentation. The space radiation environment will expose solar cells to fluxes of primarily electrons and protons. The exposure to heavier particles such as are found in cosmic rays is of sufficiently low intensity that this is not an area of concern to the solar array designer. Charged-particle accelerators are the primary sources for space-radiation simulation. The range of electron energies of interest is between 0.3 and 10 MeV. Electron or proton energies of 0.3 to 3 MeV are usually obtained with Van de Graaff and Dynamitron accelerators. Higher electron energies are available from linear electron accelerators (LINACs). Proton energies between 50 keV and 10 MeV are available from Van de Graaff or Tandem Van de Graaff accelerators. Higher proton energies are available from cyclotrons. The performance of low-energy proton irradiations (up to ≈ 10 MeV) requires that the irradiation be performed in vacuum to avoid excessive energy losses.

A successful radiation experiment must include accurate knowledge of the particle energy, measurement of the cross-sectional beam intensity at the target plane, the beam intensity (flux) during the irradiation, and the fluence (time integral of flux) given to the samples during the irradiation. The beam intensity measurements are performed with a Faraday cup, and a current monitoring and integrating instrument as discussed in reference [3.31]. The particle energy may be determined independently of the dosimetry at the accelerator facility by either using the method described in [3.31] or by using a nuclear reaction that has a sharp energy threshold.

Most radiation experiments involve the irradiation of several sample solar cells at the same time. With the ever-increasing size of modern solar cells and the high cost of performing experiments at particle accelerator facilities, it is important to utilize a particle beam that is broad in cross-sectional area. A few accelerator installations are equipped with a rastering scheme to expose a large target plane. Since most accelerators are not equipped with rastering capability, an alternate scheme of sending the beam through an appropriate scattering foil may be used. Although use of the raster method to produce a large beam gives very intense instantaneous fluxes, it has been found experimentally that for 200 keV protons incident on Si solar cells, this method produced the same results as a large DC beam produced with scattering foils. A charged-particle beam experiences multiple collisions with the electron cloud as it traverses the scattering foil. These collisions deflect the beam in such a way that the beam profile at a downstream target plane has the shape given by a circular Gaussian distribution. The shape of this Gaussian distribution can be altered significantly by the choice of the material and thickness of the scattering foil. This choice will depend on the type of charged particle and its energy. An appropriately chosen foil will produce a very uniform beam profile in the central portion of the circular Gaussian distribution. The scattering foil material and thickness must be chosen with care. If the foil is too thick, it may produce a very uniform beam at the target plane, but the straggling will be so severe that it may cause the beam intensity to be so low that the irradiation time will be prolonged beyond budgetary capacity. If the foil is too thin, the beam uniformity will be poor, and there is increased danger of the beam burning a hole through the foil. As an example, it has been found that an aluminum foil of 125 μm thickness will scatter 1 MeV electron beams to give a very uniform distribution ($\pm$ 5%) at a distance of 30 inches over a 6-inch diameter target plane. With the same geometry, a 2 MeV electron beam requires a copper foil $\approx$75 μm thick to give the same distribution. Foils for use in scattering proton beams are much thinner than those required for electron beams because of the much greater penetrating power of electron beams, and because the protons are much more difficult to deflect. The foil materials are limited to those that can be fabricated into thin, pinhole-free foils that are also fairly tough and have reasonably high heat conductivities. Gold, chromium, and titanium foils have been successfully used for scattering proton beams. For example, in a geometry where the foil-to-target distance was 240 cm, with a target area of 15 cm diameter, a chromium foil of 0.248 μm thickness was found to produce a 100 keV beam with a beam uniformity of $\pm$ 5% over the target area. A 50 μm-thick gold foil was found to be appropriate for scattering a 10 MeV proton beam in the same geometry. The methods outlined in references [3.36, 3.37, and 3.38] may be used to calculate the scattering statistics of proton beams after traversal of thin foils.

References for Chapter 3

3.1   F.S. Johnson, *Journal of Meteorology*, 11, 6,431, 1954.

3.2   M.P. Thekaekara, F.J. Drummond, D.G. Murcray, P.R. Gast, E.G. Laue, and R.C. Willson, "Solar Electromagnetic Radiation," NASA SP 8005, Revised, May 1971.

3.3   D. Labs and H. Neckel, "The Radiation of the Solar Photosphere from 2000 Angstroms to 100 Microns," *Zeitschrift für Astrophysik*, 69, 1, 1968.

3.4   H. Neckel and D. Labs, "Improved Data of Solar Spectral Irradiance from 0.33 to 1.25 μ," *Solar Physics*, 74, 231, 1981.

3.5   H. Neckel and D. Labs, "The Solar Radiation Between 3300-A and 12500-A," *Solar Physics*, 90, 205, 1984.

3.6   R.C. Willson, "Solar Total Irradiance and its Spectral Distribution," *Encyclopedia of Physics, 3rd Edition*, Robert M. Besançon, Editor, 1135, Van Nostrand Reinhold, New York, 1985.

3.7   R.F. Donnelly and J.H. Pope, "The 1-3000 A Solar Flux for a Moderate Level of Solar Activity for Use in Modeling the Ionosphere and Upper Atmosphere," NOAA Technical Report ERL 276-SEL 25, U.S. Government Printing Office, Washington DC 20402, 1973.

3.8   J.C. Arvesen, R.N. Griffin, and B.D. Pearson, "Determination of Extraterrestrial Solar Spectral Irradiance from a Research Aircraft," *Journal of Applied Optics*, 8, 2215, 1969.

3.9   C. Wehrli, "Extraterrestrial Solar Spectrum," Publication No. 615, Physikalish-Meteorologisches Observatorium & World Radiation Center, CH-7260 Davos-Dorf, Switzerland, July, 1985.

3.10   R.C. Willson and H.S. Hudson, "The Sun's Luminosity Over a Complete Solar Cycle," *Nature*, 351, 42, 1991.

3.11   R.C. Willson, "Solar Monitoring Has a Past and Present: Does it have a Future?," *Proceedings of the International Astronomical Union, Colloquium No. 143*, Boulder, Colorado, Cambridge Press, June 1993.

3.12   D. Creed, "Solar Simulator Using a Pulsed Xenon Arc Tube," *Institute of Environmental Sciences, 1969 Proceedings*, 363, April 1969.

3.13   R.W. Opjorden, "Pulsed Xenon Solar Simulator System," *Proceedings of the 8th IEEE Photovoltaic Specialists Conference*, 312, 1970.

3.14   R.L. Mueller, "Air Mass 1.5 Global and Direct Solar Simulation and Secondary Solar Cell Calibration Using a Filtered Large Area Pulsed Solar Simulator," *Proceedings of the 18th IEEE Photovoltaic Specialists Conference*, 1698, 1985.

3.15   R.L. Mueller, "The Large Area Pulsed Solar Simulator," JPL Publication 93-22, Jet Propulsion Laboratory, California Institute of Technology, Pasadena, Calif., August 15, 1993.

3.16   J. Lovelady and Y. Kohanzadeh, "Large Area Pulsed Solar Simulator (LAPSS) Testing of Back Surface Field Solar Panels," Spectrolab Inc. Report, Sept. 1985.

3.17   F.J. McKendry, H.W. Kuzminski, and C.P. Hadley, "Comparison of Flight and Terrestrial Solar Measurements on Silicon Cells," *Proceedings of the 4th Photovoltaic Specialists Conference*, II, C-3, 1964.

3.18   H.W. Brandhorst, Jr., "Airplane Testing of Solar Cells," *Proceedings of the 4th Photovoltaic Specialists Conference*, II, C-2, 1964.

3.19   M. Audibert, "Calibration of Solar Cells for Space Applications," *Solar Cells*, Gordon and Breach, London, 1971.

3.20   N.L. Thomas and D.M. Chisel, "Space Calibration of Standard Solar Cells Using High Altitude Sounding Rockets," *Proceedings of the 11th IEEE Photovoltaic Specialists Conference*, 237, 1975.

3.21   N.L. Thomas and F.W. Sarles, Jr., "High Altitude Calibration of Thirty-Three Silicon and Gallium Arsenide Solar Cells on a Sounding Rocket," *Proceedings of the 12th IEEE Photovoltaic Specialists Conference*, 560, 1976.

3.22   B.E. Anspaugh and R.S. Weiss, "Results of the 1993 NASA/JPL Balloon Flight Solar Cell Calibration Program," JPL Publication 93-30, Jet Propulsion Laboratory, California Institute of Technology, Pasadena, Calif., October 1, 1993.

3.23   B.E. Anspaugh, R.G. Downing, and L.B. Sidwell, "Solar Cell Calibration Facility Validation of Balloon Flight Data: A Comparison of Shuttle and Balloon Flight Results," JPL Publication 85-78, Jet Propulsion Laboratory, California Institute of Technology, Pasadena, Calif., Oct. 15, 1985.

3.24   H.W. Brandhorst, Jr., "Calibration of Solar Cells Using High-Altitude Aircraft," *Solar Cells*, Gordon and Breach, London, 1971.

3.25   R.J. Hardy, "Theoretical Analysis of the Series Resistance of a Solar Cell," *Solid-State Electronics*, 10, 765, 1967.

3.26   M.S. Imamura and J.I. Portscheller, "An Evaluation of the Methods of Determining Solar Cell Series Resistance," *Proceedings of the 8th IEEE Photovoltaic Specialists Conference*, 102, 1970.

3.27   G.C. Datum, Private Communication, February 1996.

3.28   L.A. Rosenberg and S.H. Gasner, "Reverse-Bias Screening of Large-Area GaAs/Ge Solar Cells at Low and High Temperatures," *Proceedings of the 23rd IEEE Photovoltaic Specialists Conference*, 1421, 1993.

3.29   P.A. Iles, H.I. Yoo, C. Chu, J. Krogen, and K-I Chang, "Reverse I-V Characteristics of GaAs Cells," *Proceedings of the 21st IEEE Photovoltaic Specialists Conference*, 448, 1990.

3.30   R.L. Mueller, "Spectral Response Measurements of Two-Terminal Triple-Junction a-Si Solar Cells," *Solar Energy Materials and Solar Cells*, 30, No. 1, 37, 1993.

3.31   H.Y. Tada, J.R. Carter, Jr., B.E. Anspaugh, and R.G. Downing, "Solar Cell Radiation Handbook," JPL Publication 82-69, Jet Propulsion Laboratory, California Institute of Technology, Pasadena, Calif., 1982.

3.32   J.A. Woolam, A.A. Khan, R.J. Soukkup, and A.M. Hermann, "Diffusion Length Measurements in Solar Cells - An Analysis and Comparison of Techniques," *Space Photovoltaic Research and Technology Conference, NASA Conference Pub. 2256*, 45, NASA Lewis, 1982.

3.33   B.T. Cavicchi, Private Communication, June 1995.

3.34   K.A. Bertness, B.T. Cavicchi, S.R. Kurtz, J.M. Olson, A.E. Kibbler, and C. Kramer, "Effect of Base Doping on Radiation Damage in GaAs Single-Junction Solar Cells," *Proceedings of the 22nd IEEE Photovoltaic Specialists Conference*, 1582, 1991.

3.35   C.C. Shen, K.P. Pande, and G.L. Pearson, "Electron Diffusion Lengths in Liquid-Phase Epitaxial p-GaAs:Ge Layers Determined by Electron-Beam-Induced Current Method," *Journal of Applied Physics*, 53, 1236, 1982.

3.36   J.B. Marion and F.C. Young, "Nuclear Reaction Analysis; Graphs and Tables," American Elsevier Publishing Company, Inc., New York, pp. 30-33, 1968.

3.37   J.B. Marion and B.A. Zimmerman, "Multiple Scattering of Charged Particles," *Nuclear Instruments and Methods*, 51, 93, 1967.

3.38   D.R. Dixon, G.L. Jensen, S.M. Morrill, C.J. Connors, R.L. Walter, C.R. Gould, and P.M. Thambiduria, "Multiple Scattering of Protons and Deuterons by Thick Foils," *IEEE Transactions on Nuclear Science*, NS-28, 1295, 1981.

Chapter 4

General Radiation Effects

In this chapter we will discuss some of the fundamental processes occurring in solar cells exposed to ionizing radiation, such as energetic electrons and protons. A detailed treatment of the interaction of radiation with solar cells can be found references [4.1, 4.2, and 4.3] and the references cited there, so the discussion here will be limited to collecting together the basic equations describing radiation damage in solar cells and certain particular aspects of damage in GaAs solar cells.

The major types of radiation damage phenomena in solids which are of interest to the solar array designer are ionization and atomic displacement. As a general statement, damage to the solar cells themselves will almost always be due to atomic displacements which disrupt the periodic structure of the lattice and interfere with the movement of minority carriers within the semiconductor. Ionization effects are important in such areas as the darkening of adhesives and the change in physical properties of materials used in solar array construction.

4.1   The Theory of Radiation Damage

The radiation that is usually of interest in the study of degradation of materials and devices consists of energetic or fast massive particles (i.e., electrons, protons, neutrons, or ions). The origin of these particles may be particle accelerators, the natural space radiation environment, nuclear reactions, radioactive sources (as in applications such as betavoltaic cells), or secondary mechanisms such as the interaction of gamma rays with electrons in a material in such a way that the electrons gain enough energy to cause damage to the material (Compton electrons). Because these particles have mass, energy, and usually charge, they can interact in several ways with materials. As an energetic charged particle enters the surface of a material, it slows down more or less continuously by interactions with the electrons and nucleii in the material. The type of interactions vary with the speed of the projectile [4.2].

There are basically two types of interactions of charged particles with matter: inelastic collisions and elastic collisions. In an inelastic collision, the projectile loses energy and the target particle gains energy, but the sum total of kinetic energies after the collision is generally less than the kinetic energies the particles had before the collision. The energy difference goes into excitation of the electrons in the material, mostly by ionization. Inelastic collisions are primarily between the incident particle and the electron cloud of the target material. Inelastic collisions are the most probable process in the interaction between space radiation and semiconductor materials, and are the primary mechanism for energy loss in the target. But once the velocity of the moving ion is much less than the velocities of the electrons at the top of the Fermi distribution, it is improbable that electrons can be excited. To within a factor of approximately two, the incident particle reaches this velocity at an energy, in keV, which is equal to its atomic weight. This estimate is independent of the target material. Thus, the limiting energy for ionization, E_i, is about 1 keV for protons, 2 keV for deuterons, 28 keV for Si, and 70 keV for a Ga atom. Below these incident energies the collisions are primarily elastic, so ionization losses may be neglected. An elastic collision occurs when the incoming particle collides with a target atom giving it a certain amount of energy and losing the same amount of energy in the process. The total kinetic energy of the projectile-target system is conserved in elastic collisions, and no energy is dissipated in electron excitation. Elastic collisions are the interactions that cause displacement damage, which is responsible for the degradation of solar cell performance.

4.1.1 Displacement Damage

We will consider the displacement of atoms by elastic processes when the incident radiation consists of heavy charged particles, and then we will see how some of the equations are modified when the incident particles are electrons with relativistic velocities. Elastic collisions are commonly classified according to the energies of the incident particles. At higher energies, the incident particle can partially penetrate the electron cloud surrounding the target atom. Such collisions are known as Rutherford collisions. At lower energies, where the electron cloud is not penetrated, the collisions are treated as though the particles were hard elastic spheres, and these interactions are known as hard sphere collisions. The displacements caused by the interaction of the incident charged-particle beam with the target atoms are known as primary displacements. It is quite probable that these "primaries" will have enough kinetic energy to produce additional displacements, and these additional displacements are known as secondary displacements or sometimes knock-ons. The secondary displacements will almost always be of the hard sphere type.

In an elastic collision the moving and stationary atoms interact with a screened Coulomb potential energy, of the form:

$$V(r) = \frac{(Z_1 Z_2 q^2)}{r} \exp(-r/a) \qquad (4\text{-}1)$$

where r is the separation between the two atoms, Z_1 and Z_2 are the atomic numbers of the moving and target particles, respectively, and a is the screening radius given by the approximate relation:

$$a \simeq \frac{a_0}{\sqrt{(Z_1^{2/3} + Z_2^{2/3})}} \qquad (4\text{-}2)$$

where a_0 is the Bohr radius of hydrogen ($\approx 5.3 \times 10^{-9}$ cm). If the energy of the incoming particle is high enough, r will be quite small and eq. (4-1) reduces to the classical Coulomb repulsion between the two charged particles. If the incident particle has sufficient energy to come closer than distance a to the target particle, the collision will be of the Rutherford type, but if it has less than this energy, the collision will be of the hard sphere type. This critical energy, E_A, will be calculated below. In the equation below, b represents the classical distance of closest approach (in the absence of screening) and is called the "collision diameter."

$$b = \frac{2 Z_1 Z_2 q^2}{\mu v^2} \qquad (4\text{-}3)$$

In the above equation, μ is the reduced mass, defined to be

$$\mu = \frac{M_1 M_2}{(M_1 + M_2)} \qquad (4\text{-}4)$$

and v is the velocity of the incoming particle.

The condition for Rutherford scattering to hold is for $b \ll a$, i.e., the distance of closest approach is $\ll$ than the screening radius. For the collisions to be of the hard sphere type, we require $b \gg a$. A critical energy, E_A, occurring at the point where $b = a$ can be calculated from (4-2) and (4-3) as follows [4.4]:

$$E_A = 2 E_R \frac{(M_1 + M_2)}{M_2} Z_1 Z_2 \sqrt{Z_1^{2/3} + Z_2^{2/3}} \tag{4-5}$$

where E_R, the Rydberg energy $= q^2/(2a_0) \approx 13.6$ eV, and M_1 and M_2 are the masses of the incident and struck particles, respectively. The collisions will become hard-sphere-like when the energy of the incident particle becomes $\ll E_A$. We find that if protons are incident on GaAs, the collisions will be hard-sphere-like if the proton energies are less than $E_A \approx 2.8$ keV, and for protons incident on Si, the collisions will be hard-sphere-like if the incident energies are less than $E_A \approx 1.03$ keV. Since we have already seen that the limiting energy for ionization with protons incident on any material is ≈ 1 keV, it is apparent that these rules of thumb defining the various energy regimes are very approximate indeed.

In radiation-damage calculations, wherein displacement damage is of primary importance, one of the most important considerations is the transfer of energy to an atom in the target. The maximum energy transfer, T_m, will occur in a head-on collision, and from classical considerations involving conservation of energy and momentum, this maximum kinetic energy transfer to the target atom can be shown in the nonrelativistic case to be

$$T_m = \frac{4 M_1 M_2}{(M_1 + M_2)^2} E \tag{4-6}$$

for an incoming particle of kinetic energy E and mass M_1, incident on a target atom of mass M_2.

When electrons are the incident particles, they must have high velocities because of their small mass in order to achieve sufficient energy to dislodge lattice atoms. A relativistic version of the above equation must be used for T_m:

$$T_m = \frac{2 m E}{M_2} \left(\frac{E}{m c^2} + 2 \right) \cos^2 \theta \tag{4-7}$$

where m = electron mass (1/1823 in atomic mass units)

     M_2 = atomic weight of target atoms (69.72 for Ga, 74.91 for As, 28 for Si)

     mc^2 = mass-energy equivalence of the electron (0.511 MeV)

θ = scattering angle of the displaced atom with respect to the incident direction of the electrons. θ is related to the deflection angle ϕ of the scattered electron by $\phi = \pi - 2\theta$

The maximum energy transfer will occur when $\theta = 0$. For example, the maximum energy a 1 MeV electron can transfer to a Si atom is 155.0 eV, and the maximum energy a 1 MeV electron can transfer to a Ga atom is 62.3 eV.

4.1.2 Atomic Displacements

Energetic electrons or protons entering an absorber lose almost all of their energy by collision with the electron "cloud" in the absorber, and these collisions with the electrons determine the range of the electrons or protons in the absorber. Only rarely does the energetic particle come close enough to a nucleus for an energy exchange to occur. This may be visualized by considering that an atom has a diameter of about 10^{-8} cm while the nucleus has a diameter of about 10^{-13} cm. If we had an atom and wished to see the nucleus, we would have to magnify it until the whole atom was the size of a large room, and then the nucleus would be a bare speck which you could just about make out with the eye [4.5]. Nevertheless, interactions with the nucleii do occur and when the incident particles come close enough to a nucleus they can give it enough energy to permanently displace it from its lattice site. These displaced atoms and their associated vacancies undergo other reactions which often involve dopant atoms, and finally form stable defects which produce significant changes in the equilibrium carrier concentrations and the minority carrier lifetime.

An energetic particle must have more than a specific threshold energy to be able to displace an atom from its lattice site. The lattice atom itself must receive a certain energy, called the displacement energy, for it to be removed sufficiently far from its site that it does not return there. In silicon, displacement energies ranging between 11.0 and 12.9 eV have been measured [4.6, 4.7, and 4.8]. In GaAs, the displacement energy is dependent on the orientation of the crystal with respect to the incident angle of the energetic particles, and has been measured to be between 7.0 and 11.0 eV, with the average displacement energy ≈ 10 eV [4.9, 4.10].

The threshold energy, E_t, and displacement energy, E_d may be calculated from eq. (4-6) for protons:

$$E_d = \left(\frac{4 M_p M_2}{(M_p + M_2)^2} \right) E_t \qquad (4\text{-}8)$$

where E_d = displacement energy (MeV)

E_t = threshold energy (MeV)

M_p = proton mass in atomic mass units

In a similar manner, and using the above definitions and eq. (4-7), the threshold energy for electrons is given by:

$$E_d = \frac{2 m E_t}{M_2} \left(\frac{E_t}{m c^2} + 2 \right) \cos^2 \theta \qquad (4\text{-}9)$$

When protons are incident on an arsenic target (AMU = 69.72 and E_d = 10 eV), eq. (4-8) tells us that the threshold energy is ≈ 179 eV. Since proton accelerators are not commonly available in this energy range, the threshold energies are more readily measured using electrons. For example, when we use eq. (4-9) to compute the electron energy threshold for a Ga target, we find that E_t = 245 keV.

As mentioned above, E_d in GaAs depends on the angle of incidence the bombarding particles have with respect to the crystalline structure. Pons and Bourgoin [4.9] have found that the principal defects caused by electron irradiation are caused by the displacement of As atoms. By an examination of Figure 4.1, we see that when the beam is incident in the [111]Ga direction, with energy just above E_t it is difficult to displace an As atom in the straight-ahead direction ($\theta \approx 0$) because a Ga atom occupies the space where the displaced As atom would

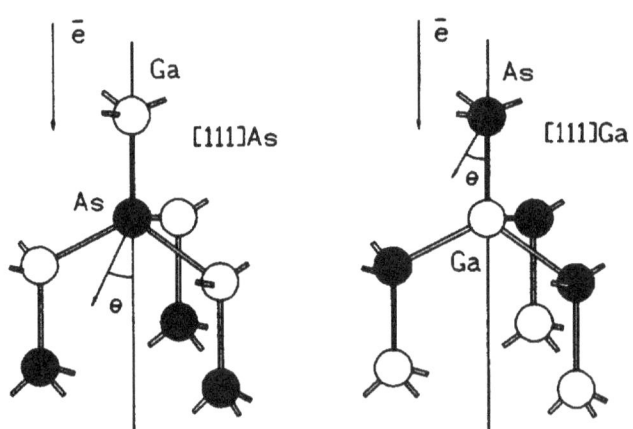

Figure 4.1.   Zinc-blende structure of GaAs (from Pons and Bourgoin)

have to go. However, when the beam is incident in the [111]As direction, there is a large empty space in the region where the displaced As atom would go, so displacements of As atoms will occur with relative ease. These observations are consistent with the measurements performed by Pons and Bourgoin, which show that the threshold energy, E_d for displacing an As atom in the [111]As direction is only 9 eV and very nearly independent of the scattering angle θ. When the bombardment occurs along the [111]Ga direction, E_d is 11 eV for small values of θ and decreases to $E_d = 7$ eV for larger values of θ. At higher electron energies, and for thick targets such as solar cells, the angular dependence is expected to disappear because the knocked-on atoms can have large scattering angles, and the empty spaces in the lattice other than those just behind the As atoms are available as landing sites for the struck atoms. Also, the electron beam loses its monodirectionality after penetrating a short distance into the lattice. Pons and Bourgoin estimate that the displacement energy for GaAs, averaged over all directions, is ≈ 10 eV.

4.1.3 Primary Displacement Cross Sections

When the collisions are in the Rutherford region (incident particle energy $> E_A$), collisions resulting in small energy transfers are the most probable. The differential cross section for kinetic energy transfer from T to (T + dT) is given by:

$$d\sigma = \frac{\pi b^2}{4} T_m \frac{dT}{T^2} = \left(4 \pi a_0^2 \frac{M_1}{M_2} Z_1^2 Z_2^2 \frac{E_R^2}{E} \right) \frac{dT}{T^2} \qquad (4\text{-}10)$$

where E is the energy of the incident particle. This equation is valid for collisions which result in the maximum energy transfer, T_m, down to some small but finite lower limit, where electronic screening cannot be neglected. In cases of present interest this lower limit lies well below that energy at which atomic displacements can be produced [4.2]. If it is assumed that a struck atom is always displaced when it receives an energy greater than E_d and is never displaced if it receives less than E_d, then the total cross section for producing a displacement can be found by integrating the above equation with the following result:

$$\sigma_d = \int_{T=T_d}^{T=T_m} d\sigma = 16\pi a_0^2 Z_1^2 Z_2^2 \frac{M_1^2}{(M_1 + M_2)^2} \frac{E_R^2}{T_m^2} \left(\frac{T_m}{E_d} - 1 \right)$$

$$or \quad \sigma_d = 4\pi a_0^2 \frac{M_1}{M_2} \frac{Z_1^2 Z_2^2 E_R^2}{E E_d}$$

(4-11)

Hard-sphere collisions occur in the energy region where the incident particle has energy $< E_A$. In this case, all energy transfers from 0 to T_m are equally probable, and the differential cross section for kinetic energy transfer from T to (T + dT) [4.2] is:

$$d\sigma = \pi a_1^2 \frac{dT}{T_m}$$

(4-12)

where a_1 is the diameter of the hard sphere, taken to be approximately the screening radius. The total cross section for producing primary displacements in the hard sphere case is:

$$\sigma_d = \frac{\pi a_1^2}{T_m} \int_{T=E_d}^{T=T_m} dT = \pi a_1^2 \frac{T_m - E_d}{T_m}$$

(4-13)

When the incident particles are electrons, the scattering involved that causes displacements is primarily due to the Coulomb interaction between the electron and the target nucleus in the Rutherford scattering regime. However, it is necessary to modify the Rutherford scattering cross-section equation to account for the relativistic velocities of the electrons. Relativistic Coulomb scattering has been treated by Mott [4.11, 4.12] and in a simplified form by McKinley and Feshbach [4.13]. The McKinley-Feshbach scattering cross section for values of kinetic energy transfer ranging from the maximum, T_m, to the minimum capable of displacing an atom, E_d, is given in the following form by reference [4.3]:

$$\sigma_d = \frac{\pi b'^2}{4} \left[\left(\frac{T_m}{E_d} - 1 \right) - \beta^2 \ln \frac{T_m}{E_d} + \pi \alpha \beta \left(2 \left[\left(\frac{T_m}{E_d} \right)^{\frac{1}{2}} - 1 \right] - \ln \frac{T_m}{E_d} \right) \right]$$

(4-14)

with $\alpha = Z_2/137$

$b' = b/\gamma$

b = distance of closest approach defined by eq. (4-3)

$$\gamma = (1 - \beta^2)^{-1/2}$$

and $\beta = $ v/c, the ratio of the electron velocity to the speed of light

The quantity $\pi b'^2/4$ may be expressed as:

$$\frac{\pi\,b'^2}{4} = \pi\,Z_2^2\left(\frac{e^2}{m\,c^2}\right)^2\frac{1}{\beta^4\,\gamma^4} = \frac{2.495\,x\,10^{-25}\,(cm^2)}{\beta^4\,\gamma^2}\,Z_2^2 \tag{4-15}$$

The total cross section for displacement rises steeply from zero at the threshold energy, to an asymptotic value as the energy increases. The calculated displacement cross section for electrons on GaAs is 98×10^{-24} cm^2 for a bombardment energy of 0.5 MeV. This cross section rises to 163×10^{-24} cm^2 for 1 MeV electrons and then reaches a maximum at ≈ 5 MeV with a cross section of 218×10^{-24} cm^2. The reader is referred to reference [4.3] for more details in making this calculation.

4.1.4 Secondary Displacements

When an atom is knocked out of its lattice site, it may have considerable kinetic energy and become a projectile itself. These energetic atoms knocked out of the lattice are known as "knock-ons" and are fully capable of producing further displacements. Such interactions, e.g., of Ga knock-ons colliding with Ga lattice atoms, will be of the hard-sphere collision type, since all Ga knock-ons will certainly have energies well below the threshold limit of $E_A = 8$ MeV given by eq. (4-5) using values suitable for Ga-Ga interactions.

For particles above the threshold energy, the probability of an atomic displacement can be described in terms of a displacement cross section along with an average number of secondary displacements induced by the primary displacement. Using this concept, the number of displacements can be estimated from the relationship:

$$N_d = n_a\,\sigma_d\,\bar{v}\,\Phi \tag{4-16}$$

where
N_d = number of displacements per unit volume

n_a = number of atoms per unit volume of absorber

= 4.42×10^{22} atoms/cm^3 in GaAs

σ_d = displacement cross section (cm^2)

$\bar{\nu}$ = average number of displacements per primary displacement including the primary, averaged over the energy spectrum of primary knock-ons

Φ = radiation fluence (particles/cm^2)

Several theories have been presented for the calculation of secondary displacements. Even though these theories use widely varying assumptions about the details of the collision process, they all produce very similar results. The theory formulated by Kinchin and Pease [4.14] is widely accepted. Assuming the secondary collisions occur by hard-sphere collisions, these authors have calculated the number of displacements produced by a knock-on of energy T to be:

$$\nu(T) = 1 \qquad\qquad 0 < T < 2E_d$$
$$\nu(T) = \frac{T}{2E_d} \qquad 2E_d < T < E_i$$
$$\nu(T) = \frac{E_i}{2E_d} \qquad\qquad T > E_i \tag{4-17}$$

The average number of displacements, $\bar{\nu}$, is calculated by averaging ν over the energy spectrum of the knock-on atoms. In a form calculated by reference [4.2], $\bar{\nu}$ is:

$$\bar{\nu} = \frac{1}{2}\left(\frac{T_m}{T_m - E_d}\right)\left(1 + \ln\frac{T_m}{2E_d}\right) \tag{4-18}$$

We can now compute the rate of displacement production for an incident particle of energy E using eqs. (4-16) and (4-18), along with eq. (4-14) if the incident particles are electrons. If the incident particles are protons we use eq. (4-11) instead of eq. (4-14) for the cross section. Some sample calculations are shown in Table 4-1 for electrons incident on Si and GaAs, and in Table 4-2 when protons are the incident particles. In these calculations we have used an E_d value of 12.9 eV for Si and 10 eV for GaAs.

It is important to recognize that the values calculated for N_d in Tables 4-1 and 4-2 are displacement <u>rates</u> produced by an incident particle of energy E. If these rates are multiplied by the

Table 4-1.   Displacement Rates for Incident Electrons

Energy (MeV)	Silicon				GaAs			
	T_m (eV)	σ_d 10^{-24} cm^2	$\bar{\nu}$	N_d Disp/cm	T_m (eV)	σ_d 10^{-24} cm^2	$\bar{\nu}$	N_d Disp/cm
0.5	58	58.4	1.16	3.40	23	92.6	1.01	4.13
1.0	154	67.8	1.52	5.16	62	153.2	1.27	8.62
2.0	462	73.6	2.00	7.35	186	187.8	1.70	14.17
3.0	922	75.3	2.32	8.74	372	197.4	2.02	17.58
5.0	2300	76.3	2.76	10.5	927	201.7	2.44	21.80
10.0	8400	76.5	3.40	13.0	3390	201.2	3.08	27.33

Table 4-2.   Displacement Rates for Incident Protons

Energy (MeV)	Silicon				GaAs			
	T_m (keV)	σ_d (cm^2)	$\bar{\nu}$	N_d Disp/cm	T_m (keV)	σ_d (cm^2)	$\bar{\nu}$	N_d Disp/cm
0.1	13.3	3.53E-19	3.63	64,012	5.6	9.00E-19	3.32	132,106
0.2	26.6	1.76E-19	3.97	35,051	11.2	4.50E-19	3.67	72,892
0.3	39.8	1.18E-19	4.17	24,558	16.7	3.00E-19	3.87	51,270
0.5	66.4	7.06E-20	4.43	15,634	27.9	1.80E-19	4.12	32,788
1.0	132.8	3.53E-20	4.77	8,429	55.8	9.00E-20	4.47	17,770
2.0	265.5	1.77E-20	5.12	4,520	111.5	4.50E-20	4.81	9,573
5.0	663.9	7.06E-21	5.58	1,969	278.8	1.80E-20	5.27	4,194
10.0	1328.0	3.53E-21	5.92	1,046	557.6	9.00E-21	5.62	2,235

incident particle fluence (particles/cm^2), then the rates will be displacements per unit volume. The displacement rate will change as the particle slows down. The total number of displacements produced by an incident particle is an integral of the displacement rates over the energy range from E to 0 if the particle stops in the material, and over E to E_0 if the particle emerges from the target with energy E_0. A beam of 1 MeV electrons will travel 0.23 cm through Si and 0.12 cm through GaAs, distances which are much greater than the thickness of modern-day solar cells. On the other hand, if the bombarding particles are 1 MeV protons, they will penetrate a distance of 15.6 μm in Si and a distance of 11.8 μm in GaAs, and will therefore come to rest inside a solar cell. The effects on the two different types of cells are expected to be different, however, because the 1 MeV proton will come to rest in a very active region of the Si solar cell, but will penetrate beyond the region of major activity in a GaAs solar cell. Tables 4-1 and 4-2 show the dramatic difference in the damage distribution produced by electrons and protons. As the energy of the electron decreases, it produces a slowly decreasing rate of displacements/cm, but as the energy of a proton decreases, it produces a very rapidly increasing displacement rate, so that most of the displacements are produced near the end of its trajectory.

Kinchin and Pease have calculated the total number of displacements, N_t, produced by light ions such as protons coming to rest in a target material. For the specific case of protons coming to rest in Si, their formula reduces to:

$$N_t = \frac{P(E \times 10^6 - 563.7) + 281.9}{12.9}$$

$$\text{with } P = \frac{6.85 \times 10^{-5}[1 + \ln(5162\,E)]}{\ln(12.68\,E)} \tag{4-19}$$

$$\text{and } E = \text{proton energy (in MeV) valid for } E \geq 0.1 \text{ MeV}$$

When the protons are incident on GaAs, the Kinchin and Pease formula for N_t is as follows:

$$N_t = \frac{P(E \times 10^6 - 503.5) + 251.8}{10.0}$$

$$\text{with } P = \frac{6.07 \times 10^{-5}[1 + \ln(2691\,E)]}{\ln(5.70\,E)} \tag{4-20}$$

$$\text{and } E = \text{proton energy (in MeV) valid for } E \geq \approx 0.3 \text{ MeV}$$

Note that these equations are not valid for low energies. Bulgakov and Kumakhov [4.15] also give a relation for the total number of displacements, which has a wider range of validity but is more complex.

4.1.5 Ionization

Ionization occurs when orbital electrons are removed from an atom or molecule in gases, liquids, or solids. Ionization is the primary mechanism for energy loss for energetic, charged particles traversing a target material. The unit measure of the <u>incident</u> intensity for ionizing radiation is the roentgen. This amount of radiation will induce a charge of 2.58 x 10^{-4} coulomb/kilogram in air. The roentgen is not a measure of the amount of energy actually absorbed by the material. The amount of <u>absorbed</u> radiation dose in a material is commonly measured in units of rads. One rad of absorbed dose occurs when an energy of 100 ergs is absorbed in one gram of material (100 ergs/gm = 0.01 joules/kg). The preferred SI unit of absorbed dose is the gray (Gy), which is defined to be 1 joule/kg, and therefore 1 Gy = 100 rads.

The absorbed dose units are most commonly used in specifying the exposure of electronic parts to gamma radiation (and also in the exposure of humans to radiation). However, by using the definition of absorbed dose, it is possible to calculate the "rad exposure" of materials to charged-particle radiation also. This is calculated by considering the radiation incident on a thin slice of material, having thickness dx. Then, using the tabulated dE/dx (stopping power) tables for the incident particle at the irradiation energy, the energy deposited in the slice of material is calculated. Finally, the deposited energy is converted into units of rads and divided by 100 times the mass of the slice of material to yield rads. For electrons or protons, this calculation and unit conversion results in the formula:

$$Dose\ (rads)\ =\ 1.6\ x\ 10^{-8}\ \frac{dE}{dx}\left(\frac{MeV-cm^2}{gm}\right)\phi\left(\frac{e}{cm^2}\right) \tag{4-21}$$

To properly use this formula, the stopping power must be in units of (MeV-cm^2)/gm. If the tabulated stopping power is given in units of MeV/cm, for example, it can be converted to the proper units by dividing by the density (in g/cm^3) of the absorber material. Note that this calculation is dependent on the particle type (electron or proton), the particle energy, and the absorber material. When the absorbed dose units are given as rads (Si) it means that the stopping powers for Si have been used in the calculation. If the dose is given as rads (water), the stopping powers for water were used. Since the particle beams lose energy as they penetrate the absorber, the value for dE/dx will change with depth, and the resultant absorbed dose rate also changes with dose. For this reason, absorbed doses are commonly calculated to be applicable to a surface absorbed dose only.

Values for the stopping power and range for electrons in various materials are tabulated in references [4.16 and 4.17] and for protons in reference [4.18]. Plots of these values for both electrons and protons in Si are given in reference [4.1]. Computer programs for calculating stopping powers and ranges in nearly all materials have recently become available. A program called EPSTAR for electron computations is available from reference [4.19] and a program for charged-ion calculations (including protons) called TRIM is available from reference [4.20]. These programs have been used to compute the stopping powers and ranges for electrons and protons in GaAs, which are plotted in Figures 4.2 and 4.3. The stopping powers are plotted in units of MeV-cm^2/gm, and in cm for the ranges. The stopping power and ranges are not a strong function of the atomic number of the absorber material. Since the stopping power plots have been normalized for density, they may be used for materials with atomic numbers similar to that of GaAs. If the range plots are to be used for materials of similar atomic number, a density correction must be made by multiplying by the density of GaAs (5.320 g/cm^3), and then dividing by the density of the similar material.

The reduction of transmittance in solar-cell coverglasses is an important effect of ionizing radiation. The darkening is caused by the formation of color centers (sometimes referred to as F-centers, from the German word "farben," meaning color) in glass or oxide materials. Color centers are formed when ionizing radiation excites an orbital electron to the conduction band. These electrons may become trapped by impurity atoms to form relatively stable charged-defect complexes which can absorb light. The use of fused silica (Corning 7940) for solar-cell coverglasses is a good choice because the color centers formed in this material absorb light at UV wavelengths where the cells have no response.

Radiation produces many ionization-related effects in organic materials. These changes all result from the production of ions, free electrons, and free radicals. As a result of these actions, transparent polymers are darkened and crosslinking between main-chain members may drastically alter the mechanical properties. The contemplated use of polymeric materials in solar arrays will require the array designer to have knowledge of the ionization-related radiation effects in those materials.

The use of silicon dioxide as a surface passivation coating and dielectric material in silicon devices results in a wide range of ionization-related radiation effects. The development of trapped charges in the silicon dioxides can cause increased leakage currents, decreased gain, and surface channel development in bipolar transistors and increased threshold voltages in metal oxide semiconductor field effect transistors (MOSFETs). Ionizing radiation in silicon excites the electrons of the valence band to the conduction band, creating electron-hole pairs in much the same way that carrier pairs are generated by visible light. Although an optical photon of energy equal to or greater than 1.1 eV will create an

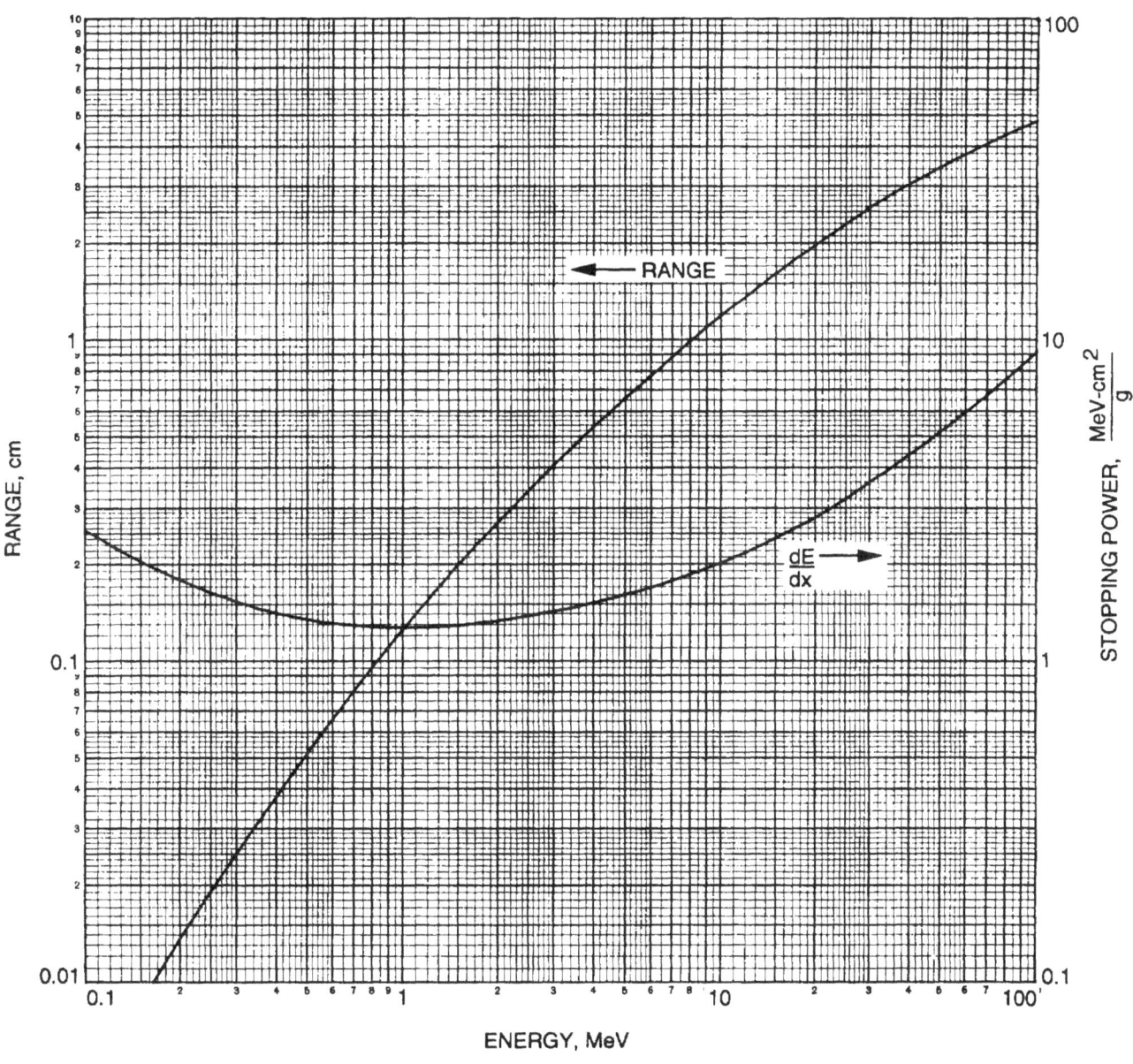

Figure 4.2.   Stopping Power and Range Curves for Electrons in GaAs
Reference: S.M. Seltzer, EPSTAR Program

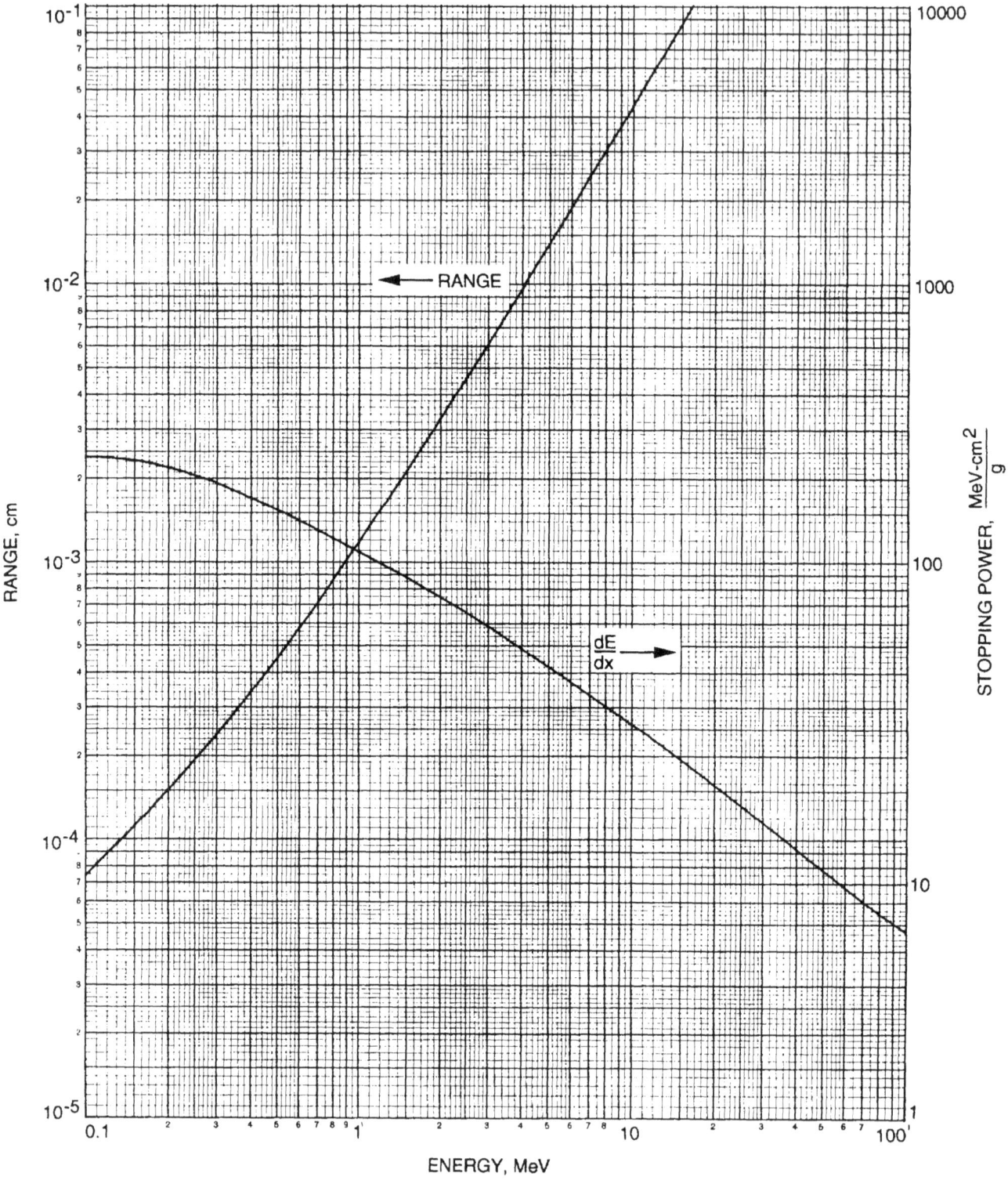

Figure 4.3.   Stopping Power and Range Curves for Protons in GaAs
Reference: J.F. Ziegler, TRIM Program

electron-hole pair, roughly three times this amount of energy must be absorbed from an ionizing high-energy particle to produce an electron-hole pair.

References for Chapter 4

4.1   H.Y. Tada, J.R. Carter, Jr., B.E. Anspaugh, and R.G. Downing, "Solar Cell Radiation Handbook," JPL Publication 82-69, Jet Propulsion Laboratory, California Institute of Technology, Pasadena, Calif., 1982.

4.2   G.J. Dienes and G.H. Vineyard, *Radiation Effects in Solids*, Interscience Publishers, Inc., New York, 1957.

4.3   F. Seitz and J.S. Koehler, "Displacement of Atoms During Irradiation," *Solid State Physics*, 2, 305, Academic Press, 1956.

4.4   N. Bohr, "The Penetration of Atomic Particles Through Matter," *Kgl. Danske Videnskab. Selskab, Mat.-fys. Medd.*, 18, 8, 1948.

4.5   R.P. Feynman, R.B. Leighton, and M. Sands, *The Feynman Lectures on Physics*, Vol. 1, Addison-Wesley, Reading, Mass., 1963.

4.6   J.J. Loferski and P. Rappaport, "Radiation Damage in Ge and Si Detected by Carrier Lifetime Changes - Damage Thresholds," *Physical Review*, 111, 2, 432, 1958.

4.7   H. Flicker, J.J. Loferski, and J. Scott-Monck, "Radiation Defect Introduction Rates in n- and p-Type Silicon in the Vicinity of the Radiation Damage Threshold," *Physical Review*, 128, 6, 2557, 1962.

4.8   N.N. Gersimenko, et al., "Threshold Energy for the Formation of Radiation Defects in Semiconductors," *Soviet Physics - Semiconductors*, 5, 8, 1439, 1972.

4.9   D. Pons and J. Bourgoin, "Anisotropic-Defect Introduction in GaAs by Electron Irradiation," *Physical Review Letters*, 47, 1293, 1981.

4.10   F.H. Eisen, "Orientation Dependence of Electron Radiation Damage in InSb," *Physical Review*, 135, No. 5A, A1394, 1964.

4.11   N.F. Mott, "The Scattering of Fast Electrons by Atomic Nuclei," *Proceedings of the Royal Society (London)*, A124, 426, 1929.

4.12   N.F. Mott, "Polarization of Electrons by Double Scattering," *Proceedings of the Royal Society (London)*, A135, 429, 1932.

4.13   W.A. McKinley, Jr. and H. Feshbach, "The Coulomb Scattering of Relativistic Electrons by Nuclei," *Physical Review*, 74, 1759, 1948.

4.14   G.W. Kinchin and R.S. Pease, "The Displacement of Atoms in Solids by Radiation," *Reports on Progress in Physics*, 18, 1, 1955.

4.15   Yu.V. Bulgakov and M.A. Kumakhov, "Spatial Distribution of Radiation Defects in Materials Irradiated with Beams of Mono-Energetic Particles," *Soviet Physics-Semiconductors*, 2, 11, 1334, 1969.

4.16   M.J. Berger and S.M. Seltzer, "Tables and Energy Losses and Ranges of Electrons and Positrons," Paper 10, NAS-NRC Publication 1133, 1964; also NASA SP-3012, 1964.

4.17   M.J. Berger and S.M. Seltzer, "Additional Stopping Power and Range Tables for Protons, Mesons, and Electrons," NASA SP-3036, 1966.

4.18   J.F. Janni, *Atomic Data and Nuclear Data Tables, Proton Range-Energy Tables, 1 keV-10 GeV, Part 1. Compounds, and Part 2. Elements*, Academic Press, New York, 1992.

4.19   S.M. Seltzer, "Electron and Positron Stopping Powers and Ranges, EPSTAR," Vers. 2.0B, National Institute of Standards and Technology, 1988.

4.20   J.F. Ziegler, J.P. Biersack, and U. Littmark, *The Stopping and Range of Ions in Solids*, Vol. 1, Pergamon Press, New York, 1985.

Chapter 5

Radiation Effects in Solar Cells

In this chapter we will discuss why radiation decreases the electrical performance of solar cells. We will review the concept of damage equivalence and show how it is used to relate the radiation damage caused by particles of one energy to that caused by particles of another energy or even by particles of a different type. We will present experimental data showing how GaAs/Ge solar cells degrade after bombardment with electrons and protons of various energies, and explain how this data is used to derive the damage coefficients needed for calculating damage equivalences in both normally incident radiation beams and in omnidirectionally incident space radiation. We will then show how to calculate the radiation damage induced in solar cells by the space environment, where the cells will be exposed to spectra of omnidirectional electrons and protons.

5.1   Dependence on Lifetime and Diffusion Length

Solar cells are usually irradiated both in the laboratory and in space at temperatures near 30°C, and the resulting defects are somewhat mobile at near this temperature. The defects will tend to bond with impurities, vacancies, or interstitials that already may be present in the semiconductor, and the resulting defects can be quite stable. A review of the defects produced in GaAs may be found in reference [5.1]. Although this review restricts itself to some of the more fundamental defects, it gives an idea of the procedures involved in measuring and identifying defects. The defects discussed are intrinsic defects, consisting of vacancies, interstitials, and antisites in both Ga and As sublattices, along with complexes formed between each other and impurities in the crystal. A Ga antisite defect is a Ga atom sitting in a lattice position that would normally be occupied by an As atom. There are shorthand labels for these defects. V_{Ga} and V_{As} are Ga and As vacancies, Ga_i and As_i are Ga and As interstitials, while Ga_{As} and As_{Ga} are Ga and As antisites. One of the most important defects formed in GaAs is known as the EL2 defect. This defect is found in all GaAs material regardless of dopant or method of growth. It appears to consist of a complex formed by an As antisite bonded with an As interstitial, i.e. As_{Ga} - As_i. The EL2 defect has two states which probably are of only academic interest to the solar array designer, but nevertheless interesting. At very low temperatures, the EL2 defect can be made to disappear by optical stimulus with a photon energy of 1.1 eV. This "disappearance" is thought to be a transformation into a metastable defect, because it does not revert if the photoexcitation is removed and the temperature remains low. However, if the crystal is heated to temperatures above 140 K, the defect reappears. Analysis suggests that the stable state consists of the existence of the As_{Ga} - As_i with the As interstitial in the second-neighbor position, while in the metastable state, the $\dot{As}_i$ moves closer to the As_{Ga}.

The following defects have also been observed in GaAs material: (1) the V_{AS}-As_i pair (or Frenkel pair) which partially anneals at $\approx 220°C$, because As interstitials begin to be mobile at that temperature, and annealing much more thoroughly at $450°C$; (2) the As vacancy, V_{AS}, which is just a special case of the V_{AS}-As_i pair with the interstitial a longer distance away; (3) the As interstitial, As_i, which becomes mobile at temperatures above $220°C$; (4) the As antisite, As_{Ga}, which is stable at temperatures up to $950°C$; (5) the Ga antisite, Ga_{As}; (6) the divacancy, appearing to be $V_{Ga} + V_{As}$ because its threshold energy of formation is twice that of the threshold for Frenkel pair formation; (7) the As antisite - As vacancy complex, As_{Ga}-V_{As}, which is probably formed when a nearest-neighbor As atom moves into a Ga vacancy. It is quite apparent that when a crystal contains two elements, each on its own sublattice, it introduces a measure of complexity into the number and types of defects that can be formed; and we have not even touched on the interaction of the above defects with the dopants commonly used in GaAs solar cells.

Defects in semiconductors are often studied by use of the deep-level transient spectroscopy (DLTS) technique. In performing these measurements, the junction capacitance of a small sample is measured during a thermal scan of the sample. Interpretation of the capacitance vs. temperature curves allows identification of the energy level and concentration of the defects within the forbidden gap. The defects can be identified as to whether they are hole traps or electron traps, but unfortunately, this technique does not identify the atomic nature of the defects. It can be useful in observing the behavior of defects as a function of doping levels in the semiconductor, radiation history, annealing, etc. An example of this technique applied to GaAs solar cells can be found in reference [5.2]. In this paper, defect levels were observed in GaAs material which was doped with Sn, and in undoped material. After the cells were irradiated with 1 MeV electrons, electron traps were found to occur at energy levels of E_c-0.31, 0.71, and 0.90 eV (labeled E3, E4, and E5) and a hole trap at energy level E_v+0.71 eV in the Sn-doped material. The authors studied the behavior of these traps as a function of electron flux rate, fluence, annealing treatment, etc.

The main effect of the displacements produced by radiation is a disruption of the periodic lattice structure, resulting in a decrease of the minority carrier lifetime. In modern GaAs solar cells that have their front surfaces passivated with an $Al_xGa_{1-x}As$ window so that front-surface recombination is not a major problem, the decrease in minority carrier lifetime is the major radiation-sensitive parameter. Since, as we saw in eq. (2.17), minority carrier lifetimes are inversely proportional to the recombination rates, the reciprocal lifetime contributions caused by various sets of recombination centers can be added to determine the inverse of the lifetime as follows:

$$\frac{1}{\tau} = \frac{1}{\tau_0} + \frac{1}{\tau_e} + \frac{1}{\tau_p} + \ldots \qquad (5\text{-}1)$$

where τ = final minority carrier lifetime

τ_0 = minority carrier lifetime before irradiation

τ_e = minority carrier lifetime due to electron irradiation

τ_p = minority carrier lifetime due to proton irradiation

One of the most commonly used analytical tools for the determination of the particle type and energy dependence of degradation in both Si and GaAs solar cells has been developed from the basic relationship for lifetime degradation:

$$\frac{1}{\tau} = \frac{1}{\tau_0} + K_\tau \phi \qquad (5\text{-}2)$$

where τ = final minority carrier lifetime

τ_0 = minority carrier lifetime before irradiation

K_τ = damage coefficient (lifetime)

Φ = radiation fluence

However, minority-carrier diffusion length is a more applicable and more easily determined parameter for solar cell analysis than minority carrier lifetime. We note from eqs. (2-34) and (2-36) that the hole and electron currents, J_p and J_n, are proportional to diffusion lengths L_p and L_n. Using $L^2 = D\tau$, the above expression becomes:

$$\frac{1}{L^2} = \frac{1}{L_0^2} + K_L \phi \tag{5-3}$$

where L = final minority carrier diffusion length

     τ_0 = minority carrier diffusion length before irradiation

     K_r = damage coefficient (diffusion length)

     $K_L = K_r/D$

     Φ = radiation fluence

But the degradation of solar cells induced by radiation is most commonly measured in terms of the common electrical parameters such as I_{sc}, V_{oc}, and P_{max}, since most laboratories are not equipped to measure lifetimes or diffusion lengths. The situation is even more complicated for GaAs cells because radiation degrades both the p and n parts of the cell, so diffusion lengths would have to be measured in both parts to adequately characterize the damage to the cell.

Experience has shown that the degradation of solar cell electrical parameters due to radiation can usually be expressed as follows for the case of I_{sc}:

$$I_{sc} = I_{sco} - C \log\left(1 + \frac{\phi}{\phi_x}\right) \tag{5-4}$$

The Φ_x term represents the radiation fluence at which I_{sc} starts to change to a linear function of the logarithm of the fluence. The constant C represents the decrease in I_{sc} per decade in radiation fluence in the logarithmic region. As discussed in reference [5.3], the degradation in I_{sc} may be expressed as a function of L (through K_L) in an equation which has the same form as eq. (5-4) as follows:

$$I_{sc} = A - B \log(1 + K_L L_0^2 \phi) \tag{5-5}$$

Similar expressions may be obtained for V_{oc} and P_{max}, but their applicability to GaAs cells may be limited because their derivation rests on an expression between a single diffusion length and the short circuit current, $I_{sc} = A \ln L + B$, which has questionable validity for cells having diffusion lengths degrading at different rates on each side of the junction.

5.2   The Concept of Damage Equivalence

The wide range of electron and proton energies present in the space environment necessitates some method of describing the effects of various types of radiation in terms of a radiation environment which can be produced under laboratory conditions. The concept of damage equivalence may be discussed in terms of the variation of ordinary solar cell electrical parameters as a function of radiation fluence. In Figure 5.1, we plot the degradation of normalized I_{sc} of GaAs solar cells after irradiation with both protons and electrons of various energies. The radiation was produced by electron and proton accelerators and entered the cells at normal incidence. It is clear that if we bombard GaAs cells with a fluence of 3.8×10^{11} p/cm^2 of 1 MeV protons, we will reduce the cell's I_{sc} to 80% of its starting value (20% degradation); if we bombard a similar cell with a fluence of 3.6×10^{12} p/cm^2 of 10 MeV protons, we can reduce the cell's output to the same I_{sc} value. Or, we could have used a fluence of 6×10^{14} e/cm^2 of 1 MeV electrons to produce the equivalent degradation. The fluences required to reduce a solar cell's electrical performance to some specified level (such as 80% of I_{sc0}) are sometimes referred to as critical fluences, Φ_c. We determine the equivalent fluence for 1 MeV protons relative to 10 MeV protons in the first case by ratioing the 10 MeV fluence required to produce the given degradation to the 1 MeV fluence required to produce the same degradation. This is known as the relative-damage coefficient for 1 MeV protons and, for the case cited, equals 9.5. Similarly, we may ratio the 1 MeV electron fluence required to produce 20% degradation in I_{sc} to the 10 MeV proton fluence required to produce the same amount of damage. This ratio will define a damage coefficient of 10 MeV protons relative to 1 MeV electrons. For the example described above, this ratio is 167.

As is evident from Figure 5.1, this procedure may be carried out for all the degradation curves for each energy. The same procedure may be carried out for P_{max} and V_{oc}. We may also choose another degradation level at which to pick the fluences for the ratio calculation, and we would get exactly the same results as we did with the 20% degradation level, but only if all the degradation curves such as those shown in Figure 5.1 were parallel to each other. Since this is not the case, and particularly at low proton energies, the calculated damage coefficients will depend on the reference degradation level chosen for the calculation. In practice these coefficients are derived by performing the calculation at 4 or 5 different degradation levels, then estimating averages from these values.

The difficulty with the low energy protons in establishing damage equivalency arises because the range of protons below $\approx$ 1 MeV in GaAs is less than the active solar cell thickness. For this reason,

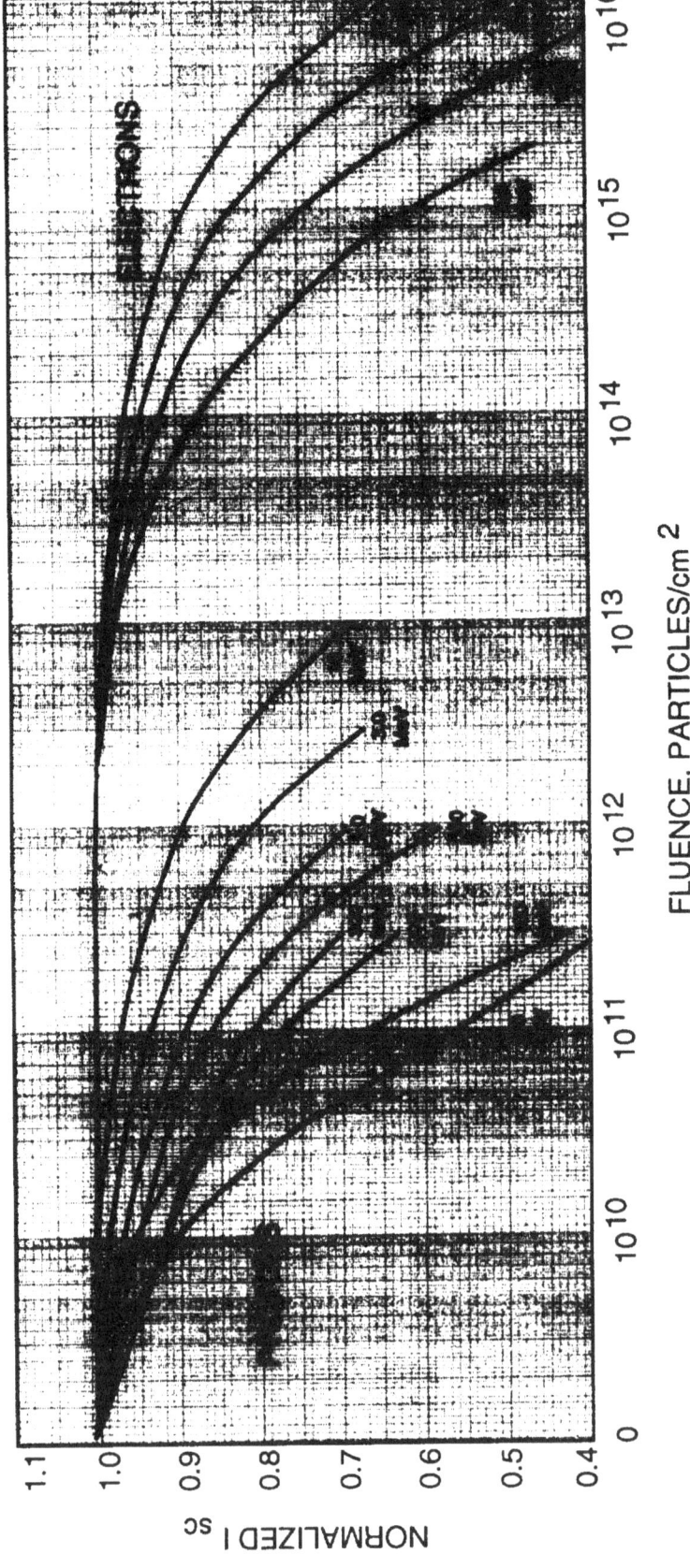

Figure 5.1   Normalized Degradation of I_{sc} for GaAs/Ge Solar Cells vs. Proton and Electron Fluence

low energy protons produce nonuniform damage. This situation is further complicated by the fact that the damage produced per unit pathlength is very high when proton energies are low. As a result, when a low energy proton is stopped in a solar cell, a large amount of damage is concentrated at the end of the track. The derivation of equivalent damage coefficients based on the critical fluence method outlined above relates back to eq. (5-3), where it is assumed that the degradation of diffusion length is uniform throughout the entire thickness of the cell. This will be true for electrons and protons with sufficient energy to penetrate the cell ($E \geq 45$ keV for electrons, and $E \geq 1$ MeV for protons in GaAs solar cells). This observation seems to be borne out by the curves of Figure 5.1, where the degradation curves induced by particles that penetrate the solar cell are very nearly parallel.

By use of the critical-fluence damage coefficients, it is possible to construct a model in which the various components of a combined radiation environment can be described in terms of a damage-equivalent fluence produced by energetic particles of a single energy. 1 MeV electrons are a common and significant component of space radiation, and 1 MeV is an energy which is easily achievable by most electron accelerators. For this reason, 1 MeV electron fluence has been used as a basis of the damage-equivalent fluences which describe solar cell degradation. Similarly, proton damage can be normalized to the damage produced by protons of a specific energy. The proton energy employed for normalization of relative damage should be an energy that occurs in the space environment, an energy that produces relatively uniform damage, and an energy that is commonly available in accelerator laboratories. 10 MeV protons fulfill these requirements and are therefore commonly used as a basis for calculating equivalent proton damage in solar cells.

5.3   Calculation of Damage Coefficients for Space Radiation

The laboratory data used to calculate damage coefficients for normally incident beams of electrons and protons is not directly applicable to space radiation effects because space radiation is omnidirectional and because of the shielding provided by coverglasses mounted on the solar cells. In this section, the analytical methods of calculating the damage effectiveness of each component of the space radiation environment will be detailed. The damage effectiveness of space radiation is calculated relative to normal incidence 1 MeV electrons and to 10 MeV protons on unshielded GaAs solar cells. It will allow the reduction of all components of space radiation to an equivalent laboratory (normal incidence, monoenergetic) irradiation. In this way, laboratory data can be used to predict the behavior of shielded solar arrays in space.

An omnidirectional flux is defined as the number of radiation particles of a particular type and energy that isotropically traverse a test sphere which has a cross-sectional area of 1 cm^2 (radius = $1/\sqrt{\pi}$ cm) per unit time. (The units do not have to be cm, but cm is a commonly used unit for radiation calculations). The commonly used sources of space radiation literature tabulate the environment in terms of omnidirectional fluxes, with units of particles/(cm^2-day). We now follow reference [5.4] in deriving the conversion of omnidirectional fluxes to unidirectional fluxes. Assume a plane of area dA in space, with an incident omnidirectional flux of particles, Φ_0. The portion of omnidirectional flux that arrives at our test sphere per unit solid angle is $\Phi_0/4\pi$, and $(\Phi_0/4\pi)d\Omega$ is the number of particles arriving at the test sphere within solid angle $d\Omega$. The number of particles incident on our plane at angle of incidence θ is:

$$d\phi_n = \frac{\phi_0}{4\pi} (\cos\theta \, dA) \, d\Omega$$

$$and \ \ \phi_n = 2 \, \frac{\phi_0}{4\pi} \, dA \int_0^{\frac{\pi}{2}} 2\pi \sin\theta \cos\theta \, d\theta \tag{5-6}$$

$$giving \ \ \phi_n = \frac{\phi_0}{2}$$

where the solid angle $d\Omega = 2\pi \sin\theta \, d\theta$. We have used integration limits of 0 to $\pi/2$ and multiplied the result by 2 to account for the fact that flux incident on the top half of the plane should not cancel out the flux incident on the bottom half.

The above derivation implies that the unidirectional flux is equal in intensity or "equivalent" to the omnidirectional flux divided by 2. Likewise, if the unit plane area has infinite back shielding, the unidirectional flux will be equal to 1/4 the omnidirectional flux. The above expression determines the normal component of an omnidirectional fluence, that is, the fluence that would pass through the unit plane area. This result is a common derivation found in discussions of omnidirectional vs. normally incident fluxes, but in order for the calculation to be useful for solar cells, we must also properly weigh the damage effectiveness of all angular components.

The expression for the effectiveness weighted for all angular components of an omnidirectional monoenergetic flux and assuming infinite back shielding, is as follows:

$$D(E,t) = \frac{1}{4\pi} \int_0^{\pi/2} D(E_0,\theta) \, 2\pi \sin\theta \cos\theta \, d\theta \qquad (5\text{-}7)$$

where $D(E,t)$ = relative damage coefficient of omnidirectional radiation particles with energy E, relative to unidirectional 1 MeV electrons or 10 MeV protons for a cell protected by a coverglass of thickness t.

$D(E_0,\theta)$ = damage coefficient of unidirectional radiation particles with angle of incidence θ and energy E_0 relative to unidirectional 1 MeV electrons or 10 MeV protons.

t = shielding thickness. When $t = 0$, $E = E_0$

E = proton energy incident on the coverglass

E_0 = proton energy as it enters the solar cell

Equation (5-7) must be further modified to reflect the energy loss in the coverglass.

A common solar panel configuration involves infinite back shielding provided by the solar panel or spacecraft body and an optically transparent, finite shield covering the front surface of the cell. The assumption of infinite back shielding is not valid for those designs where the solar panel consists of very light substrates; and those cases will require separate treatment for the radiation incident on the rear surface of the solar panels. If an omnidirectional flux of radiation particles with energy E is incident on a solar cell shield of thickness t, the particles not stopped in the shielding will emerge from the shielding with an energy of E_0 and in our case, will enter the solar cell with that energy. The emerging energy will be a strong function of the angle of incidence of the particle. The particle track length in the shield is equal to $t/\cos\theta$. The energy E_0 of the emerging particle is found in a slightly roundabout manner. First, we use tables of range vs. energy to find the range of the particle incident on the coverglass, R(E). Subtract the pathlength of the particle in the glass, $t/\cos\theta$ to find the residual range $R(E_0)$ the particle has as it emerges from the coverglass and enters the cell. Again using the range-energy tables, find the energy E_0 corresponding to $R(E_0)$. This procedure is summarized as follows:

$$E_0(E,\theta,t) = R^{-1}\left[R(E) - \frac{t}{\cos\theta} \right] \qquad (5\text{-}8)$$

where R^{-1} is a convenient form used to represent an inverse function of the range-energy relation R(E). Proton and electron range-energy data suitable for this calculation have been tabulated by Janni [5.5, 5.6],

Berger and Seltzer [5.7, 5.8] and more recently in the form of computer programs named TRIM [5.9] and EPSTAR [5.10].

5.4   Electron Space Radiation Effects

The evaluation of $D(E_0,\theta)$ is necessary to complete the integration of eq. (5-7). Recently, an extensive series of electron irradiations of GaAs/Ge solar cells was undertaken at JPL. The experimentally derived values for $D(E_0,\theta)$ resulting from this work are shown in Figure 5.2. The top curve shows the damage coefficients for normally incident electrons. The lower curves, depicting damage coefficients a propos to omnidirectionally incident radiation, are calculated from the experimental curve as described below. Electrons in the MeV energy range produce uniform damage along their tracks and many of them will easily penetrate the active volume of a GaAs solar cell. For this reason, the amount of displacement damage produced by a high energy electron is proportional to the total track length produced in the cell. The length of an individual electron track in a solar cell is proportional to $1/\cos\theta$, therefore:

$$D(E_0,\theta) = \frac{D(E_0,0)}{\cos\theta} \qquad (5-9)$$

However, the number of electrons intercepted by the cell is proportional to its projected area normal to the direction of the incident radiation, i.e., $\propto \cos\theta$, therefore the cos terms cancel out and the damage induced in the solar cell is independent of θ. The fact that MeV electron damage of unshielded Si solar cells is independent of the angle of incidence was confirmed experimentally by Barrett [5.11]. The same arguments hold for high energy protons, which easily penetrate solar cells, and the independence of damage to angle of incidence was confirmed by Anspaugh and Downing [5.12] for both Si and GaAs solar cells.

Equation (5-7) may therefore be written as follows for electrons:

$$D(E,t) = \frac{1}{4\pi} \int_0^{\pi/2} D(E_0,0)\, 2\pi \sin\theta\, d\theta \qquad (5-10)$$

Equation (5-10) can be evaluated with the aid of eq. (5-8) to find E_0, and the upper curve of Figure 5.2 to evaluate $D(E_0,0)$. Equation (5-10) has been integrated numerically for various values of shielding

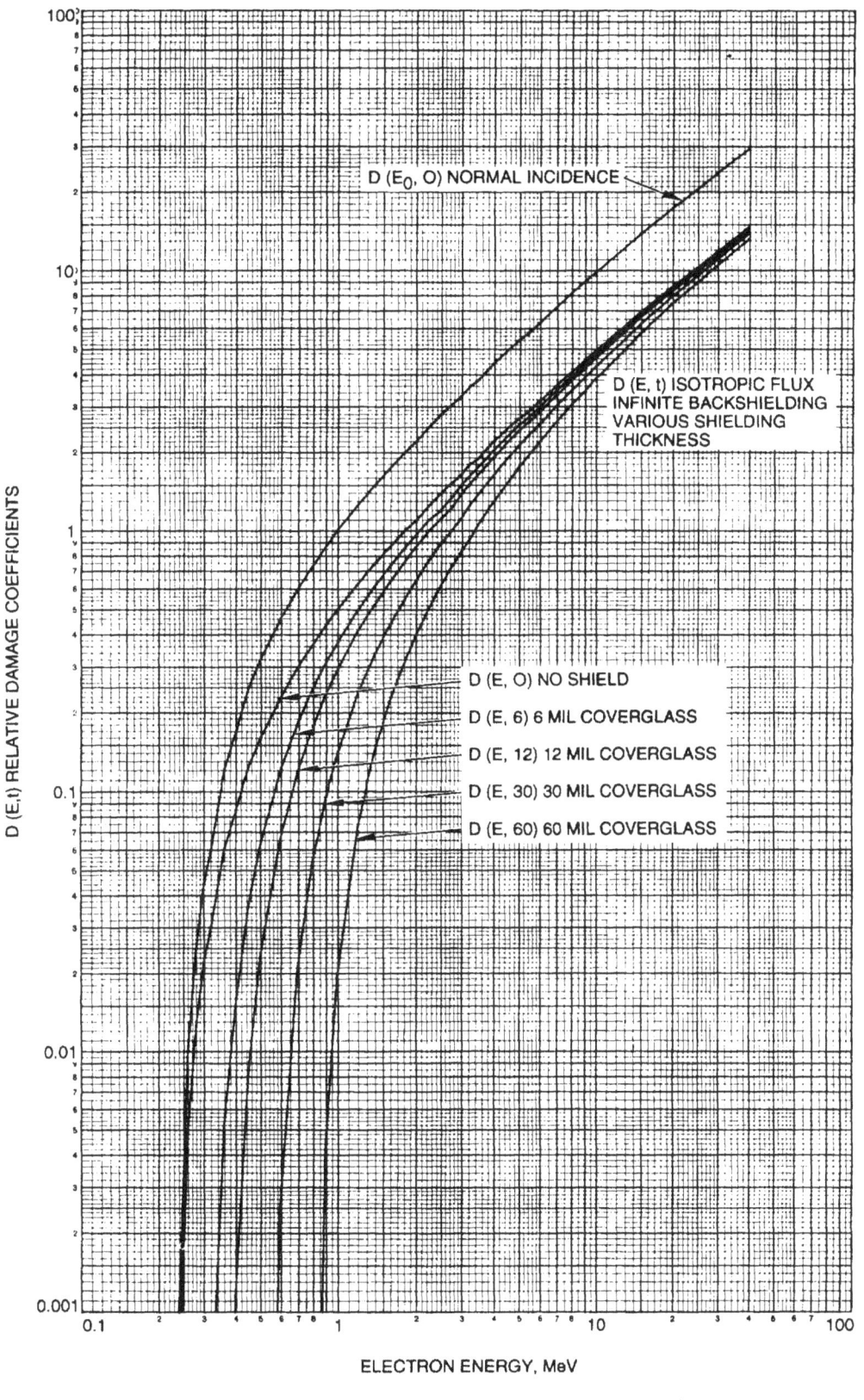

Figure 5.2.   Relative Damage Coefficients for Electron Irradiation of GaAs/Ge Solar Cells

thickness, t, and the results are plotted as the lower curves in Figure 5.2. This process was performed separately for each of the parameters I_{sc}, V_{oc}, and P_{max}. The results are also tabulated in Table 5-1. It was found that the same set of damage coefficients was applicable to all 3 parameters, so there is only one set of damage coefficients for electrons on GaAs.

5.5   Proton Space Radiation Effects

For proton space radiation, the evaluation of eq. (5-7) is more complex than for electrons. Two problems arise in the treatment of space protons with energies less than that required to penetrate the active volume of the solar cell. One problem exists because the relative damage constants based on I_{sc} differ quite markedly from those based on V_{oc} and P_{max} at low proton energies, so we now have two sets of damage coefficients to compute. The second problem is that at low proton energies the degradation curves vs. proton fluence (e.g., Figure 5.1) are not parallel with the degradation curves at higher energies, thereby ensuring some degree of inaccuracy. The proton damage coefficients for highly penetrating energies of 10 MeV or higher can be assumed to be independent of the angle of radiation incidence as discussed above.

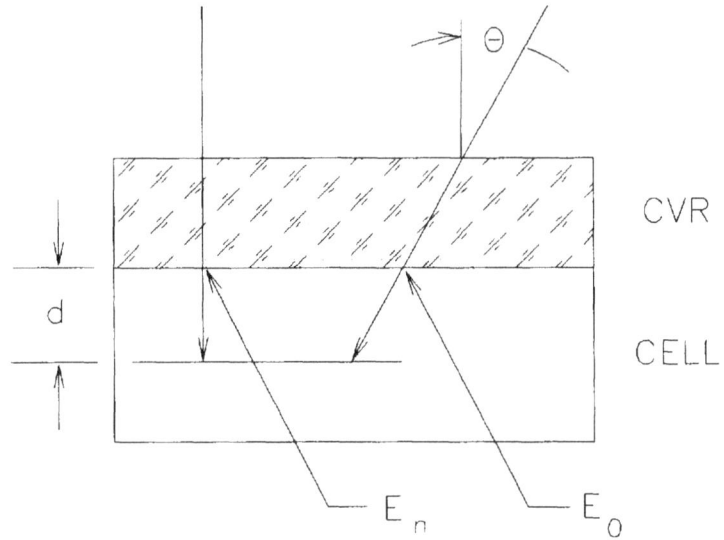

The physical distribution of low energy proton damage is known to be non-linear along the path of the bombarding proton, with most of the damage concentrated near the end of the track. The high damage concentration near the end of the proton track allows the construction

Figure 5.3.   Geometries Used in Omnidirectional Proton Damage Coefficient Calculation

of a simple damage model for the prediction of the effect of angle of incidence on low energy proton damage in Si or GaAs solar cells. It is assumed that the effect of a low energy proton, of arbitrary angle of incidence and energy, is roughly equal to that of a normally incident proton with a range equal to the perpendicular penetration of the non-normally incident proton, as illustrated in Figure 5.3. Therefore, to a first approximation, we can use the known damage coefficient for the normally incident proton of energy E_n for the proton entering with energy E_0 at angle θ. But the proton incident at the slant angle will produce more damage in the cell because its track is longer, so it should have a larger coefficient than the normally incident proton. To compensate for this, the total number of displacements $N_{td}(E_0)$ for the off-normal proton, and $N_{td}(E_n)$, for the normally incident proton, are calculated using the Kinchin and Pease model (discussed in Chapter 4). The ratio, $N_{td}(E_0)/N_{td}(E_n)$, is used as a compensation factor to improve on the estimate for $D(E_0,\theta)$, as follows:

$$D(E_0,\theta) = D(E_n,0) \frac{N_{td}(E_0)}{N_{td}(E_n)} \qquad (5\text{-}11)$$

where $D(E_0,\theta) =$ relative damage coefficient for protons entering a solar cell with energy E_0 at an angle θ

$D(E_n,0) =$ relative damage coefficient for a proton of normal incidence ($\theta = 0$) with range E_n (range equal to $R(E_0 \cos\theta)$

$N_{td}(E_0) =$ the total number of displacements created by a proton entering the solar cell with energy E_0

$N_{td}(E_n) =$ the total number of displacements created by a proton entering the solar cell with energy E_n

$\cos\theta =$ unit cell projected area

$E_n =$ $R^{-1}[R(E_0) \cos\theta]$

$E_0 =$ proton energy as it emerges from the coverglass and enters the solar cell

When the range of a proton of energy E_0 incident on a solar cell at angle θ exceeds the cell thickness divided by $\cos\theta$, the proton will penetrate the cell. This case is entirely analogous to the case previously discussed for high energy electrons, so that:

$$D(E_0,0) = \frac{D(E_0,0)}{\cos\theta} \qquad (5\text{-}12)$$

Equations (5-11) and (5-12) allow the evaluation of eq. (5-7) for infinite backshielding as follows:

$$D(E,t) = \frac{1}{4\pi} \int_0^{\theta_p} D(E_0,0) \, 2\pi \sin\theta \, d\theta$$

$$+ \frac{1}{4\pi} \int_{\theta_p}^{\frac{\pi}{2}} D(E_n,0) \, \frac{N_{td}(E_0)}{N_{td}(E_n)} \, 2\pi \sin\theta \, \cos\theta \, d\theta \qquad (5\text{-}13)$$

where θ_p =   the angle of incidence for which a proton of energy E will just penetrate both the coverglass and the solar cell.

The first term in eq. (5-13) represents the case when the proton completely penetrates the coverglass and the solar cell, while the second term applies when the proton penetrates the coverglass but stops in the cell. This integration has been done by machine using the experimentally derived $D(E_0,0)$ values for I_{sc} and for V_{oc}, P_{max} shown in Figure 5.4. Separate integrations were done for $D(E_0,0)$ values based on I_{sc} and on V_{oc}, P_{max}. $D(E,t)$ values calculated by eq. (5-13) are a function of solar cell thickness. However, evaluations made with several solar cell thicknesses revealed that the dependence on cell thickness is very slight, and for practical purposes the results can be considered independent of thickness. The results of numerical integrations of eq. (5-13) are shown in Figures 5.5 and 5.6. The same data is displayed in tabular form in Tables 5-2 and 5-3.

The values of relative damage constants for omnidirectional fluences of protons on shielded solar cells allow a space proton environment to be reduced to an equivalent fluence of normally incident 10 MeV protons on unshielded GaAs solar cells. Experimental studies of GaAs solar cells have indicated that a fluence of normally incident 10 MeV protons produces damage in I_{sc} that can be approximated by a fluence of 1 MeV electrons, which is 400 times that of the 10 MeV proton fluence. A similar factor for degradation of V_{oc} was found to be 1400 and for P_{max} the factor is 1000. Note that these values are quite different than was found for Si where all three factors for all three cases were equal to 3000.

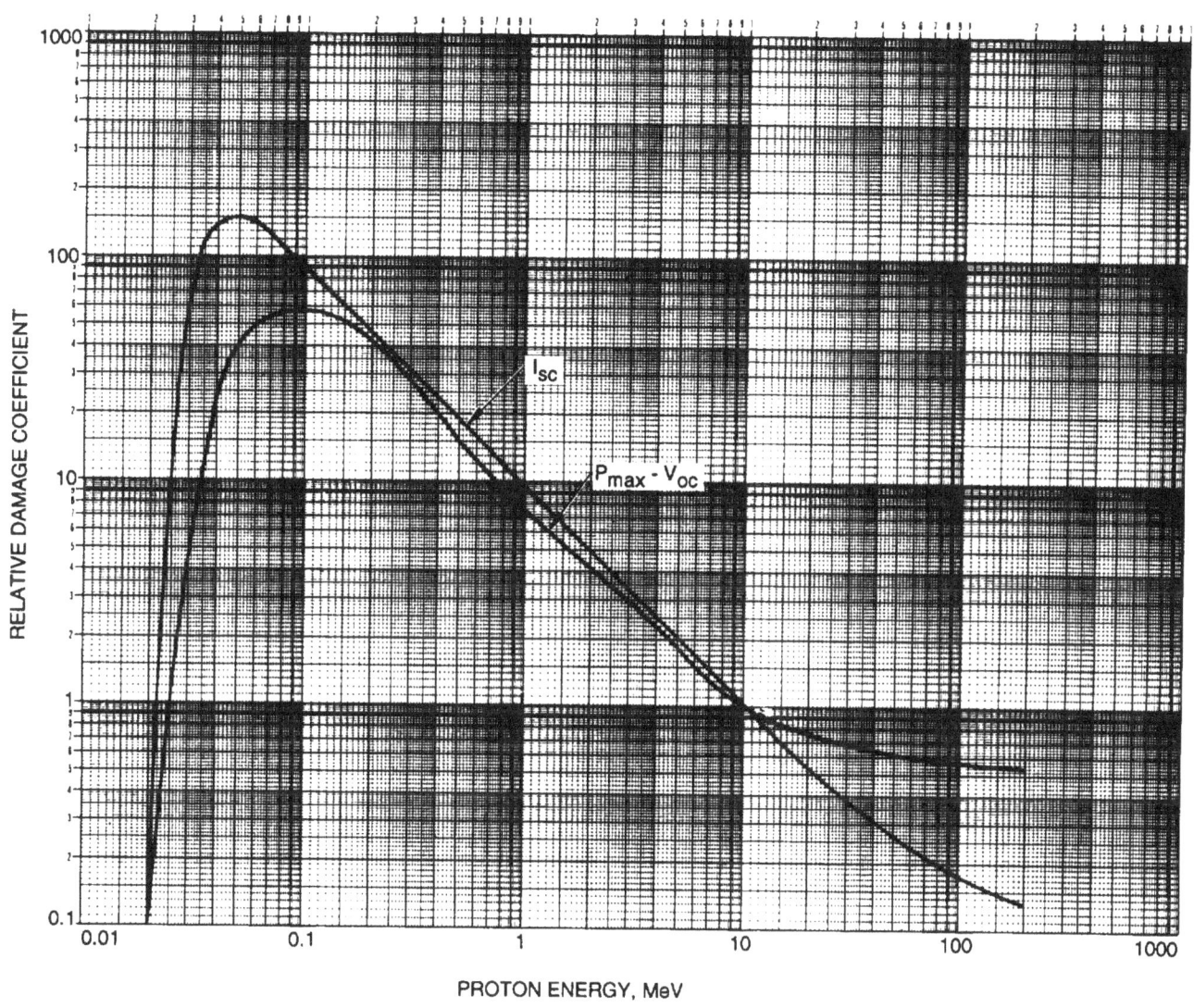

Figure 5.4.   Experimental I_{sc} and V_{oc}, P_{max} Relative Damage Coefficients for GaAs/Ge Solar
Cells and Normally Incident Proton Irradiation, Normalized to 10 MeV

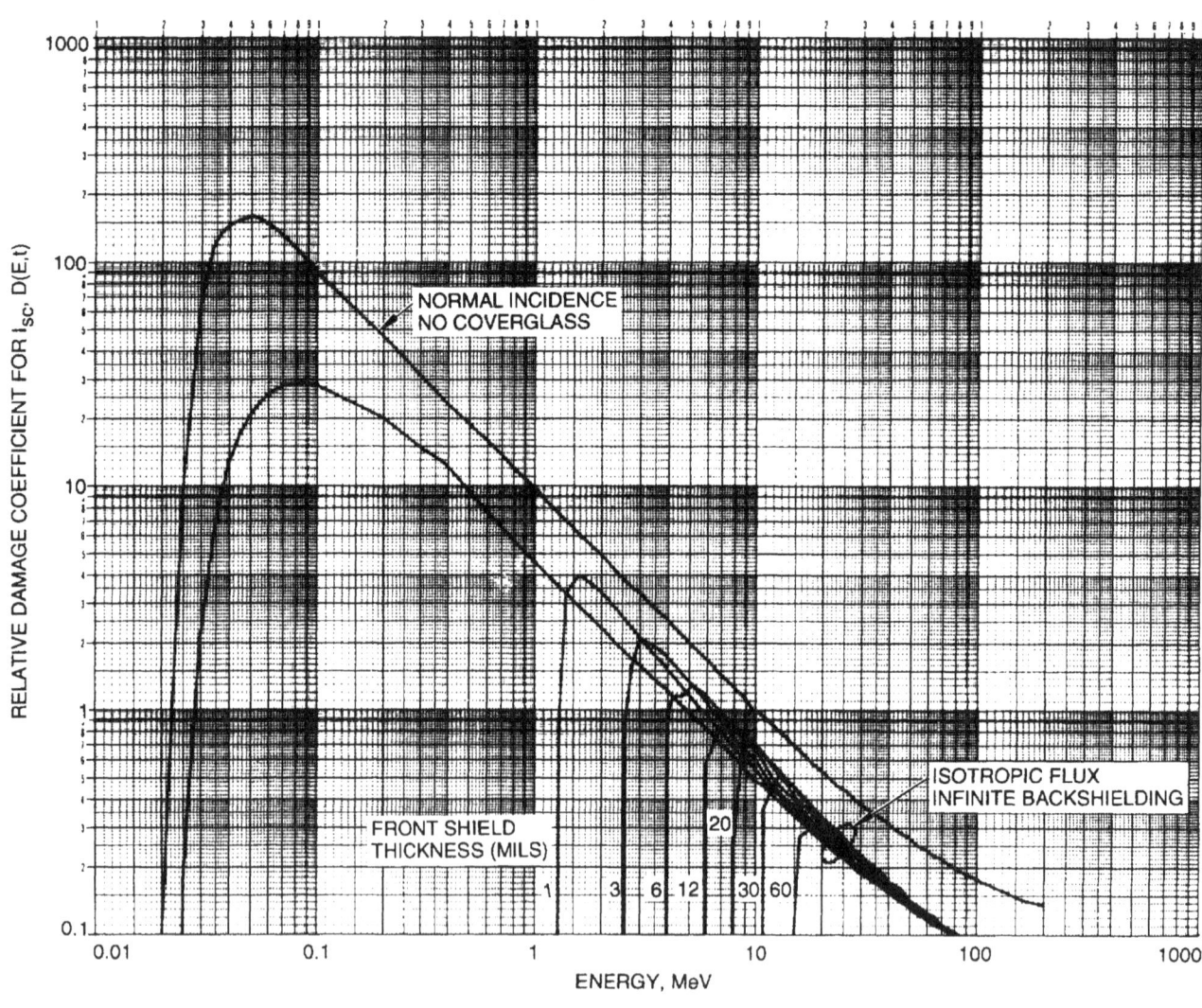

Figure 5.5. I_{sc} Relative Damage Coefficients for Omnidirectional Space Proton Irradiation of Shielded GaAs/Ge Solar Cells

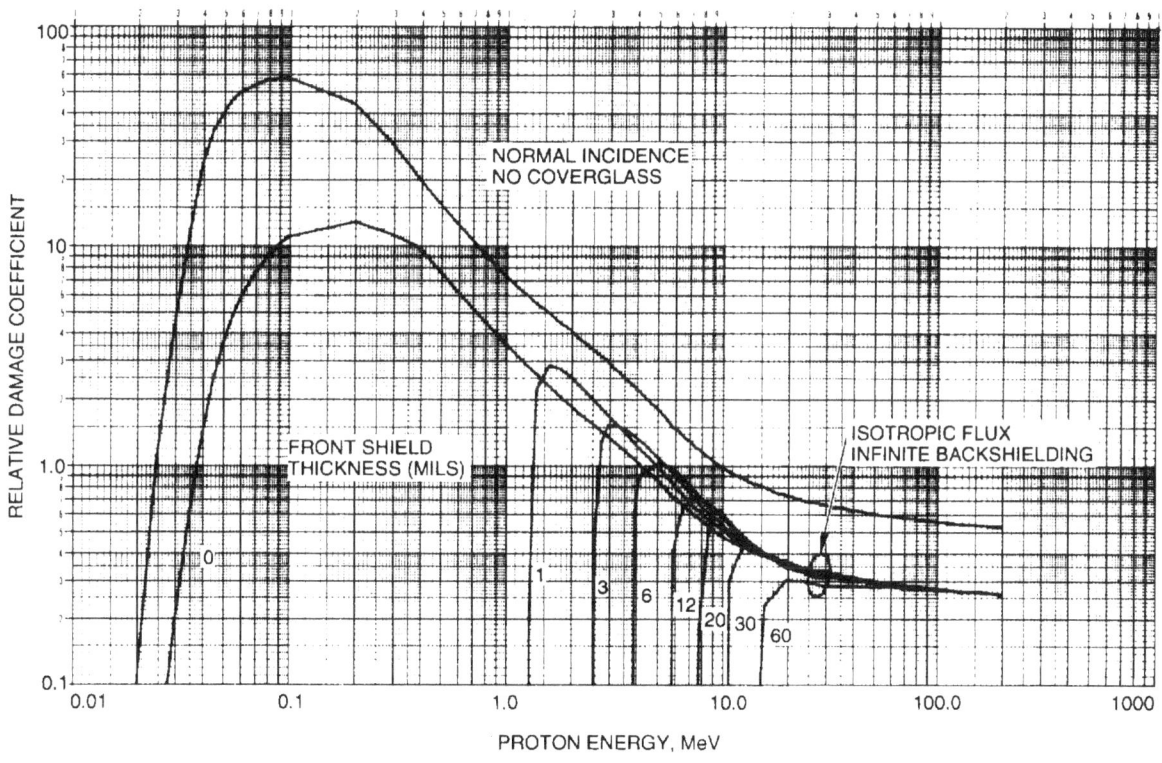

Figure 5.6. V_{oc} or P_{max} Relative Damage Coefficients for Omnidirectional Space Proton
Irradiation of Shielded GaAs/Ge Solar Cells

Table 5-1.   Electron Damage Coefficients for GaAs Solar Cells

Shield Thickness, Mils

Energy (Mev)	0 Nor	0	1	3	6	12	20	30	60
0.15	5.00E-06	2.500E-06	0.0	0.0	0.0	0.0	0.0	0.0	0.0
0.16	1.00E-05	5.000E-06	0.0	0.0	0.0	0.0	0.0	0.0	0.0
0.17	2.00E-05	1.000E-05	9.582E-07	0.0	0.0	0.0	0.0	0.0	0.0
0.18	5.00E-05	2.500E-05	2.932E-06	0.0	0.0	0.0	0.0	0.0	0.0
0.19	1.00E-04	5.000E-05	6.978E-06	0.0	0.0	0.0	0.0	0.0	0.0
0.20	2.00E-04	1.000E-04	1.679E-05	7.415E-07	0.0	0.0	0.0	0.0	0.0
0.22	5.00E-04	2.500E-04	6.805E-05	5.762E-06	0.0	0.0	0.0	0.0	0.0
0.24	1.00E-03	5.000E-04	1.818E-04	2.974E-05	6.795E-07	0.0	0.0	0.0	0.0
0.26	1.00E-02	5.000E-03	5.399E-04	9.434E-05	5.553E-06	0.0	0.0	0.0	0.0
0.28	2.50E-02	1.250E-02	3.622E-03	2.535E-04	2.769E-05	0.0	0.0	0.0	0.0
0.30	4.40E-02	2.200E-02	9.504E-03	1.794E-03	8.541E-05	1.232E-07	0.0	0.0	0.0
0.32	6.00E-02	3.000E-02	1.693E-02	5.463E-03	3.417E-04	1.916E-06	0.0	0.0	0.0
0.36	1.20E-01	6.000E-02	3.612E-02	1.724E-02	5.163E-03	4.519E-05	0.0	0.0	0.0
0.40	1.70E-01	8.500E-02	6.181E-02	3.674E-02	1.533E-02	1.154E-03	1.186E-06	0.0	0.0
0.45	2.50E-01	1.250E-01	9.577E-02	6.445E-02	3.643E-02	8.242E-03	5.982E-05	0.0	0.0
0.50	3.20E-01	1.600E-01	1.318E-01	9.744E-02	6.232E-02	2.290E-02	2.608E-03	1.527E-06	0.0
0.60	4.60E-01	2.300E-01	1.999E-01	1.619E-01	1.215E-01	6.664E-02	2.393E-02	2.558E-03	0.0
0.70	6.00E-01	3.000E-01	2.693E-01	2.286E-01	1.835E-01	1.190E-01	6.257E-02	2.076E-02	0.0
0.80	7.30E-01	3.650E-01	3.348E-01	2.940E-01	2.467E-01	1.756E-01	1.088E-01	5.335E-02	1.348E-05
0.90	8.60E-01	4.300E-01	3.993E-01	3.574E-01	3.083E-01	2.332E-01	1.595E-01	9.337E-02	4.126E-03
1.00	1.00E+00	5.000E-01	4.677E-01	4.227E-01	3.704E-01	2.908E-01	2.117E-01	1.381E-01	1.951E-02
1.20	1.25E+00	6.250E-01	5.943E-01	5.486E-01	4.947E-01	4.090E-01	3.198E-01	2.330E-01	7.296E-02
1.40	1.50E+00	7.500E-01	7.185E-01	6.715E-01	6.146E-01	5.246E-01	4.295E-01	3.345E-01	1.429E-01
1.60	1.74E+00	8.700E-01	8.384E-01	7.907E-01	7.334E-01	6.396E-01	5.394E-01	4.369E-01	2.222E-01
1.80	1.99E+00	9.950E-01	9.615E-01	9.112E-01	8.516E-01	7.539E-01	6.482E-01	5.402E-01	3.063E-01
2.00	2.20E+00	1.100E+00	1.069E+00	1.022E+00	9.643E-01	8.673E-01	7.590E-01	6.445E-01	3.939E-01
2.25	2.50E+00	1.250E+00	1.215E+00	1.165E+00	1.103E+00	9.999E-01	8.875E-01	7.711E-01	5.040E-01
2.50	2.80E+00	1.400E+00	1.279E+00	1.249E+00	1.203E+00	1.119E+00	1.017E+00	8.984E-01	6.175E-01
2.75	3.05E+00	1.525E+00	1.478E+00	1.417E+00	1.341E+00	1.226E+00	1.111E+00	1.002E+00	7.295E-01
3.00	3.30E+00	1.650E+00	1.614E+00	1.568E+00	1.501E+00	1.390E+00	1.262E+00	1.123E+00	8.334E-01
3.25	3.60E+00	1.800E+00	1.760E+00	1.710E+00	1.637E+00	1.523E+00	1.395E+00	1.260E+00	9.354E-01
3.50	3.80E+00	1.900E+00	1.867E+00	1.825E+00	1.760E+00	1.654E+00	1.530E+00	1.389E+00	1.058E+00
3.75	4.10E+00	2.050E+00	2.008E+00	1.959E+00	1.885E+00	1.770E+00	1.644E+00	1.509E+00	1.174E+00
4.00	4.40E+00	2.200E+00	2.156E+00	2.106E+00	2.030E+00	1.909E+00	1.771E+00	1.628E+00	1.289E+00
4.50	4.90E+00	2.450E+00	2.408E+00	2.362E+00	2.291E+00	2.172E+00	2.033E+00	1.882E+00	1.514E+00
5.00	5.40E+00	2.700E+00	2.656E+00	2.610E+00	2.540E+00	2.416E+00	2.276E+00	2.122E+00	1.747E+00
5.50	5.80E+00	2.900E+00	2.858E+00	2.818E+00	2.755E+00	2.638E+00	2.502E+00	2.352E+00	1.970E+00
6.00	6.30E+00	3.150E+00	3.102E+00	3.056E+00	2.987E+00	2.859E+00	2.716E+00	2.561E+00	2.182E+00
7.00	7.20E+00	3.600E+00	3.550E+00	3.506E+00	3.440E+00	3.307E+00	3.166E+00	3.005E+00	2.600E+00
8.00	8.10E+00	4.050E+00	3.996E+00	3.951E+00	3.884E+00	3.748E+00	3.601E+00	3.433E+00	3.011E+00
9.00	9.00E+00	4.500E+00	4.442E+00	4.396E+00	4.328E+00	4.191E+00	4.033E+00	3.856E+00	3.418E+00
10.00	9.80E+00	4.900E+00	4.840E+00	4.798E+00	4.733E+00	4.602E+00	4.444E+00	4.269E+00	3.824E+00
15.00	1.38E+01	6.900E+00	6.823E+00	6.777E+00	6.709E+00	6.574E+00	6.393E+00	6.203E+00	5.704E+00
20.00	1.73E+01	8.650E+00	8.560E+00	8.516E+00	8.450E+00	8.318E+00	8.142E+00	7.926E+00	7.408E+00
25.00	2.05E+01	1.025E+01	1.015E+01	1.010E+01	1.004E+01	9.909E+00	9.735E+00	9.514E+00	8.976E+00
30.00	2.37E+01	1.185E+01	1.173E+01	1.169E+01	1.161E+01	1.147E+01	1.129E+01	1.106E+01	1.047E+01
40.00	2.95E+01	1.475E+01	1.461E+01	1.456E+01	1.449E+01	1.434E+01	1.415E+01	1.392E+01	1.327E+01

Table 5-2. Proton Damage Coefficients for GaAs Solar Cells -- I_{sc}

Energy (Mev)	0 Nor	0	1	3	6	12	20	30	60
0.020	1.00E-01	0.00	0.00	0.00	0.00	0.00	0.00	0.00	0.00
0.025	1.00E+01	1.885E-01	0.00	0.00	0.00	0.00	0.00	0.00	0.00
0.030	6.50E+01	2.186E+00	0.00	0.00	0.00	0.00	0.00	0.00	0.00
0.035	1.25E+02	7.022E+00	0.00	0.00	0.00	0.00	0.00	0.00	0.00
0.040	1.45E+02	1.246E+01	0.00	0.00	0.00	0.00	0.00	0.00	0.00
0.045	1.52E+02	1.701E+01	0.00	0.00	0.00	0.00	0.00	0.00	0.00
0.050	1.56E+02	2.066E+01	0.00	0.00	0.00	0.00	0.00	0.00	0.00
0.055	1.52E+02	2.349E+01	0.00	0.00	0.00	0.00	0.00	0.00	0.00
0.060	1.48E+02	2.560E+01	0.00	0.00	0.00	0.00	0.00	0.00	0.00
0.070	1.30E+02	2.806E+01	0.00	0.00	0.00	0.00	0.00	0.00	0.00
0.080	1.13E+02	2.887E+01	0.00	0.00	0.00	0.00	0.00	0.00	0.00
0.090	1.00E+02	2.877E+01	0.00	0.00	0.00	0.00	0.00	0.00	0.00
0.100	9.00E+01	2.821E+01	0.00	0.00	0.00	0.00	0.00	0.00	0.00
0.200	4.65E+01	1.997E+01	0.00	0.00	0.00	0.00	0.00	0.00	0.00
0.300	3.10E+01	1.477E+01	0.00	0.00	0.00	0.00	0.00	0.00	0.00
0.400	2.32E+01	1.221E+01	0.00	0.00	0.00	0.00	0.00	0.00	0.00
0.600	1.60E+01	7.742E+00	0.00	0.00	0.00	0.00	0.00	0.00	0.00
0.800	1.20E+01	5.749E+00	0.00	0.00	0.00	0.00	0.00	0.00	0.00
1.000	9.80E+00	4.610E+00	0.00	0.00	0.00	0.00	0.00	0.00	0.00
1.200	8.10E+00	3.851E+00	0.00	0.00	0.00	0.00	0.00	0.00	0.00
1.300	7.50E+00	3.560E+00	8.283E-01	0.00	0.00	0.00	0.00	0.00	0.00
1.400	7.00E+00	3.315E+00	3.344E+00	0.00	0.00	0.00	0.00	0.00	0.00
1.600	6.10E+00	2.914E+00	4.011E+00	0.00	0.00	0.00	0.00	0.00	0.00
1.800	5.45E+00	2.597E+00	3.695E+00	0.00	0.00	0.00	0.00	0.00	0.00
2.000	4.95E+00	2.347E+00	3.384E+00	0.00	0.00	0.00	0.00	0.00	0.00
2.200	4.50E+00	2.139E+00	3.070E+00	0.00	0.00	0.00	0.00	0.00	0.00
2.400	4.10E+00	1.961E+00	2.784E+00	0.00	0.00	0.00	0.00	0.00	0.00
2.600	3.80E+00	1.816E+00	2.538E+00	8.575E-01	0.00	0.00	0.00	0.00	0.00
2.800	3.50E+00	1.683E+00	2.325E+00	1.604E+00	0.00	0.00	0.00	0.00	0.00
3.000	3.35E+00	1.596E+00	2.138E+00	2.063E+00	0.00	0.00	0.00	0.00	0.00
3.200	3.15E+00	1.503E+00	1.975E+00	2.039E+00	0.00	0.00	0.00	0.00	0.00
3.400	2.95E+00	1.413E+00	1.837E+00	1.973E+00	0.00	0.00	0.00	0.00	0.00
3.600	2.80E+00	1.342E+00	1.722E+00	1.899E+00	0.00	0.00	0.00	0.00	0.00
3.800	2.65E+00	1.273E+00	1.620E+00	1.811E+00	0.00	0.00	0.00	0.00	0.00
4.000	2.55E+00	1.233E+00	1.521E+00	1.724E+00	8.558E-01	0.00	0.00	0.00	0.00
4.200	2.40E+00	1.157E+00	1.439E+00	1.635E+00	1.134E+00	0.00	0.00	0.00	0.00
4.400	2.30E+00	1.110E+00	1.370E+00	1.564E+00	1.160E+00	0.00	0.00	0.00	0.00
4.600	2.20E+00	1.063E+00	1.298E+00	1.487E+00	1.153E+00	0.00	0.00	0.00	0.00
4.800	2.12E+00	1.025E+00	1.227E+00	1.414E+00	1.175E+00	0.00	0.00	0.00	0.00
5.200	1.95E+00	9.450E-01	1.123E+00	1.298E+00	1.298E+00	0.00	0.00	0.00	0.00
5.600	1.82E+00	8.838E-01	1.028E+00	1.188E+00	1.230E+00	0.00	0.00	0.00	0.00
6.000	1.70E+00	8.269E-01	9.618E-01	1.095E+00	1.158E+00	5.747E-01	0.00	0.00	0.00
6.400	1.60E+00	7.792E-01	8.899E-01	1.007E+00	1.086E+00	6.661E-01	0.00	0.00	0.00
6.800	1.50E+00	7.318E-01	8.381E-01	9.438E-01	1.017E+00	7.922E-01	0.00	0.00	0.00
7.200	1.42E+00	6.928E-01	7.731E-01	8.831E-01	9.575E-01	8.016E-01	0.00	0.00	0.00
7.600	1.35E+00	6.594E-01	7.402E-01	8.238E-01	8.971E-01	8.778E-01	0.00	0.00	0.00
8.000	1.28E+00	6.262E-01	6.977E-01	7.624E-01	8.465E-01	8.545E-01	2.305E-01	0.00	0.00
9.000	1.14E+00	5.586E-01	6.011E-01	6.640E-01	7.347E-01	7.787E-01	7.459E-01	0.00	0.00
10.000	1.00E+00	4.911E-01	5.594E-01	6.148E-01	6.800E-01	7.238E-01	7.284E-01	0.00	0.00
11.000	9.40E-01	4.613E-01	4.922E-01	5.299E-01	5.909E-01	6.401E-01	6.329E-01	3.661E-01	0.00
12.000	8.60E-01	4.226E-01	4.427E-01	4.696E-01	5.055E-01	5.652E-01	5.839E-01	4.471E-01	0.00
13.000	8.00E-01	3.931E-01	4.188E-01	4.347E-01	4.642E-01	5.124E-01	5.359E-01	5.136E-01	0.00
14.000	7.50E-01	3.686E-01	3.847E-01	3.989E-01	4.295E-01	4.701E-01	4.918E-01	4.895E-01	0.00
15.000	7.00E-01	3.441E-01	3.551E-01	3.799E-01	4.067E-01	4.570E-01	4.636E-01	8.524E-02
16.000	6.60E-01	3.246E-01	3.431E-01	3.526E-01	3.722E-01	3.914E-01	4.230E-01	4.316E-01	2.732E-01
18.000	5.90E-01	2.902E-01	3.018E-01	3.115E-01	3.263E-01	3.531E-01	3.709E-01	3.827E-01	3.019E-01
20.000	5.40E-01	2.656E-01	2.736E-01	2.851E-01	2.913E-01	3.085E-01	3.242E-01	3.417E-01	3.367E-01
22.000	5.00E-01	2.459E-01	2.507E-01	2.577E-01	2.715E-01	2.827E-01	3.012E-01	3.108E-01	3.115E-01
24.000	4.60E-01	2.264E-01	2.304E-01	2.412E-01	2.439E-01	2.538E-01	2.643E-01	2.753E-01	2.895E-01
26.000	4.40E-01	2.165E-01	2.186E-01	2.242E-01	2.301E-01	2.349E-01	2.415E-01	2.545E-01	2.683E-01
28.000	4.10E-01	2.018E-01	2.043E-01	2.092E-01	2.133E-01	2.258E-01	2.262E-01	2.356E-01	2.454E-01
30.000	3.85E-01	1.895E-01	1.916E-01	2.006E-01	2.016E-01	2.126E-01	2.162E-01	2.228E-01	2.294E-01
34.000	3.55E-01	1.747E-01	1.758E-01	1.801E-01	1.882E-01	1.912E-01	1.929E-01	1.972E-01	2.067E-01
38.000	3.30E-01	1.624E-01	1.633E-01	1.653E-01	1.683E-01	1.736E-01	1.718E-01	1.777E-01	1.887E-01
42.000	3.05E-01	1.501E-01	1.509E-01	1.527E-01	1.535E-01	1.568E-01	1.625E-01	1.642E-01	1.685E-01
46.000	2.85E-01	1.403E-01	1.409E-01	1.422E-01	1.462E-01	1.470E-01	1.504E-01	1.518E-01	1.607E-01
50.000	2.70E-01	1.329E-01	1.333E-01	1.342E-01	1.369E-01	1.370E-01	1.382E-01	1.407E-01	1.440E-01
55.000	2.50E-01	1.230E-01	1.234E-01	1.243E-01	1.264E-01	1.285E-01	1.290E-01	1.308E-01	1.348E-01
60.000	2.40E-01	1.181E-01	1.183E-01	1.187E-01	1.196E-01	1.209E-01	1.230E-01	1.232E-01	1.266E-01
65.000	2.25E-01	1.107E-01	1.110E-01	1.115E-01	1.124E-01	1.165E-01	1.149E-01	1.178E-01	1.207E-01
70.000	2.18E-01	1.073E-01	1.074E-01	1.077E-01	1.082E-01	1.104E-01	1.115E-01	1.102E-01	1.117E-01
80.000	2.00E-01	9.843E-02	9.856E-02	9.883E-02	9.926E-02	1.008E-01	1.003E-01	1.017E-01	1.036E-01
90.000	1.87E-01	9.203E-02	9.212E-02	9.230E-02	9.259E-02	9.326E-02	9.529E-02	9.541E-02	9.542E-02
100.000	1.78E-01	8.760E-02	8.766E-02	8.778E-02	8.796E-02	8.839E-02	8.961E-02	8.880E-02	9.020E-02
130.000	1.58E-01	7.776E-02	7.779E-02	7.785E-02	7.795E-02	7.816E-02	7.846E-02	7.893E-02	8.036E-02
160.000	1.44E-01	7.087E-02	7.089E-02	7.093E-02	7.099E-02	7.112E-02	7.129E-02	7.153E-02	7.375E-02
200.000	1.35E-01	6.644E-02	6.645E-02	6.646E-02	6.649E-02	6.654E-02	6.661E-02	6.671E-02	6.707E-02

Table 5-3.   Proton Damage Coefficients for GaAs Solar Cells -- P_{max}, V_{oc}

Shield Thickness, Mils

Energy (Mev)	0 Nor	0	1	3	6	12	20	30	60
0.020	1.00E-01	0.00	0.00	0.00	0.00	0.00	0.00	0.00	0.00
0.025	1.10E+00	3.504E-02	0.00	0.00	0.00	0.00	0.00	0.00	0.00
0.030	4.40E+00	1.900E-01	0.00	0.00	0.00	0.00	0.00	0.00	0.00
0.035	1.15E+01	5.793E-01	0.00	0.00	0.00	0.00	0.00	0.00	0.00
0.040	2.35E+01	1.323E+00	0.00	0.00	0.00	0.00	0.00	0.00	0.00
0.045	3.35E+01	2.376E+00	0.00	0.00	0.00	0.00	0.00	0.00	0.00
0.050	4.00E+01	3.512E+00	0.00	0.00	0.00	0.00	0.00	0.00	0.00
0.055	4.60E+01	4.631E+00	0.00	0.00	0.00	0.00	0.00	0.00	0.00
0.060	5.00E+01	5.690E+00	0.00	0.00	0.00	0.00	0.00	0.00	0.00
0.070	5.40E+01	7.508E+00	0.00	0.00	0.00	0.00	0.00	0.00	0.00
0.080	5.70E+01	8.966E+00	0.00	0.00	0.00	0.00	0.00	0.00	0.00
0.090	5.80E+01	1.013E+01	0.00	0.00	0.00	0.00	0.00	0.00	0.00
0.100	5.80E+01	1.104E+01	0.00	0.00	0.00	0.00	0.00	0.00	0.00
0.200	4.40E+01	1.283E+01	0.00	0.00	0.00	0.00	0.00	0.00	0.00
0.300	2.90E+01	1.123E+01	0.00	0.00	0.00	0.00	0.00	0.00	0.00
0.400	2.00E+01	9.781E+00	0.00	0.00	0.00	0.00	0.00	0.00	0.00
0.600	1.27E+01	6.271E+00	0.00	0.00	0.00	0.00	0.00	0.00	0.00
0.800	9.30E+00	4.608E+00	0.00	0.00	0.00	0.00	0.00	0.00	0.00
1.000	7.40E+00	3.647E+00	0.00	0.00	0.00	0.00	0.00	0.00	0.00
1.200	6.30E+00	3.034E+00	0.00	0.00	0.00	0.00	0.00	0.00	0.00
1.300	5.90E+00	2.807E+00	1.954E-01	0.00	0.00	0.00	0.00	0.00	0.00
1.400	5.50E+00	2.611E+00	2.227E+00	0.00	0.00	0.00	0.00	0.00	0.00
1.600	5.00E+00	2.305E+00	2.866E+00	0.00	0.00	0.00	0.00	0.00	0.00
1.800	4.50E+00	2.070E+00	2.771E+00	0.00	0.00	0.00	0.00	0.00	0.00
2.000	4.20E+00	1.890E+00	2.553E+00	0.00	0.00	0.00	0.00	0.00	0.00
2.200	3.80E+00	1.735E+00	2.334E+00	0.00	0.00	0.00	0.00	0.00	0.00
2.400	3.60E+00	1.632E+00	2.139E+00	0.00	0.00	0.00	0.00	0.00	0.00
2.600	3.35E+00	1.527E+00	1.973E+00	4.618E-01	0.00	0.00	0.00	0.00	0.00
2.800	3.15E+00	1.442E+00	1.825E+00	1.306E+00	0.00	0.00	0.00	0.00	0.00
3.000	3.00E+00	1.373E+00	1.705E+00	1.544E+00	0.00	0.00	0.00	0.00	0.00
3.200	2.80E+00	1.294E+00	1.598E+00	1.528E+00	0.00	0.00	0.00	0.00	0.00
3.400	2.65E+00	1.230E+00	1.505E+00	1.507E+00	0.00	0.00	0.00	0.00	0.00
3.600	2.50E+00	1.168E+00	1.423E+00	1.468E+00	0.00	0.00	0.00	0.00	0.00
3.800	2.40E+00	1.123E+00	1.346E+00	1.420E+00	0.00	0.00	0.00	0.00	0.00
4.000	2.30E+00	1.072E+00	1.273E+00	1.370E+00	5.910E-01	0.00	0.00	0.00	0.00
4.200	2.20E+00	1.028E+00	1.212E+00	1.321E+00	8.959E-01	0.00	0.00	0.00	0.00
4.400	2.10E+00	9.870E-01	1.159E+00	1.269E+00	9.438E-01	0.00	0.00	0.00	0.00
4.600	2.00E+00	9.431E-01	1.109E+00	1.223E+00	9.566E-01	0.00	0.00	0.00	0.00
4.800	1.93E+00	9.120E-01	1.064E+00	1.169E+00	9.768E-01	0.00	0.00	0.00	0.00
5.200	1.80E+00	8.527E-01	9.783E-01	1.084E+00	1.034E+00	0.00	0.00	0.00	0.00
5.600	1.65E+00	7.880E-01	9.081E-01	1.006E+00	1.002E+00	0.00	0.00	0.00	0.00
6.000	1.50E+00	7.209E-01	8.357E-01	9.363E-01	9.588E-01	3.840E-01	0.00	0.00	0.00
6.400	1.45E+00	6.966E-01	7.743E-01	8.714E-01	9.109E-01	5.509E-01	0.00	0.00	0.00
6.800	1.35E+00	6.512E-01	7.290E-01	8.104E-01	8.622E-01	6.679E-01	0.00	0.00	0.00
7.200	1.30E+00	6.263E-01	6.838E-01	7.615E-01	8.161E-01	6.843E-01	0.00	0.00	0.00
7.600	1.23E+00	5.946E-01	6.469E-01	7.141E-01	7.704E-01	7.207E-01	0.00	0.00	0.00
8.000	1.19E+00	5.764E-01	6.241E-01	6.746E-01	7.266E-01	7.110E-01	1.844E-01	0.00	0.00
9.000	1.08E+00	5.246E-01	5.557E-01	5.987E-01	6.423E-01	6.603E-01	6.049E-01	0.00	0.00
10.000	1.00E+00	4.856E-01	5.278E-01	5.587E-01	5.979E-01	6.216E-01	6.348E-01	0.00	0.00
11.000	9.40E-01	4.578E-01	4.808E-01	5.039E-01	5.371E-01	5.580E-01	5.368E-01	3.042E-01	0.00
12.000	9.00E-01	4.397E-01	4.488E-01	4.608E-01	4.808E-01	5.104E-01	5.037E-01	3.847E-01	0.00
13.000	8.70E-01	4.249E-01	4.327E-01	4.364E-01	4.539E-01	4.698E-01	4.738E-01	4.361E-01	0.00
14.000	8.40E-01	4.101E-01	4.160E-01	4.213E-01	4.350E-01	4.442E-01	4.452E-01	4.244E-01	0.00
15.000	8.20E-01	4.002E-01	4.030E-01	4.091E-01	4.171E-01	4.177E-01	4.214E-01	4.091E-01	7.178E-02
16.000	8.00E-01	3.924E-01	4.051E-01	4.050E-01	3.995E-01	4.014E-01	4.042E-01	3.941E-01	2.328E-01
18.000	7.60E-01	3.728E-01	3.771E-01	3.778E-01	3.778E-01	3.871E-01	3.834E-01	3.712E-01	2.652E-01
20.000	7.40E-01	3.629E-01	3.660E-01	3.733E-01	3.681E-01	3.598E-01	3.553E-01	3.459E-01	3.028E-01
22.000	7.20E-01	3.530E-01	3.555E-01	3.581E-01	3.517E-01	3.586E-01	3.430E-01	3.378E-01	3.005E-01
24.000	7.00E-01	3.445E-01	3.454E-01	3.461E-01	3.470E-01	3.394E-01	3.332E-01	3.319E-01	2.995E-01
26.000	6.90E-01	3.396E-01	3.394E-01	3.405E-01	3.376E-01	3.366E-01	3.245E-01	3.195E-01	2.964E-01
28.000	6.80E-01	3.347E-01	3.343E-01	3.348E-01	3.323E-01	3.300E-01	3.201E-01	3.110E-01	2.896E-01
30.000	6.70E-01	3.297E-01	3.306E-01	3.347E-01	3.325E-01	3.262E-01	3.197E-01	3.123E-01	2.851E-01
34.000	6.50E-01	3.199E-01	3.206E-01	3.213E-01	3.223E-01	3.268E-01	3.188E-01	3.126E-01	2.827E-01
38.000	6.40E-01	3.150E-01	3.153E-01	3.150E-01	3.149E-01	3.128E-01	3.070E-01	3.007E-01	2.832E-01
42.000	6.30E-01	3.100E-01	3.104E-01	3.111E-01	3.095E-01	3.064E-01	3.082E-01	2.989E-01	2.840E-01
46.000	6.20E-01	3.051E-01	3.054E-01	3.061E-01	3.064E-01	3.054E-01	3.039E-01	2.989E-01	2.830E-01
50.000	6.10E-01	3.002E-01	3.005E-01	3.011E-01	3.022E-01	3.014E-01	2.985E-01	2.951E-01	2.804E-01
55.000	6.00E-01	2.953E-01	2.955E-01	2.959E-01	2.967E-01	2.955E-01	2.921E-01	2.891E-01	2.827E-01
60.000	5.95E-01	2.928E-01	2.929E-01	2.931E-01	2.936E-01	2.919E-01	2.943E-01	2.906E-01	2.835E-01
65.000	5.90E-01	2.904E-01	2.905E-01	2.906E-01	2.910E-01	2.914E-01	2.905E-01	2.865E-01	2.793E-01
70.000	5.80E-01	2.854E-01	2.856E-01	2.860E-01	2.865E-01	2.876E-01	2.862E-01	2.825E-01	2.762E-01
80.000	5.75E-01	2.830E-01	2.830E-01	2.831E-01	2.832E-01	2.837E-01	2.817E-01	2.827E-01	2.748E-01
90.000	5.62E-01	2.766E-01	2.767E-01	2.769E-01	2.771E-01	2.777E-01	2.787E-01	2.769E-01	2.748E-01
100.000	5.60E-01	2.756E-01	2.756E-01	2.756E-01	2.757E-01	2.758E-01	2.762E-01	2.741E-01	2.706E-01
130.000	5.42E-01	2.667E-01	2.668E-01	2.668E-01	2.669E-01	2.671E-01	2.674E-01	2.678E-01	2.664E-01
160.000	5.40E-01	2.658E-01	2.658E-01	2.658E-01	2.658E-01	2.658E-01	2.658E-01	2.659E-01	2.654E-01
200.000	5.25E-01	2.584E-01	2.584E-01	2.584E-01	2.585E-01	2.585E-01	2.587E-01	2.588E-01	2.593E-01

5.6   Miscellaneous Radiation Effects

5.6.1 Low Energy Proton Irradiation of Partially Covered Solar Cells

Some early satellites were launched into synchronous orbit with coverslides that were not quite large enough to completely cover the solar cells. Two such satellites, ATS-F1 and Intelsat II-F4, experienced immediate, unpredicted solar panel degradation soon after reaching synchronous altitude. Subsequent investigations revealed that the degradation was due to irradiation of the small exposed areas by the intense low energy protons existing at synchronous altitude [5.3, 5.13-5.17]. It was found that cells irradiated with low energy protons through narrow spaces in their shielding (gaps) exhibited large losses in P_{max} and V_{oc}, but their I-V curves near I_{sc} remained unchanged. This behavior was explained by a model consisting of two solar cells in parallel, wherein one cell is large and shielded, while the other is small and exposed to radiation. The irradiated cell may be damaged heavily in the junction region by the creation of a large number of recombination centers. This induces a large recombination-generation current, I_{02}, causing a loss in V_{oc} (cf. eq. 2-3). The I-V curve resulting from this model matched the experimental data quite well. It also showed that if the onboard power management system was designed to operate on the V_{oc} side of maximum power, this type of cell damage could cause large losses in power available to the spacecraft.

There is nothing in the above model which would indicate that GaAs cells should be immune from this phenomenon, and several workers did indeed perform low energy proton irradiations of GaAs solar cells through small gaps [5.18 - 5.21]. The first test was performed at JPL [5.18] on four GaAs cells from the Mantech line. These cells were placed in fixtures which shielded all the cell surface except for the busbar and ≈ 125 μm of GaAs just inside the busbar. The cells were exposed to 200 keV protons in two steps, to fluences of 1 x 10^{11} and 1 x 10^{12} p/cm^2, with I-V characteristics measured after each step. Following this, the cells were exposed to the same two fluences with 500 keV protons, then again with 1 MeV protons. The cells exhibited $\approx 5\%$ degradation in both P_{max} and V_{oc}, but not in I_{sc}, thus duplicating the behavior of Si cells. Approximately 3% of the degradation occurred after the 200 keV exposure and an additional 1% after each of the higher energy irradiations.

Another test of this phenomenon was performed at JPL in 1991 [5.19] using GaAs/Ge cells from ASEC. These cells were protected by 12-mil thick coverglasses that entirely covered the cell and part of the busbar. As shown in Figure 5.7, the silver busbar is 6 μm thick, which is slightly less than the range of 1 MeV protons. There is also a small exposed area of GaAS just outside the busbar, ≈ 2 mils wide. This is deliberately designed in the cell geometry to leave clearance for a saw cut after processing. In this experiment, eight samples were irradiated with 500 keV protons to a fluence of 1 x 10^{13} p/cm^2. Although these protons could not penetrate the busbar, they nevertheless produced an average degradation in cell power of 11%, with a loss in V_{oc} of 3.5%, but no loss in I_{sc}, (see Figure 5.8) typical of low energy proton gap irradiations. It is clear that the exposed gap is a vulnerable area.

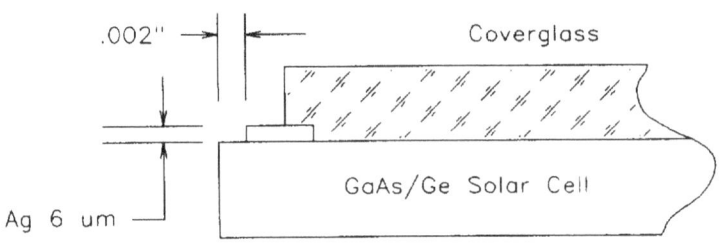

Figure 5.7.   GaAs/Ge Solar Cell and Coverglass Geometry Used in Low Energy Proton Gap Irradiations

More recently, Scott and Marvin [5.20] performed similar low energy proton irradiations on ASEC cells similar to those used in the 1991 JPL irradiations. These cells also had the 2-mil wide gaps outside the busbar. They irradiated the cells to successively higher fluences, up to a total of 1 x 10^{14} p/cm^2. As the irradiation proceeded, they found that P_{max} at first degraded very rapidly, then stopped degrading, and after this point began to improve with irradiation. The fluences at which minima occurred varied with proton energy, varying from 1 x 10^{13} to 3 x 10^{13} p/cm^2, for 50 keV to 200 keV protons. The initial power loss is explained by an increase in leakage current through the small shunt diode, through the creation of recombination centers in the junction region. At higher doses, the resistance of the heavily damaged material increases and reduces the current flow through the leakage area. These authors also found that the cells' power degradation increased as the length of the exposed gap was increased. The decrease is rapid for small gap lengths, but as the length increases the damage produced by a constant incident fluence is proportional to exp($-\alpha t$) and effectively saturates for gap lengths over ≈ 2 cm. Since the power loss due to the presence of gaps is proportional to the cell perimeter, and the cell power is proportional to cell area, the authors concluded that the use of larger area cells would be advantageous because the ratio of cell area to perimeter length increases with cell size. The results from all these studies strongly suggest that the unprotected area near the busbar is very vulnerable to low energy proton damage and steps should be taken to protect it when launching into a proton-rich environment.

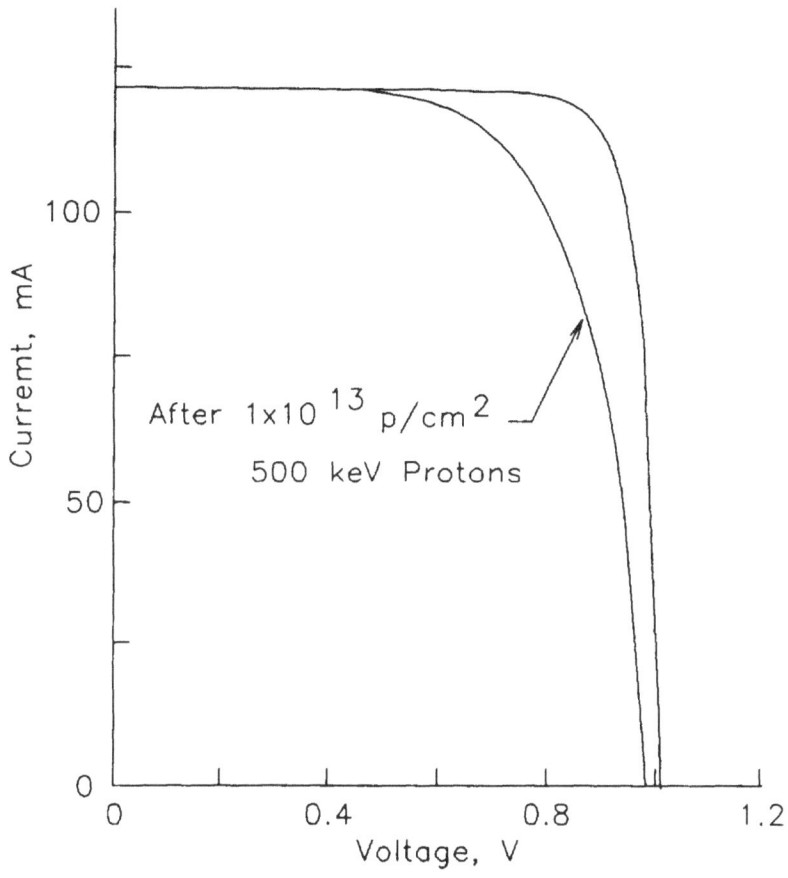

Figure 5.8.   GaAs/Ge Solar Cell Performance after Irradiation with 500 keV Protons to a
Fluence of 1 x 10^{13} p/cm^2

5.6.2 Radiation Rate Effects

When solar cells are irradiated with accelerators, they are always irradiated at a much greater flux rate than experienced in space. Cells irradiated with electrons to fluences between 1 x 10^{15} and 1 x 10^{16} e/cm^2 are usually irradiated with fluxes in the order of 1 x 10^{11} e/cm^2-sec, so that the radiation times are no more than a few hours. A beam of 1 MeV electrons at a flux of 1 x 10^{11} e/cm^2-sec will cause the temperature of a GaAs solar cell to rise at a rate of $\approx 0.2°$C/sec if the cell is not attached to a thermal sink (assuming half the electrons stop in a 2 x 2 cm cell, 0.02 cm thick). Even when the cell is thermally attached to a heat sink, it is possible that some internal heating may occur inside the cell during the irradiation, perhaps annealing some of the radiation damage during the irradiation. This could then turn out to be a poor simulation of the behavior to be expected in space.

Three different experiments were performed at JPL to investigate the possibility of a rate effect in the electron irradiation of GaAs cells. The first experiment [5.22], performed in 1978 on LPE cells made by Hughes Research Lab, involved the slow irradiation of two sets of cells to fluences of 1×10^{15} e/cm^2. One set was irradiated at a temperature of 28°C, and the other set was held at a temperature of 126°C. The flux rates were low, 2×10^{10} e/cm^2-sec, so that the total irradiation time was 13.5 hours. At the end of these irradiations, the electrical characteristics of each set of cells were found to be identical. The cells irradiated at 28°C were annealed at 129°C for 15 hours and remeasured. The cell characteristics did not improve. All the above I-V characteristics also agreed with cells irradiated at the usual laboratory flux rates of ≈ 2 to 5×10^{15} e/cm^2, so the possibility of internal cell heating (and annealing) during laboratory irradiation testing was concluded to be unlikely. The second experiment also involved Hughes LPE cells. Two cells, with 12 mil coverglasses attached, were irradiated at 80°C to 5 successive fluences, to a total 1 MeV electron fluence of 6×10^{14} e/cm^2 [5.23]. One cell was irradiated at a flux rate of 2×10^9 e/cm^2-sec, and the other at 2×10^{11} e/cm^2-sec. No difference was observed in the electrical performance.

The third JPL study was performed in 1991 [5.24], using GaAs/Ge solar cells of then-current vintage from the ASEC. During this investigation, 36 cells were divided into two groups. The first group of cells was irradiated with 1 MeV electrons to three successive fluences of 1×10^{14}, 5×10^{14}, and 1×10^{15} e/cm^2, at a flux rate of 5×10^{10} e/cm^2-sec for the first irradiation, and at 2×10^{11} e/cm^2-sec for the second two fluences. The second group was irradiated at 1/10 the above flux rates. In each case, the cells were attached to a temperature-controlled heat sink with Apiezon H vacuum grease, and the heat sink was maintained at 28°C during all the irradiations. The cells were measured after each set of fluence levels. All electrical parameters of the solar cells, namely I_{sc}, V_{oc}, I_{mp}, V_{mp}, P_{max}, and fill factor degraded identically (to within 1%) in each case. Although even the slow rate used in this experiment is one or two orders of magnitude higher than flux rates experienced in space, the results confirm that rate effects are not a prime issue of concern.

A related study was reported by Loo et al. [5.25], in which LPE GaAs solar cells were irradiated to four fluences up to a maximum of 1×10^{15} e/cm^2 at low flux rates, but here the cells were irradiated at much higher temperatures, 150°C and 200°C. Flux rates used were 2×10^9 and 4×10^{10} e/cm^2-sec, so that the irradiation at the low flux rate required an irradiation time of 139 hours. Their data shows that at fluences up to 1×10^{14} e/cm^2 there was not much difference in the power degradation, regardless of the flux or cell temperature used, and the results also compared favorably with irradiations performed at room temperature and higher flux rates. However, the cells irradiated for 139 hours to 1×10^{15} e/cm^2 at 200°C degraded to a P/P_0 of 90%, while all cells irradiated at higher fluxes and lower temperatures degraded to a P/P_0 of 80%. But after a post-irradiation anneal at 200°C for 40 hours, all cells ended up with identical P/P_0 of 90%. The authors concluded that the radiation damage in the cells was due to annealing at 200°C, and the much longer annealing time associated with the low fluxes is primarily responsible for the apparently lower degradation rate.

5.6.3 Annealing of Irradiated GaAs Cells

At an early stage of development, electron-irradiated GaAs solar cells were found to be capable of annealing at temperatures on the order of 200°C to 300°C [5.26 - 5.29]. In a radiation study of the early Hughes LPE cells, Loo et al. [5.28] found that cells irradiated with 0.7, 1.0, and 1.9 MeV electrons recovered significantly at annealing temperatures of 210°C in time periods on the order of 10 hours. They found that annealing improved the cell's spectral response, indicating that there was recovery of the diffusion length. The dark I-V characteristics of the irradiated cells showed a significant increase in current due to production of recombination centers in the junctions, making them somewhat leaky. The leakage current recovered to very nearly its pre-irradiation value after the annealing. The behavior of the defect levels in GaAs cells irradiated with 1 MeV electrons and subsequently annealed was reported in reference [5.2].

Heinbockel et al. [5.30] proposed that the simultaneous irradiation and annealing of GaAs solar cells could be very advantageous. Soon thereafter, Loo et al. [5.31] reported the annealing characteristics of Hughes LPE GaAs cells irradiated with 200 keV protons, along with several annealing treatments. 200 keV was selected because the damage induced in GaAs solar cells in this particular cell structure ($x_w = xj = 0.5 \mu$m) had been found to be particularly severe. In their continuous annealing experiments, where the cells were held at temperatures as low as 150°C during irradiation, they found the radiation-induced power loss was greatly decreased. These workers found, that in contrast to electron-irradiated GaAs cells, there was very little damage recovery at post-irradiation annealing temperatures of 200°C. However, in a subsequent paper [5.32] these same workers reported that annealing of proton damage was quite effective. In these annealing experiments, a periodic annealing schedule was used, wherein irradiations were interspersed with annealing, as well as irradiation to the total cumulative fluence given in the periodic annealing experiments, culminating with a one-shot anneal at the prescribed temperature.

These workers found that: (1) the same P_{max}/P_{max0} was obtained from the periodic annealing treatment as obtained if the cells are irradiated to the cumulative fluence, then annealed; (2) there was substantial improvement in cell performance when the annealing temperature was as low as 200°C; (3) there was a moderate improvement in performance when the annealing temperatures were raised to 300°C and 400°C; and (4) performance recovery could be enhanced by the injection of minority carriers into the cells during annealing, either by photoinjection (concentrated light at 5 suns) or dark forward-current biasing at 125 mA/cm². Injection-assisted annealing appears to be confirmed by a study of the defect levels in electron irradiated GaAs by Stievenard and Bourgoin [5.33], who found significant annealing of one of the electron traps (E5 at E_c-0.83 eV) at an injection current of 1.3 A/cm², but at a very low temperature of 57°C. The authors did not report whether there was an associated improvement in the I-V characteristic.

A study of GaAs solar cell annealing after irradiation with 8.3 MeV protons and 1 MeV electrons was performed at JPL using Hughes LPE cells [5.34]. In this study the solar cells were annealed after irradiation using isochronal anneals of 30 minutes in a nitrogen atmosphere. The annealing results are plotted as a function of unannealed fraction defined as:

$$f(x) = \frac{X_0 - X_t}{X_0 - X_i} \tag{5-14}$$

where X_0 is the pre-irradiation value of the photovoltaic parameter, X_i is the value after irradiation, and X_t is the value after annealing at temperature t. In plots of this type, improvement of cell performance after annealing causes the value of f(x) to decrease. The results of the annealing experiments are shown in Figure 5.9. The figure shows the results after irradiation with 8.3 MeV protons to a fluence of 1×10^{13} p/cm² and after an irradiation with 1 MeV electrons to a fluence of 1×10^{15} e/cm². This data shows that recovery from electron-induced damage has a very rapid onset at $\approx$ 200°C, but proton-induced damage seems to have a much more gradual onset at about the same temperature. After treatment to temperatures above 375°C all GaAs cells showed deterioration in I_{sc}, V_{oc}, and P_{max}. Since other investigators have not reported such severe degradation at these temperatures, the cell deterioration may be due in part to the nitrogen atmosphere used during the annealing.

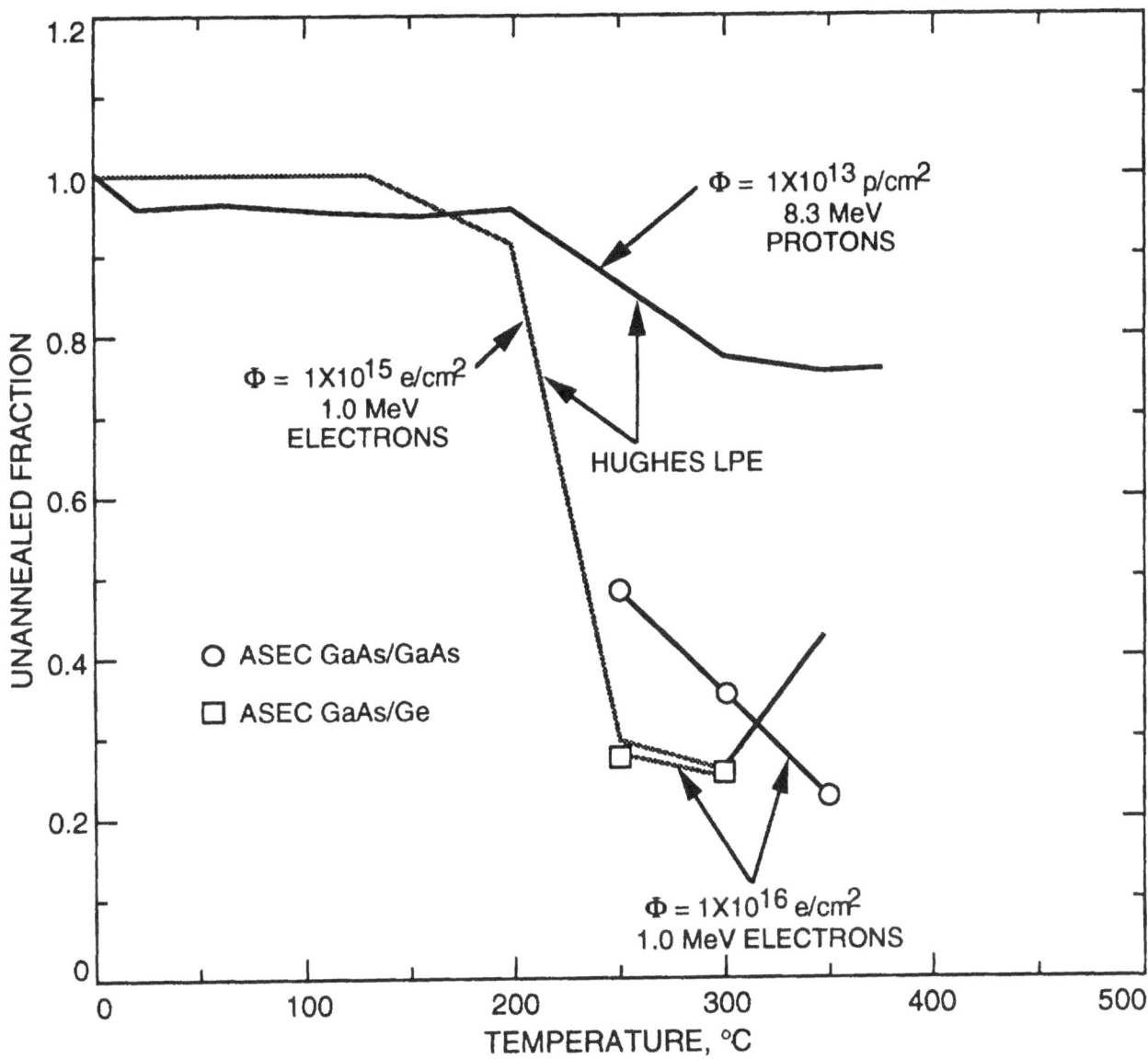

Figure 5.9. P_{max} Annealing of Electron and Proton Irradiated Hughes LPE GaAs Solar Cells

All the annealing studies discussed above were performed on early vintage GaAs solar cells manufactured by Hughes Research Lab with the LPE process. A study performed in 1988 by Chung et al. [5.35] examined the annealing of some early GaAs cells made by ASEC on both GaAs and Ge substrates. The cells were irradiated in steps of 1×10^{16} e/cm^2, to a total fluence of 1×10^{17} e/cm^2, with annealing between each step at 250°C for 1 hour in a nitrogen atmosphere. Parallel experiments were performed at annealing temperatures of 300°C and 350°C. These early cells did not have high efficiencies by today's standards (16.5% for the GaAs/GaAs cells and 13.6% for the GaAs/Ge cells), but the results should be indicative of modern cell behavior. The 1×10^{16} e/cm^2 irradiation/anneal results have been converted to unannealed fractions and plotted on Figure 5.9. Their results appear to parallel the earlier results for LPE cells. Since GaAs cells are annealable at relatively low temperatures, it may be possible to design annealable panels for some space missions.

References for Chapter 5

5.1   J.C. Bourgoin, H.J. von Bardeleben, and D. Stiévenard, "Native Defects in Gallium Arsenide," *Journal of Applied Physics*, 64, R65, November 1, 1988.

5.2   S.S. Li, W.L. Wang, R.Y. Loo, and W.P. Rahilly, "Deep Level Defects and Annealing Studies in 1-MeV Electron Irradiated (AlGa)As-GaAs Solar Cells," *Proceedings of the 16th IEEE Photovoltaic Specialists Conference*, 211, 1982.

5.3   H.Y. Tada, J.R. Carter, Jr., B.E. Anspaugh, and R.G. Downing, *Solar Cell Radiation Handbook*, JPL Publication 82-69, Jet Propulsion Laboratory, California Institute of Technology, Pasadena, Calif., 1982.

5.4   D.P. LeGalley and A. Rosen (Eds.), *Space Physics*, John Wiley and Sons, New York, p. 693, 1964.

5.5   J.E. Janni, "Calculations of Energy Loss, Range, Pathlength, Straggling, Multiple Scattering, and the Probability of Inelastic Nuclear Collisions for 0.1 to 1000 MeV Protons," AFWL-TR-65-150, September, 1966.

5.6   J.F. Janni, "Proton Range-Energy Tables, 1 keV-10 GeV," *Atomic Data and Nuclear Data Tables*, 27, Nos. 2/3, Academic Press, New York, March/May 1982.

5.7   M.J. Berger and S.M. Seltzer, "Tables of Energy Losses and Ranges of Electrons and Positrons," Paper 10, NASA, NRC Pub. 1133, 1964; also NASA SP-3012, 1964.

5.8   M.J. Berger and S.M. Seltzer, "Additional Stopping Power and Range Tables for Protons, Mesons, and Electrons," NASA SP-3036, 1966.

5.9   J.F. Ziegler, "TRIM-91 The Transport of Ions in Matter," Version 5.3, IBM Research Center, Yorktown, New York.

5.10   S.M. Seltzer, "Electron and Positron Stopping Powers and Ranges, EPSTAR," Vers. 2.0B, National Institute of Standards and Technology, 1988.

5.11   M.J. Barrett, "Electron Damage Coefficients in P-Type Silicon," *IEEE Transactions on Nuclear Science*, NS-14, 6, 82, 1967.

5.12   B.E. Anspaugh and R.G. Downing, "Radiation Effects in Silicon and Gallium Arsenide Solar Cells Using Isotropic and Normally Incident Radiation," JPL Publication 84-61, Jet Propulsion Laboratory, California Institute of Technology, Pasadena, Calif., September 1, 1984.

5.13   G. J. Brucker, W. Dennehy, A.G. Holmes-Siedle, and J.J. Loferski, "Low Energy Proton Damage in Partially Shielded Solar Cells," *Proceedings of the IEEE*, 54, 798, 1966.

5.14   W.D. Brown, "ATS Power Subsystem Radiation Effects Study," Hughes Aircraft Company, NAS 5-3823, SSD-80089R, 1968.

5.15 L.J. Goldhammer, "Solar Cell Radiation Flight Report," Hughes Aircraft Co., Report SSD-90329R, May 1969.

5.16 E. Stofel and D. Joslin, "Low Energy Proton Damage to Silicon Solar Cells," *IEEE Transactions on Nuclear Science*, NS-17, 250, 1970.

5.17 R.L. Statler and D.J. Curtin, "Radiation Damage in Silicon Solar Cells from Low Energy Protons," *IEEE Transactions on Electron Devices*, ED-18, 412, 1971.

5.18 B.E. Anspaugh, "Unpublished JPL Data," 1986.

5.19 B.E. Anspaugh, "Unpublished JPL Data," 1991.

5.20 D.M. Scott and D.C. Marvin, "The Effects of Low-Energy Proton Irradiation on Partially Shielded GaAS Solar Cells," *Proceedings of the 23rd IEEE Photovoltaic Specialists Conference*, 1338, 1993.

5.21 M.T. Gates, "Predicted Solar Cell Edge Radiation Effects," *Proceedings of the 23rd IEEE Photovoltaic Specialists Conference*, 1404, 1993.

5.22 B.E. Anspaugh, "Unpublished JPL Data," 1978.

5.23 B.E. Anspaugh, "The Evaluation of Production-Type GaAs Solar Cells for Space," JPL D-466, Internal Document, Jet Propulsion Laboratory, California Institute of Technology, Pasadena, Calif., January 17, 1983.

5.24 B.E. Anspaugh, "Unpublished JPL Data," 1991.

5.25 R.Y. Loo, G.S. Kamath, and R.C. Knechtli, "Effect of Electron Flux in Radiation Damage in GaAs Solar Cells," *Proceedings of the 16th IEEE Photovoltaic Specialists Conference*, 307, 1982.

5.26 R.S. Miller and J.S. Harris, "Gallium Arsenide Concentration System," *AIAA Conference on the Future of Aerospace Power Systems*, March 1977.

5.27 G.H. Walker and E.J. Conway, "Annealing of GaAs Solar Cells Damaged by Electron Irradiation," *Journal of the Electrical Chemical Society*, 125, No. 4, 676, 1978.

5.28 R. Loo, L. Goldhammer, B. Anspaugh, R.C. Knechtli, and G.S. Kamath, "Electron and Proton Degradation in (AlGa)As-GaAs Solar Cells," *Proceedings of the 13th IEEE Photovoltaic Specialists Conference*, 562, 1978.

5.29 G.H. Walker and E.J. Conway, "Annealing Kinetics of Electron-Irradiated GaAs Heteroface Solar Cells in the Range 175° to 200°C," *Applied Physics Letters*, 35, 459, September 15, 1979.

5.30   J.H. Heinbockel, E.J. Conway, and G.H. Walker, "Simultaneous Radiation Damage and Annealing of GaAs Solar Cells," *Proceedings of the 14th IEEE Photovoltaic Specialists Conference*, 1085, 1980.

5.31   R.Y. Loo, R.C. Knechtli, and G.S. Kamath, "Periodic Annealing of Radiation Damage in GaAs Solar Cells," *Proceedings of the Space Photovoltaic Research and Technology Conference, NASA Lewis Conference Publication 2169*, 249, October 1990.

5.32   R. Loo, R.C. Knechtli, and G.S. Kamath, "Enhanced Annealing of GaAs Solar Cell Radiation Damage," *Proceedings of the 15th IEEE Photovoltaic Specialists Conference*, 33, 1981.

5.33   D. Stievenard and J.C. Bourgoin, "Degradation and Recovery of GaAs Solar Cells Under Electron Irradiation," *Proceedings of the 17th IEEE Photovoltaic Specialists Conference*, 1103, 1984.

5.34   B.E. Anspaugh and R.G. Downing, "Damage Coefficients and Thermal Annealing of Irradiated Silicon and GaAs Solar Cells," *Proceedings of the 15th IEEE Photovoltaic Specialists Conference*, 499, 1981.

5.35   M.A. Chung, D.L. Meier, J.R. Szedon, and J. Bartko, "Electron Radiation and Annealing of MOCVD GaAs and GaAs/Ge Solar Cells," *Proceedings of the 20th IEEE Photovoltaic Specialists Conference*, 924, 1988.

Chapter 6

Electrical Performance of GaAs Solar Cells

In this chapter we will present data showing the degradation of GaAs solar cells from various manufacturers as a function of 1 MeV electron fluence. We will present the behavior of GaAs solar cells as a function of solar intensity, cell temperature, and irradiation, and will also report some GaAs solar cell flight data.

6.1   1 MeV Electron Radiation Data

All the irradiations and electrical measurements that follow were performed at the JPL Dynamitron/Solar Cell Laboratory using the techniques outlined in Chapter 2. The cells were held at a temperature of 28°C during the irradiation. Even though we observed negligible change in performance due to annealing at temperatures of 60°C, the measurements reported here were made after a 20-hour anneal at 60°C. Figures 6.1 through 6.10 are plots of the behavior of ASEC GaAs/Ge cells manufactured in 1990. Ten cells were irradiated and measured in cumulative fluences. Averages of the electrical parameters were calculated. Cubic spline fits to these average values are plotted in the figures. These cells had an AlGaAs window thickness of 0.05 μm, a junction depth of 0.40 μm, a GaAs buffer layer of 8 μm, and were constructed with dual antireflection coatings. The busbar consisted of a thin strip along one edge and two contact pads. The current and maximum power values have been divided by the total cell area (including busbar) for the plots to give I_{xx}/cm^2 and P_{max}/cm^2. The busbar area on these 2 x 2 cm^2 cells was 0.13 cm^2; therefore, a slight adjustment may be desirable when applying these values to samples with different cell and busbar areas.

Using least-square polynomial fits, the values of normalized cell performance taken from Figures 6.6 through 6.10 were fit to equations of the form:

$$Y = \sum_{i=0}^{m} a_i \left[\ln(\phi) \right]^i \qquad (6\text{-}1)$$

where Y is a solar cell electrical parameter, ln(Φ) is the logarithm (base e) of the total 1 MeV electron fluence (e/cm^2), a_i are the parameters determined by the least square fits, and m is the degree of polynomial which was determined to give the best fit for parameter Y. The values found for the a_i

parameters are tabulated in Table 6-1. These fits are not the same fits used in the cubic spline fitting used in the plotting; however, the agreement between the two fitting techniques turns out to be surprisingly good. As a cautionary note, these fits have been made for 1 MeV electron fluences out to a maximum fluence of 1 x 10^{16} e/cm^2, and should not under any circumstances be used to calculate the degradation at higher fluences.

As GaAs cells manufactured by Spectrolab became available, they were also irradiated and measured. Their performance as a function of 1 MeV electron fluence is plotted in Figures 6.11 through 6.20. These cells were made in 1993 with an AlGaAs window thickness of 0.05 μm, a junction depth of 0.40 μm, a buffer layer of 1 μm, and dual antireflection coatings. Sample size for these measurements was five cells. The busbars on the Spectrolab cells consisted of a wide strip along one edge. The cell area was 4 cm^2 and the busbar area was 0.18 cm^2. Least-square fits to the normalized data for the Spectrolab cells are listed in Table 6-2.

The data in Figures 6.1 through 6.20 is intended to be used with the calculations of total 1 MeV equivalent fluence, computed for various space radiation environments, to predict on-orbit solar cell performance. The manufacturers are constantly improving their cells, and the data shown here may or may not be representative of the cells currently in production. It is highly recommended that samples be taken from current production lines and checked at one or two fluences to assess how well their behavior is represented by these curves.

Much of the data presented in this handbook is based on GaAs cells made by Hughes Research Laboratory using the LPE process. We have also presented a great deal of data on cells made by ASEC with the MOCVD process. This has primarily been because these cells were available in quantity for performance of the experiments. Figures 6.21 through 6.30 are plots which compare the performance of the later Hughes LPE cells (vintage $\approx$ 1984) with some early (vintage $\approx$ 1984) ASEC MOCVD cells. These figures also show the performance at fluence levels as high as 1 x 10^{17} e/cm^2. The current and power performance at extremely high fluences does not continue to decrease linearly with the logarithm of the fluence, but levels off at fluences higher than 10^{16} e/cm^2, perhaps indicating a saturating of defect levels which affect the minority carrier diffusion lengths.

Table 6-1.   Least-Square Fits to the Normalized Electrical Parameters of ASEC GaAs/Ge
              Cells Irradiated with 1 MeV Electrons (see Eq. 6-1)

Fit Coeff.	I_{sc}/I_{sc0}	V_{oc}/V_{oc0}	P_{max}/P_{max0}	I_{mp}/I_{mp0}	V_{mp}/V_{mp0}
a_0	-3.21012E+1	-6.93638E+0	-2.63101E+2	-9.23637E+1	1.40839E+2
a_1	2.87190E+0	1.06168E+0	4.08797E+1	1.27738E+1	-2.28925E+1
a_2	-7.16821E-3	-5.38317E-2	-2.50093E+0	-6.54682E-1	1.49427E+0
a_3	-6.15155E-3	1.23296E-3	7.53995E-2	1.49083E-2	-4.86444E-2
a_4	2.13215E-4	-1.08219E-5	-1.11568E-3	-1.27378E-4	7.90489E-4
a_5	-2.19063E-6		6.43817E-6		-5.13603E-6

Table 6-2.   Least-Square Fits to the Normalized Electrical Parameters of Spectrolab GaAs/Ge
              Cells Irradiated with 1 MeV Electrons (see Eq. 6-1)

Fit Coeff.	I_{sc}/I_{sc0}	V_{oc}/V_{oc0}	P_{max}/P_{max0}	I_{mp}/I_{mp0}	V_{mp}/V_{mp0}
a_0	-2.61758E+2	2.59327E+2	-1.46904E+2	-3.94407E+2	4.60191E+2
a_1	4.08147E+1	-4.19118E+1	2.39582E+1	6.25491E+1	-7.37150E+1
a_2	-2.50241E+0	2.70466E+0	-1.53609E+0	-3.91923E+0	4.70944E+0
a_3	7.55032E-2	-8.67942E-2	4.86061E-2	1.21426E-1	-1.49715E-1
a_4	-1.11673E-3	1.38569E-3	-7.56180E-4	-1.85668E-3	2.36935E-3
a_5	6.43804E-6	-8.81256E-6	4.59927E-6	1.11787E-5	-1.49430E-5

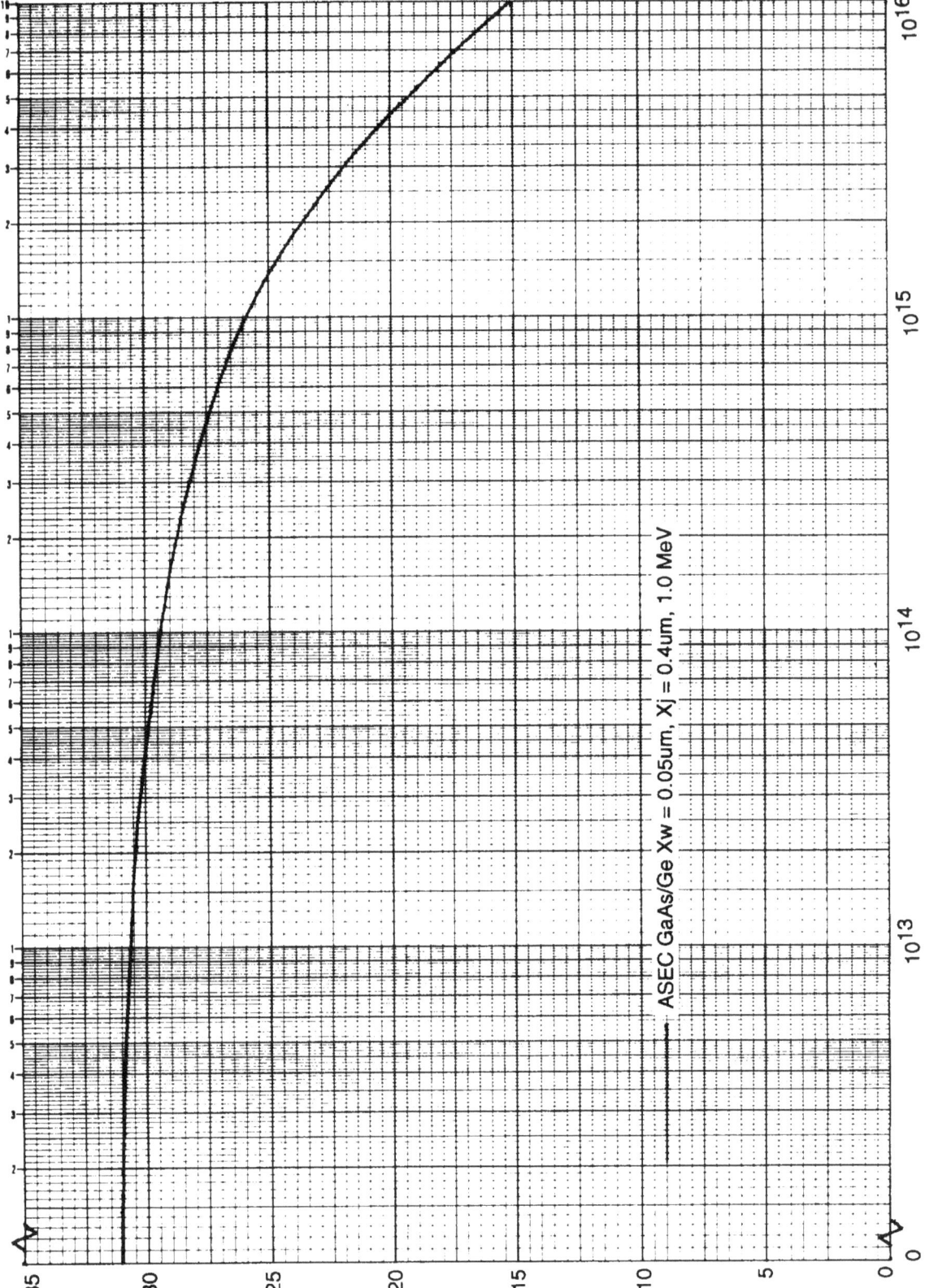

SHORT CIRCUIT CURRENT, mA/cm²

ELECTRON FLUENCE, e/cm²

ASEC GaAs/Ge Xw = 0.05um, Xj = 0.4um, 1.0 MeV

Figure 6.1. I_{sc}/cm² vs. 1 MeV Electron Fluence for ASEC GaAs/Ge Solar Cells

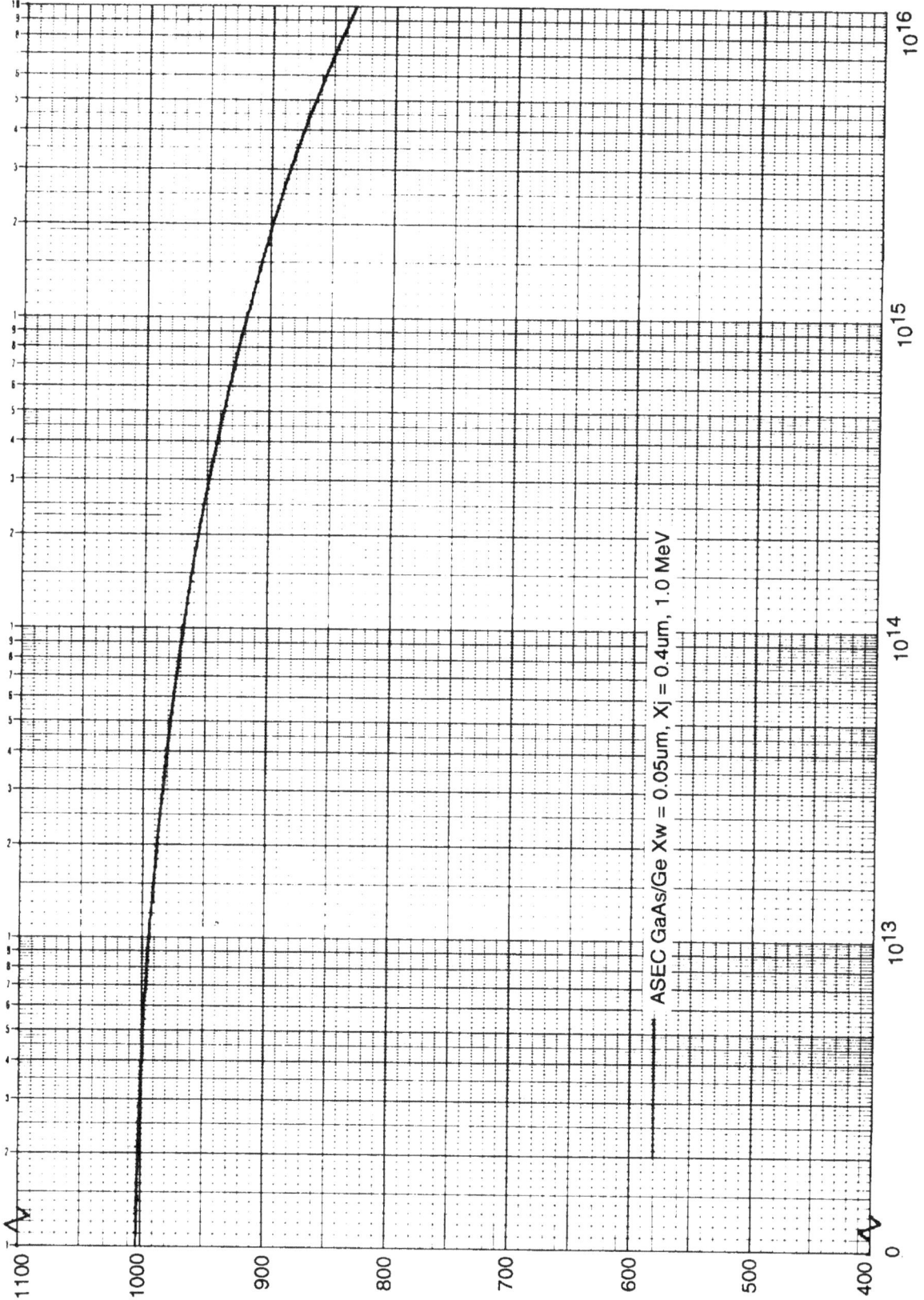

ELECTRON FLUENCE, e/cm²

OPEN CIRCUIT VOLTAGE, mV

ASEC GaAs/Ge Xw = 0.05um, Xj = 0.4um, 1.0 MeV

Figure 6.2. V_{oc} vs. 1 MeV Electron Fluence for ASEC GaAs/Ge Solar Cells

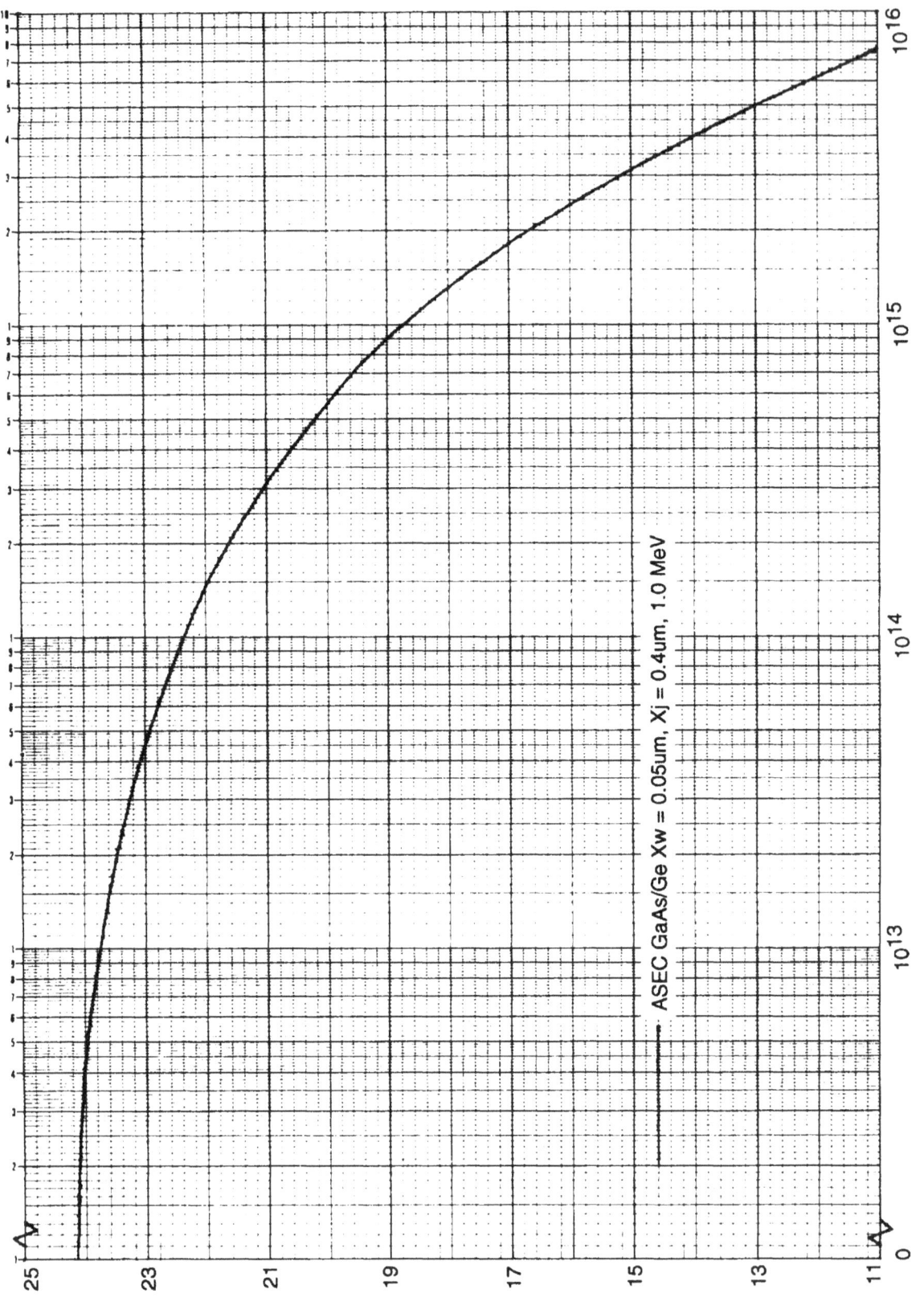

ELECTRON FLUENCE, e/cm²

MAXIMUM POWER, mW/cm²

ASEC GaAs/Ge Xw = 0.05um, Xj = 0.4um, 1.0 MeV

Figure 6.3. P_{max}/cm^2 vs. 1 MeV Electron Fluence for ASEC GaAs/Ge Solar Cells

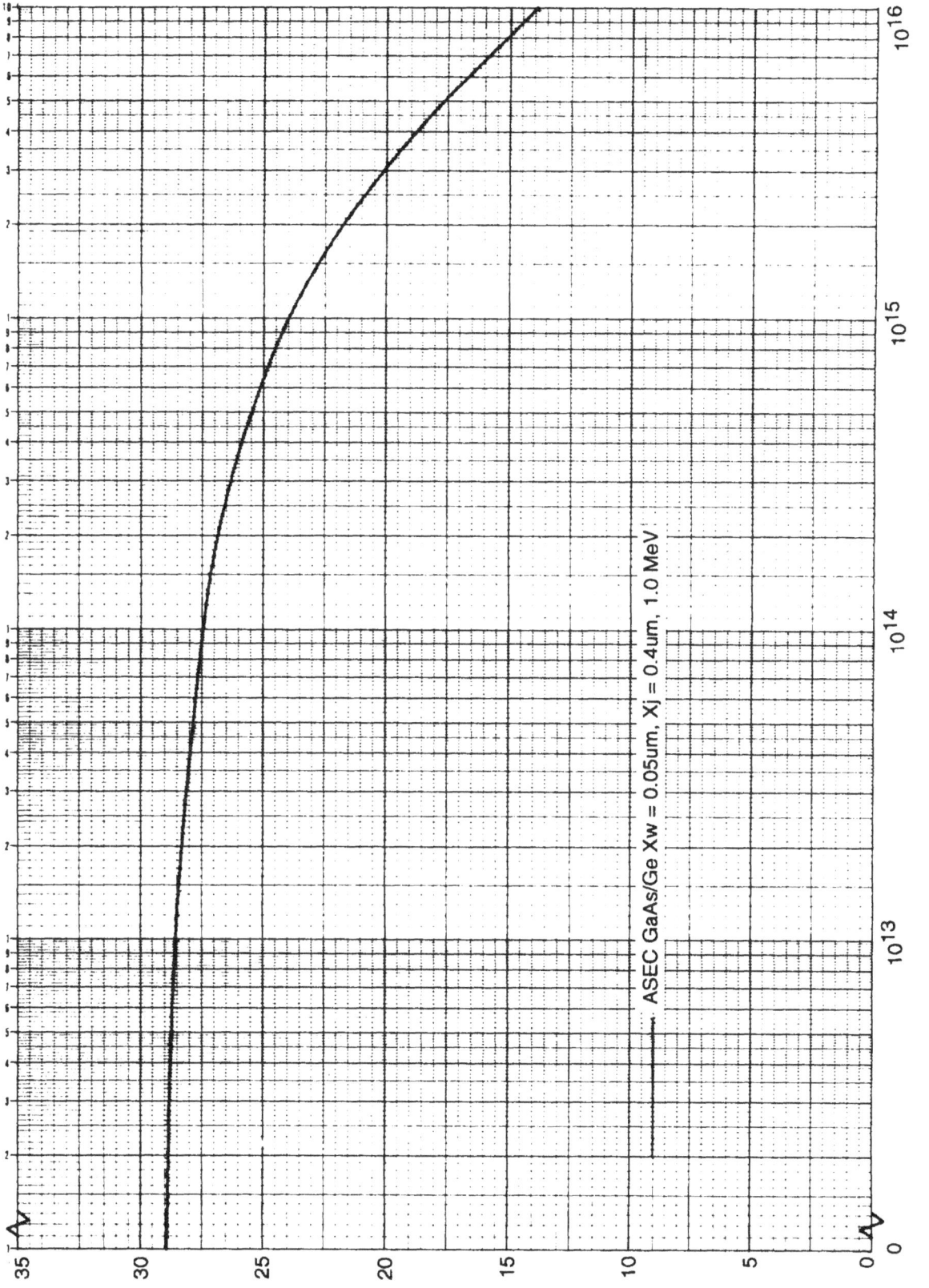

ELECTRON FLUENCE, e/cm²

CURRENT AT MAXIMUM POWER, mA/cm²

ASEC GaAs/Ge Xw = 0.05um, Xj = 0.4um, 1.0 MeV

Figure 6.4. I_{mp}/cm^2 vs. 1 MeV Electron Fluence for ASEC GaAs/Ge Solar Cells

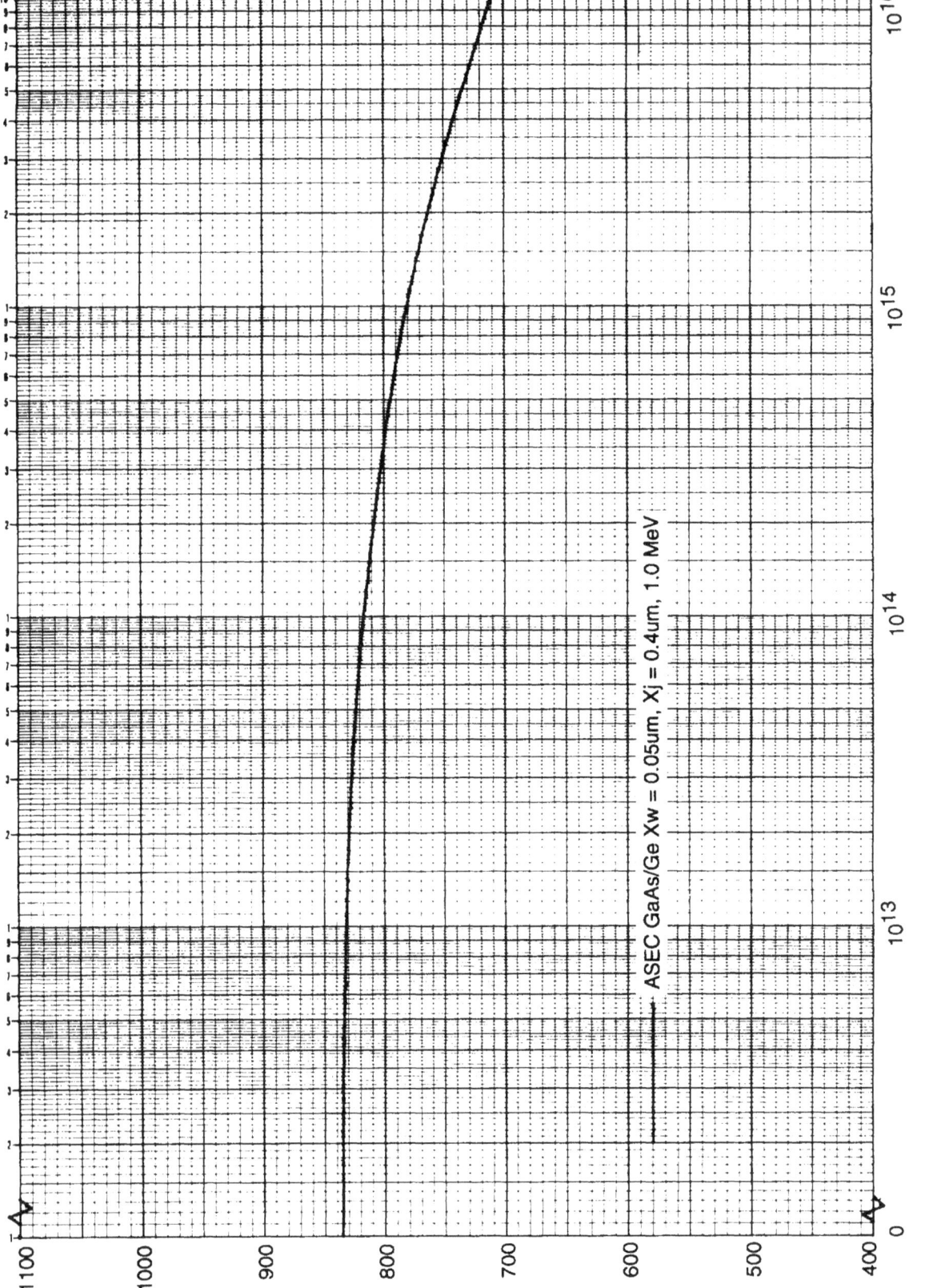

ELECTRON FLUENCE, e/cm²

VOLTAGE AT MAXIMUM POWER, mV

ASEC GaAs/Ge Xw = 0.05um, Xj = 0.4um, 1.0 MeV

Figure 6.5. V_{mp} vs. 1 MeV Electron Fluence for ASEC GaAs/Ge Solar Cells

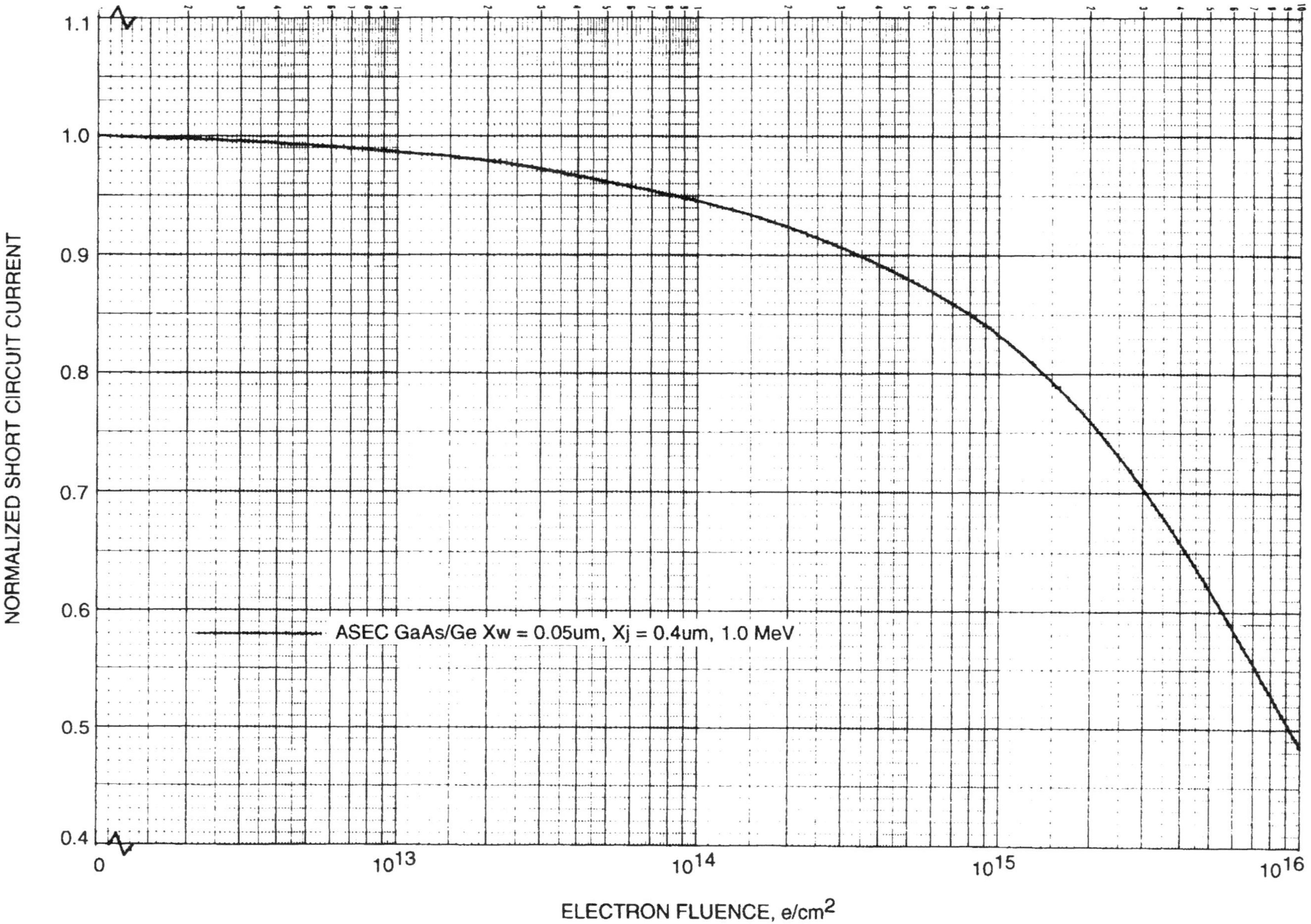

ELECTRON FLUENCE, e/cm^2

NORMALIZED SHORT CIRCUIT CURRENT

ASEC GaAs/Ge Xw = 0.05um, Xj = 0.4um, 1.0 MeV

Figure 6.6.   Normalized I$_{sc}$ vs. 1 MeV Electron Fluence for ASEC GaAs/Ge Solar Cells

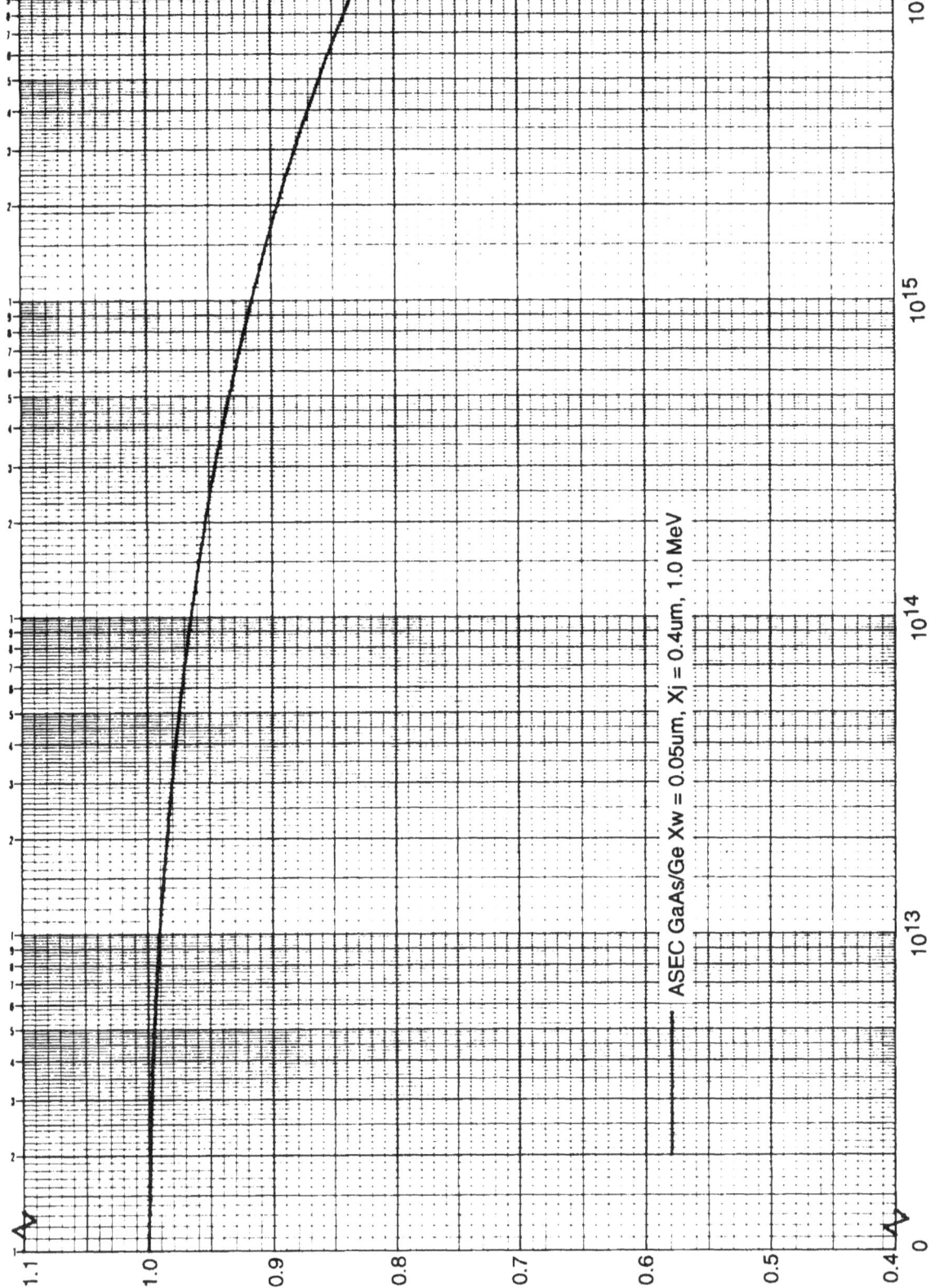

ELECTRON FLUENCE, e/cm²

NORMALIZED OPEN CIRCUIT VOLTAGE

ASEC GaAs/Ge Xw = 0.05um, Xj = 0.4um, 1.0 MeV

Figure 6.7.   Normalized V∝ vs. 1 MeV Electron Fluence for ASEC GaAs/Ge Solar Cells

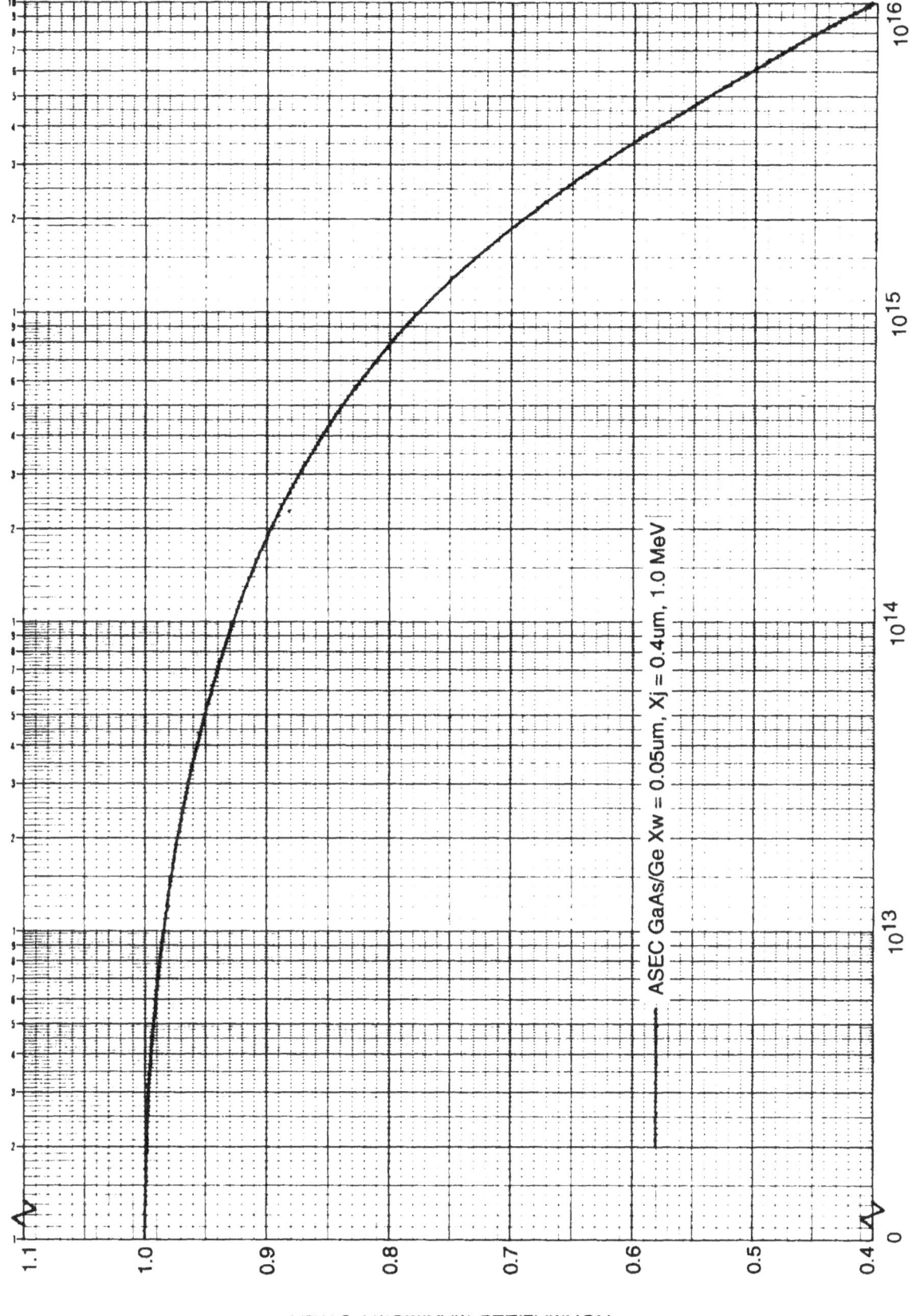

ELECTRON FLUENCE, e/cm²

NORMALIZED MAXIMUM POWER

ASEC GaAs/Ge Xw = 0.05um, Xj = 0.4um, 1.0 MeV

Figure 6.8.   Normalized P_{max} vs. 1 MeV Electron Fluence for ASEC GaAs/Ge Solar Cells

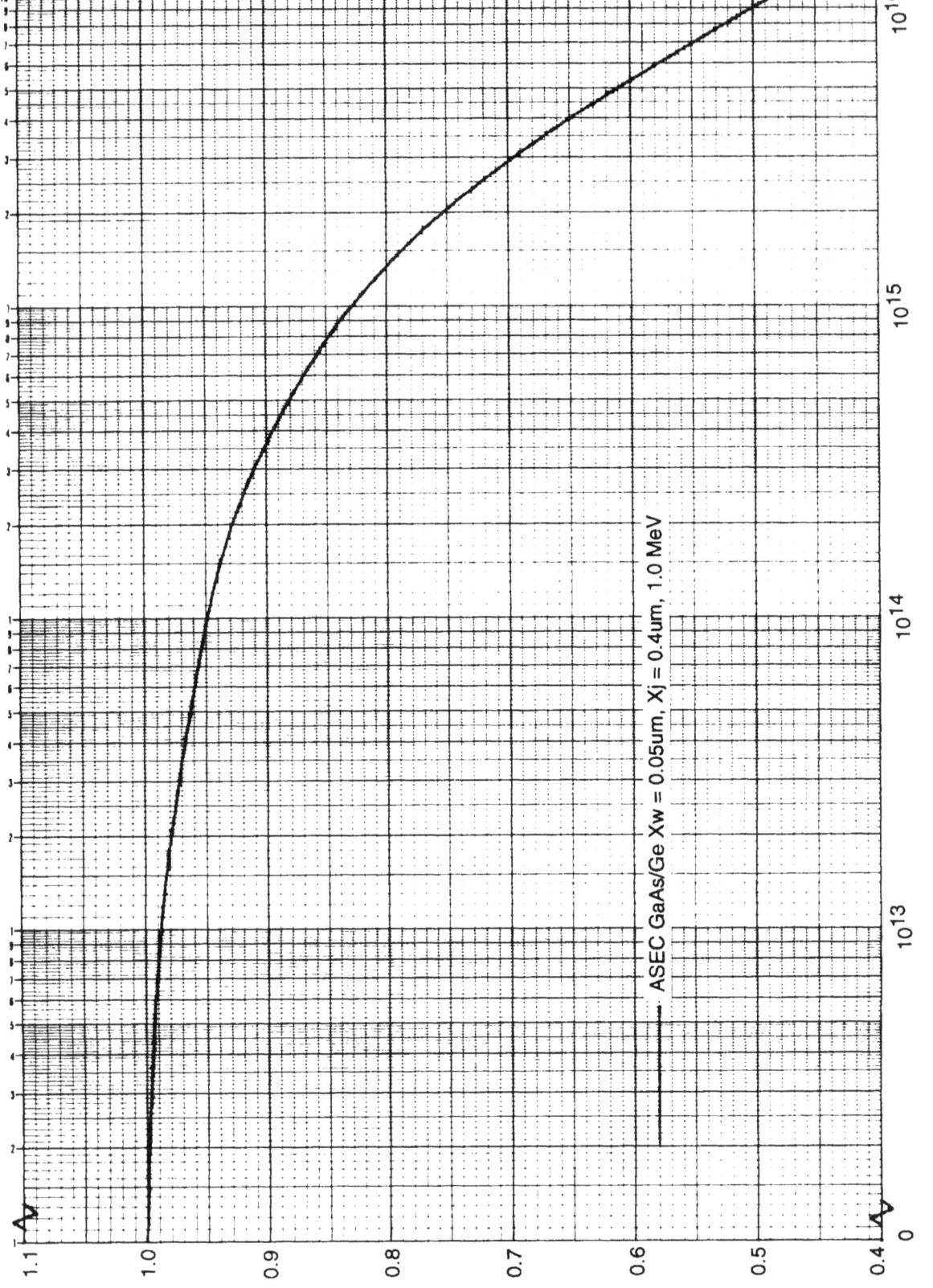

NORMALIZED CURRENT AT MAXIMUM POWER

ELECTRON FLUENCE, e/cm²

ASEC GaAs/Ge Xw = 0.05um, Xj = 0.4um, 1.0 MeV

Figure 6.9.   Normalized I_{mp} vs. 1 MeV Electron Fluence for ASEC GaAs/Ge Solar Cells

6-12

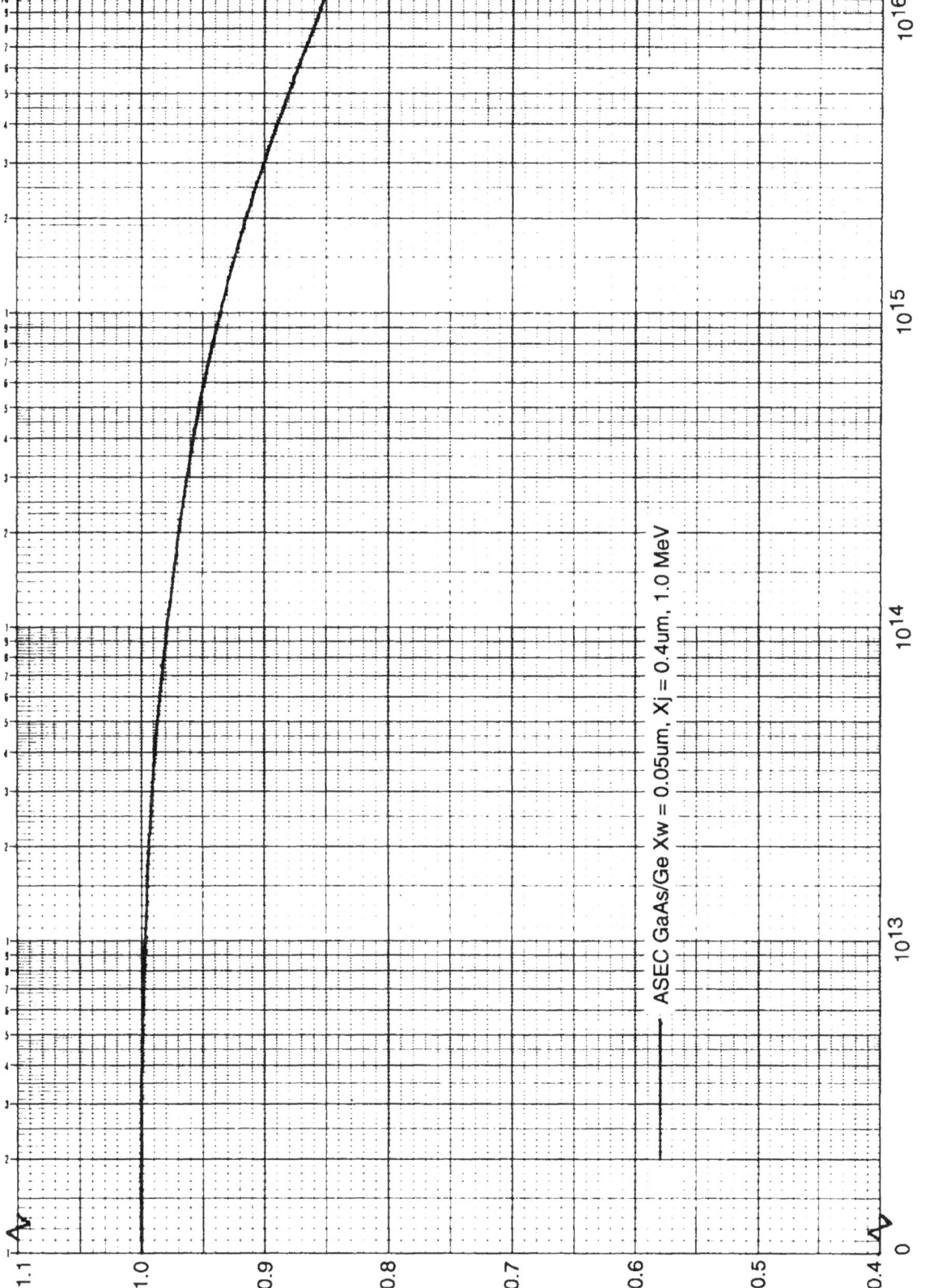

ELECTRON FLUENCE, e/cm²

NORMALIZED VOLTAGE AT MAXIMUM POWER

ASEC GaAs/Ge Xw = 0.05um, Xj = 0.4um, 1.0 MeV

Figure 6.10   Normalized V_{mp} vs. 1 MeV Electron Fluence for ASEC GaAs/Ge Solar Cells

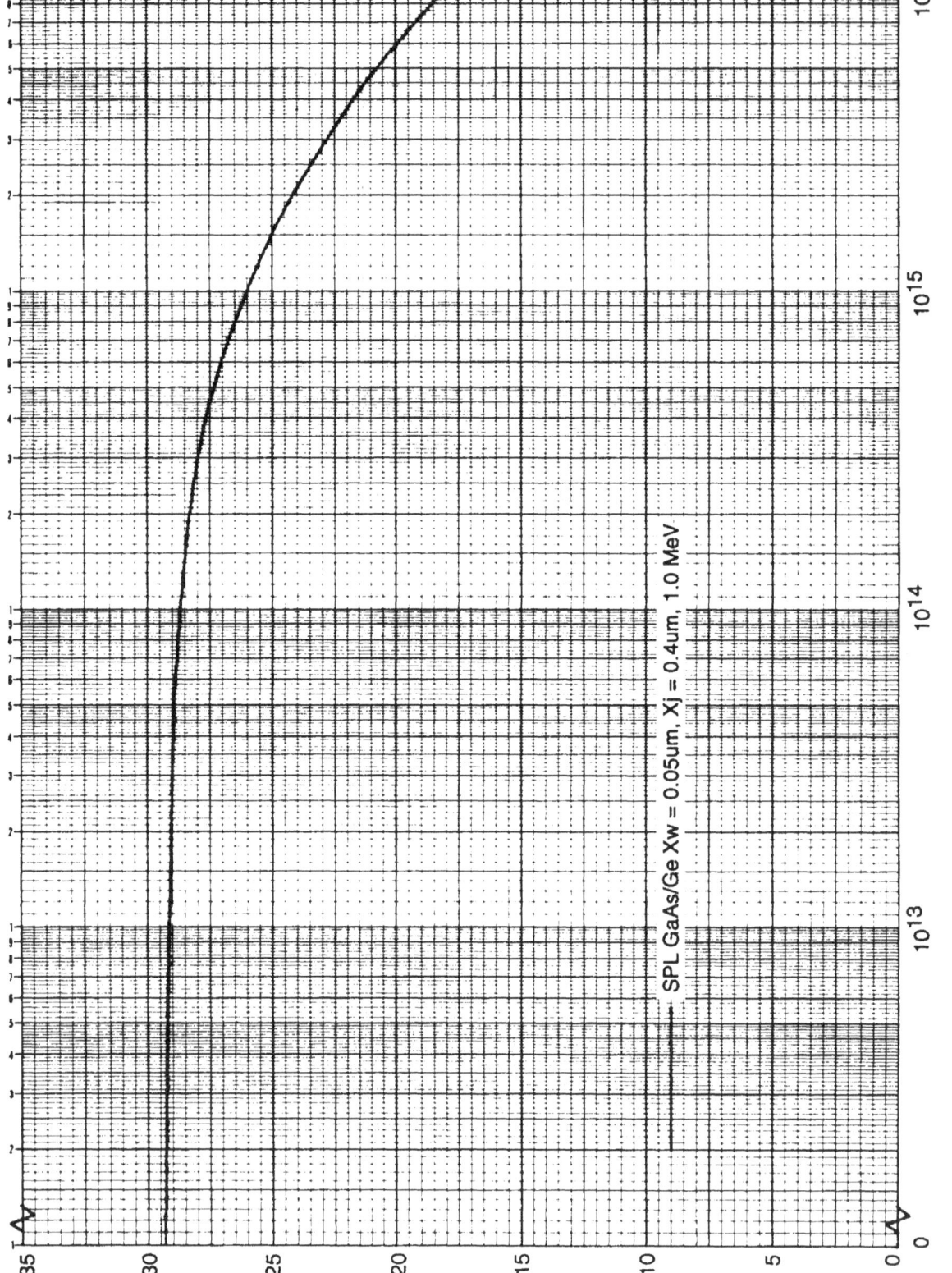

ELECTRON FLUENCE, e/cm^2

SHORT CIRCUIT CURRENT, mA/cm^2

SPL GaAs/Ge Xw = 0.05um, Xj = 0.4um, 1.0 MeV

Figure 6.11. I_{sc}/cm^2 vs. 1 MeV Electron Fluence for Spectrolab GaAs/Ge Solar Cells

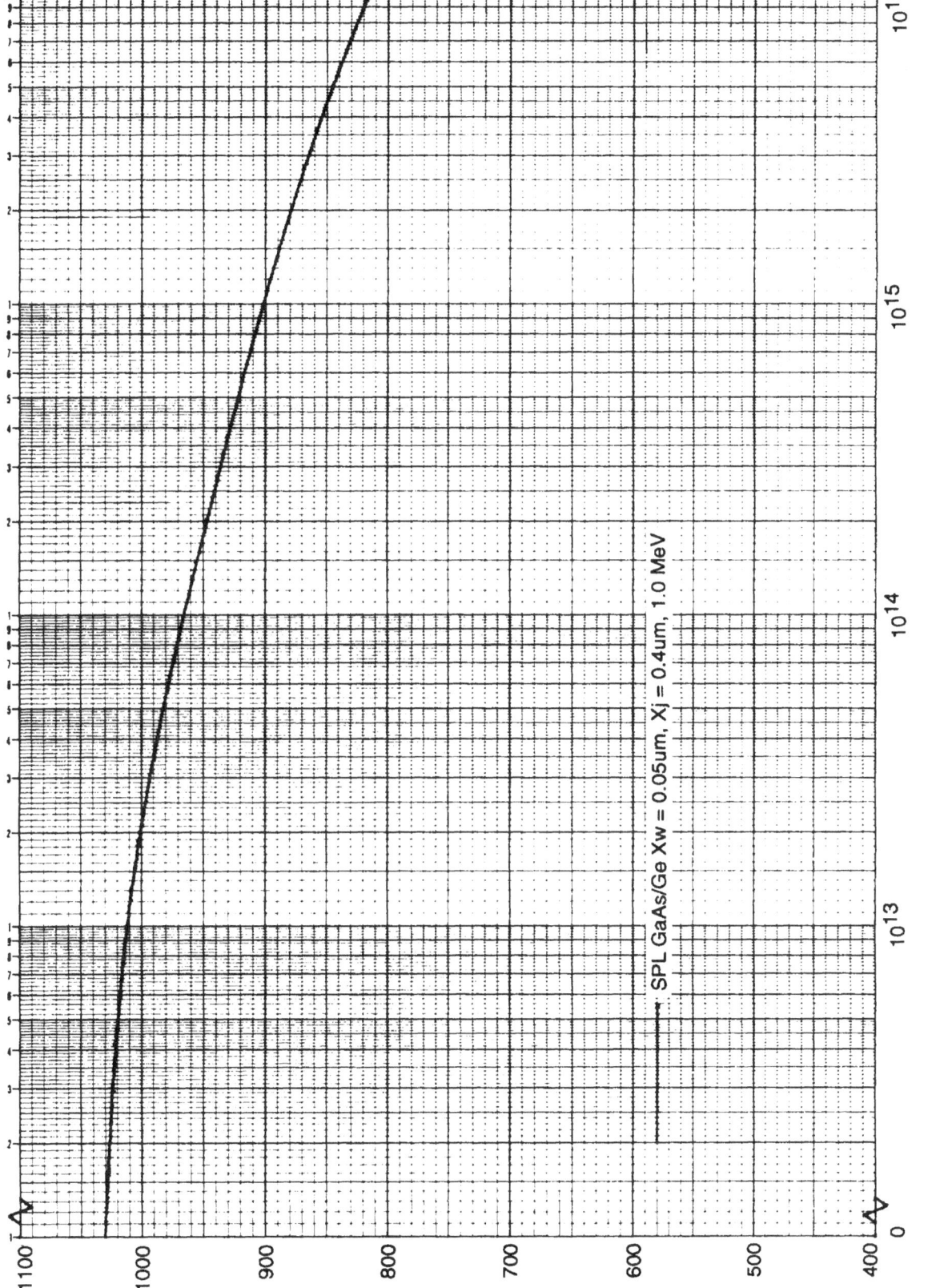

ELECTRON FLUENCE, e/cm^2

OPEN CIRCUIT VOLTAGE, mV

SPL GaAs/Ge Xw = 0.05um, Xj = 0.4um, 1.0 MeV

Figure 6.12. V_{oc} vs. 1 MeV Electron Fluence for Spectrolab GaAs/Ge Solar Cells

6-15

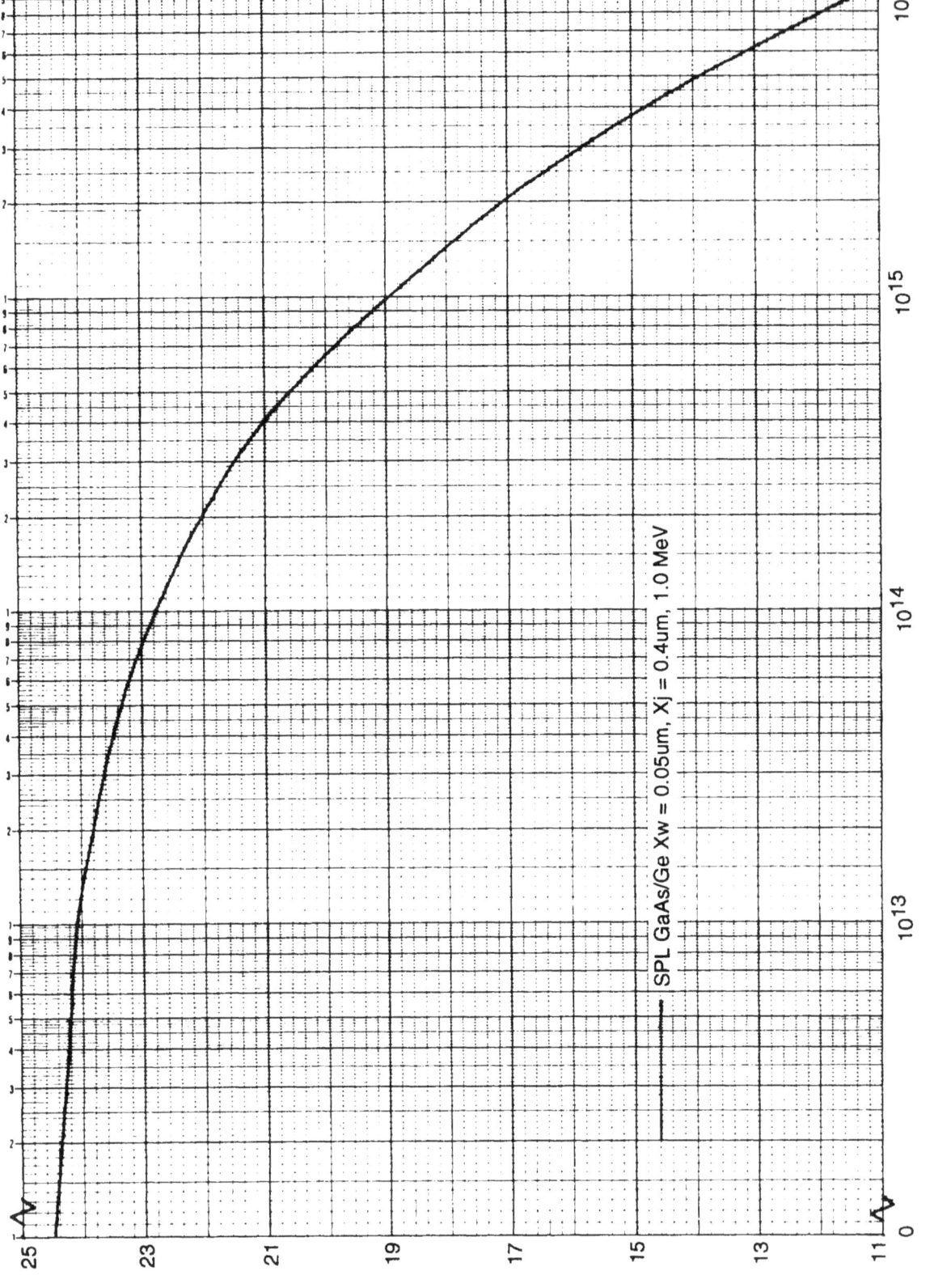

ELECTRON FLUENCE, e/cm²

MAXIMUM POWER, mW/cm²

SPL GaAs/Ge Xw = 0.05um, Xj = 0.4um, 1.0 MeV

Figure 6.13. P_{max}/cm^2 vs. 1 MeV Electron Fluence for Spectrolab GaAs/Ge Solar Cells

6-16

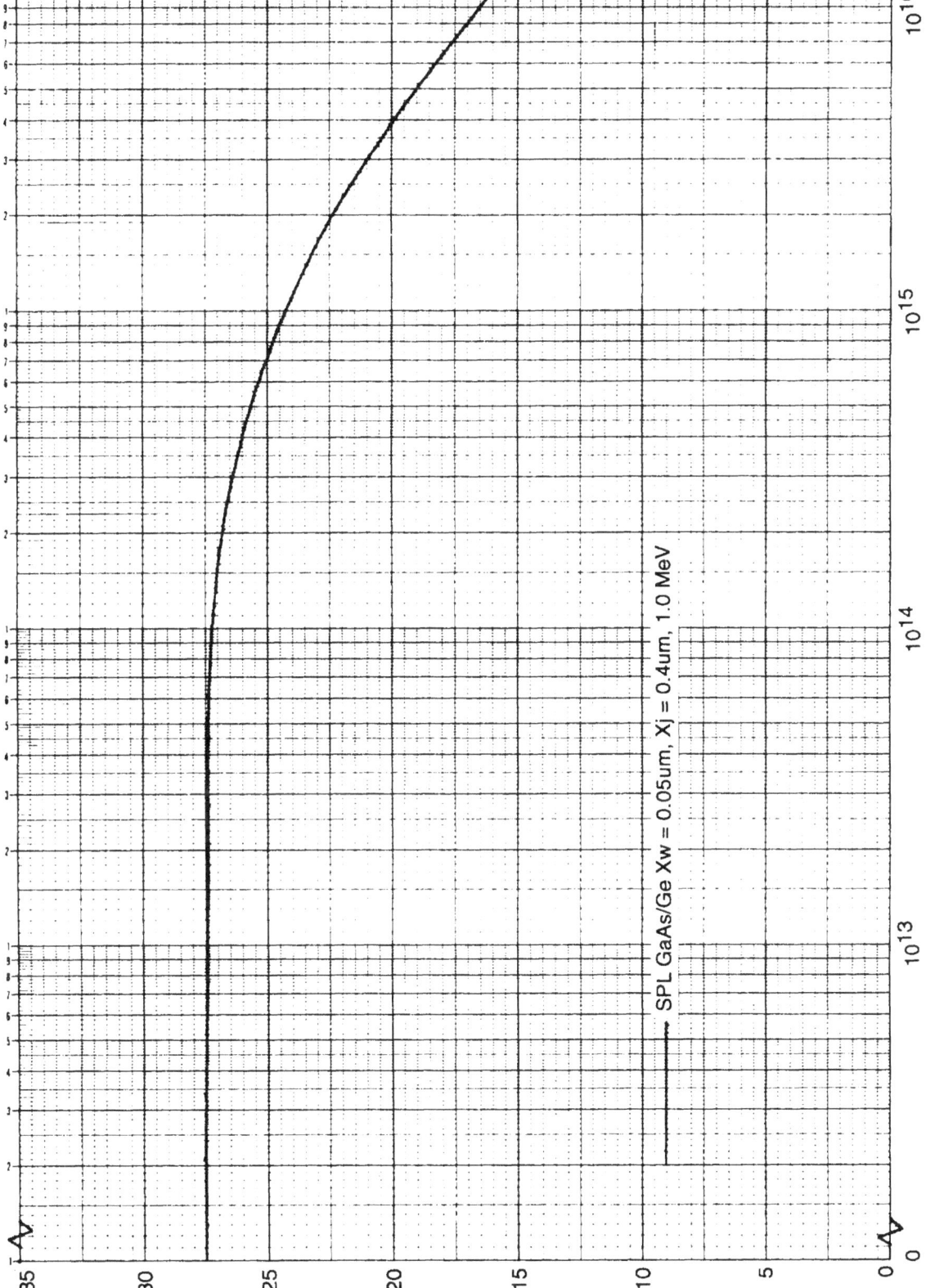

ELECTRON FLUENCE, e/cm²

SPL GaAs/Ge Xw = 0.05um, Xj = 0.4um, 1.0 MeV

CURRENT AT MAXIMUM POWER, mA/cm²

Figure 6.14. I_{mp}/cm² vs. 1 MeV Electron Fluence for Spectrolab GaAs/Ge Solar Cells

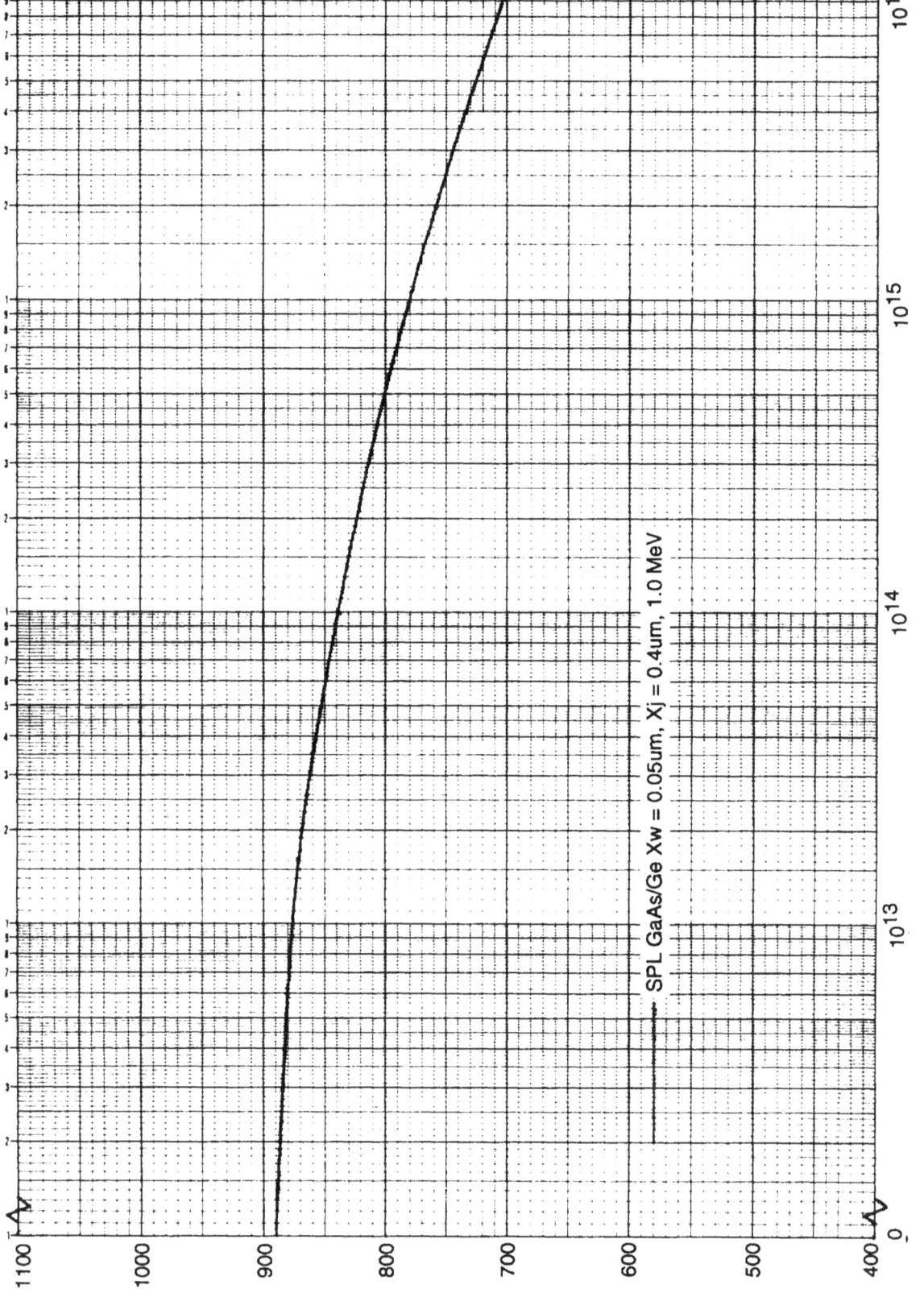

ELECTRON FLUENCE, e/cm²

VOLTAGE AT MAXIMUM POWER, mV

SPL GaAs/Ge Xw = 0.05um, Xj = 0.4um, 1.0 MeV

Figure 6.15.   V$_{mp}$ vs. 1 MeV Electron Fluence for Spectrolab GaAs/Ge Solar Cells

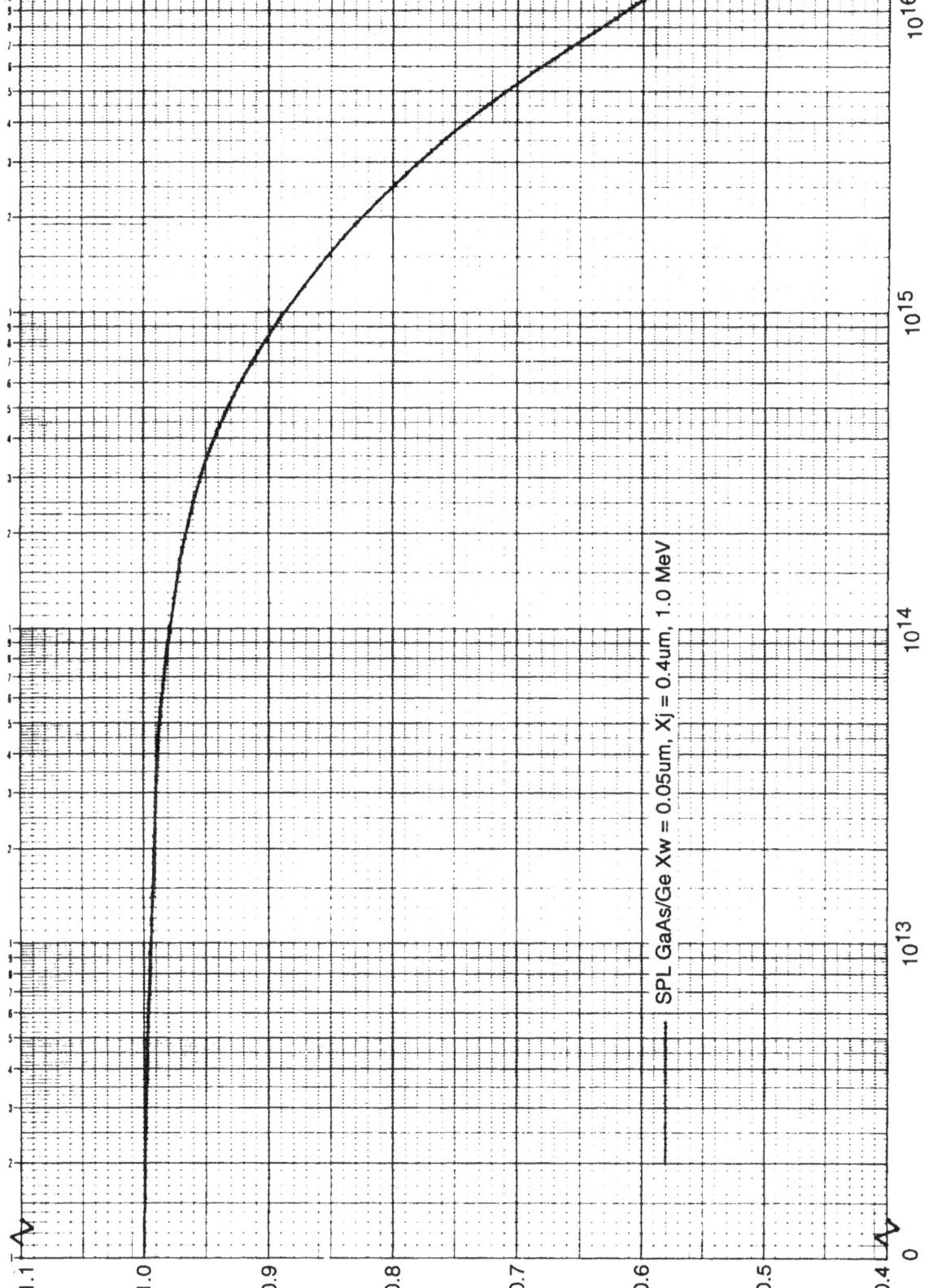

NORMALIZED SHORT CIRCUIT CURRENT

ELECTRON FLUENCE, e/cm^2

SPL GaAs/Ge Xw = 0.05um, Xj = 0.4um, 1.0 MeV

Figure 6.16. Normalized I$_{sc}$ vs. 1 MeV Electron Fluence for Spectrolab

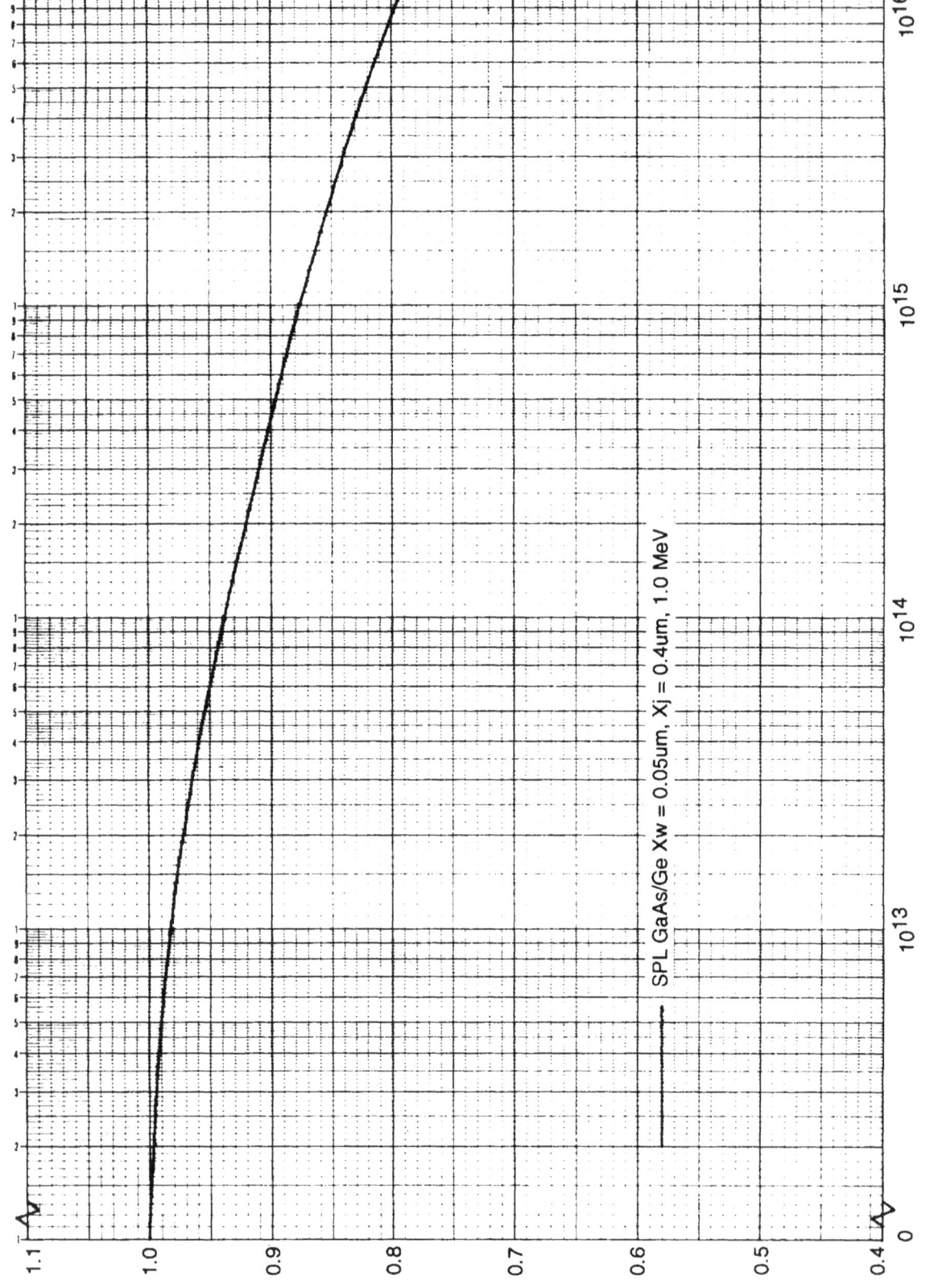

ELECTRON FLUENCE, e/cm²

NORMALIZED OPEN CIRCUIT VOLTAGE

SPL GaAs/Ge Xw = 0.05um, Xj = 0.4um, 1.0 MeV

Figure 6.17.   Normalized $V_{\propto}$ vs. 1 MeV Electron Fluence for Spectrolab GaAs/Ge Solar Cells

6-20

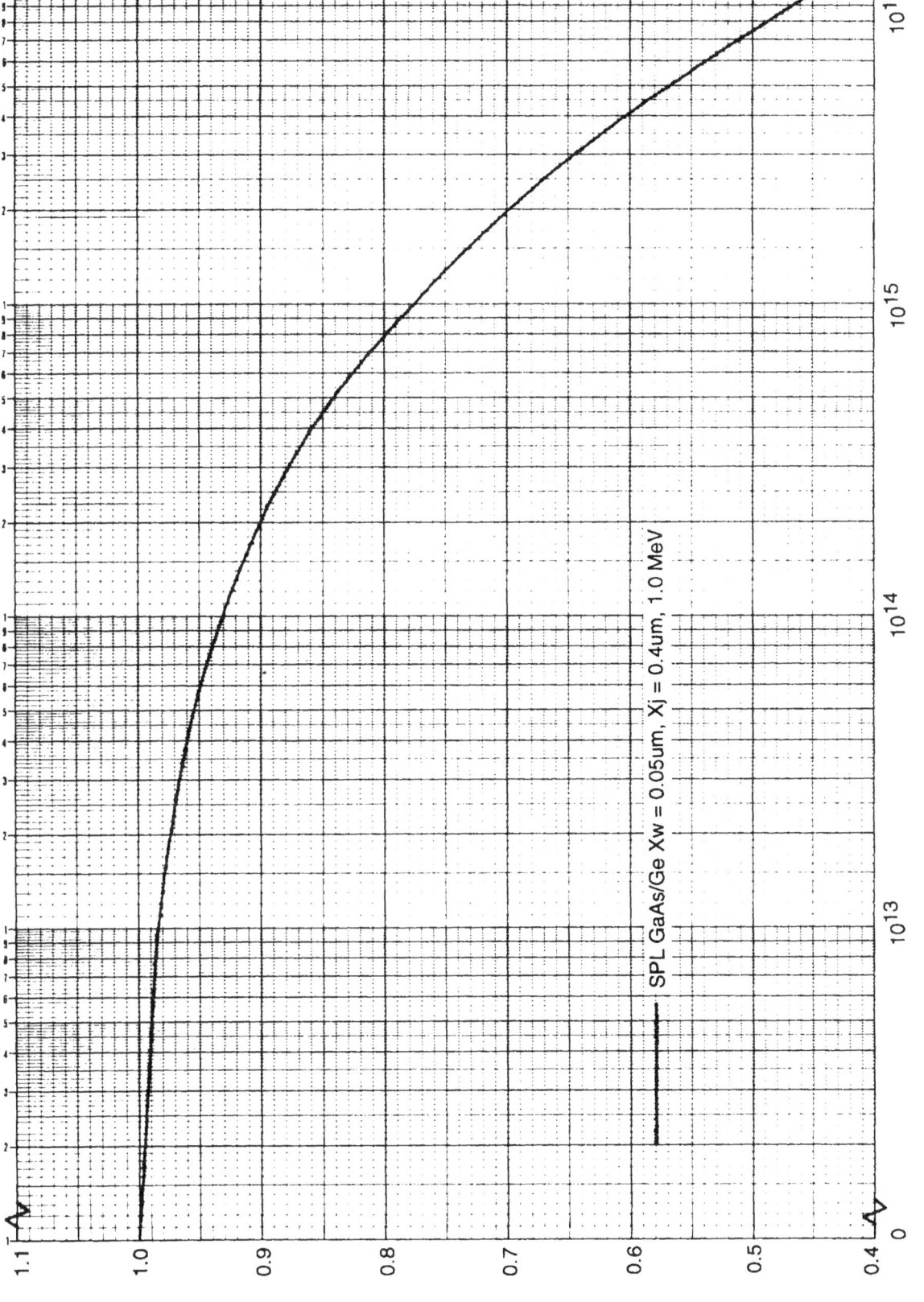

ELECTRON FLUENCE, e/cm²

NORMALIZED MAXIMUM POWER

SPL GaAs/Ge Xw = 0.05um, Xj = 0.4um, 1.0 MeV

Figure 6.18.   Normalized P_{max} vs. 1 MeV Electron Fluence for Spectrolab GaAs/Ge Solar Cells

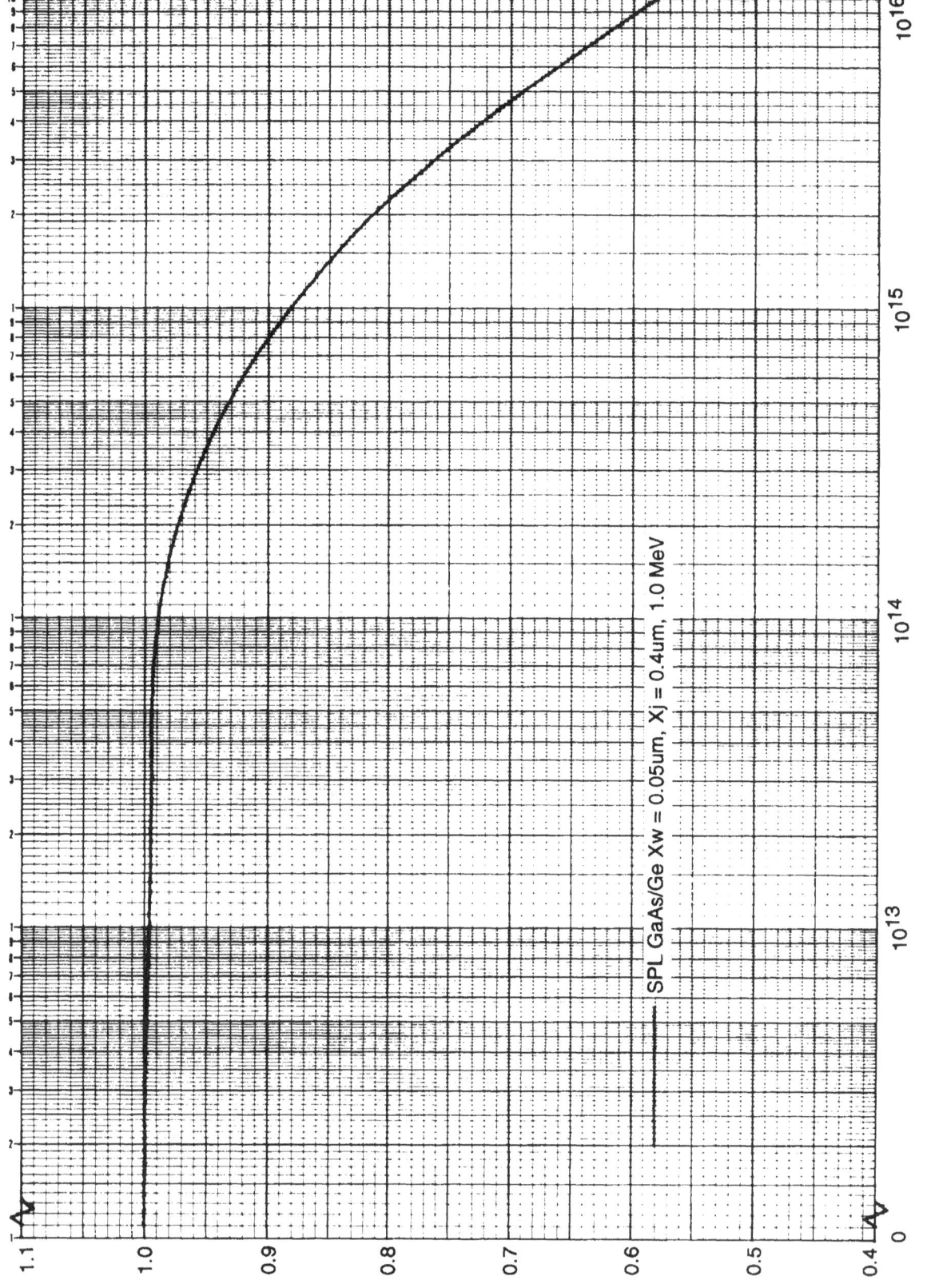

Figure 6.19. Normalized I_{mp} vs. 1 MeV Electron Fluence for Spectrolab GaAs/Ge Solar Cells

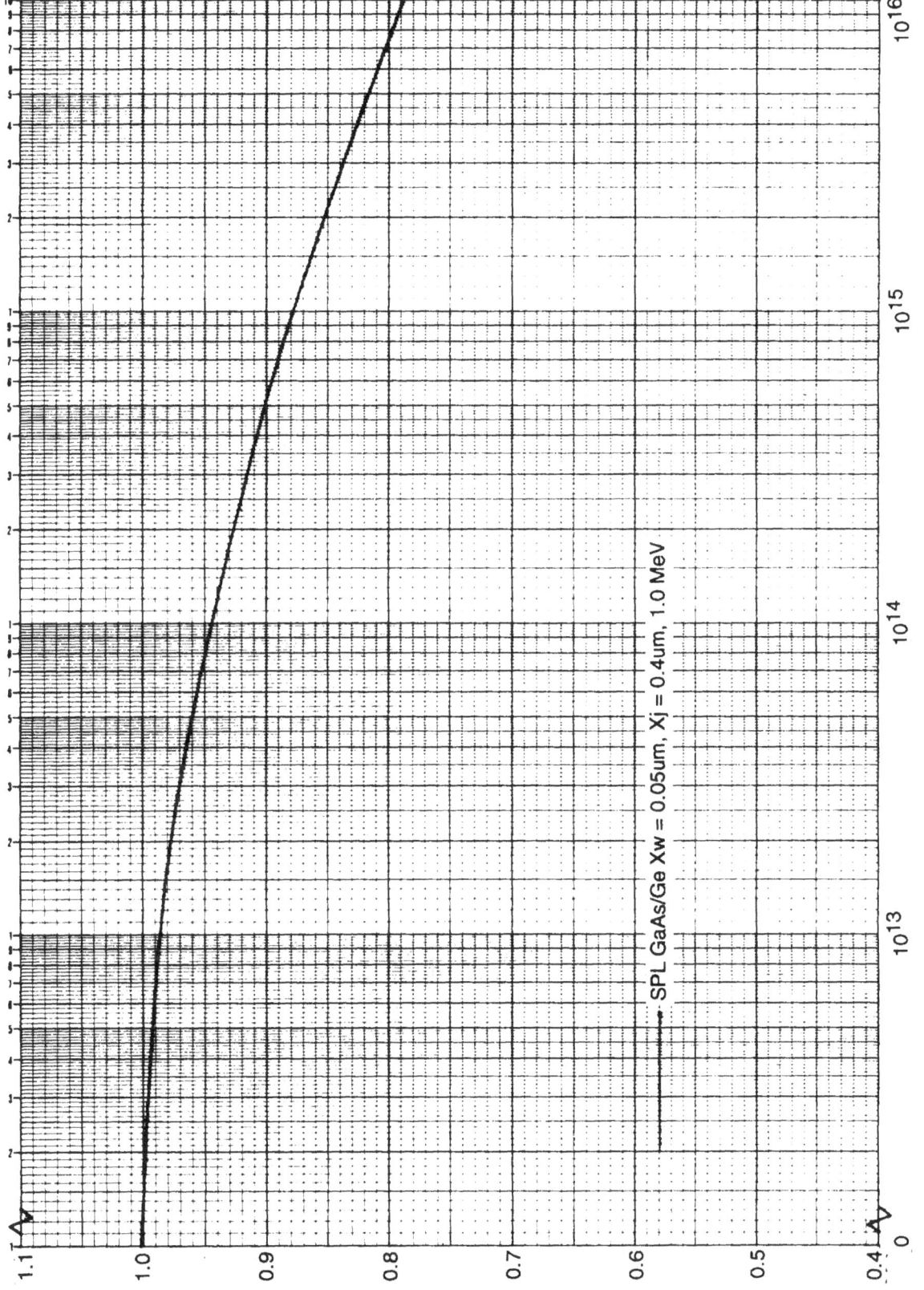

ELECTRON FLUENCE, e/cm²

NORMALIZED VOLTAGE AT MAXIMUM POWER

SPL GaAs/Ge Xw = 0.05um, Xj = 0.4um, 1.0 MeV

Figure 6.20   Normalized V_{mp} vs. 1 MeV Electron Fluence for Spectrolab GaAs/Ge Solar Cells

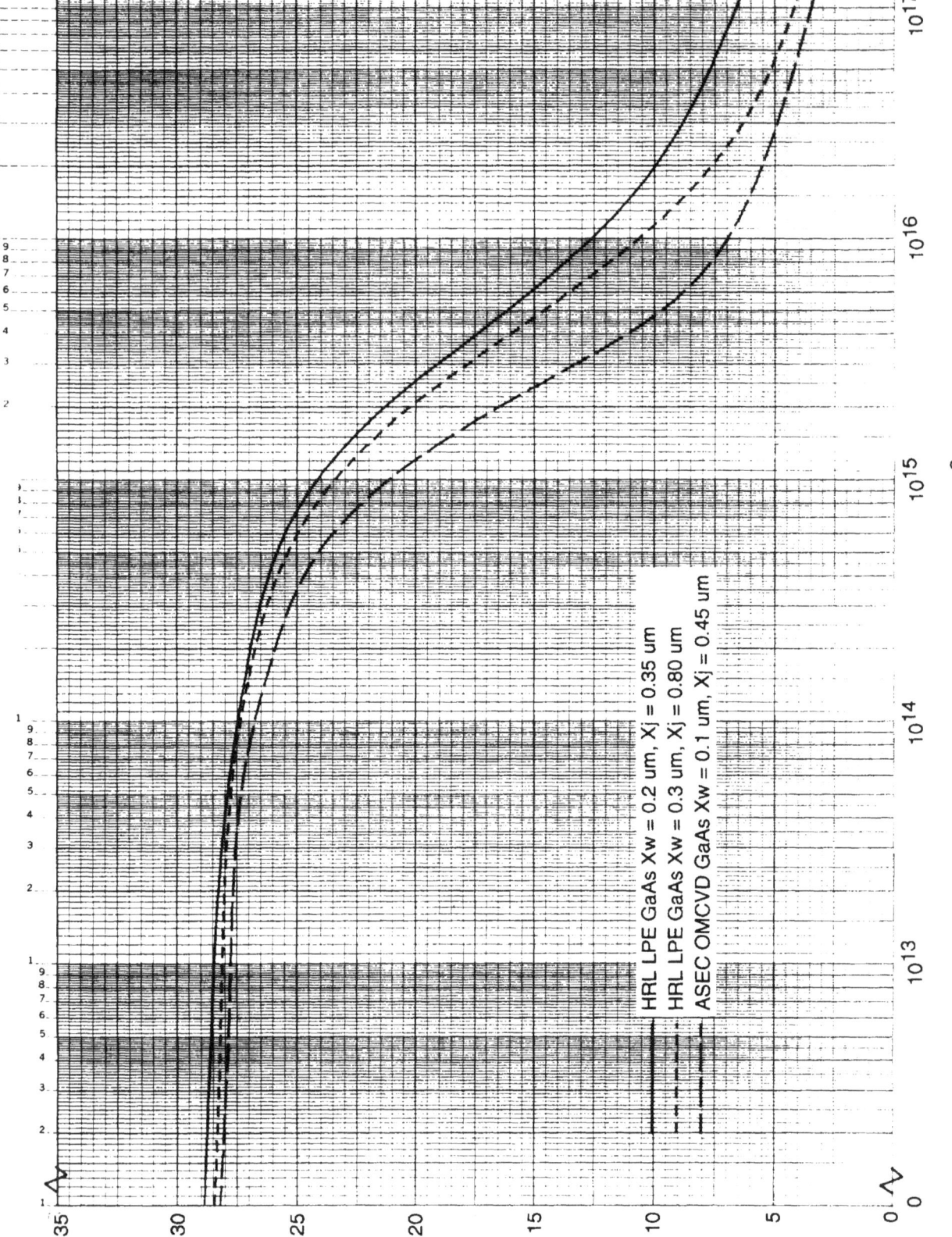

ELECTRON FLUENCE, e/cm²

SHORT CIRCUIT CURRENT, mA/cm²

HRL LPE GaAs Xw = 0.2 um, Xj = 0.35 um
HRL LPE GaAs Xw = 0.3 um, Xj = 0.80 um
ASEC OMCVD GaAs Xw = 0.1 um, Xj = 0.45 um

Figure 6.21. I_{sc}/cm² vs. 1 MeV Electron Fluence for Hughes LPE and ASEC OMCVD GaAs/Ge Solar Cells

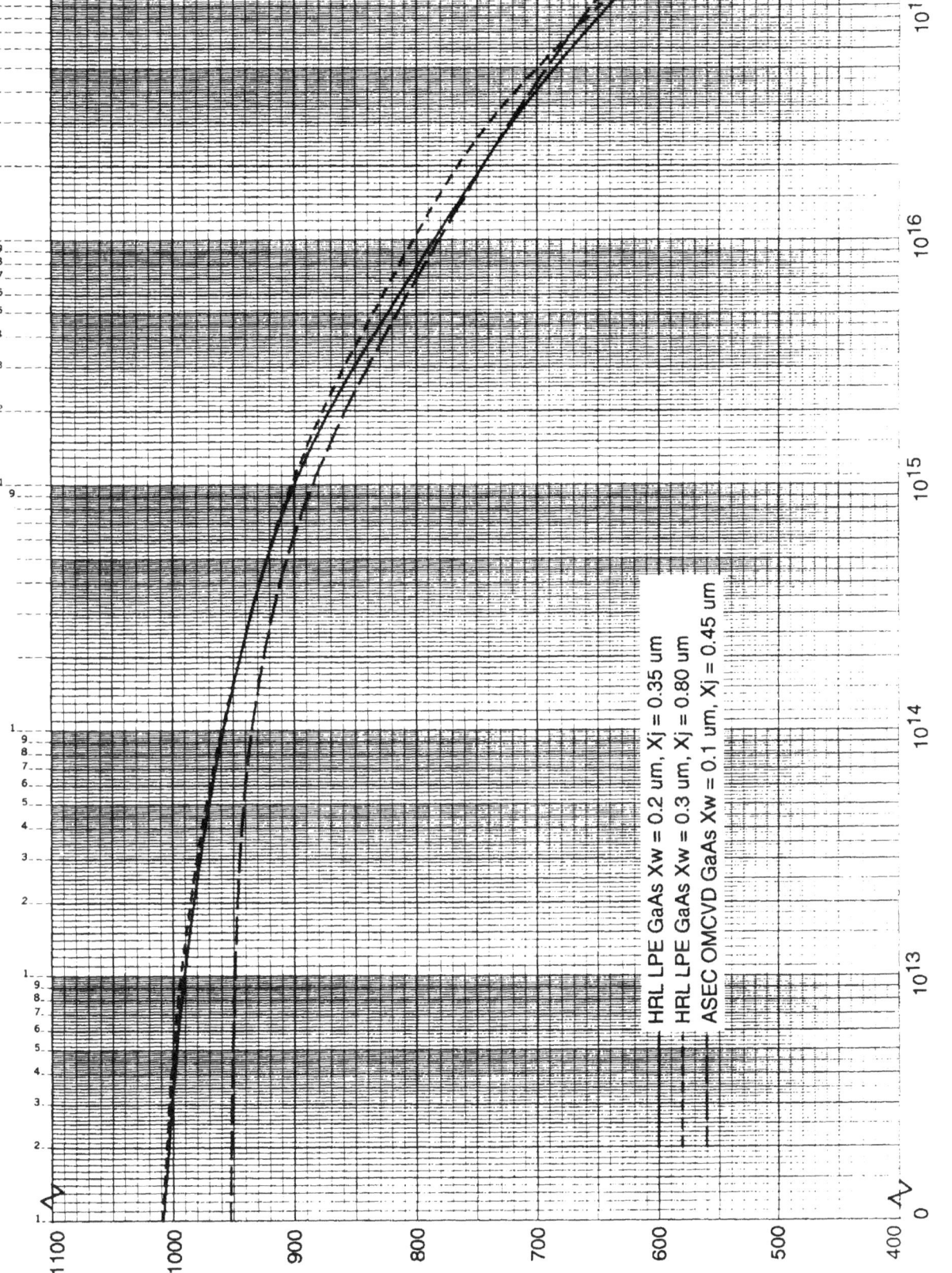

ELECTRON FLUENCE, e/cm²

OPEN CIRCUIT VOLTAGE, mV

Legend:
— HRL LPE GaAs Xw = 0.2 um, Xj = 0.35 um
–– HRL LPE GaAs Xw = 0.3 um, Xj = 0.80 um
ASEC OMCVD GaAs Xw = 0.1 um, Xj = 0.45 um

Figure 6.22. V_{oc} vs. 1 MeV Electron Fluence for Hughes LPE and ASEC OMCVD GaAs/Ge Solar Cells

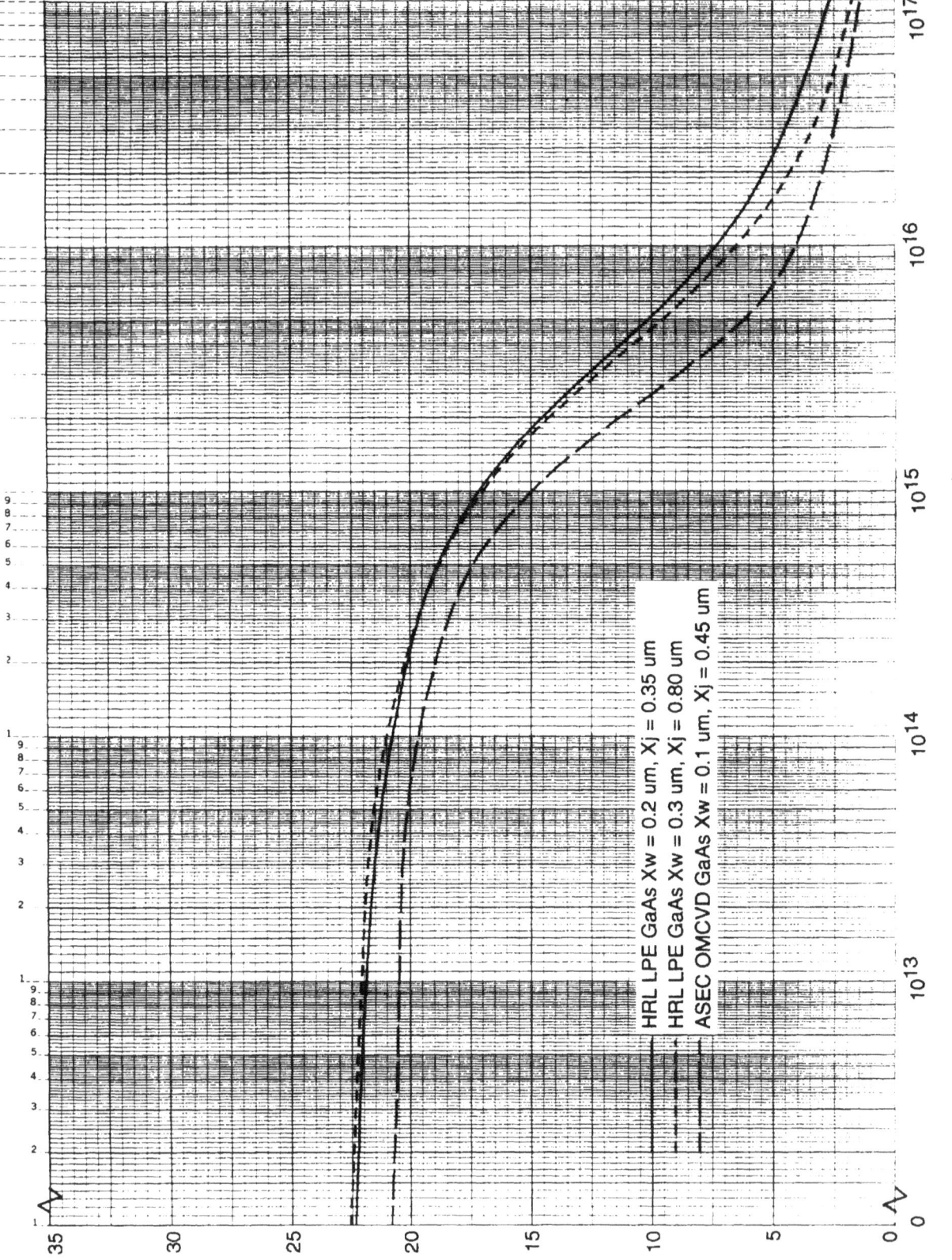

ELECTRON FLUENCE, e/cm²

MAXIMUM POWER, mW/cm²

HRL LPE GaAs Xw = 0.2 um, Xj = 0.35 um
HRL LPE GaAs Xw = 0.3 um, Xj = 0.80 um
ASEC OMCVD GaAs Xw = 0.1 um, Xj = 0.45 um

Figure 6.23. P_{max}/cm² vs. 1 MeV Electron Fluence for Hughes LPE and ASEC OMCVD GaAs/Ge Solar Cells

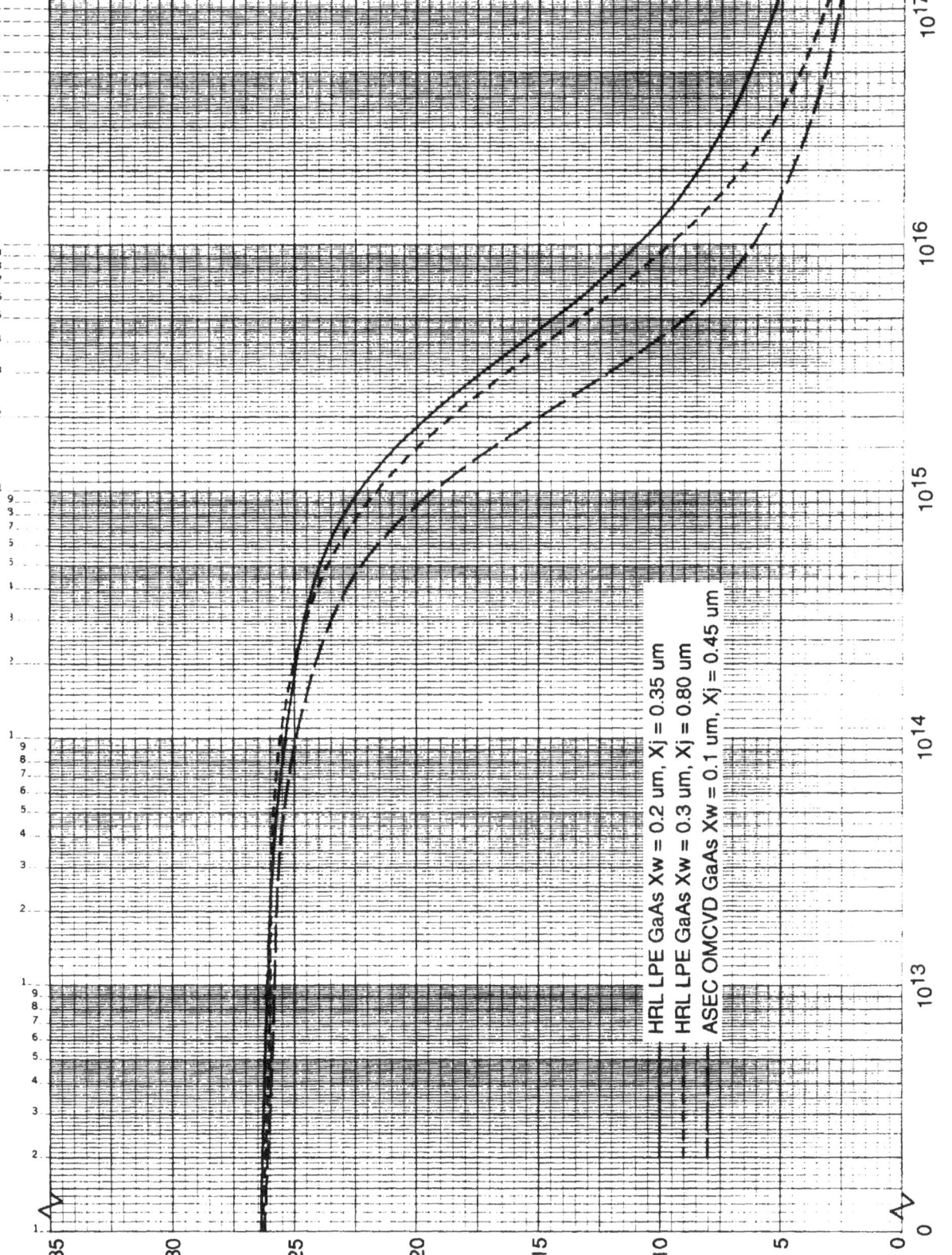

ELECTRON FLUENCE, e/cm²

CURRENT AT MAXIMUM POWER, mA/cm²

HRL LPE GaAs Xw = 0.2 um, Xj = 0.35 um
HRL LPE GaAs Xw = 0.3 um, Xj = 0.80 um
ASEC OMCVD GaAs Xw = 0.1 um, Xj = 0.45 um

Figure 6.24. I_{mp}/cm² vs. 1 MeV Electron Fluence for Hughes LPE and ASEC OMCVD GaAs/Ge Solar Cells

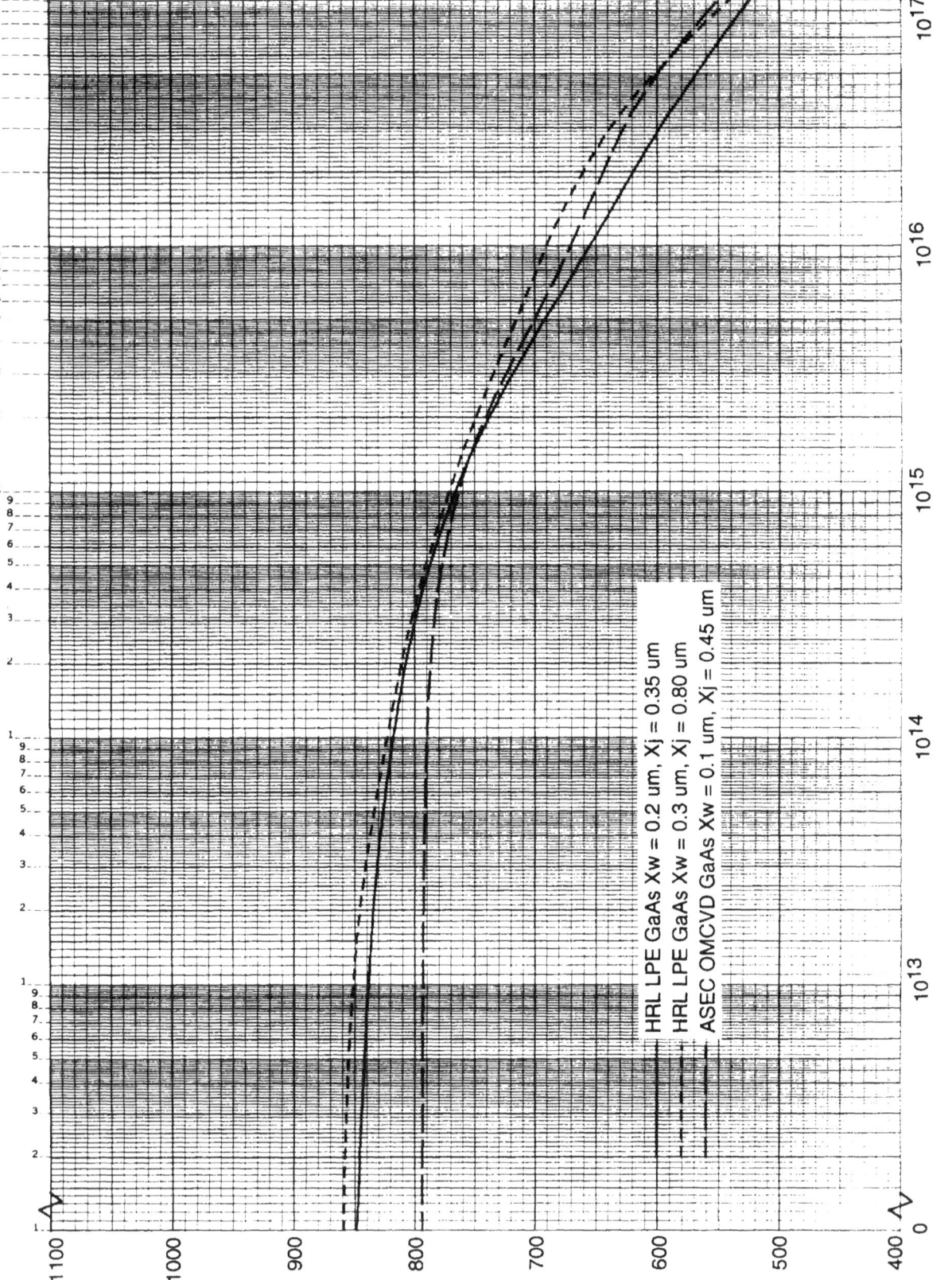

Figure 6.25. V_{mp} vs. 1 MeV Electron Fluence for Hughes LPE and ASEC OMCVD GaAs/Ge Solar Cells

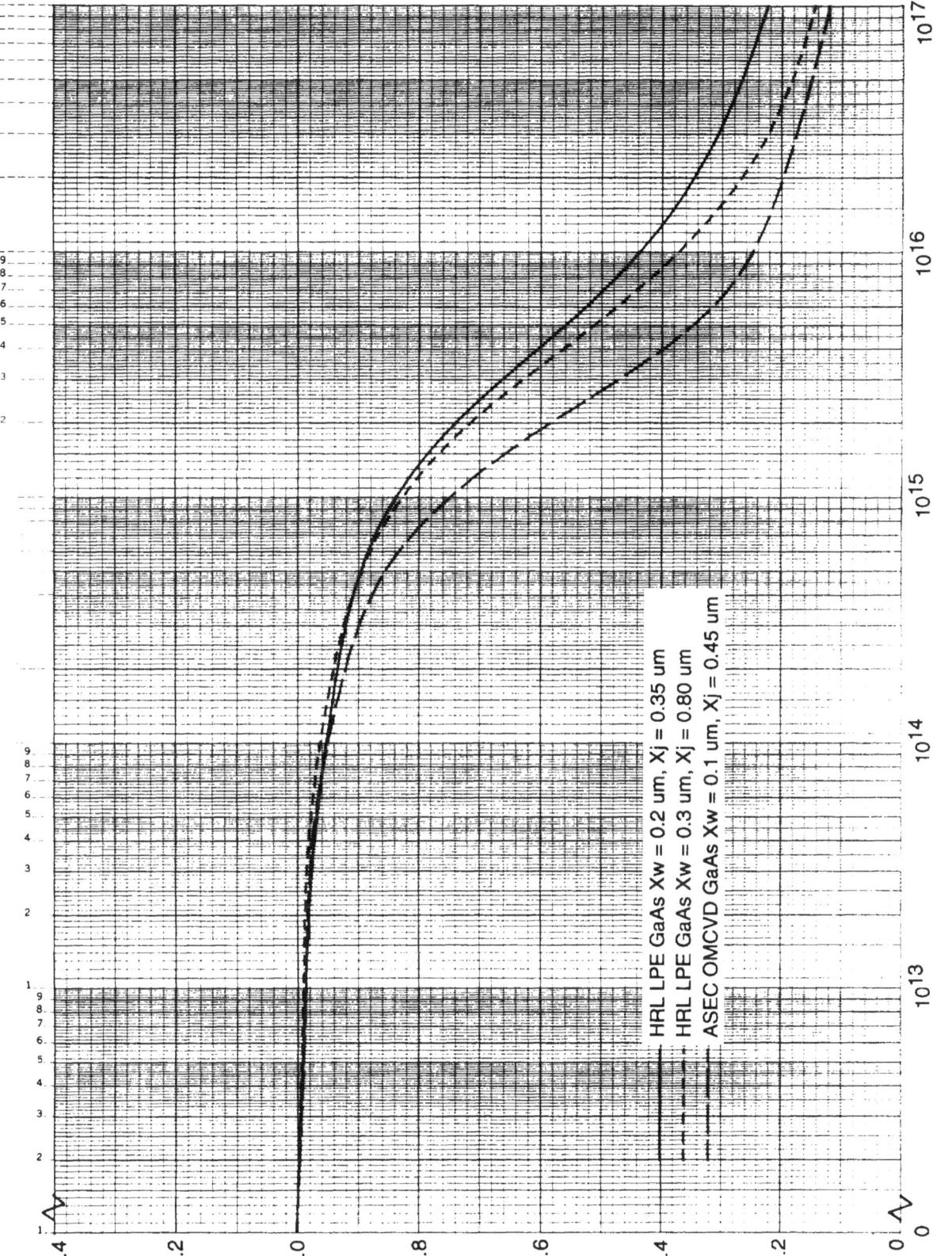

NORMALIZED SHORT CIRCUIT CURRENT

ELECTRON FLUENCE, e/cm²

HRL LPE GaAs Xw = 0.2 um, Xj = 0.35 um
HRL LPE GaAs Xw = 0.3 um, Xj = 0.80 um
ASEC OMCVD GaAs Xw = 0.1 um, Xj = 0.45 um

Figure 6.26.   Normalized I_{sc} vs. 1 MeV Electron Fluence for Hughes LPE and ASEC OMCVD GaAs/Ge Solar Cells

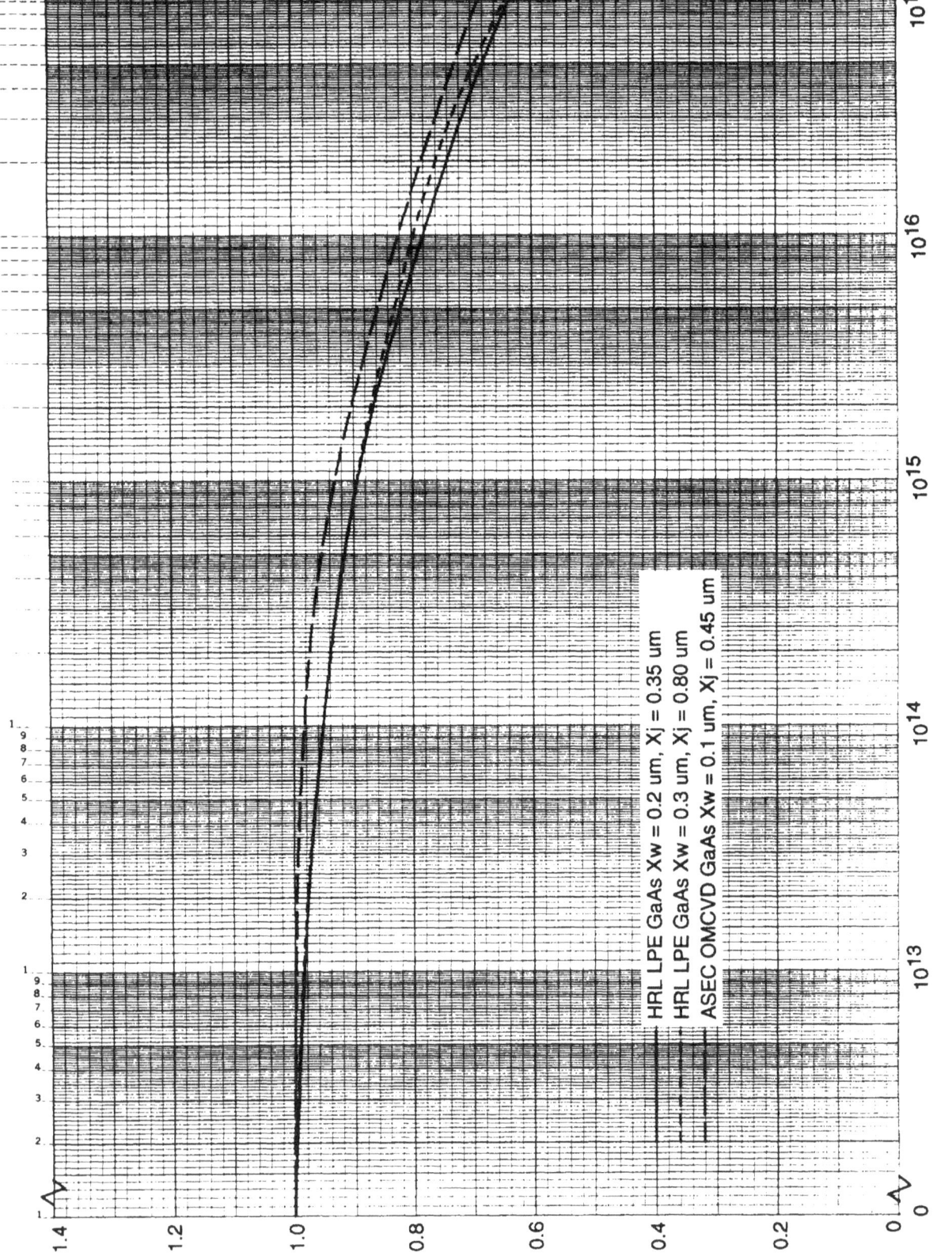

ELECTRON FLUENCE, e/cm^2

NORMALIZED OPEN CIRCUIT VOLTAGE

HRL LPE GaAs Xw = 0.2 um, Xj = 0.35 um
HRL LPE GaAs Xw = 0.3 um, Xj = 0.80 um
ASEC OMCVD GaAs Xw = 0.1 um, Xj = 0.45 um

Figure 6.27.   Normalized V_{oc} vs. 1 MeV Electron Fluence for Hughes LPE and ASEC OMCVD GaAs/Ge Solar Cells

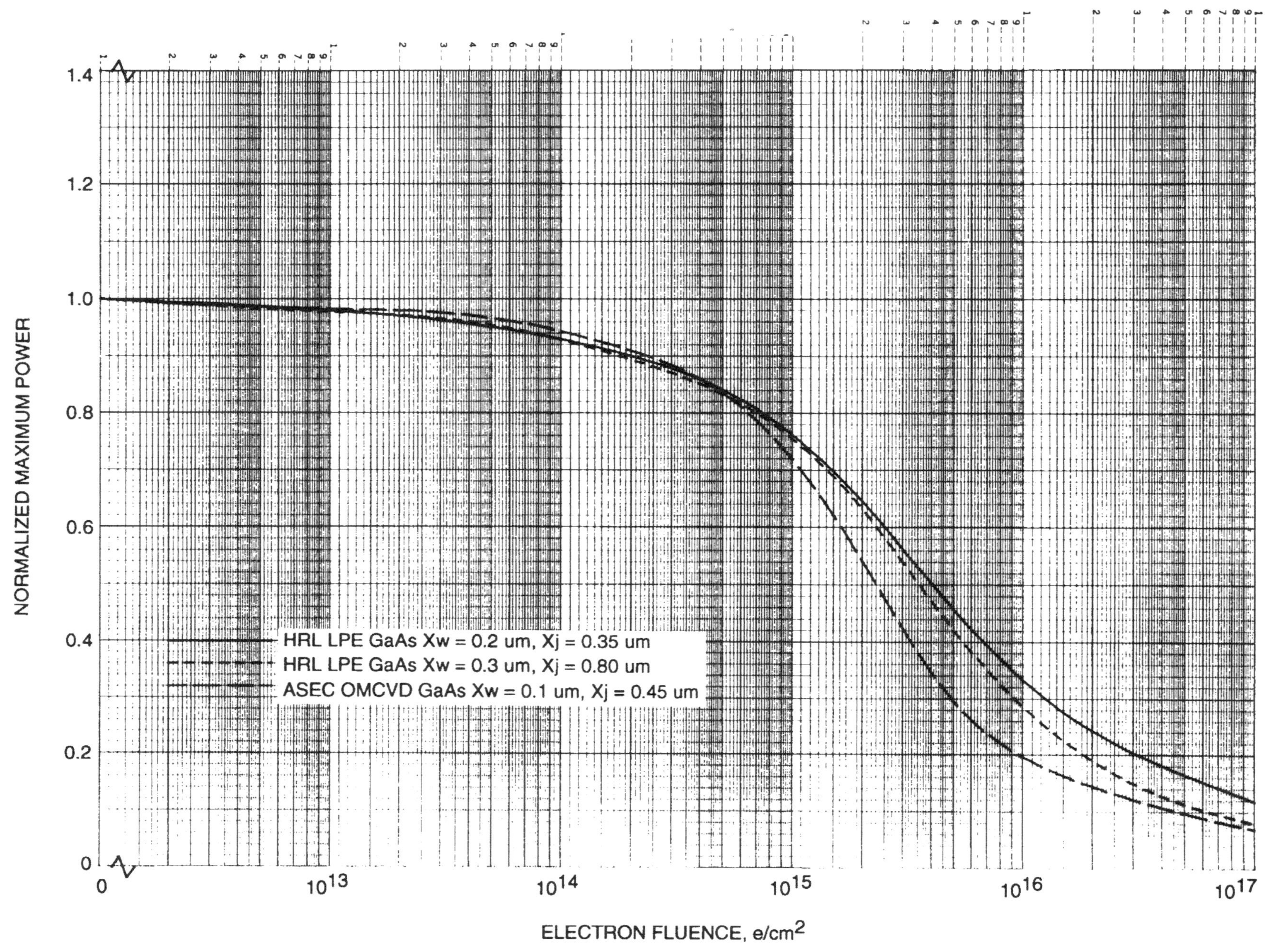

Figure 6.28.   Normalized P_{max} vs. 1 MeV Electron Fluence for Hughes LPE and ASEC OMCVD GaAs/Ge Solar Cells

Legend (within figure):
- HRL LPE GaAs Xw = 0.2 um, Xj = 0.35 um
- HRL LPE GaAs Xw = 0.3 um, Xj = 0.80 um
- ASEC OMCVD GaAs Xw = 0.1 um, Xj = 0.45 um

X-axis: ELECTRON FLUENCE, e/cm^2
Y-axis: NORMALIZED MAXIMUM POWER

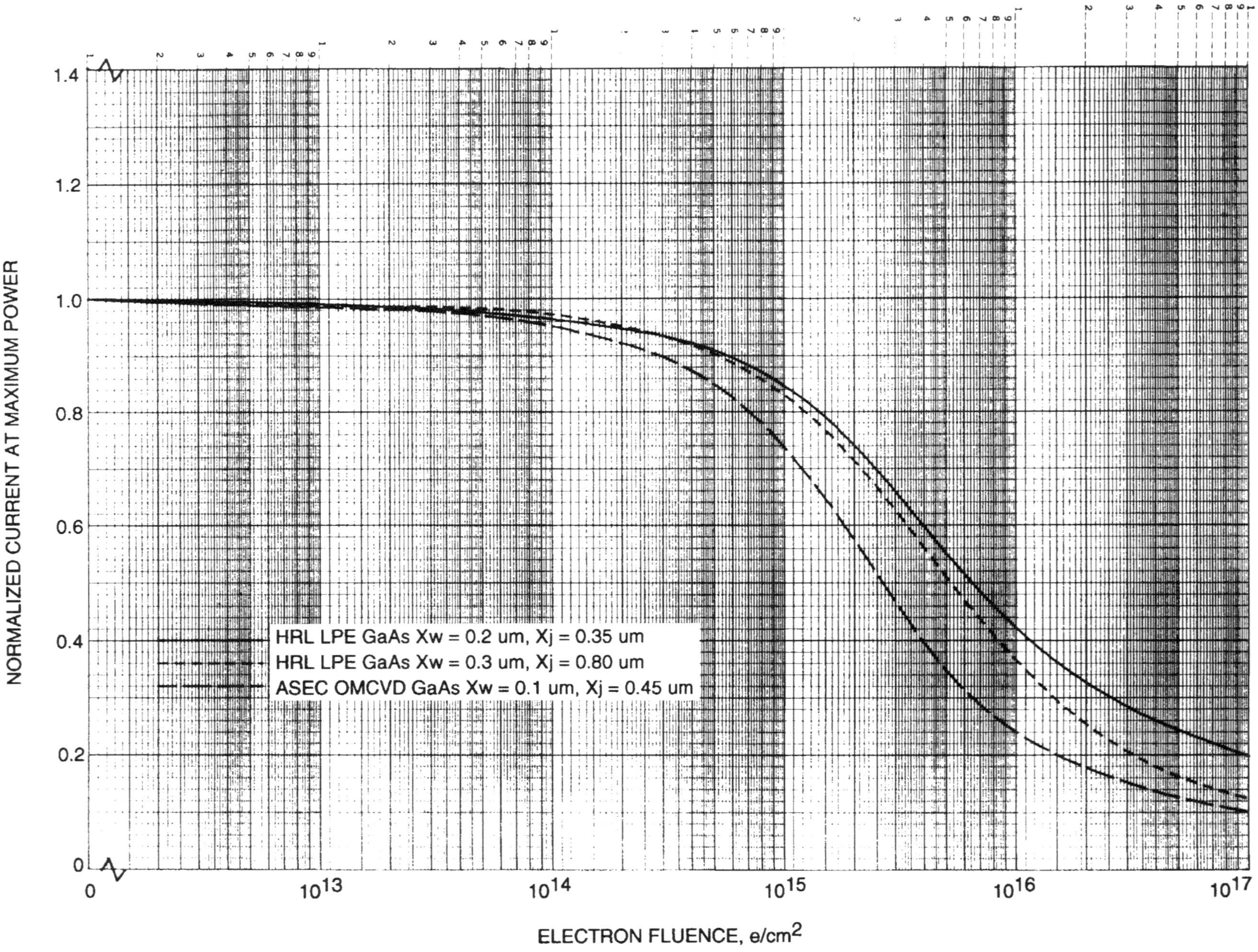
1.4
1.2
1.0
0.8
0.6
0.4
0.2
0
NORMALIZED CURRENT AT MAXIMUM POWER
HRL LPE GaAs Xw = 0.2 um, Xj = 0.35 um
HRL LPE GaAs Xw = 0.3 um, Xj = 0.80 um
ASEC OMCVD GaAs Xw = 0.1 um, Xj = 0.45 um
0
10^13
10^14
10^15
10^16
10^17
ELECTRON FLUENCE, e/cm^2

Figure 6.29. Normalized I_{mp} vs. 1 MeV Electron Fluence for Hughes LPE and ASEC OMCVD GaAs/Ge Solar Cells

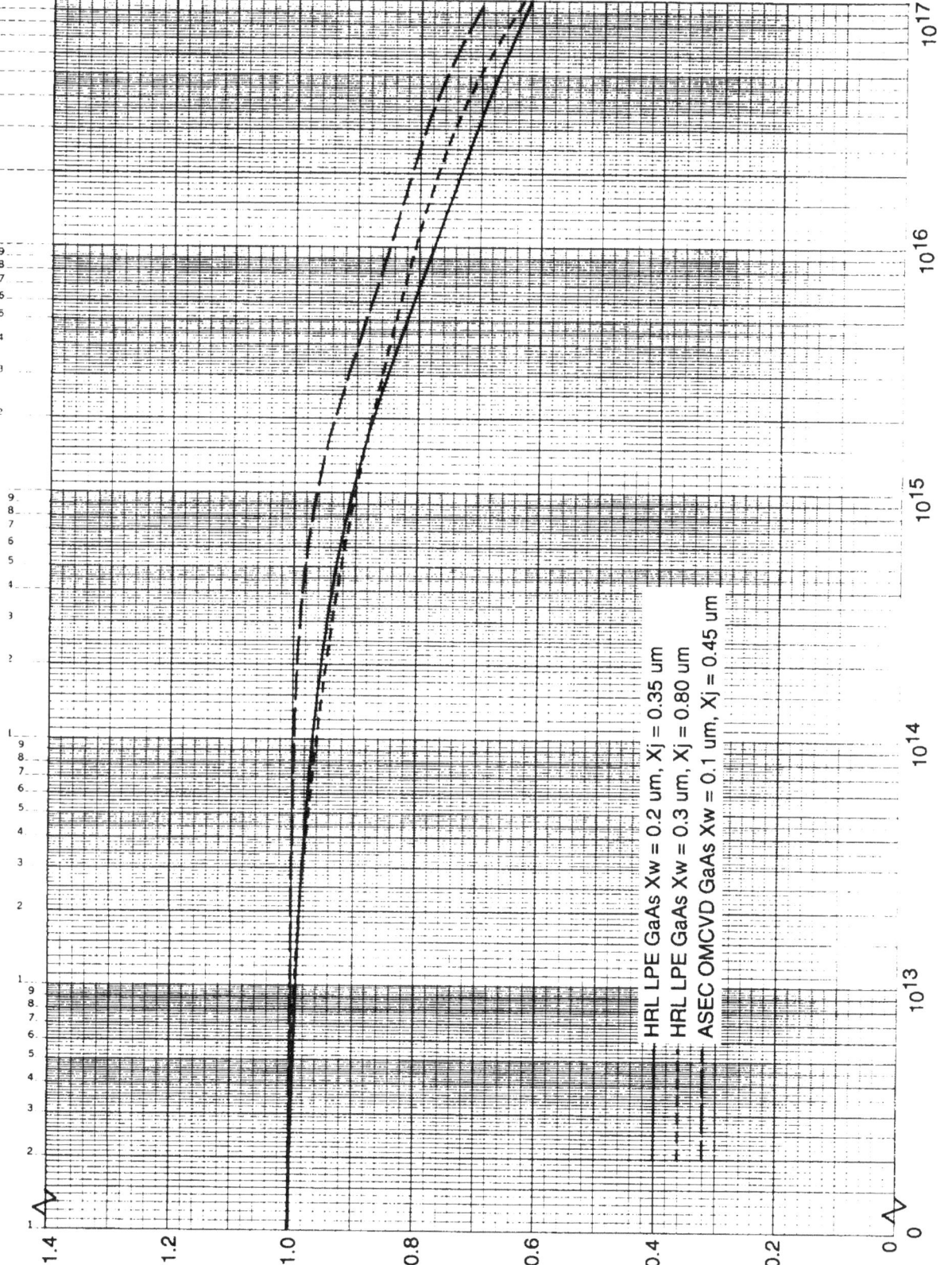

ELECTRON FLUENCE, e/cm²

NORMALIZED VOLTAGE AT MAXIMUM POWER

HRL LPE GaAs Xw = 0.2 um, Xj = 0.35 um
HRL LPE GaAs Xw = 0.3 um, Xj = 0.80 um
ASEC OMCVD GaAs Xw = 0.1 um, Xj = 0.45 um

Figure 6.30   Normalized V_{mp} vs. 1 MeV Electron Fluence for Hughes LPE and ASEC OMCVD GaAs/Ge Solar Cells

6.2   GaAs Solar Cell Behavior with Temperature and Solar Intensity

In this section we will show the behavior of ASEC GaAs/Ge cells as a function of temperature ($-120°C$ to $+140°C$), incident solar intensity (5 to 250 mW/cm^2), and 1 MeV electron irradiation (after fluences of 0, 1 x 10^{14} and 1.1 x 10^{15} e/cm^2). These cells were of 1992 vintage and had a window thickness of 0.06 μm, a junction thickness of 0.45 μm, and a dual antireflection (AR) coating of TiO$_x$ - Al$_2$O$_3$. The cell size was 2 x 2 x 0.02 cm. For these experiments, seven cells were bonded to a copper plate with RTV 560, a silicone rubber compound made by General Electric. A thermocouple was bonded to the busbar of a dummy cell mounted on the plate to enable temperature measurements. During the irradiations, the copper plate was mounted on a temperature-controlled target plane and held at a temperature of 28°C during the exposures. The plate was mounted on another temperature-controlled surface in a vacuum system for the temperature/intensity measurements. An X-25L simulator was used as the light source and illuminated the cells through a 7940 fused-silica window in the vacuum chamber. A balloon flight standard cell was also mounted in the chamber for setting the simulator intensity. The standard cell was mounted on its own temperature block and held at 28°C throughout the measurements.

The results of the electrical measurements prior to irradiation are tabulated in Tables 6-3 through 6-9. The numbers in the tables are average values of the seven cells measured. The intensities were measured with reference to the Thekaekara standard of 135.3 mW/cm^2, so for increased accuracy, the reported intensity values should be adjusted upward by the ratio of 136.8/135.3 (about a 1% adjustment). The current and maximum power values, I_{sc}, I_{mp}, and P_{max} have all been divided by the cell area, which in this case was 4 cm^2. Figures 6.31 through 6.33 are plots of I_{sc}, V_{oc}, and P_{max} as a function of temperature, and Figures 6.34 through 6.36 are plots of the same parameters as a function of intensity. The continuous curves are least-square fits to the data using either second or third degree polynomials. The data for these same cells, after the 1 MeV electron irradiation to 1 x 10^{14} e/cm^2 is shown in Tables 6-10 through 6-16, and in Figures 6.37 through 6.42; the data following the 1 MeV electron irradiation to 1.1 x 10^{15} e/cm^2 is shown in Tables 6-17 through 6-23, and in Figures 6.43 through 6.48.

Table 6-3. I_{sc}/cm^2 vs. Temperature and Intensity, Pre-Irradiation

Solar Intensity, mW/cm^2

Temp, °C	5.0	15.0	25.0	50.0	100.0	135.3	250.0
-120	1.05	3.13	5.18	10.35	20.43	28.80	51.10
-100	1.03	3.15	5.28	10.38	20.70	28.28	52.00
-80	1.03	3.18	5.40	10.50	21.10	28.43	52.83
-60	1.05	3.25	5.48	10.70	21.23	28.73	53.55
-40	1.10	3.25	5.53	10.83	21.58	28.95	53.90
-20	1.10	3.33	5.55	10.90	21.75	29.20	54.48
0	1.15	3.30	5.58	11.00	21.68	29.05	54.15
20	1.15	3.43	5.63	11.10	21.93	29.60	54.83
40	1.18	3.45	5.75	11.43	22.48	30.58	55.83
60	1.15	3.50	5.83	11.55	22.93	31.08	56.90
80	1.15	3.58	5.95	11.75	23.43	31.63	57.95
100	1.20	3.65	6.10	12.00	23.85	32.33	58.95
120	1.23	3.70	6.20	12.18	24.15	32.90	59.98
140	1.25	3.75	6.28	12.35	24.55	33.45	61.08

Table 6-4. V_{oc} vs. Temperature and Intensity, Pre-Irradiation

Solar Intensity, mW/cm^2

Temp, °C	5.0	15.0	25.0	50.0	100.0	135.3	250.0
-120	1066.9	1144.9	1174.4	1213.3	1249.3	1259.6	1287.9
-100	1042.1	1116.1	1146.5	1183.5	1218.7	1227.9	1254.6
-80	1013.5	1086.2	1116.1	1152.5	1186.2	1195.5	1221.7
-60	982.3	1053.2	1082.9	1118.7	1151.4	1161.8	1186.3
-40	950.4	1017.6	1047.2	1082.8	1115.2	1125.8	1150.8
-20	912.1	980.0	1009.5	1045.2	1077.8	1089.3	1114.2
0	874.5	939.7	970.3	1006.6	1039.0	1050.8	1076.3
20	832.5	900.6	930.2	966.7	1000.2	1013.0	1038.6
40	791.0	858.8	889.8	928.0	961.7	975.9	1000.5
60	742.9	815.7	847.8	886.7	921.5	935.7	962.0
80	696.5	773.2	805.3	845.2	881.5	895.9	922.9
100	650.3	728.6	762.3	803.4	840.4	855.9	882.8
120	602.2	683.6	718.6	760.2	798.6	814.6	842.7
140	552.5	637.9	673.4	716.9	756.6	772.6	801.4

Table 6-5.　　I_{mp}/cm^2 vs. Temperature and Intensity, Pre-Irradiation

Solar Intensity, mW/cm^2

Temp, °C	5.0	15.0	25.0	50.0	100.0	135.3	250.0
-120	0.93	2.88	4.78	9.58	19.08	25.60	48.28
-100	0.93	2.90	4.88	9.65	19.40	26.08	48.90
-80	0.93	2.90	4.98	9.83	19.85	26.45	50.33
-60	0.93	3.00	5.05	9.93	20.03	26.88	50.78
-40	0.98	3.03	5.08	10.05	20.28	26.98	50.70
-20	0.98	3.00	5.08	10.20	20.40	27.03	51.48
0	1.00	3.00	5.08	10.23	20.33	27.25	50.75
20	1.00	3.13	5.18	10.20	20.35	27.53	51.35
40	1.03	3.15	5.28	10.55	20.78	28.03	51.93
60	1.00	3.15	5.25	10.60	21.00	28.68	52.75
80	1.00	3.18	5.38	10.70	21.48	28.98	53.28
100	1.03	3.18	5.43	10.75	21.63	29.28	53.53
120	1.03	3.23	5.43	10.83	21.70	29.55	54.18
140	1.03	3.18	5.50	10.80	21.75	29.40	55.03

Table 6-6.　　V_{mp} vs. Temperature and Intensity, Pre-Irradiation

Solar Intensity, mW/cm^2

Temp, °C	5.0	15.0	25.0	50.0	100.0	135.3	250.0
-120	878.7	958.6	998.2	1046.7	1091.1	1110.0	1129.6
-100	855.5	940.0	975.0	1019.0	1063.6	1075.8	1107.4
-80	832.3	921.5	947.6	985.5	1024.4	1040.9	1062.4
-60	820.7	883.5	917.1	960.9	991.7	1000.5	1033.5
-40	782.5	847.4	886.9	924.0	959.2	971.5	1003.2
-20	751.8	828.1	859.2	884.8	921.0	940.0	960.7
0	715.0	785.5	817.1	843.5	882.2	898.5	927.9
20	678.7	733.8	770.3	813.0	846.3	856.3	887.7
40	632.8	694.0	730.4	766.2	807.3	820.7	850.4
60	587.7	654.8	696.7	726.3	770.2	779.2	809.9
80	536.5	619.8	646.1	684.4	723.9	739.4	773.7
100	497.8	577.8	607.6	647.7	686.5	706.5	735.5
120	450.0	527.0	567.7	604.1	646.3	665.8	695.9
140	407.7	491.6	515.5	566.8	607.4	631.4	650.0

Table 6-7. P_{max}/cm^2 vs. Temperature and Intensity, Pre-Irradiation

Solar Intensity, mW/cm^2

Temp, °C	5.0	15.0	25.0	50.0	100.0	135.3	250.0
-120	0.81	2.76	4.77	10.04	20.80	28.41	54.54
-100	0.79	2.72	4.75	9.85	20.64	28.06	54.16
-80	0.77	2.68	4.71	9.68	20.35	27.54	53.48
-60	0.76	2.65	4.63	9.54	19.85	26.89	52.48
-40	0.76	2.57	4.50	9.29	19.45	26.22	50.88
-20	0.73	2.49	4.36	9.01	18.77	25.41	49.44
0	0.72	2.36	4.15	8.63	17.94	24.49	47.08
20	0.68	2.30	3.98	8.29	17.23	23.56	45.58
40	0.65	2.19	3.85	8.09	16.77	23.00	44.15
60	0.59	2.06	3.66	7.70	16.18	22.35	42.72
80	0.55	1.96	3.48	7.32	15.56	21.41	41.23
100	0.51	1.83	3.29	6.96	14.85	20.69	39.36
120	0.46	1.70	3.07	6.55	14.02	19.68	37.70
140	0.41	1.57	2.83	6.12	13.21	18.58	35.78

Table 6-8.   Fill Factor vs. Temperature and Intensity, Pre-Irradiation

Solar Intensity, mW/cm^2

Temp, °C	5.0	15.0	25.0	50.0	100.0	135.3	250.0
-120	0.731	0.771	0.784	0.798	0.815	0.809	0.829
-100	0.733	0.772	0.783	0.801	0.817	0.808	0.830
-80	0.732	0.773	0.784	0.800	0.813	0.810	0.828
-60	0.732	0.772	0.783	0.796	0.812	0.806	0.826
-40	0.727	0.774	0.779	0.793	0.808	0.805	0.820
-20	0.724	0.768	0.777	0.791	0.801	0.799	0.815
0	0.717	0.760	0.766	0.779	0.797	0.790	0.808
20	0.708	0.748	0.760	0.773	0.785	0.786	0.800
40	0.703	0.739	0.751	0.763	0.776	0.777	0.790
60	0.688	0.721	0.741	0.751	0.766	0.768	0.780
80	0.671	0.708	0.726	0.736	0.753	0.755	0.771
100	0.650	0.690	0.709	0.722	0.741	0.748	0.756
120	0.627	0.671	0.689	0.707	0.727	0.734	0.746
140	0.600	0.651	0.669	0.691	0.711	0.718	0.731

Table 6-9.   Efficiency vs. Temperature and Intensity, Pre-Irradiation

Temp, °C	Solar Intensity, mW/cm²						
	5.0	15.0	25.0	50.0	100.0	135.3	250.0
-120	16.23	18.39	19.09	20.07	20.80	21.00	21.82
-100	15.80	18.13	18.98	19.69	20.64	20.74	21.66
-80	15.30	17.84	18.85	19.36	20.35	20.36	21.39
-60	15.08	17.63	18.51	19.08	19.84	19.87	20.99
-40	15.11	17.11	18.00	18.59	19.45	19.38	20.35
-20	14.49	16.61	17.43	18.03	18.77	18.78	19.77
0	14.38	15.73	16.59	17.26	17.94	18.10	18.83
20	13.59	15.33	15.93	16.59	17.23	17.41	18.23
40	13.03	14.56	15.41	16.18	16.77	17.00	17.66
60	11.80	13.75	14.63	15.40	16.18	16.51	17.09
80	10.88	13.08	13.91	14.64	15.56	15.83	16.49
100	10.19	12.21	13.17	13.92	14.85	15.29	15.75
120	9.18	11.30	12.29	13.10	14.02	14.54	15.08
140	8.27	10.43	11.33	12.24	13.21	13.73	14.31

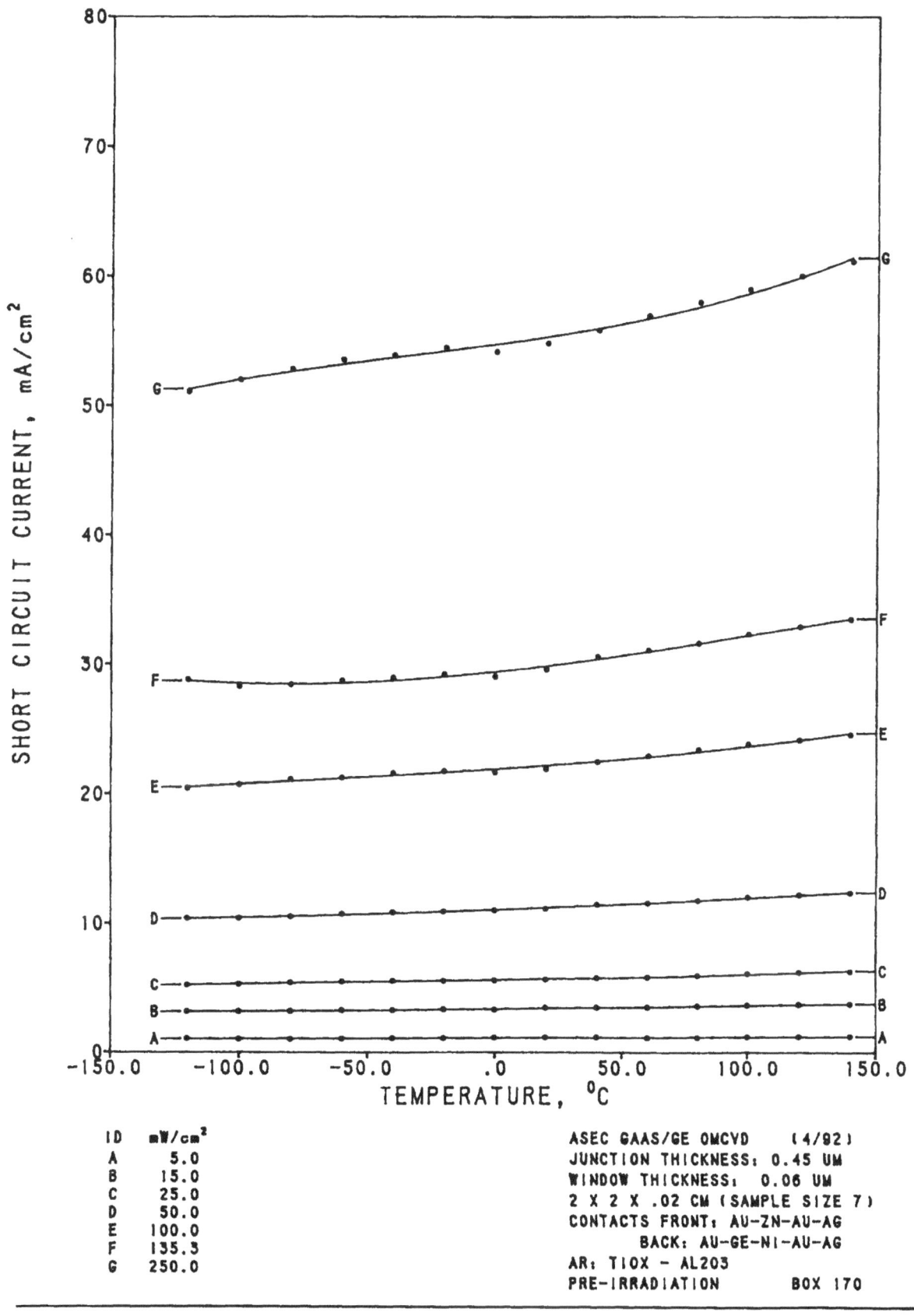

Figure 6.31.   Average I_{sc}/cm^2 as a Function of Temperature, Pre-Irradiation

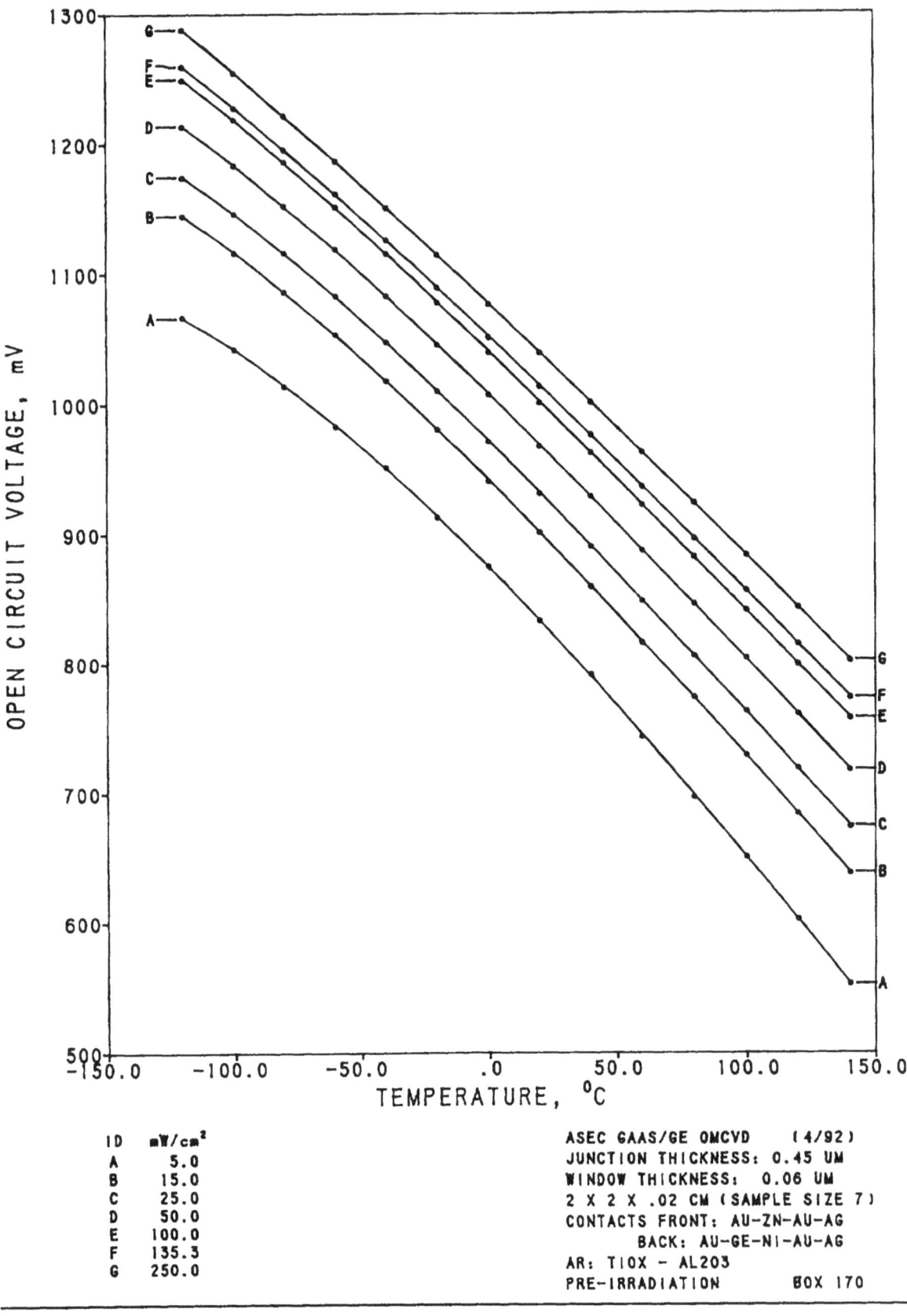

ID   mW/cm²
A   5.0
B   15.0
C   25.0
D   50.0
E   100.0
F   135.3
G   250.0

ASEC GAAS/GE OMCVD   (4/92)
JUNCTION THICKNESS: 0.45 UM
WINDOW THICKNESS: 0.06 UM
2 X 2 X .02 CM (SAMPLE SIZE 7)
CONTACTS FRONT: AU-ZN-AU-AG
           BACK: AU-GE-NI-AU-AG
AR: TIOX - AL2O3
PRE-IRRADIATION   BOX 170

Figure 6.32.   Average V_{oc} as a Function of Temperature, Pre-Irradiation

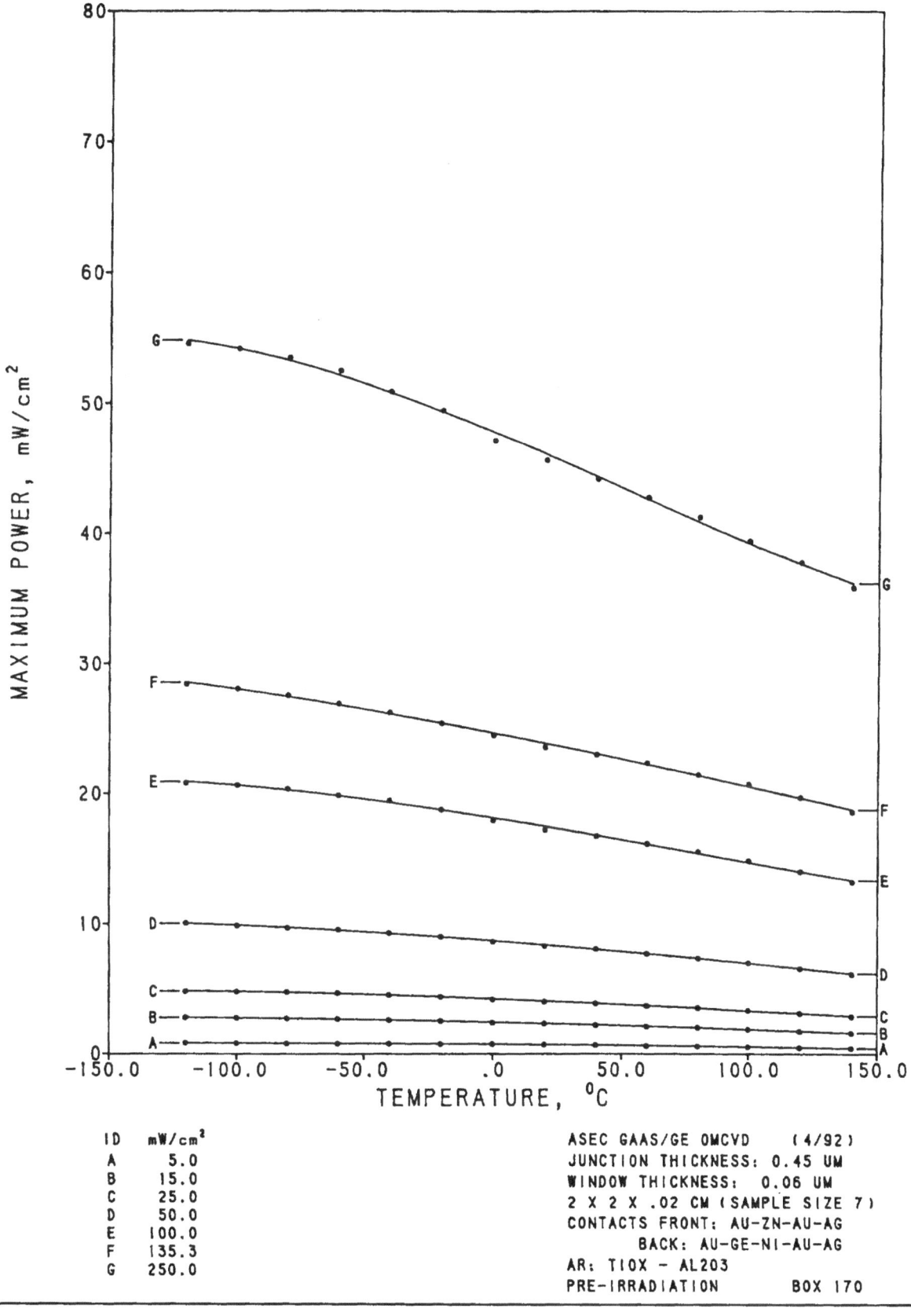

ID	mW/cm^2
A	5.0
B	15.0
C	25.0
D	50.0
E	100.0
F	135.3
G	250.0

ASEC GAAS/GE OMCVD   (4/92)
JUNCTION THICKNESS: 0.45 UM
WINDOW THICKNESS: 0.06 UM
2 X 2 X .02 CM (SAMPLE SIZE 7)
CONTACTS FRONT: AU-ZN-AU-AG
            BACK: AU-GE-NI-AU-AG
AR: TIOX - AL203
PRE-IRRADIATION   BOX 170

Figure 6.33.   Average P_{max}/cm^2 as a Function of Temperature, Pre-Irradiation

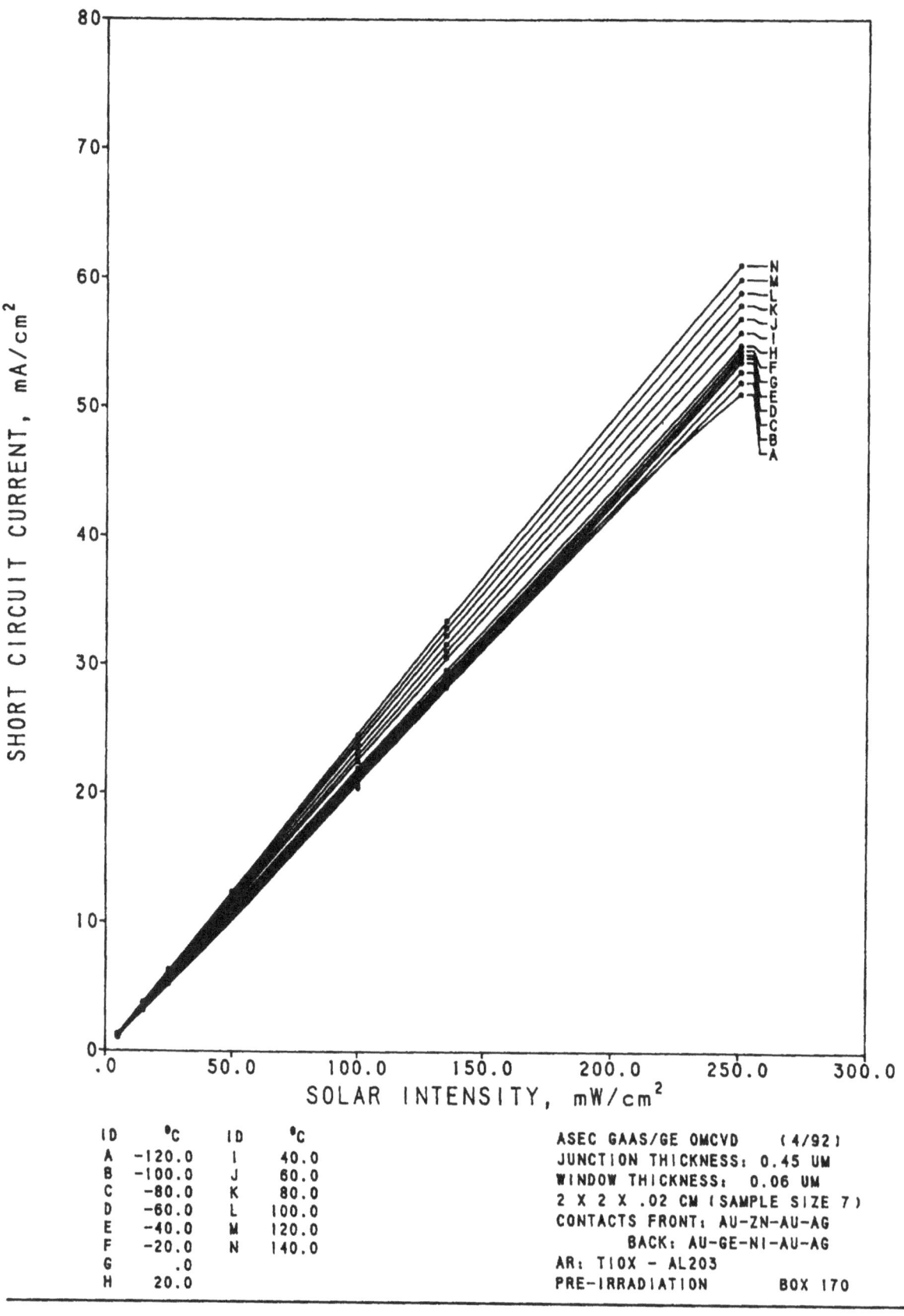

Figure 6.34.   Average I_{sc}/cm^2 as a Function of Intensity, Pre-Irradiation

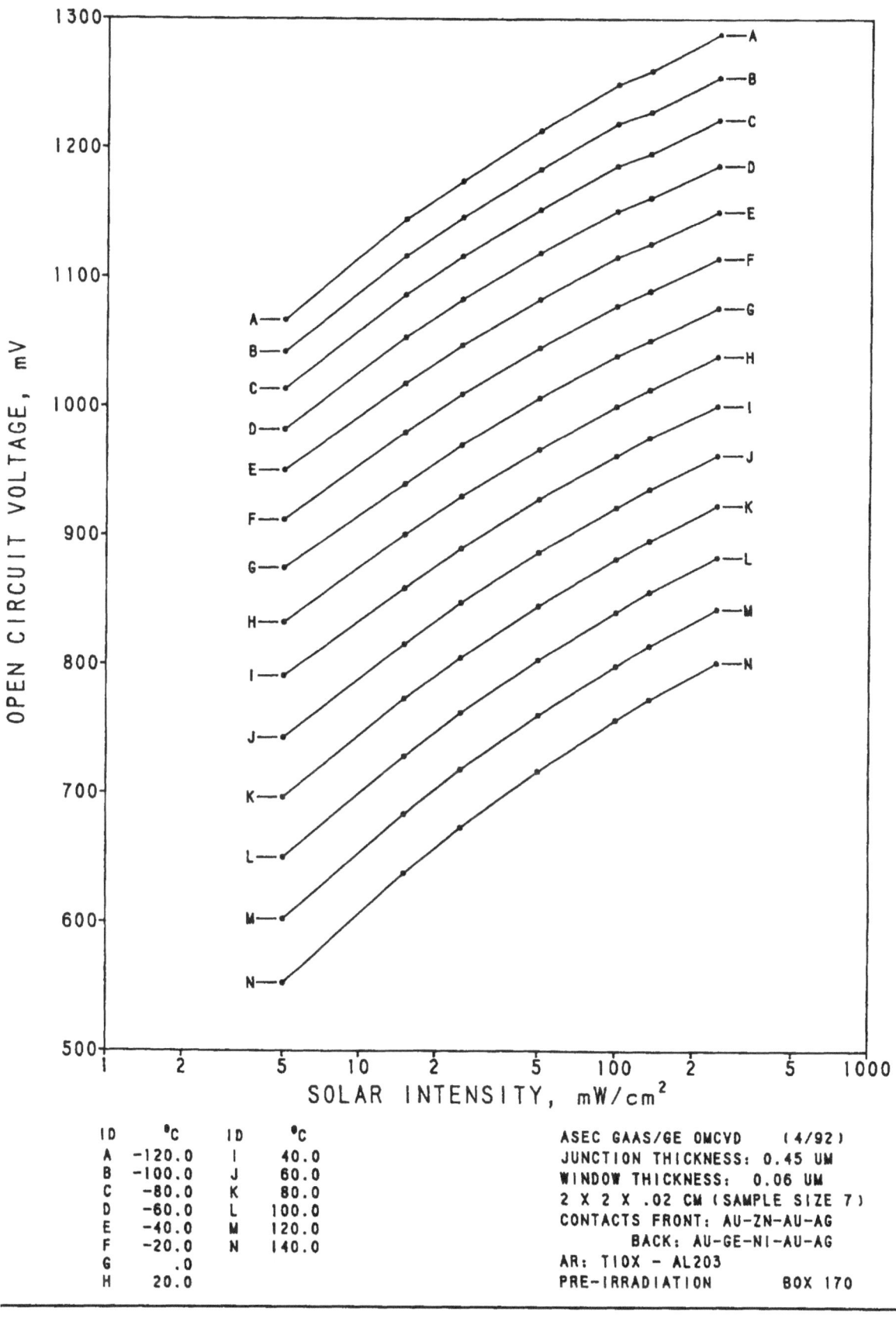

Figure 6.35.   Average V_{oc} as a Function of Intensity, Pre-Irradiation

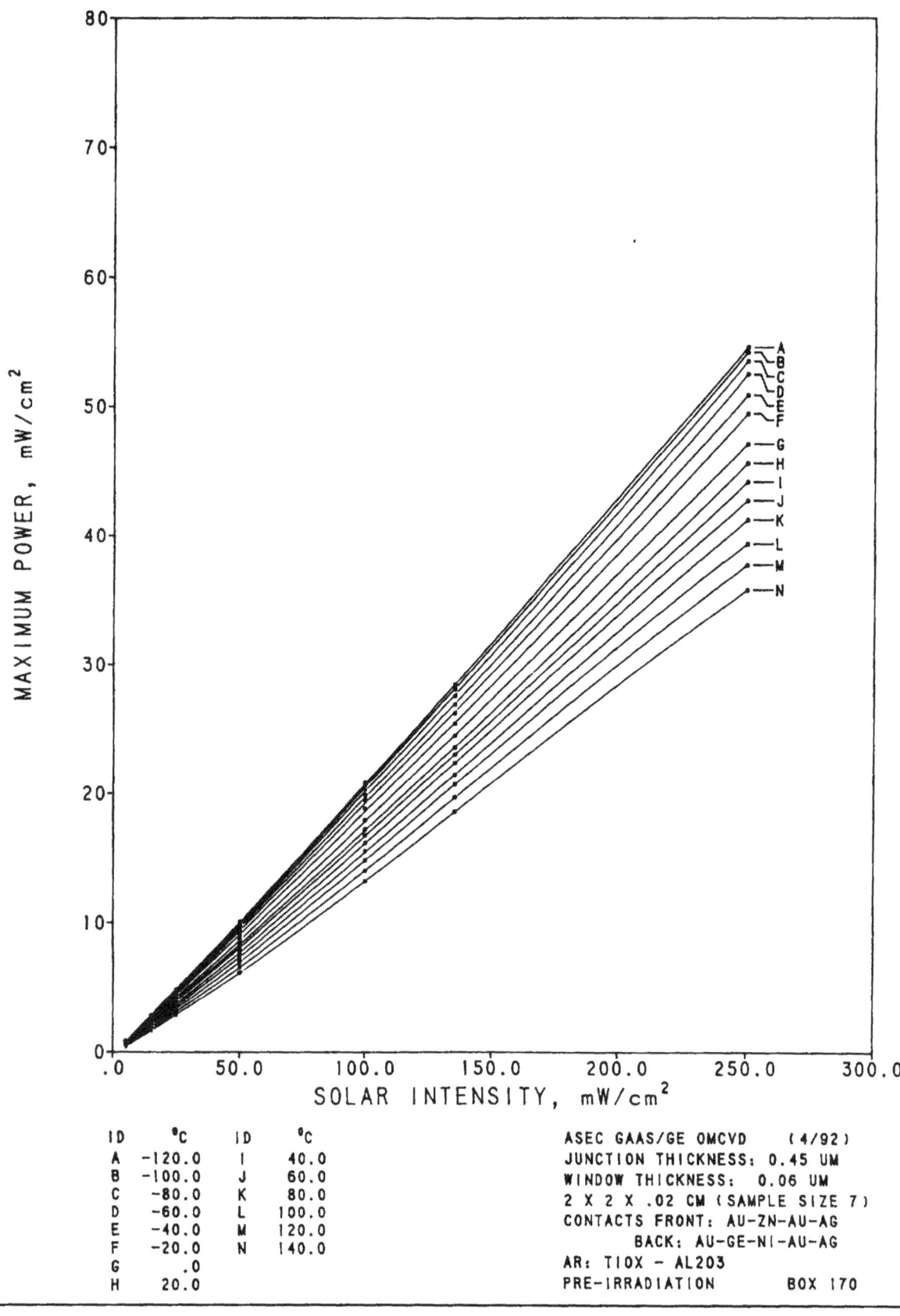

ID	°C	ID	°C
A	-120.0	I	40.0
B	-100.0	J	60.0
C	-80.0	K	80.0
D	-60.0	L	100.0
E	-40.0	M	120.0
F	-20.0	N	140.0
G	.0		
H	20.0		

ASEC GAAS/GE OMCVD (4/92)
JUNCTION THICKNESS: 0.45 UM
WINDOW THICKNESS: 0.06 UM
2 X 2 X .02 CM (SAMPLE SIZE 7)
CONTACTS FRONT: AU-ZN-AU-AG
 BACK: AU-GE-NI-AU-AG
AR: TIOX - AL203
PRE-IRRADIATION BOX 170

Figure 6.36.   Average P_{max}/cm^2 as a Function of Intensity, Pre-Irradiation

6-44

Table 6-10. I_{sc}/cm^2 vs. Temperature and Intensity, after 1×10^{14} e/cm^2

Solar Intensity, mW/cm^2

Temp, °C	5.0	15.0	25.0	50.0	100.0	135.3	250.0
-120	0.93	2.83	4.80	9.35	18.43	25.20	46.70
-100	0.95	2.93	4.80	9.60	18.93	25.45	47.83
-80	0.98	2.95	4.93	9.93	19.13	26.38	48.85
-60	1.00	3.00	5.00	9.95	19.55	26.85	50.08
-40	1.03	3.05	5.08	10.13	20.23	27.25	50.53
-20	1.05	3.08	5.13	10.25	20.53	27.60	51.10
0	1.05	3.18	5.20	10.35	20.53	27.83	51.35
20	1.05	3.18	5.28	10.28	20.68	27.98	51.45
40	1.08	3.20	5.33	10.55	21.05	28.60	52.48
60	1.10	3.25	5.48	10.73	21.58	29.38	53.85
80	1.13	3.30	5.58	11.00	22.03	29.93	54.63
100	1.13	3.43	5.68	11.30	22.43	30.73	55.50
120	1.18	3.48	5.80	11.50	22.90	31.48	56.40
140	1.20	3.55	5.85	11.70	23.43	32.03	57.68

Table 6-11. V_{oc} vs. Temperature and Intensity, after 1×10^{14} e/cm^2

Solar Intensity, mW/cm^2

Temp, °C	5.0	15.0	25.0	50.0	100.0	135.3	250.0
-120	1054.8	1132.8	1161.8	1195.2	1225.3	1239.0	1261.4
-100	1031.6	1104.5	1130.3	1162.4	1191.9	1204.7	1227.5
-80	1004.5	1072.0	1097.5	1128.8	1157.1	1169.4	1193.0
-60	974.8	1038.8	1062.7	1092.7	1121.1	1133.4	1156.7
-40	940.8	1004.1	1026.5	1056.4	1085.0	1096.9	1119.9
-20	904.1	964.3	989.0	1019.5	1047.7	1059.6	1083.1
0	864.0	925.8	950.0	981.0	1010.9	1021.6	1045.6
20	821.4	884.8	910.3	941.6	971.8	984.7	1007.8
40	778.3	843.0	869.3	902.6	932.5	945.4	969.7
60	733.9	800.0	829.1	862.7	893.4	906.0	931.2
80	688.6	755.9	785.3	820.6	852.9	865.6	892.1
100	640.1	711.8	742.0	778.9	811.8	825.6	851.8
120	593.0	665.5	697.2	735.7	770.3	783.9	811.8
140	543.7	619.6	651.6	691.9	727.4	742.4	770.4

Table 6-12. I_{mp}/cm^2 vs. Temperature and Intensity, after 1×10^{14} e/cm^2

Solar Intensity, mW/cm^2

Temp, °C	5.0	15.0	25.0	50.0	100.0	135.3	250.0
-120	0.80	2.55	4.40	8.55	17.08	23.50	43.65
-100	0.83	2.68	4.40	8.83	17.60	23.88	45.10
-80	0.85	2.68	4.53	9.20	17.90	24.68	46.23
-60	0.88	2.73	4.58	9.25	18.35	25.25	47.53
-40	0.90	2.78	4.70	9.45	18.85	25.45	47.75
-20	0.90	2.80	4.73	9.48	19.25	25.80	48.23
0	0.93	2.88	4.75	9.60	19.18	26.03	48.20
20	0.90	2.88	4.80	9.45	19.28	26.00	48.40
40	0.93	2.88	4.80	9.68	19.48	26.50	49.35
60	0.93	2.90	4.95	9.83	19.88	29.50	49.90
80	0.98	2.90	4.98	9.93	20.13	27.60	50.33
100	0.95	2.98	5.05	10.08	20.33	27.98	50.98
120	0.98	3.00	5.08	10.30	20.53	28.53	51.20
140	0.98	3.03	5.08	10.30	20.95	28.60	51.88

Table 6-13. V_{mp} vs. Temperature and Intensity, after 1×10^{14} e/cm^2

Solar Intensity, mW/cm^2

Temp, °C	5.0	15.0	25.0	50.0	100.0	135.3	250.0
-120	870.8	948.9	981.5	1037.7	1077.6	1092.1	1122.5
-100	850.7	927.0	961.1	1010.0	1049.2	1057.3	1086.2
-80	824.1	908.1	938.4	979.4	1014.3	1030.1	1051.3
-60	805.6	878.3	911.6	946.6	977.6	993.4	1013.7
-40	779.3	849.9	874.6	908.7	950.2	964.0	981.0
-20	745.2	813.1	842.8	880.1	907.7	924.2	944.4
0	706.7	777.0	806.2	838.7	871.5	882.9	910.3
20	667.3	736.9	764.4	800.9	831.5	847.6	866.3
40	625.7	696.5	725.9	759.8	793.2	805.6	824.2
60	582.3	652.6	681.4	717.4	752.4	769.2	792.4
80	534.8	612.7	641.3	680.0	712.7	722.0	750.7
100	491.9	572.3	594.1	640.4	670.8	685.8	708.4
120	447.0	520.5	552.9	587.1	632.4	642.0	670.2
140	401.0	474.9	508.4	549.6	585.2	604.5	629.1

Table 6-14. P_{max}/cm^2 vs. Temperature and Intensity, after 1×10^{14} e/cm^2

Solar Intensity, mW/cm^2

Temp, °C	5.0	15.0	25.0	50.0	100.0	135.3	250.0
-120	0.71	2.43	4.32	8.88	18.42	25.67	49.00
-100	0.71	2.48	4.24	8.92	18.46	25.24	48.99
-80	0.71	2.44	4.24	9.02	18.16	25.43	48.59
-60	0.70	2.41	4.19	8.77	17.95	25.09	48.16
-40	0.70	2.37	4.10	8.58	17.91	24.54	46.86
-20	0.68	2.28	3.97	8.33	17.47	23.86	45.56
0	0.65	2.24	3.83	8.05	16.73	22.99	43.90
20	0.61	2.11	3.67	7.58	16.03	22.04	41.94
40	0.58	2.00	3.50	7.35	15.45	21.36	40.68
60	0.55	1.90	3.37	7.04	14.96	20.78	39.55
80	0.52	1.78	3.19	6.74	14.35	19.94	37.78
100	0.47	1.70	3.00	6.45	13.63	19.19	36.12
120	0.44	1.57	2.81	6.05	12.99	18.32	34.32
140	0.39	1.44	2.58	5.66	12.27	17.29	32.64

Table 6-15.   Fill Factor vs. Temperature and Intensity, after 1×10^{14} e/cm^2

Solar Intensity, mW/cm^2

Temp, °C	5.0	15.0	25.0	50.0	100.0	135.3	250.0
-120	0.720	0.759	0.774	0.795	0.816	0.822	0.832
-100	0.721	0.765	0.780	0.800	0.818	0.823	0.834
-80	0.723	0.769	0.786	0.804	0.820	0.824	0.833
-60	0.725	0.770	0.788	0.805	0.819	0.824	0.831
-40	0.724	0.771	0.786	0.803	0.816	0.821	0.828
-20	0.718	0.767	0.782	0.798	0.812	0.816	0.823
0	0.715	0.762	0.776	0.792	0.805	0.809	0.817
20	0.706	0.754	0.765	0.783	0.798	0.799	0.809
40	0.696	0.742	0.754	0.772	0.786	0.790	0.799
60	0.683	0.729	0.741	0.760	0.776	0.781	0.788
80	0.670	0.716	0.729	0.747	0.764	0.769	0.775
100	0.649	0.695	0.713	0.733	0.749	0.756	0.764
120	0.628	0.677	0.694	0.715	0.736	0.742	0.749
140	0.600	0.656	0.676	0.699	0.720	0.727	0.734

Table 6-16.   Efficiency vs. Temperature and Intensity, after 1 x 10^{14} e/cm^2

Solar Intensity, mW/cm^2

Temp, °C	5.0	15.0	25.0	50.0	100.0	135.3	250.0
-120	14.14	16.20	17.27	17.76	18.42	18.97	19.60
-100	14.26	16.54	16.96	17.84	18.46	18.66	19.60
-80	14.20	16.26	16.96	18.03	18.16	18.79	19.44
-60	13.99	16.03	16.74	17.53	17.95	18.54	19.27
-40	13.89	15.76	16.41	17.17	17.91	18.14	18.74
-20	13.56	15.19	15.89	16.66	17.47	17.63	18.22
0	12.96	14.93	15.33	16.09	16.73	16.99	17.56
20	12.13	14.08	14.66	15.15	16.03	16.29	16.77
40	11.53	13.33	13.97	14.69	15.44	15.79	16.27
60	10.91	12.63	13.48	14.08	14.96	15.36	15.82
80	10.31	11.87	12.75	13.48	14.35	14.74	15.11
100	9.41	11.30	12.00	12.90	13.63	14.18	14.45
120	8.77	10.43	11.22	12.11	12.99	13.54	13.73
140	7.79	9.59	10.32	11.32	12.27	12.78	13.06

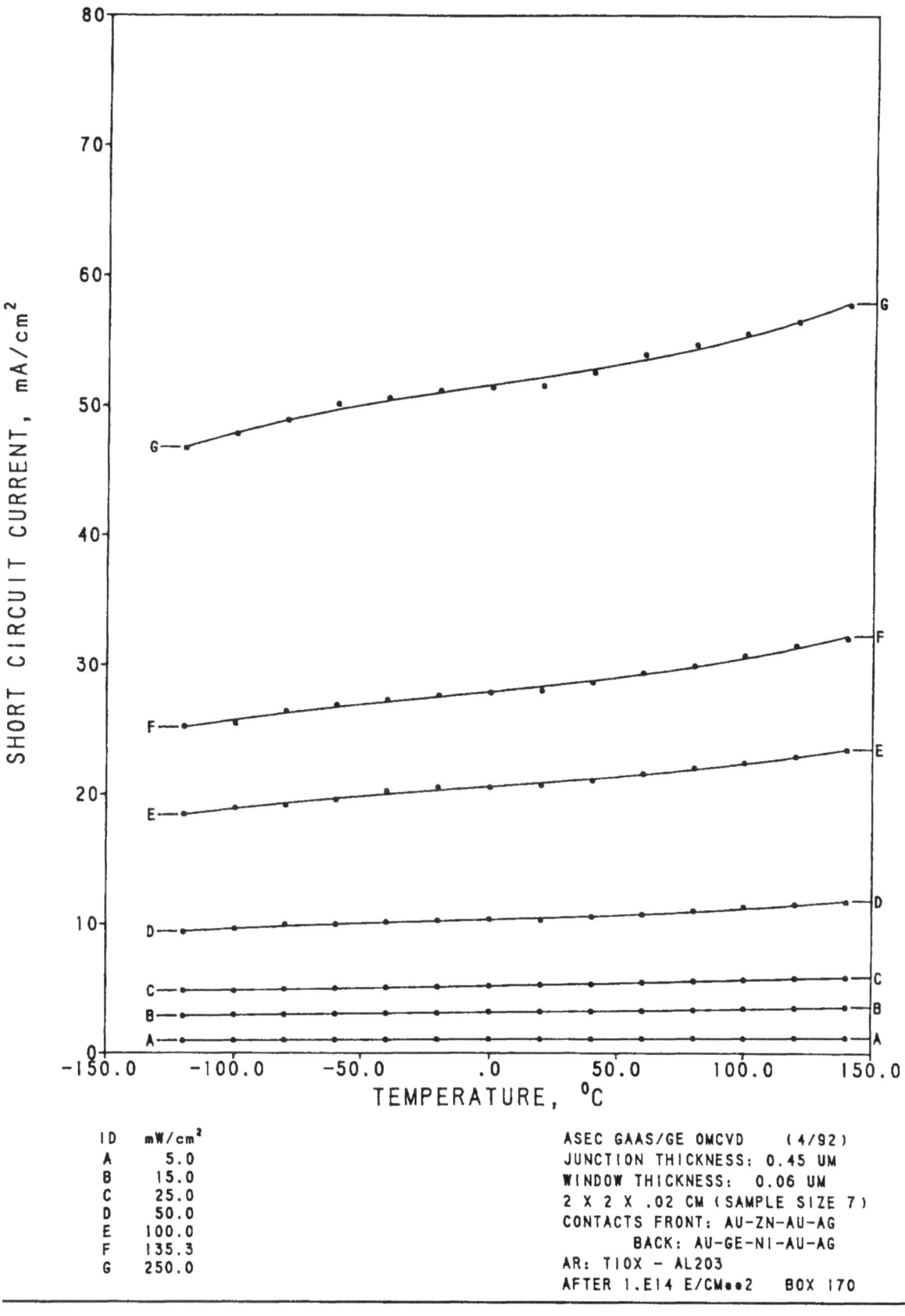

ID	mW/cm²
A	5.0
B	15.0
C	25.0
D	50.0
E	100.0
F	135.3
G	250.0

ASEC GAAS/GE OMCVD   (4/92)
JUNCTION THICKNESS: 0.45 UM
WINDOW THICKNESS: 0.06 UM
2 X 2 X .02 CM (SAMPLE SIZE 7)
CONTACTS FRONT: AU-ZN-AU-AG
         BACK: AU-GE-NI-AU-AG
AR: TIOX - AL2O3
AFTER 1.E14 E/CM**2   BOX 170

Figure 6.37.   Average I_{sc}/cm² as a Function of Temperature, after 1×10^{14} e/cm²

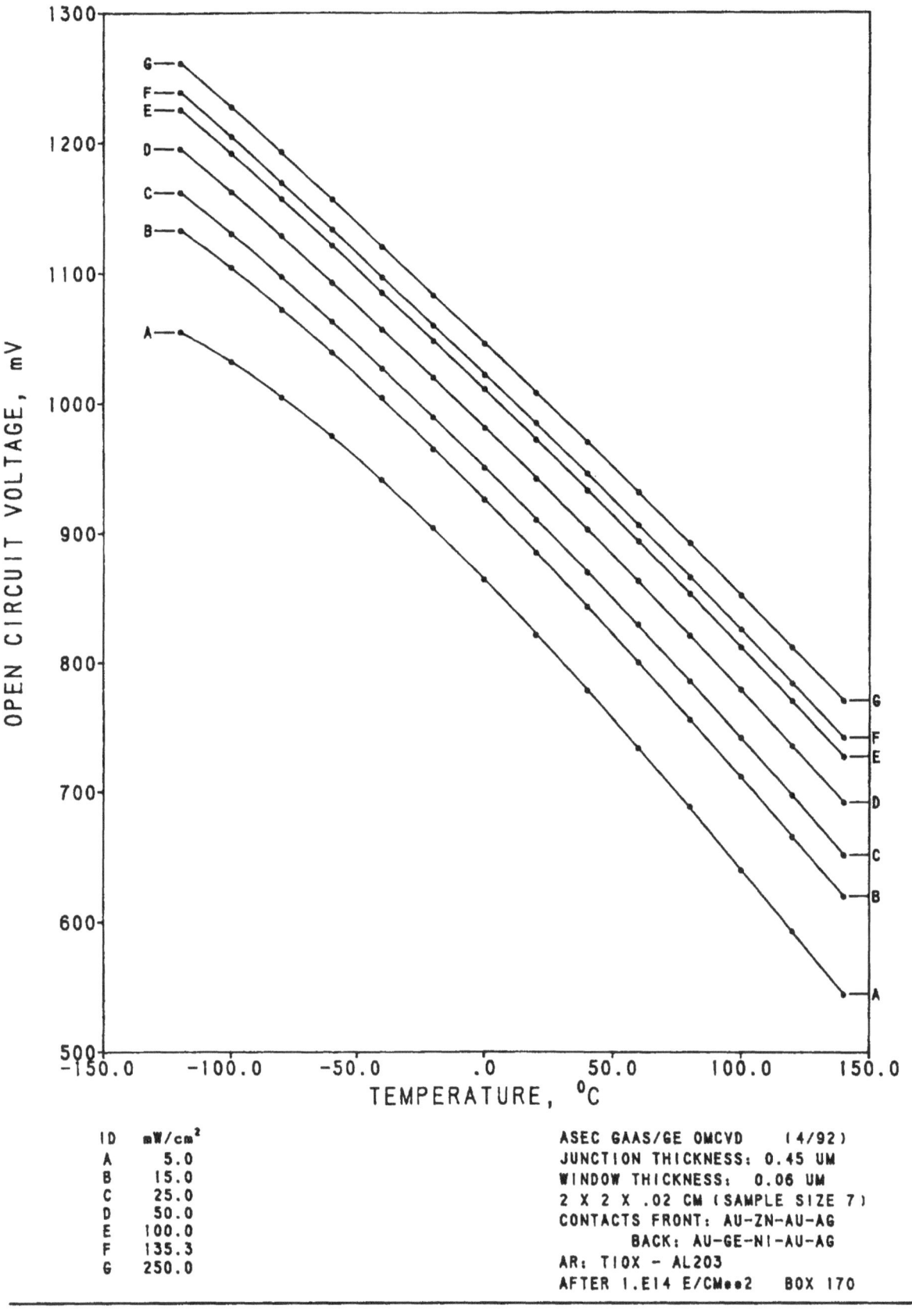

Figure 6.38. Average V_{oc} as a Function of Temperature, after 1×10^{14} e/cm^2

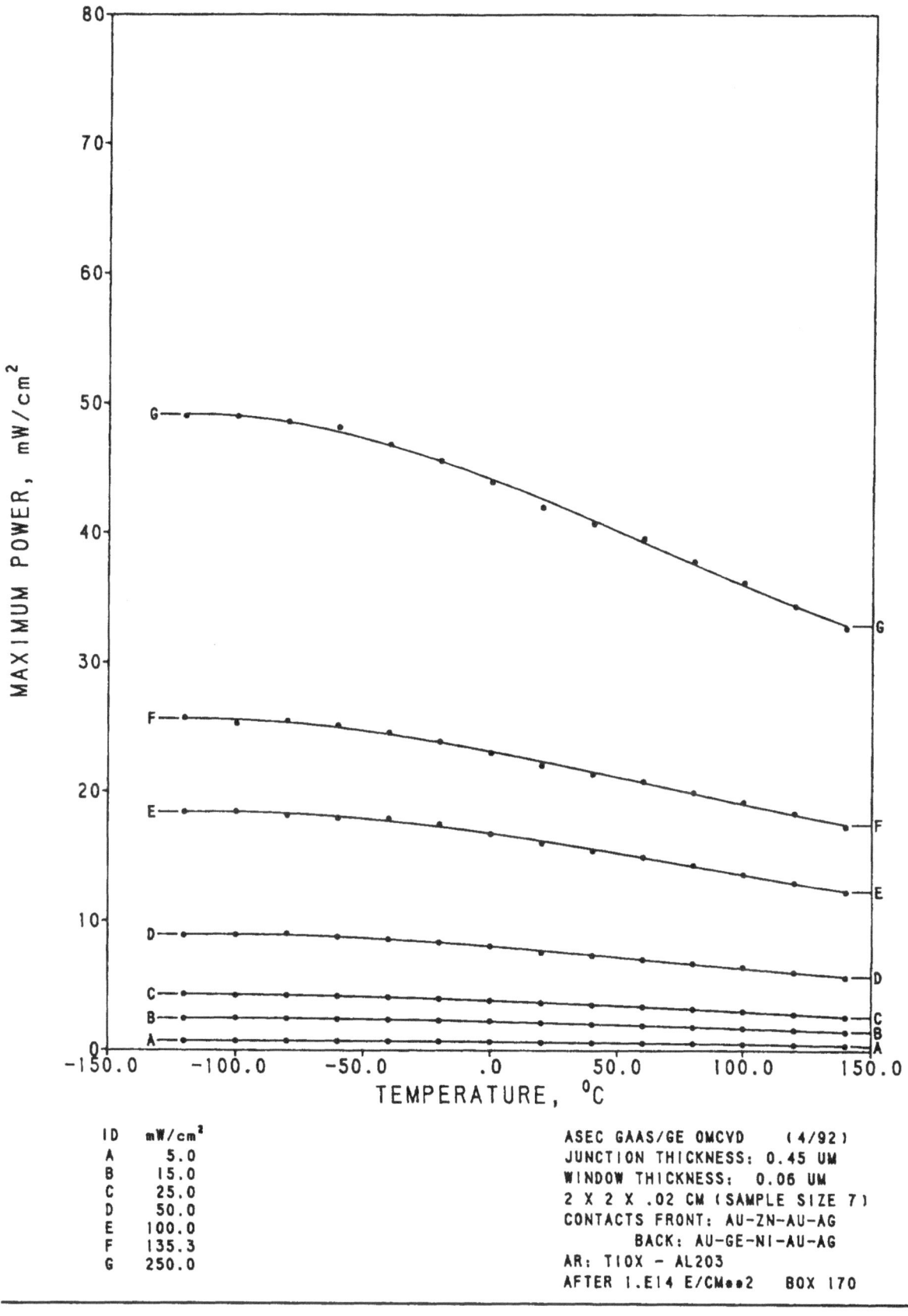

Figure 6.39. Average P_{max}/cm^2 as a Function of Temperature, after 1×10^{14} e/cm^2

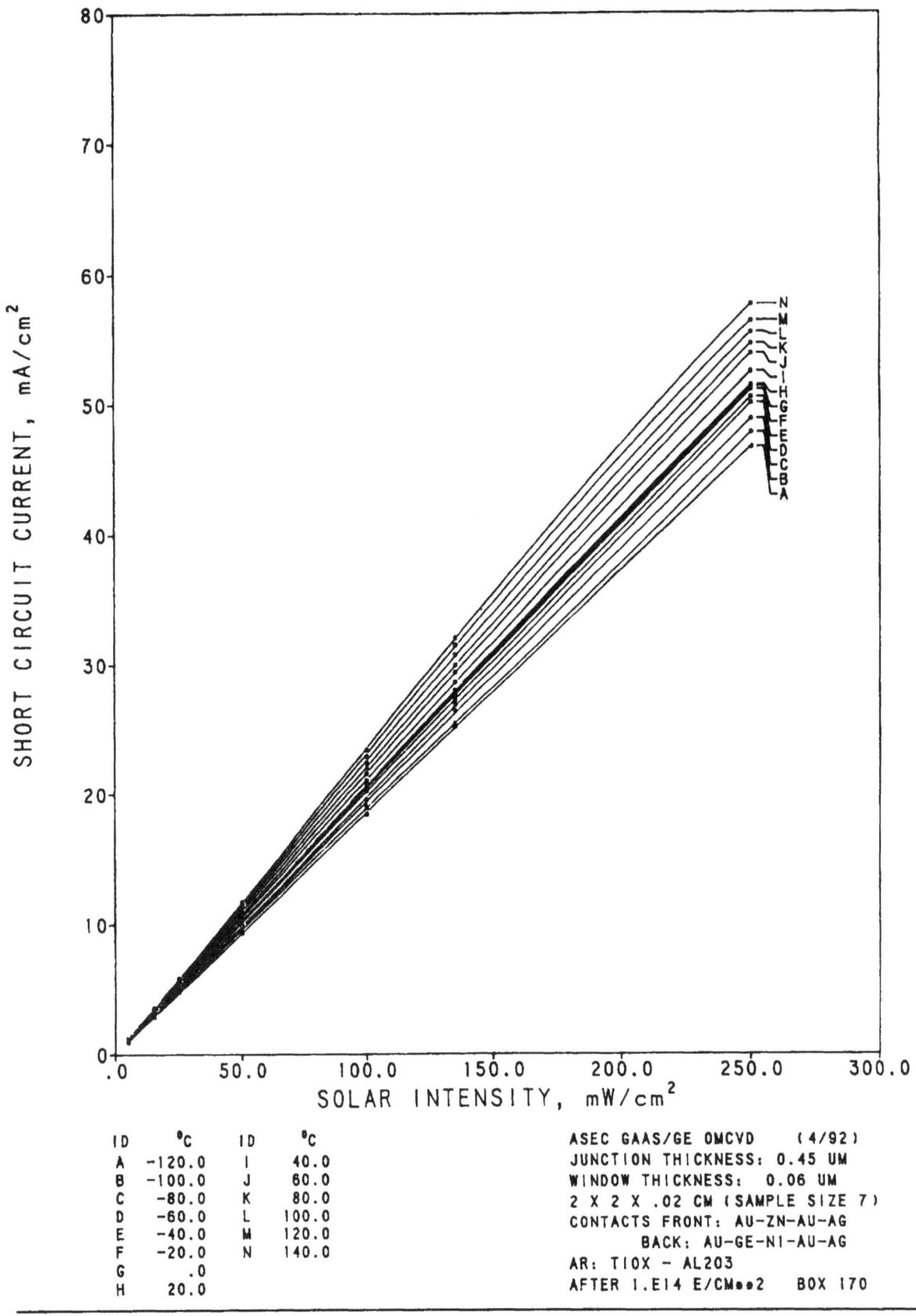

ID	°C	ID	°C
A	-120.0	I	40.0
B	-100.0	J	60.0
C	-80.0	K	80.0
D	-60.0	L	100.0
E	-40.0	M	120.0
F	-20.0	N	140.0
G	.0		
H	20.0		

ASEC GAAS/GE OMCVD   (4/92)
JUNCTION THICKNESS: 0.45 UM
WINDOW THICKNESS: 0.06 UM
2 X 2 X .02 CM (SAMPLE SIZE 7)
CONTACTS FRONT: AU-ZN-AU-AG
           BACK: AU-GE-NI-AU-AG
AR: TIOX - AL203
AFTER 1.E14 E/CM**2   BOX 170

Figure 6.40.   Average I_{sc}/cm^2 as a Function of Intensity, after 1×10^{14} e/cm^2

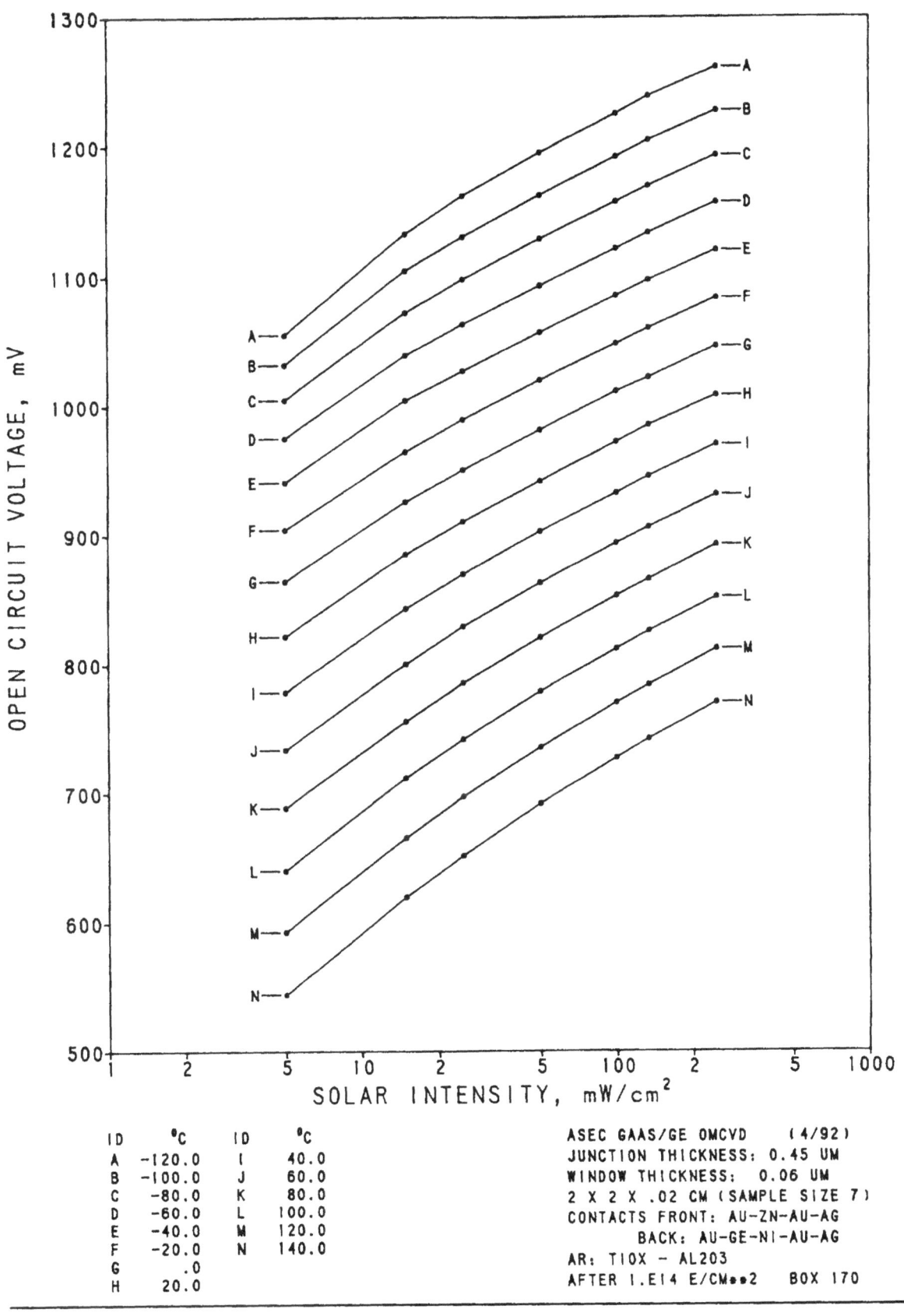

Figure 6.41.   Average V_{oc} as a Function of Intensity, after 1×10^{14} e/cm²

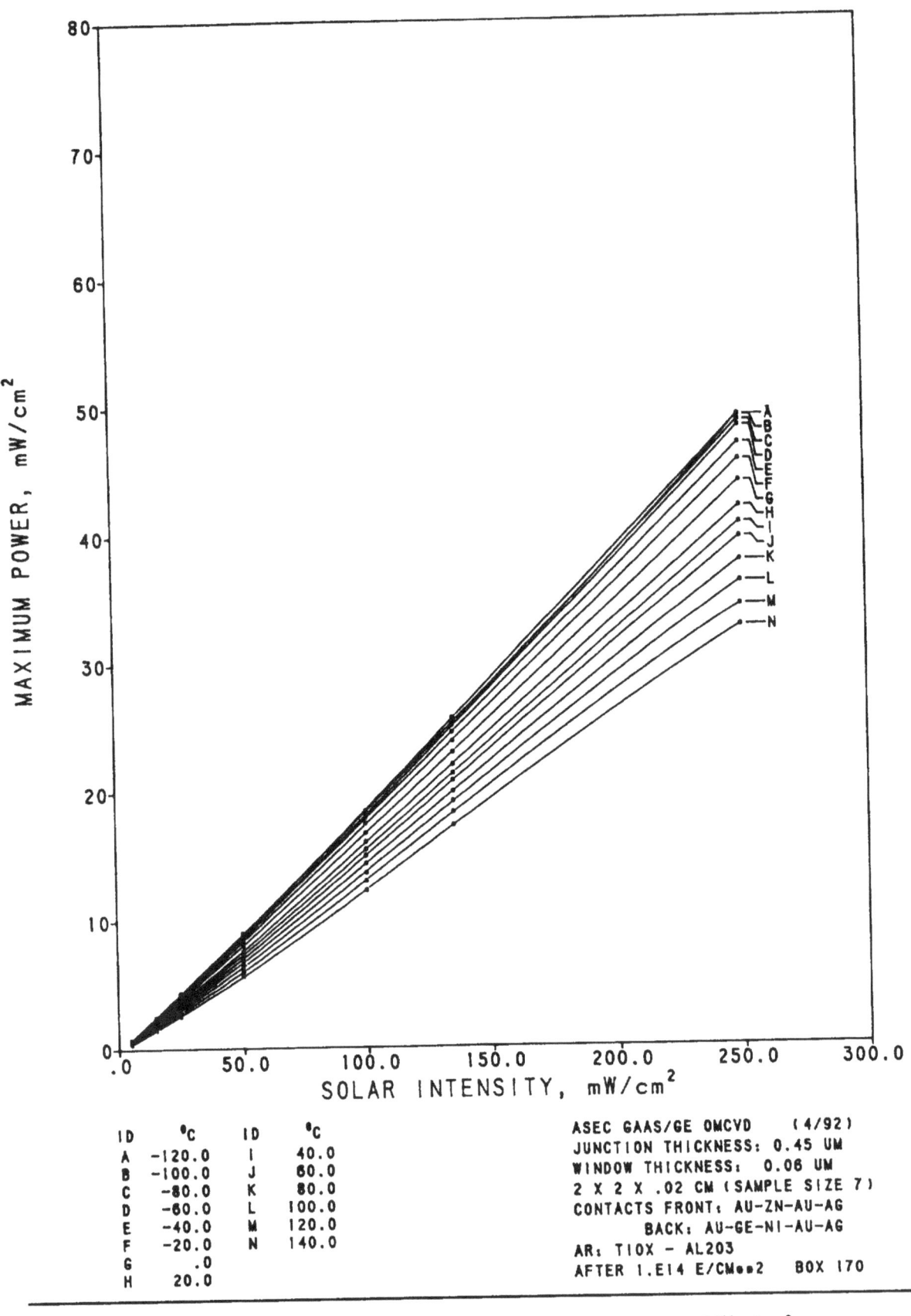

ASEC GAAS/GE OMCVD   (4/92)
JUNCTION THICKNESS: 0.45 UM
WINDOW THICKNESS: 0.06 UM
2 X 2 X .02 CM (SAMPLE SIZE 7)
CONTACTS FRONT: AU-ZN-AU-AG
BACK: AU-GE-NI-AU-AG
AR: TIOX - AL203
AFTER 1.E14 E/CM**2   BOX 170

ID	°C	ID	°C
A	-120.0	I	40.0
B	-100.0	J	60.0
C	-80.0	K	80.0
D	-60.0	L	100.0
E	-40.0	M	120.0
F	-20.0	N	140.0
G	.0		
H	20.0		

Figure 6.42.   Average P_{max}/cm^2 as a Function of Intensity, after 1×10^{14} e/cm²

Table 6-17. I_{sc}/cm^2 vs. Temperature and Intensity, after 1.1×10^{15} e/cm^2

Solar Intensity, mW/cm^2

Temp, °C	5.0	15.0	25.0	50.0	100.0	135.3	250.0
-120	0.80	2.48	4.15	8.20	16.30	21.60	41.88
-100	0.83	2.48	4.10	8.20	16.70	22.28	42.58
-80	0.83	2.53	4.20	8.28	17.08	22.63	42.93
-60	0.85	2.58	4.25	8.35	17.33	23.20	43.88
-40	0.88	2.63	4.35	8.65	17.68	23.55	44.73
-20	0.88	2.68	4.45	8.75	17.85	23.80	45.75
0	0.90	2.70	4.48	8.78	17.88	23.70	46.23
20	0.90	2.73	4.55	8.80	17.83	24.08	46.30
40	0.93	2.78	4.65	8.93	18.38	24.55	46.93
60	0.93	2.85	4.73	9.18	18.70	25.00	47.93
80	0.95	2.90	4.80	9.33	19.05	25.55	49.08
100	0.98	2.93	4.90	9.50	19.38	26.13	49.88
120	1.00	3.00	5.00	9.70	19.88	26.73	51.40
140	1.03	3.05	5.10	9.90	20.28	27.48	52.20

Table 6-18. V_{oc} vs. Temperature and Intensity, after 1.1×10^{15} e/cm^2

Solar Intensity, mW/cm^2

Temp, °C	5.0	15.0	25.0	50.0	100.0	135.3	250.0
-120	1034.8	1106.4	1129.6	1154.6	1178.6	1187.6	1205.9
-100	1010.2	1072.8	1093.4	1117.4	1140.9	1150.6	1168.6
-80	981.5	1038.4	1058.1	1081.1	1104.2	1113.3	1131.1
-60	950.3	1002.8	1021.7	1044.1	1067.6	1077.0	1094.3
-40	914.3	965.5	984.4	1007.9	1031.1	1039.9	1058.1
-20	876.9	926.2	945.9	970.1	994.7	1003.8	1022.9
0	836.3	886.3	906.2	931.2	956.7	965.6	985.8
20	793.4	845.6	866.4	892.2	918.1	927.8	948.6
40	749.3	803.5	825.1	852.2	879.6	889.6	910.8
60	703.2	759.9	782.7	811.5	840.2	850.1	872.9
80	655.8	715.2	739.9	769.9	799.5	810.0	833.2
100	607.2	669.8	695.7	727.2	757.5	769.6	793.1
120	557.3	623.7	650.3	683.3	715.7	728.2	753.2
140	506.9	576.1	604.3	638.9	672.6	686.8	712.3

Table 6-19. I_{mp}/cm^2 vs. Temperature and Intensity, after 1.1×10^{15} e/cm^2

Solar Intensity, mW/cm^2

Temp, °C	5.0	15.0	25.0	50.0	100.0	135.3	250.0
-120	0.73	2.20	3.75	7.50	15.08	20.10	39.38
-100	0.73	2.20	3.73	7.48	15.50	20.85	40.18
-80	0.73	2.25	3.83	7.65	15.90	21.23	40.65
-60	0.75	2.30	3.88	7.70	16.23	21.80	41.50
-40	0.75	2.35	3.98	8.00	16.55	22.05	42.35
-20	0.78	2.43	4.03	8.10	16.65	22.35	43.13
0	0.78	2.43	4.05	8.10	16.73	22.28	43.60
20	0.80	2.45	4.15	8.15	16.55	22.48	43.65
40	0.80	2.53	4.23	8.25	17.10	22.93	43.85
60	0.80	2.55	4.25	8.40	17.30	23.00	44.70
80	0.80	2.58	4.35	8.45	17.40	23.60	45.40
100	0.80	2.58	4.38	8.58	17.68	23.90	45.98
120	0.83	2.60	4.40	8.60	17.85	24.13	46.93
140	0.80	2.58	4.48	8.78	17.98	24.60	47.35

Table 6-20. V_{mp} vs. Temperature and Intensity, after 1.1×10^{15} e/cm^2

Solar Intensity, mW/cm^2

Temp, °C	5.0	15.0	25.0	50.0	100.0	135.3	250.0
-120	852.6	937.2	963.6	1013.3	1055.8	1064.4	1082.8
-100	837.4	915.0	938.4	993.0	1023.5	1032.7	1048.1
-80	817.5	889.2	922.1	954.4	989.5	998.1	1010.0
-60	799.9	861.0	888.1	922.7	950.6	962.2	976.4
-40	759.4	830.1	855.4	886.8	914.9	927.5	937.7
-20	729.4	791.7	819.6	850.1	880.9	886.2	902.4
0	690.3	751.6	781.2	809.0	835.6	842.2	860.4
20	648.9	712.8	735.6	766.0	799.5	805.7	818.3
40	614.9	664.3	693.2	721.1	752.9	761.8	783.2
60	572.9	620.4	654.2	684.2	712.9	726.6	740.2
80	530.4	577.2	604.4	642.7	673.9	677.5	699.6
100	484.4	533.2	559.5	596.3	627.3	638.4	656.7
120	429.4	493.1	518.9	555.4	588.6	601.5	617.0
140	387.6	446.6	469.2	507.0	547.8	557.4	573.7

Table 6-21. P_{max}/cm^2 vs. Temperature and Intensity, after 1.1 x 10^{15} e/cm^2

Solar Intensity, mW/cm^2

Temp, °C	5.0	15.0	25.0	50.0	100.0	135.3	250.0
-120	0.61	2.06	3.63	7.60	15.91	21.40	42.65
-100	0.61	2.02	3.50	7.43	15.87	21.53	42.12
-80	0.60	2.02	3.53	7.31	15.75	21.19	41.06
-60	0.59	1.99	3.45	7.12	15.43	20.98	40.55
-40	0.58	1.96	3.40	7.11	15.14	20.45	39.71
-20	0.57	1.92	3.31	6.88	14.68	19.81	38.93
0	0.54	1.83	3.17	6.56	13.98	18.76	37.53
20	0.52	1.75	3.06	6.24	13.24	18.11	35.72
40	0.49	1.68	2.92	5.96	12.88	17.48	34.35
60	0.45	1.59	2.78	5.74	12.33	16.71	33.10
80	0.42	1.49	2.62	5.44	11.73	15.99	31.77
100	0.39	1.38	2.45	5.11	11.09	15.27	30.20
120	0.35	1.28	2.29	4.78	10.51	14.52	28.97
140	0.31	1.16	2.10	4.45	9.86	13.71	27.18

Table 6-22.   Fill Factor vs. Temperature and Intensity, after 1.1 x 10^{15} e/cm^2

Solar Intensity, mW/cm^2

Temp, °C	5.0	15.0	25.0	50.0	100.0	135.3	250.0
-120	0.730	0.754	0.775	0.803	0.828	0.834	0.844
-100	0.732	0.763	0.782	0.810	0.833	0.839	0.846
-80	0.736	0.769	0.791	0.816	0.836	0.841	0.845
-60	0.734	0.771	0.792	0.817	0.834	0.839	0.844
-40	0.732	0.771	0.792	0.816	0.831	0.835	0.839
-20	0.728	0.772	0.788	0.810	0.826	0.829	0.832
0	0.723	0.764	0.782	0.803	0.818	0.820	0.823
20	0.716	0.758	0.774	0.793	0.809	0.811	0.813
40	0.705	0.748	0.762	0.783	0.796	0.800	0.803
60	0.692	0.736	0.751	0.771	0.784	0.786	0.791
80	0.674	0.719	0.737	0.756	0.770	0.773	0.777
100	0.652	0.701	0.719	0.740	0.756	0.759	0.763
120	0.628	0.683	0.702	0.722	0.738	0.745	0.748
140	0.602	0.658	0.680	0.703	0.723	0.726	0.730

Table 6-23.   Efficiency vs. Temperature and Intensity, after 1.1×10^{15} e/cm^2

Solar Intensity, mW/cm^2

Temp, °C	5.0	15.0	25.0	50.0	100.0	135.3	250.0
-120	12.20	13.75	14.51	15.20	15.91	15.81	17.06
-100	12.15	13.49	14.01	14.87	15.87	15.91	16.85
-80	12.05	13.43	14.11	14.62	15.75	15.66	16.42
-60	11.85	13.28	13.80	14.23	15.42	15.50	16.22
-40	11.56	13.04	13.60	14.21	15.14	15.12	15.88
-20	11.31	12.78	13.24	13.76	14.67	14.64	15.57
0	10.81	12.16	12.67	13.11	13.98	13.87	15.01
20	10.29	11.68	12.24	12.47	13.24	13.39	14.29
40	9.75	11.17	11.69	11.92	12.88	12.92	13.74
60	9.07	10.59	11.11	11.48	12.33	12.35	13.24
80	8.44	9.92	10.49	10.87	11.73	11.82	12.71
100	7.73	9.19	9.79	10.23	11.09	11.29	12.08
120	6.98	8.53	9.14	9.57	10.51	10.73	11.59
140	6.20	7.70	8.40	8.89	9.86	10.13	10.87

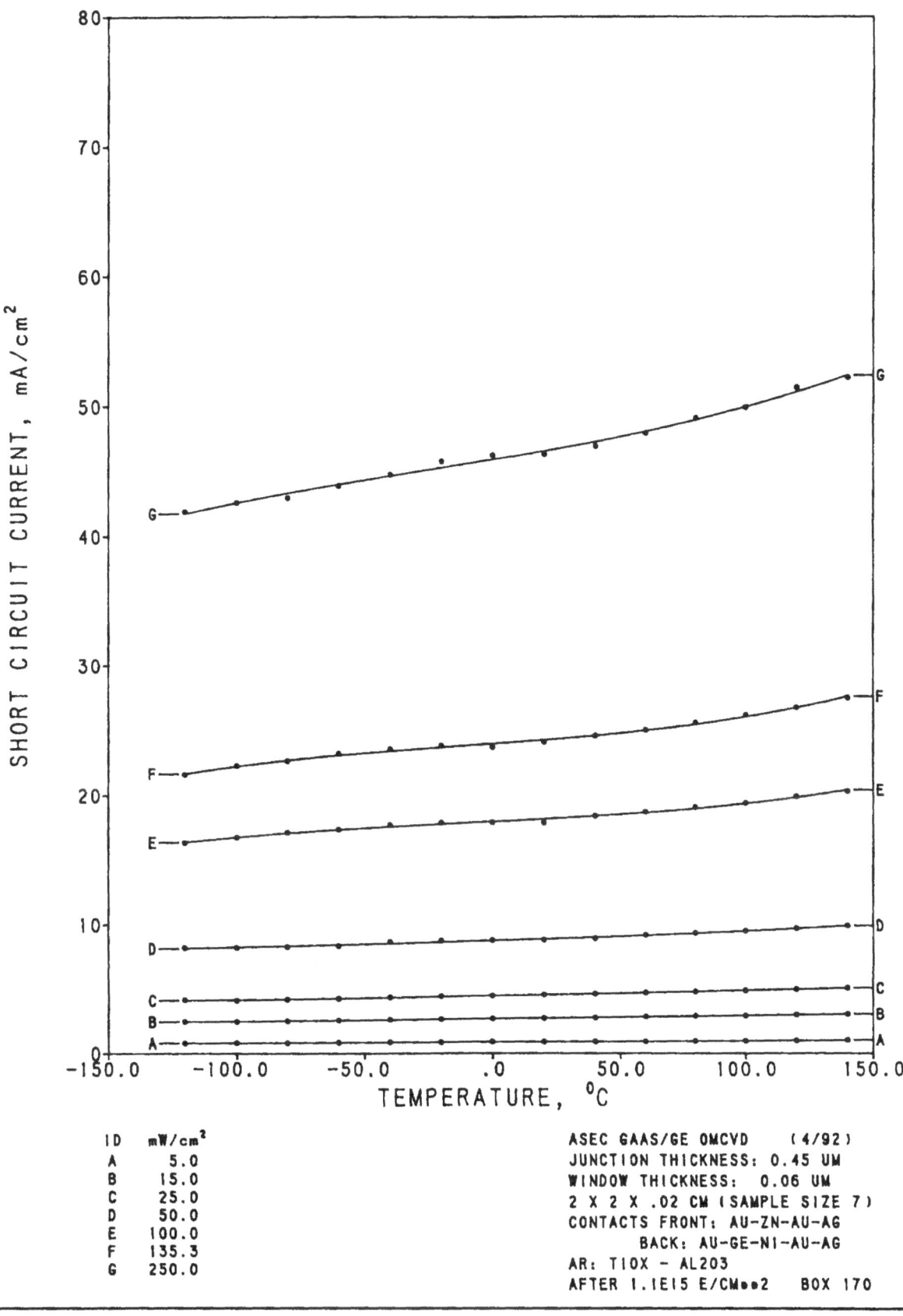

ID	mW/cm²
A	5.0
B	15.0
C	25.0
D	50.0
E	100.0
F	135.3
G	250.0

ASEC GAAS/GE OMCVD   (4/92)
JUNCTION THICKNESS: 0.45 UM
WINDOW THICKNESS: 0.06 UM
2 X 2 X .02 CM (SAMPLE SIZE 7)
CONTACTS FRONT: AU-ZN-AU-AG
          BACK: AU-GE-NI-AU-AG
AR: TIOX - AL2O3
AFTER 1.1E15 E/CM**2   BOX 170

Figure 6.43.   Average I_{sc}/cm² as a Function of Temperature, after 1.1×10^{15} e/cm²

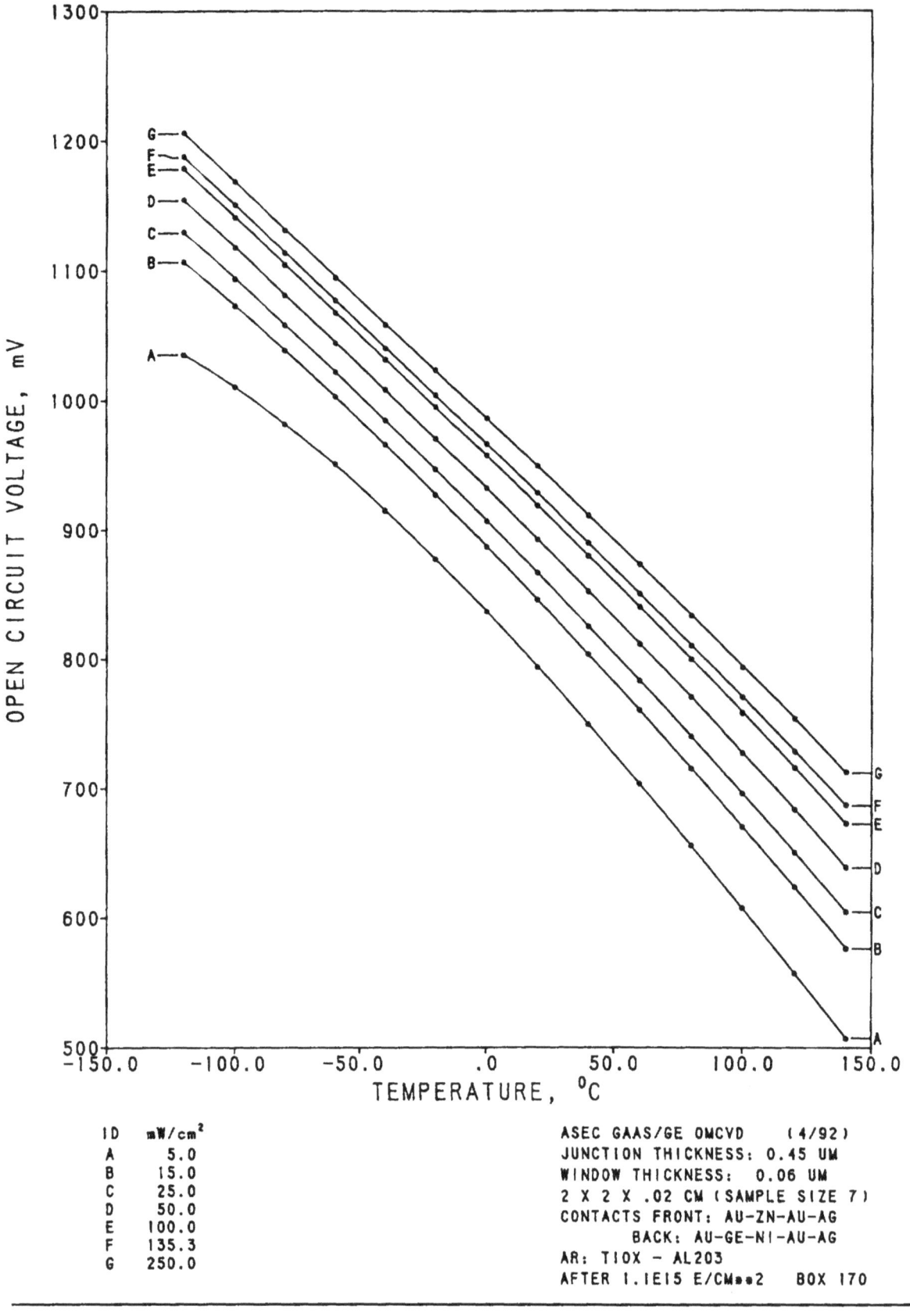

Figure 6.44. Average V_{oc} as a Function of Temperature, after 1.1×10^{15} e/cm^2

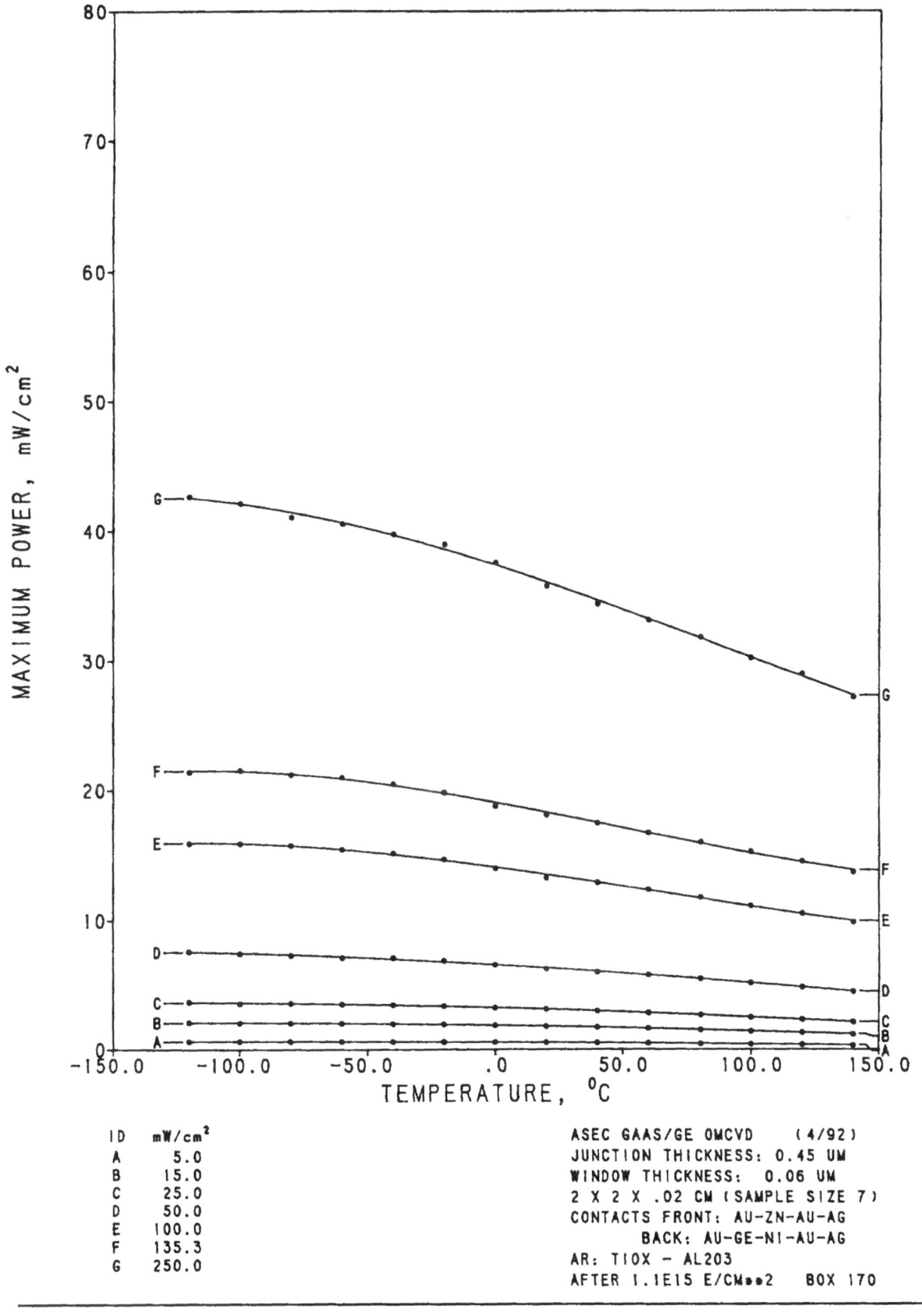

Figure 6.45.   Average P_{max}/cm^2 as a Function of Temperature, after 1.1×10^{15} e/cm²

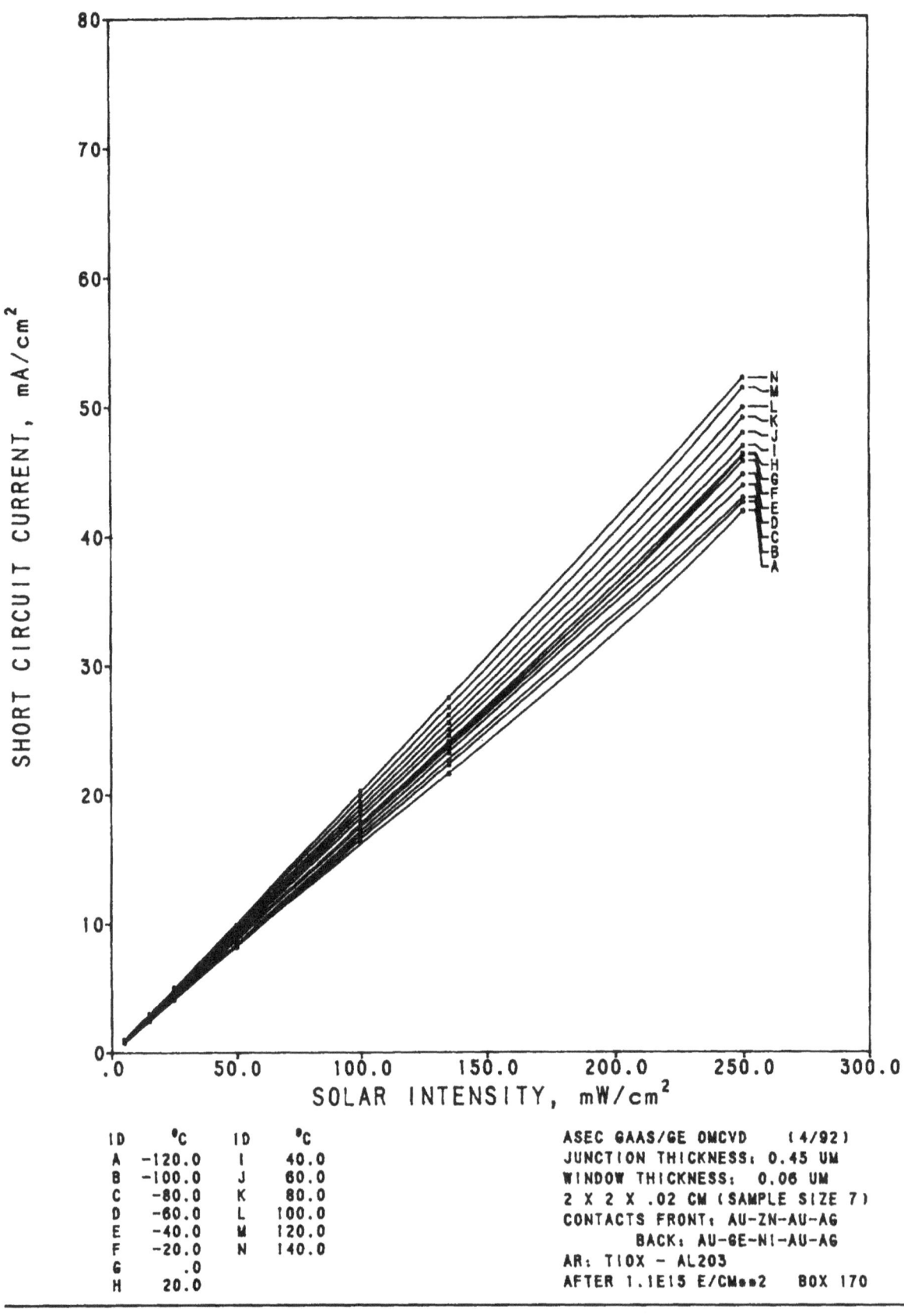

ID	°C	ID	°C
A	-120.0	I	40.0
B	-100.0	J	60.0
C	-80.0	K	80.0
D	-60.0	L	100.0
E	-40.0	M	120.0
F	-20.0	N	140.0
G	.0		
H	20.0		

ASEC GAAS/GE OMCVD   (4/92)
JUNCTION THICKNESS: 0.45 UM
WINDOW THICKNESS: 0.06 UM
2 X 2 X .02 CM (SAMPLE SIZE 7)
CONTACTS FRONT: AU-ZN-AU-AG
            BACK: AU-GE-NI-AU-AG
AR: TIOX - AL2O3
AFTER 1.1E15 E/CM**2   BOX 170

Figure 6.46.   Average I_{sc}/cm^2 as a Function of Intensity, after 1.1×10^{15} e/cm^2

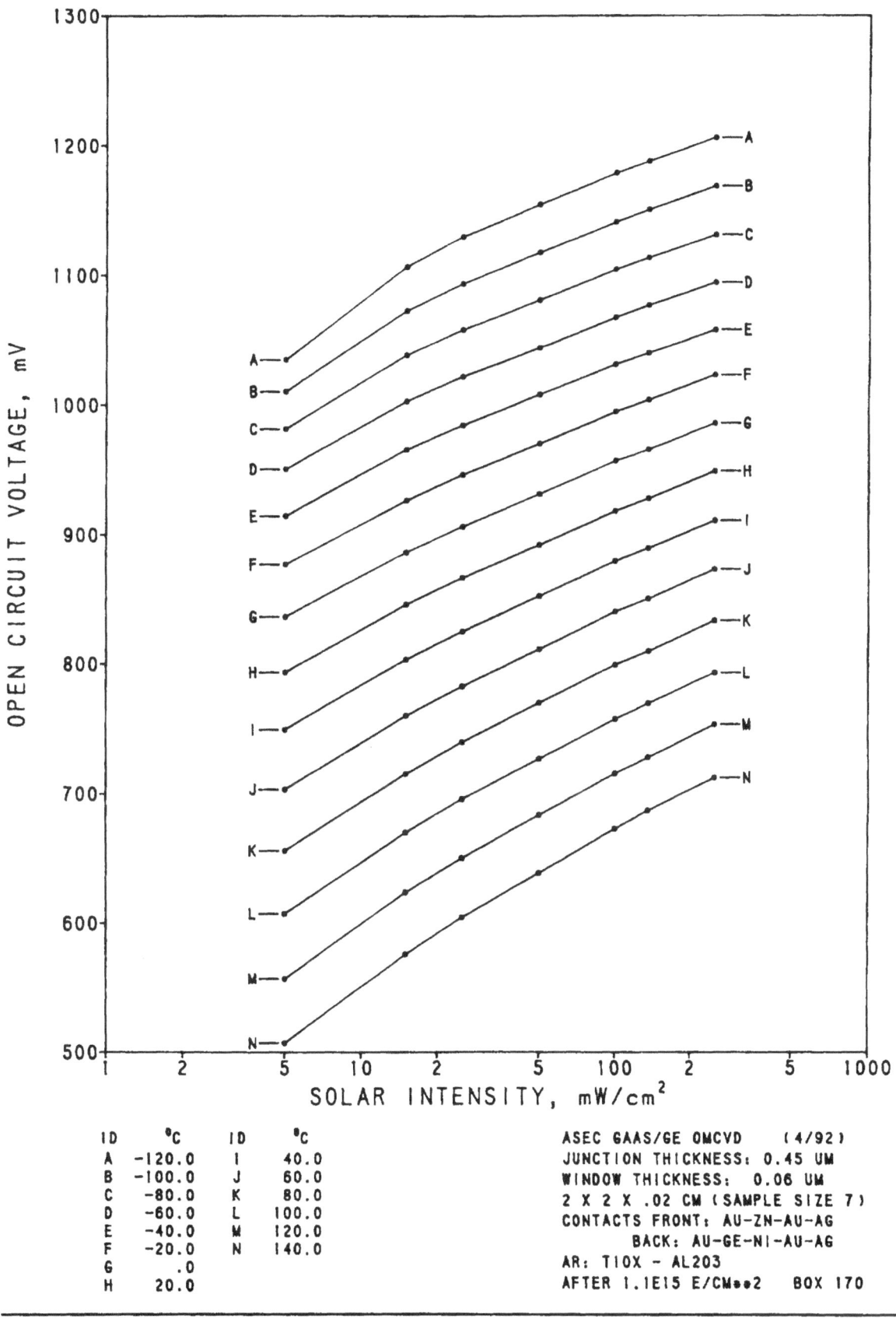

Figure 6.47.   Average V_{oc} as a Function of Intensity, after 1.1×10^{15} e/cm^2

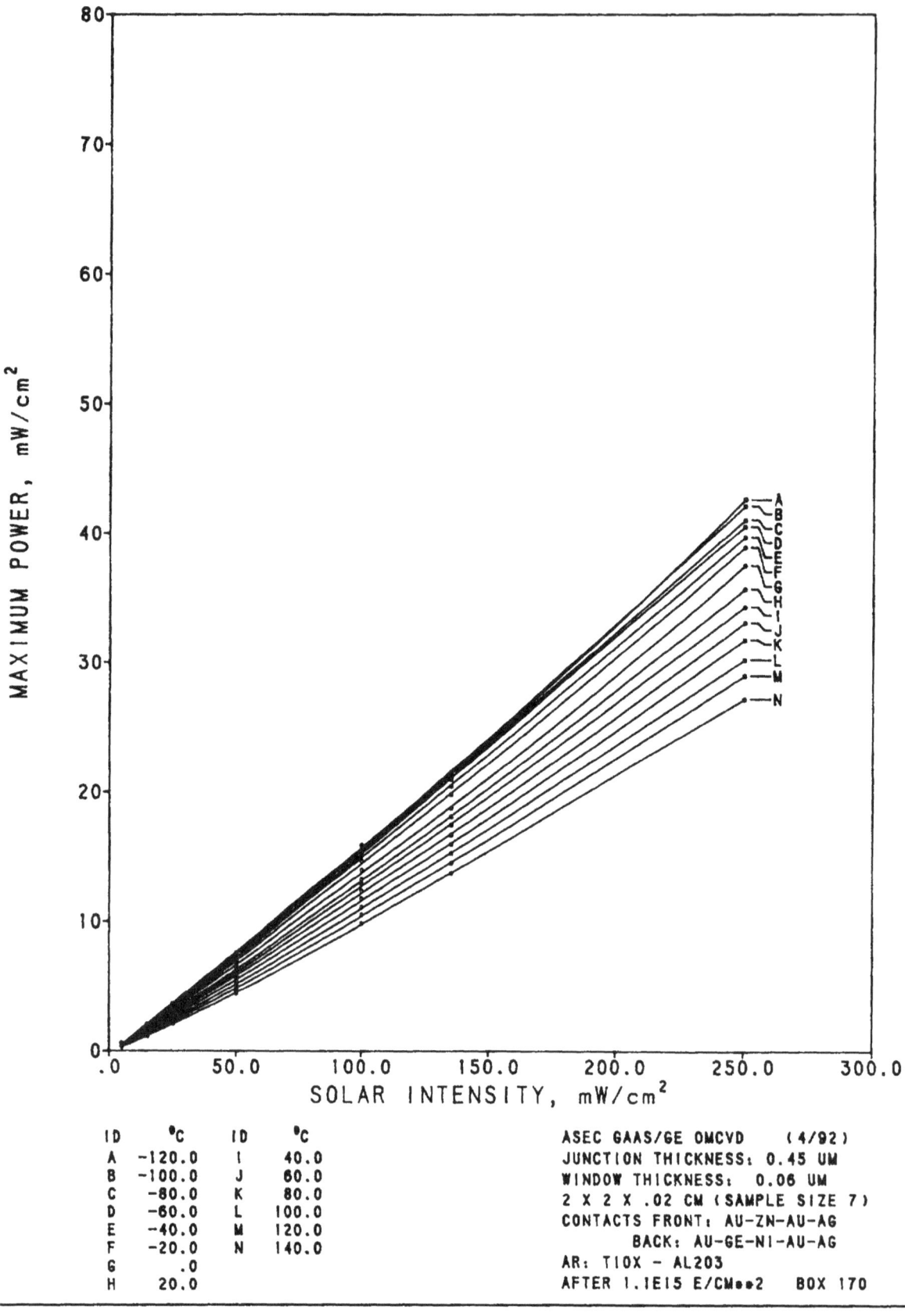

Figure 6.48.   Average P_{max}/cm^2 as a Function of Intensity, after 1.1×10^{15} e/cm^2

Chapter 7

Spacecraft Flight Data for GaAs Solar Arrays

7.1   NTS-II

A very brief reference to GaAs solar cells flown on this satellite was mentioned in reference [7.1]. At the time of that writing (May 1981), NTS-II had been on-orbit for more than 3 years with LPE GaAs solar cells made by Hughes Research Lab (HRL). The reference stated that the power loss was much less than predicted, even though the cell junctions were deep ($\approx 1\ \mu$m). It was speculated that on-panel annealing may have occurred because the panel temperatures were running $\approx 100°$C. At that time, it was too early to compare the degradation rate of the cells with damage coefficient theory.

7.2   LIPS-II

This satellite was launched by the Naval Research Laboratory (NRL) in February 1983 into a 600 nmi circular orbit with 63° inclination. It carried a panel of 300 LPE GaAs solar cells made by HRL. The cells were arranged in 3 parallel strings of 100 cells each. Each string consisted of 25 cells in series by 4 cells in parallel. The performance of the panel was discussed in references [7.2] and [7.3]. The initial on-orbit measurement of this panel showed a power loss of $\approx 7\%$ as compared to the prelaunch measurement; this could be accounted for by the loss of one of the parallel strings. (Damage to only one cell in the string could cause this loss). Other difficulties included the ability to take measurements only sporadically, the absence of an accurate measurement of the panel's temperature (it was known to vary between -45°C and +45°C), and the possibility of reflections off part of the spacecraft body (the plume shield). No comparisons of actual performance with predicted performance could be made with confidence.

7.3   LIPS-III

LIPS-III was launched in the spring of 1987 into a circular orbit with an altitude of 1100 km and an inclination of 63°. There were two experiments which measured GaAs solar cells on this spacecraft. One was a joint experiment by the Applied Solar Energy Corp. (ASEC) and the U. S. Air Force, described in reference [7.4], and the other was an experiment by the Boeing company, which measured GaAs and CuInSe$_2$ cells described in reference [7.5].

The ASEC/Air Force experiment utilized both GaAs and GaAs/Ge solar cells. The GaAs cells had deep junctions and the GaAs/Ge cells were from the early efforts at ASEC to make such cells. The experiment also incorporated some Si solar cells for reference. The results of the ASEC/Air Force experiment were disappointing. The initial on-orbit performance did not agree with the prelaunch data. Indeed, there were two sets of prelaunch data, and neither of these agreed with the on-orbit data, nor did they agree with each other, so there was some uncertainty in establishing initial I-V curves. The reported data collection was infrequent, that is, 9 times between flight days 12 and 175. Panel temperatures ran between 80°C and 90°C. The GaAs cells degraded to values of 0.913, 1.0, and 0.93 times the "initial" values of I_{sc}, V_{oc}, and P_{max} respectively, whereas degradation predictions based on the methods in this handbook predict degradations of 1% or less in all parameters. In view of the fact that 1) the panels were apparently assembled in haste with possible attendant mechanical damage, 2) the prelaunch data is somewhat uncertain, 3) and the GaAs/Ge cells may have had unmatched, leaky GaAs/Ge junctions, it is hard to draw solid conclusions from this experiment. Even so, the experimenters have posed the possibility that there may be more radiation in this orbit than predicted by the radiation models.

The Boeing GaAs panel had three n/p concentrator cells manufactured by the MOCVD process and one planar p/n cell made with the LPE process. No data on junction depths was reported. All cells were protected with ceria doped microsheet (CMX) coverglasses, 12 mils thick. The on-orbit behavior of two of the concentrator cells and the planar cell was measured over a 965-day period. The experimenters measured a degradation in power of ≈ 2.2 to 4.5% per year, whereas the predictions are for only $\approx 1\%$ per year. This reinforces the ASEC/Air Force group's postulate that the actual environment may be more severe than the modeling predictions.

7.4   High Efficiency Solar Panel (HESP) on the Combined Release and Radiation Effects Satellite (CRRES)

The results of this experiment were reported in references [7.6 and 7.7]. The CRRES satellite was launched on July 25, 1990 into an elliptical orbit with an apogee of 35,000 km, a perigee of 350 km, and an inclination of 18.2°. The spacecraft ceased functioning on October 12, 1991 after 1067 orbits.

The solar cell radiation damage experiment consisted of two panels of cells. One panel, called the ambient panel, left the panel at the equilibrium temperature it reached in space. The annealing panel consisted of four annealing experiments. Each experiment consisted of 2 strings of cells, 4 cells per string. Three of these experiments were programmed to a specific annealing schedule. The fourth experiment was allowed to reach and maintain space ambient temperature. The ambient panel incorporated 10 strings (4 cells per string) of GaAs/Ge cells with coverglasses of two materials (CMX and quartz) ranging in thickness from 3.5 to 30 mils. The cells were of the ASEC MOCVD type with junction depths of 0.47 μm. The ambient panel also incorporated two Si cell strings, (5 cells per string). One string was made of Spectrolab K4-3/4 cells which were 200 μm thick, and the other string was made of Spectrolab K7-3/4 cells which were 62 μm thick. The ambient panel reached temperatures of $\approx 90°C$ while in the Sun and dropped to -50°C during each eclipse.

The annealing panel utilized LPE cells with junction depths between 0.5 and 0.6 μm. The cells were made by HRL in the 1983 to 1984 time period. Interconnections on these cells were made by ultrasonic welding, and all the cells had quartz coverglasses except for one string, which utilized ceria-doped microsheet (CMX) coverglasses. The first group, consisting of two strings of LPE cells, was held at a constant 170°C. The second group of two strings was designed to be heated to a temperature of 250°C for 30 minutes, once each week. The third group of two strings was designed to allow heating of the cells to a temperature of $\approx 190°C$ by means of forward-current biasing for 30 minutes once each week. The fourth group of two strings remained at space ambient temperature. Each group had a set of GaAs/Ge cells protected by 6-mil (150 μm) thick coverglasses and another set of GaAs/Ge cells protected by 12-mil (300 μm) thick coverglasses.

Spectrometers aboard this spacecraft measured the differential proton energy spectrum in 24 energy channels logarithmically spaced from 1 to 100 MeV, and the electron fluence in the energy range between 1 and 8.3 MeV. The data from these spectrometers permitted the correlation of observed solar cell degradation to the radiation levels they actually experienced.

The results after 30 days in orbit were reported by reference [7.6], and after mission termination in reference [7.7]. The authors emphasize that the proton and electron radiation belts are very dynamic, and they can change dramatically when solar flares and geomagnetic storms occur. The accurate prediction of solar cell degradation over short periods of time is therefore all but impossible. The period from the beginning of the mission until orbit 587 was a quiet solar period. During the early part of this quiet time (orbits 0 through 400), the solar cell performance of all strings degraded approximately monotonically, but the degradation leveled off between orbits 400 and 600. This levelling off was matched by a levelling off of the integrated proton and electron fluences measured during this time. Then, a large solar proton event occurred on March 22, 1991 (orbit no. 584). Soon thereafter a storm began and actually created a second inner proton radiation belt [7.8]. After this event, the levelling off of the solar cell degradation ceased, and the cells began degrading at approximately the same rate as observed during orbits 0 through 400.

Other observations reported by these authors are as follows. The GaAs/Ge solar cells degraded at lower rates than did either type of Si cell. The thin K7-3/4 cells degraded initially much faster than the K4-3/4 cells, due to degradation of the back surface field. Even so, the thin cells always had a higher P_{max} than the thicker Si cells, throughout the duration of the flight. Annealing of the GaAs/Ge cells showed mixed results. In all cases, annealing of the cells with 150-μm-thick coverglasses was more effective than it was for cells with the 300-μm-thick coverglasses. The most effective annealing method was the weekly, forward-bias annealing at 190°C, followed by the weekly, thermal annealing at 250°C, then the constant thermal annealing at 170°C. The 170°C thermal annealing only gave a marginal improvement in cell behavior over the cells maintained at the ambient temperature of ≈ 100°C. There was no difference in power loss in cells protected by quartz or CMX coverglasses, as long as the coverglass thicknesses were the same. There was also no difference in cell performance observed between soldered and welded interconnects. No attempt was made to calculate cell degradation from the measured radiation levels.

7.5   EURECA

This satellite was launched on July 31, 1992 into a 508-km circular orbit and was retrieved in August 1993 [7.9, 7.10]. The spacecraft carried a flight experiment named the Advanced GaAs Solar Array (ASGA) that was primarily made of GaAs and GaAs/Ge solar cells manufactured by CISE (Italy) and LPE GaAs cells by EEV (United Kingdom). The experiment consisted of two arrays; a planar array and a concentrator array. The planar array had six strings of various types of cells, coverglasses, and interconnects. Each string had three 2 x 2 cm cells, connected in series. The concentrator panel consisted of three Cassegrainian concentrators, each focusing on one 0.25 cm^2 round GaAs cell.

According to theory, there should be almost no cell degradation in this orbit. After 239 days in orbit, the most degradation observed in the planar cells was 2.2% in a 300-μm-thick LPE cell with a deep junction (X_j = 0.8 μm) and a very thin (50 μm) coverglass. The concentrator cells degraded significantly. The CISE cells lost I_{sc} very rapidly during the first 60 days and ended up with I_{sc}/I_{sc0} values of 0.729 and 0.796 after 226 days in orbit. The EEV concentrator cell displayed much less current degradation, measuring I_{sc}/I_{sc0} of 0.933 after 226 days. The rapid decrease in response of the CISE cells was possibly due to the fact that the cell's active region was 4 mm in diameter, while the concentrator spot diameter was 4.5 mm in diameter. The concentrator spot would have strongly illuminated the interconnectors and may have contributed to cell contamination. Both the CISE cells and the EEV cell may have lost output because of contamination of the mirrors.

7.6   STRV-1 A and B

These spacecraft were launched together into a geostationary transfer orbit, (nominally 200 x 36000 km, with an inclination of 7°) on June 17, 1994. A complete description of the experiment was given in reference [7.11]. The spacecraft carried a flight experiment designed to measure the degradation of many types of solar cells, including several GaAs cells. Unfortunately, the experiment experienced a failure and no data is available directly from the experiment. However, four of the power producing solar panels for the STRV-1B were populated with GaAs solar cells and it was possible to measure the panel current at 28 V as a function of time, from each panel [7.12]. The panels are made up of 40 cells in a series string, hence 28 V across the panels corresponds to a voltage drop of 700 mV across each cell. For cells with high shunt resistances, the current at 700 mV can be assumed to be a reasonably accurate approximation for short circuit current, I_{sc}. Figure 7.1 is a plot of one such panel powered by Spectrolab GaAs/Ge cells. These cells measured 1.8 x 6.3 cm x 200-μm thick, and were covered with 500-μm-thick coverglasses. The experimental data is plotted as open diamonds, and the predicted behavior is plotted as the solid line. The prediction is based on the AP8 environment together with an equivalent fluence calculation based on the methods outlined in this book [7.13]. Figure 7.2 is a similar plot for a panel powered by ASEC GaAs/Ge cells. The ASEC cells measured 4 x 4 cm x 90-μm thick, and were covered with 500 μm-thick-coverglasses. As the figures show, the observed cell degradation matches the predicted degradation very well.

7.7   UoSAT-5

The UoSAT-5 spacecraft was launched into a 770-km Sun-synchronous orbit on July 16, 1991 [7.14]. It carried a solar cell experiment consisting of various types of Si, GaAs, and InP solar cells from the United Kingdom, Europe, and the United States. This experiment had difficulties with measuring cell temperatures and the I-V curve near V_{oc}; also the I_{sc} measurements were confounded by varying contributions from the Earth's albedo. The experimental measurements after 1 year in orbit were reported [7.14]. The largest degradation observed in GaAs cell short circuit current was 0.5% and in maximum power was 1.5%. Theory predicts that there should be no degradation of GaAs cells in this orbit.

7.8   PASP-PLUS Experiment on the Advanced Photovoltaic and Electronic Experiments (APEX) Spacecraft

The APEX spacecraft was launched on August 3, 1994 into an elliptical orbit with an apogee of 2537 km, perigee 363 km, and an inclination of 69.9°. The PASP-PLUS experiment aboard this spacecraft carried 16 photovoltaic experiments, several of which were designed to measure the performance of GaAs solar cells. A detailed description of the cells and modules is presented in reference [7.15]. The spacecraft also had its own onboard dosimeters, so the actual radiation environment could be compared to the environment predicted by the models. The radiation environment in this highly elliptical orbit is dominated by protons and the cell degradation induced by electrons, in comparison, is negligible for all practical purposes. The results after the first 3 months in orbit were reported in reference [7.16] and results after 6 months in orbit were reported in references [7.17, 7.18, and 7.19]. Experiments 4 and 6 had identical GaAs/Ge solar cells with coverglasses which were 3.5-mils thick. Experiment 4 had a string configuration of 5 cells in series and 4 in parallel. Experiment 6 had a string of 4 in series and 3 in parallel. Figures 7.3 and 7.4 are plots of the normalized maximum power, Pmax, of these two experiments vs. time. The circles are the experimental measurements and the solid line is the predicted P_{max} behavior using AP8 and an equivalent fluence calculation. The beginning efficiencies noted on the figures are derived from ground measurements prior to launch. Figure 7.5 is a plot of the normalized P_{max} vs. time observed for the GaAs/Ge cells on Experiment 11. The cells in this experiment were 178-μm thick and used 150-μm-thick CMX coverglasses. The eight cells in this experiment were connected with 2 cells in parallel and 4 in series. The plots from all three of these experiments show that the prediction tends to be slightly conservative, but deviates no more than 2-3% from the observed behavior. All three experiments show a rapid power loss in the first few days after launch followed by what appears to be a slight recovery. The reason for this behavior is, as yet, unknown.

7.9   Advanced Solar Cell Orbit Test (ASCOT) Flight Experiment

The ASCOT flight experiment was launched into a highly elliptical orbit in 1995. The purpose of ASCOT was to flight test six advanced solar cell types in a high-radiation, proton-dominated space environment. The test cells included three different types of GaAs/Ge cells, a GaAs/CIS design, and two types of Si cells, one 200-μm thick and the other 50-μm thick. The GaAs cells were protected by 300-μm coverglasses. Each Si cell type was protected by a mix of both 300-μm and 760-μm coverglasses. Four modules of five series-connected cells were flown for each of the six cell types, so there were a total of 24 modules on the spacecraft. Details of the spacecraft and the orbit are not available, but it has been calculated that one year in the ASCOT orbit would produce an equivalent fluence of 2.2 x 10^{15} e/cm^2 in Si cells with 300 μm coverglasses. [7.20].

The measured behavior of one of the GaAs/Ge cell types after part of one year in orbit is presented in Figure 7.6 [7.21]. The figure depicts normalized maximum power (P_{max}) of both the GaAs/Ge cells on the ASCOT experiment (open diamonds), and the power degradation of GaAs/Ge cells flown on the PASP-Plus experiment (open squares). The PASP-Plus data is the same data shown in Figure 7.3, with the abscissa scale changed from days in orbit to 1 MeV equivalent electron fluence. The solid line is a prediction of the degradation behavior of these cells based on a radiation calculation using the AP8 model and an equivalent fluence calculation. Figures 7.7 and 7.8 are similar plots for normalized open circuit voltage (V_{oc}) and normalized short circuit current for the same cells. In addition, Figure 7.8 includes the behavior of one of the STRV-1B solar panels. It would appear that the data from the space-based experiments is in satisfactory agreement with the theoretical predictions.

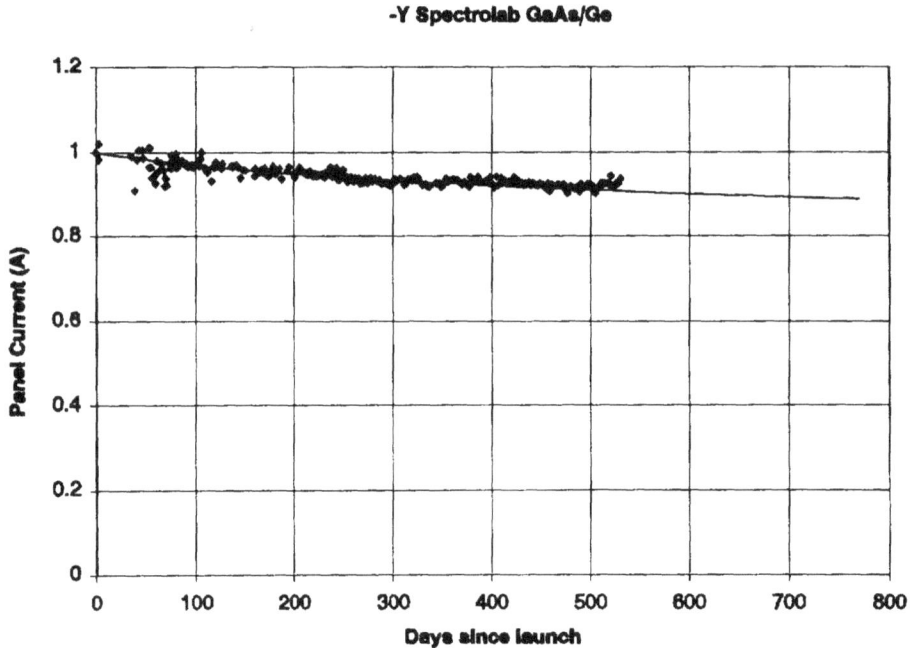

Figure 7.1. I_{sc} vs. Time in Orbit for the Spectrolab 200-μm GaAs/Ge Solar Cells with 500-μm Coverglasses on STRV-1B

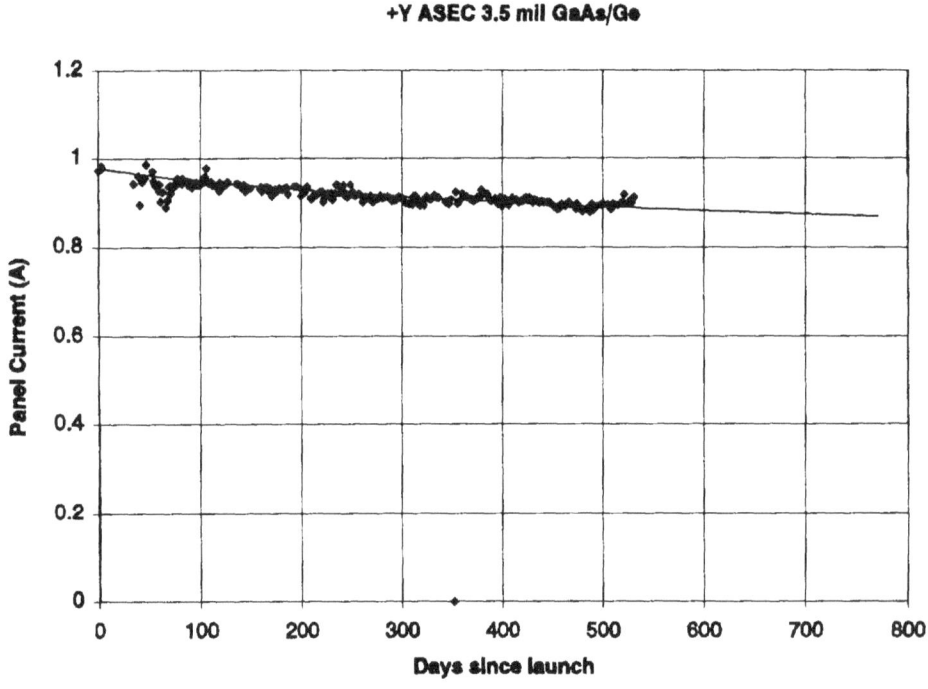

Figure 7.2. I_{sc} vs. Time in Orbit for the ASEC 90-μm GaAs/Ge Solar Cells with 500-μm Coverglasses on STRV-1B

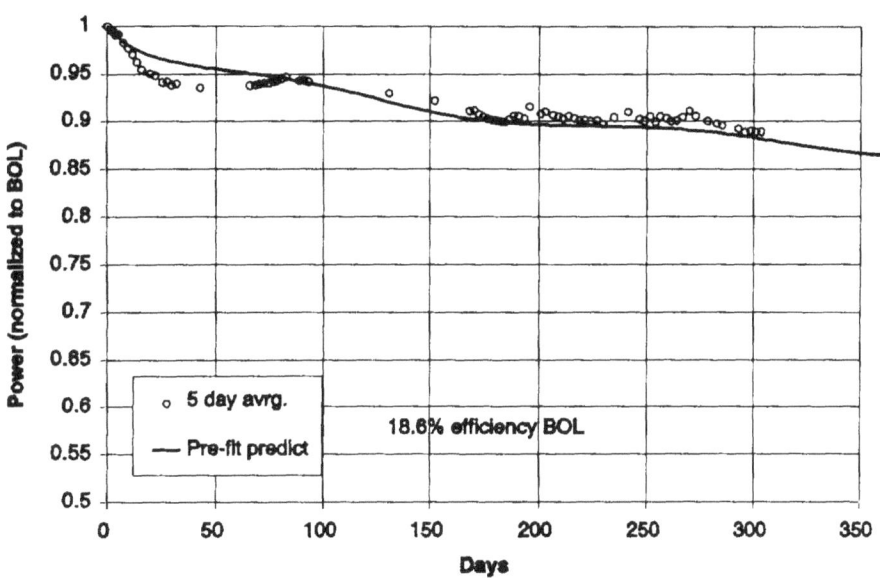

Figure 7.3.   Normalized P_{max} vs. Time in Orbit for the 90-μm GaAs/Ge Solar Cells with
100-μm-CMX Coverglasses on PASP-Plus Experiment 4

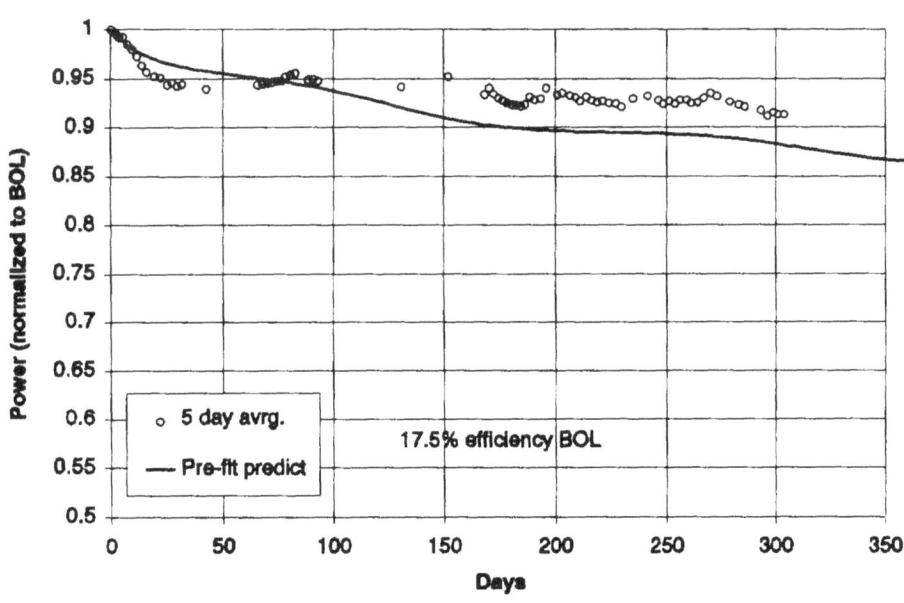

Figure 7.4.   Normalized P_{max} vs. Time in Orbit for the 90-μm GaAs/Ge Solar Cells with
100-μm-CMX Coverglasses on PASP-Plus Experiment 6

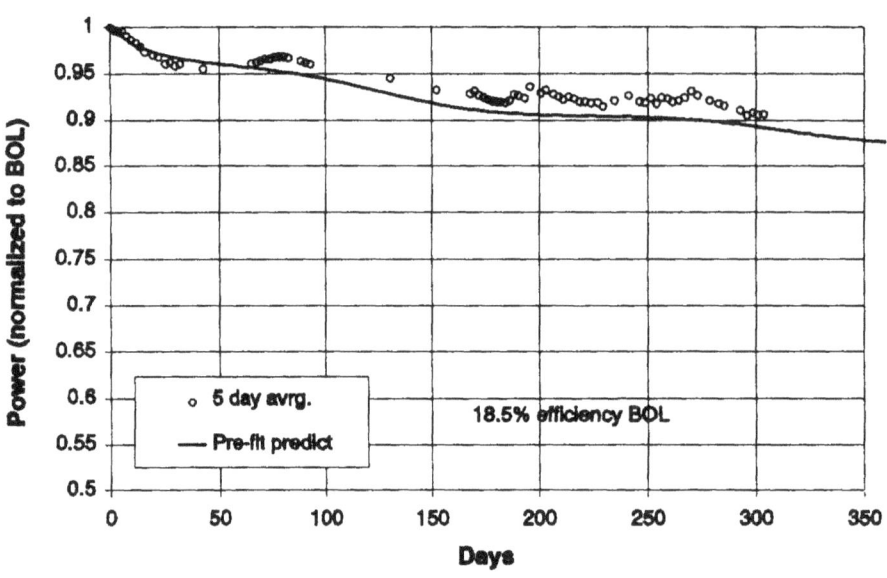

Figure 7.5. Normalized P_{max} vs. Time in Orbit for the 178-μm GaAs/Ge Solar Cells with 150-μm-CMX Coverglasses on PASP-Plus Experiment 11

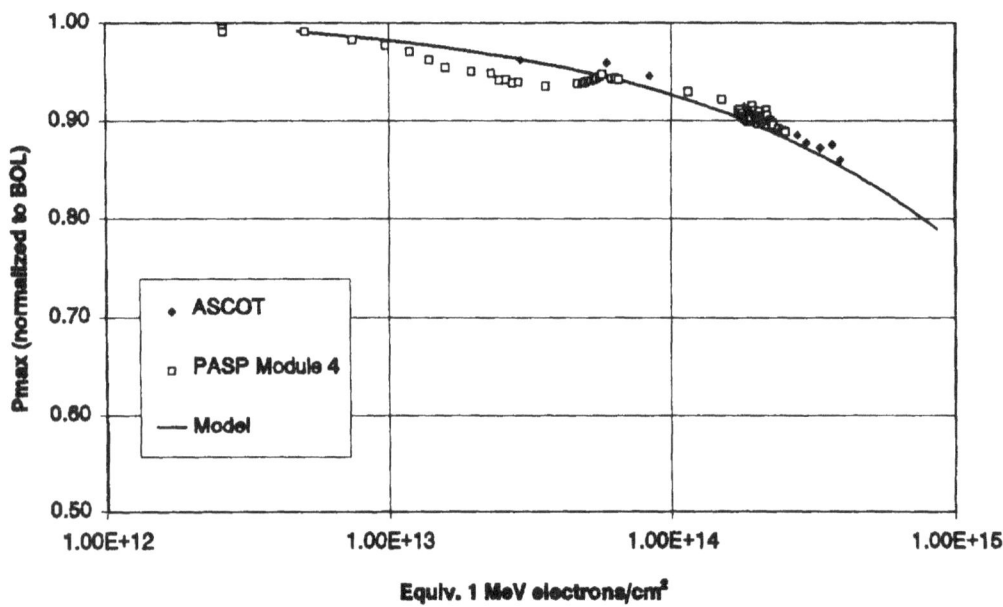

Figure 7.6. Normalized P_{max} vs. Equivalent 1 MeV Electron Fluence for GaAs/Ge Solar Cells on ASCOT and on PASP-Plus Module 4

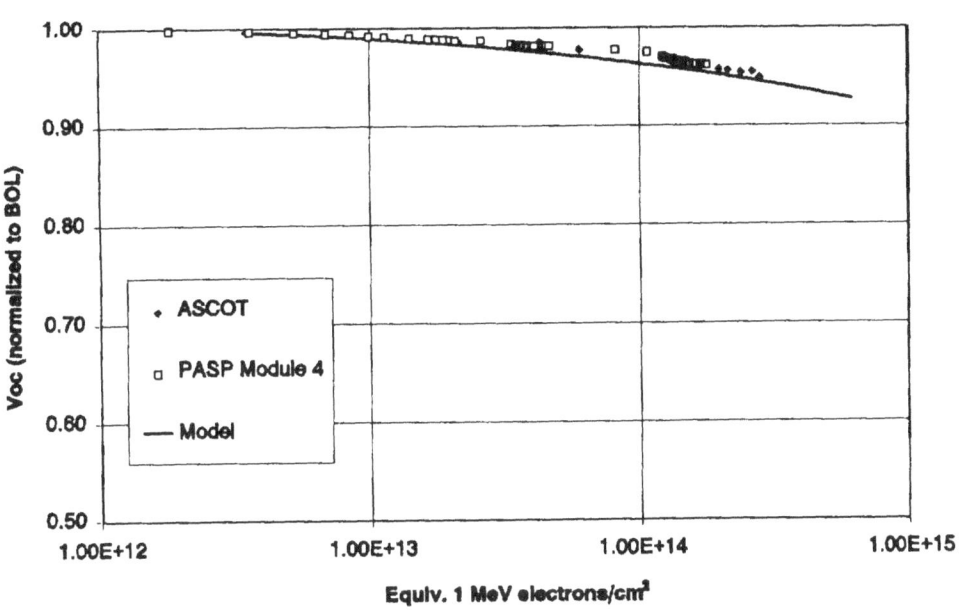

Figure 7.7.   Normalized V_{oc} vs. Equivalent 1 MeV Electron Fluence for GaAs/Ge Solar Cells on ASCOT and on PASP-Plus Module 4

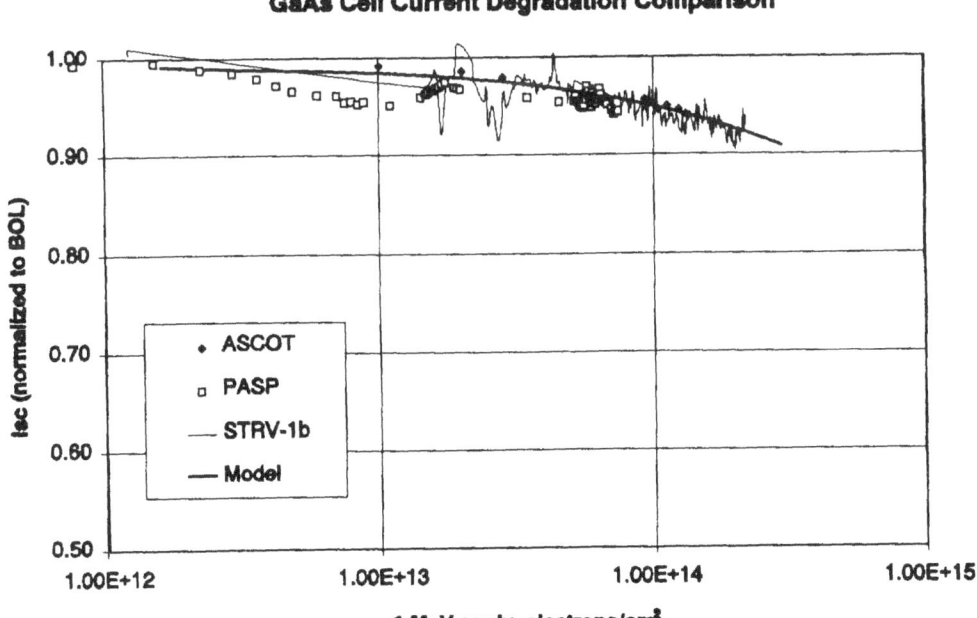

Figure 7.8.   Normalized I_{sc} vs. Equivalent 1 MeV Electron Fluence for GaAs/Ge Solar Cells on ASCOT, PASP-Plus Module 4, and on STRV-1B

References for Chapter 7

7.1   R. Loo, R.C. Knechtli, and G.S. Kamath, "Enhanced Annealing of GaAs Solar Cell Radiation Damage," *Proceedings of the 15th IEEE Photovoltaic Specialists Conference*, 33, 1981.

7.2   R.W. Francis and F.E. Betz, "Two Years of On-Orbit Gallium Arsenide Performance from the LIPS Solar Cell Panel Experiment," *Space Photovoltaic Research and Technology Conference, NASA Conference Publication 2408*, 203, 1985.

7.3   R.K. Morris, "New Insight Into the LIPS-II GaAs Solar Panel Performance," *Proceedings of the 18th IEEE Photovoltaic Specialists Conference*, 688, 1985.

7.4   P.A. Iles, K.I. Chang, K.S. Ling, C. Chu, J. Wise, and R.K. Morris, "ASEC/Air Force LIPS-3 Test Panel Results," *Proceedings of the 20th IEEE Photovoltaic Specialists Conference*, 826, 1988.

7.5   R.M. Burgess, W.E. Devaney, and W.S. Chen, "Performance Analysis of CuInSe$_2$ and GaAs Solar Cells Aboard the LIPS-III Flight Boeing Lightweight Panel," *Proceedings of the 23rd IEEE Photovoltaic Specialists Conference*, 1465, 1993.

7.6   J.S. Powe, E.L. Ralph, G. Wolff, and T.M. Trumble, "On-Orbit Performance of CRRES/HESP Experiment," *Proceedings of the 22nd IEEE Photovoltaic Specialists Conference*, 1409, 1991.

7.7   K.P. Ray, E.G. Mullen, and T.M. Trumble, "Results from the High Efficiency Solar Panel Experiment Flown on CRRES," *IEEE Transactions on Nuclear Science*, 40, 6, 1505, December 1993.

7.8   E.G. Mullen, M.S. Gussenhoven, K. Ray, and M. Violet, "A Double-Peaked Inner Radiation Belt: Cause and Effect as Seen on CRRES," *IEEE Transactions on Nuclear Science*, NS-38, 1713, December 1991.

7.9   C. Flores, R. Campesato, F. Paletta, G.L. Timo, E. Rossi, L. Brambilla, A. Caon, R. Contini, and F. Svelto, "The Flight Data of the GaAs Solar Cell Experiment on EURECA," *Proceedings of the 23rd IEEE Photovoltaic Specialists Conference*, 1369, 1993.

7.10   L. Gerlach, "Post-Flight Investigation Programmes of Recently Retrieved Solar Generators," *Space Research and Technology Conference Publication 3278*, 269, 1994.

7.11   C. Goodbody and N. Monekosso, "The STRV-1 A & B Solar Cell Experiments," *Proceedings of the 23rd IEEE Photovoltaic Specialists Conference*, 1459, 1993.

7.12   C. Goodbody, Private Communication, February, 1996.

7.13   D.C. Marvin, Private Communication, February, 1996.

7.14 C. Goodbody, "The UoSAT-5 Solar Cell Experiment - First Year in Orbit," *Space Research and Technology Conference Publication 3210*, 280, 1992.

7.15 W.T. Cooley, S.F. Adams, K.C. Reinhardt, and M.F. Piszczor, "Photovoltaic Array Space Power Flight Experiment Plus Diagnostics (PASP+) Modules," *Proceedings of the Intersociety Energy Conversion Conference (IECEC)*, 1, 1.295, 1992.

7.16 H. Curtis, M. Piszczor, P. Severance, D. Guidice, and D. Olson, "Early Results from the PASP Plus Flight Experiment," *Proceedings of the 24th IEEE Photovoltaic Specialists Conference*, 2169, 1994.

7.17 D.C. Marvin, H. Curtis, M. Piszczor, K.P. Ray, D. Guidice, D. Hardy, and W.R. James, "Flight Data from the PASP-PLUS Experiment," *Proceedings of the Fourth European Space Power Conference*, Poitiers, France, September 1995.

7.18 D.C. Marvin, Private Communication, February 1996.

7.19 D.C. Marvin, W.R. James, D. Guidice, D. Hardy, K.P. Ray, V. Davis, J. Soldi, D. Ferguson, M. VanRiet, M. Piszczor, and H. Curtis, "PASP-Plus Experiment Final Report," *U.S. Air Force Phillips Laboratory Technical Report*, 1966 (to be Published).

7.20 D.C. Marvin, M. Gates, "The Advanced Solar Cell Orbit Test (ASCOT) Flight Experiment," *Proceedings of the 22nd IEEE Photovoltaic Specialists Conference*, 1491, 1991.

7.21 D.C. Marvin and M. Gates, "Results of Three Recent GaAs Flight Experiments: ASCOT, PASP-Plus, and STRV-1b," *Proceedings of the 25th IEEE Photovoltaic Specialists Conference*, 1996, to be published.

Chapter 8

The Space Radiation Environment

The radiation environment near the Earth consists of electrons and protons trapped in the Earth's geomagnetic field and protons produced by solar flares. There is a minor component of radiation received from galactic cosmic rays. Although these highly energetic, sometimes heavy ions are quite important in inducing single-event upsets in integrated circuits, their influence on solar panels is negligible. In interplanetary space, only the solar flare protons are of importance in affecting solar panel performance. Jupiter, Saturn, and Uranus also have magnetic fields and trapped radiation belts which must be taken into account for missions that approach these planets. A great deal of the text for this chapter was taken from chapter 5 of reference [8.1]. A wealth of information on this subject is also available from recent reviews by Garrett and coworkers [8.2, 8.3].

8.1   Geomagnetically Trapped Radiation

The Earth's magnetic field is essentially that of a dipole. The dipole may be thought of as arising from a bar magnet lying near the center of the Earth, but displaced ≈ 436 km from the center, in the direction of the Pacific Ocean [8.4]. It has a magnetic moment of $\approx 8 \times 10^{25}$ Gauss-cm^3, and the magnetic field has a maximum value of ≈ 0.6 Gauss near the polar cap and a minimum value of ≈ 0.3 Gauss near the equator at the Earth's surface. It is tilted $\approx 11.5°$ with respect to the Earth's axis of rotation, and an extension of the bar magnet axis would intersect the surface of the Earth near Greenland at 78.5° N, 70.1° W (the geomagnetic north pole) in the northern hemisphere, according to the 1965 IGRF magnetic field model of the Earth. This offset and tilt of the magnetic field also explains why the geomagnetic equator does not coincide with the Earth's geographical equator. The position of the imaginary bar magnet changes fairly rapidly with time. The geomagnetic North Pole is currently drifting westward at a rate approaching 0.1 degree per year. The imaginary bar magnet is thought to have been displaced some 285 km from the Earth's rotational axis since 1845, as compared to the 436-km displacement today. These changes are extremely rapid on a geologic time scale. To make matters even more complicated, reference [8.2] shows that at an altitude of 400 km above the Earth there are two maxima in the magnetic field in the northern hemisphere and two minima near the geomagnetic equator. The largest of these minima occurs in the South Atlantic Ocean and is responsible for high values of the trapped radiation in this region, thus the origin of the term "South Atlantic Anomaly." Models of the Earth's magnetic field, including the time variation (the term EPOCH is used to specify the year described by the model), have been developed under the auspices of the International Union of Geodesy and Geophysics. They are updated periodically and are available from the National Space Science Data Center [8.5]. These magnetic field models play a very important role in the modeling of the Earth's radiation belts. Because of the rapid time variation of the magnetic field, it is important to use the proper magnetic field EPOCH when making calculations with the models of the Earth's trapped radiation environment.

The planetary magnetic fields are entirely responsible for the presence of the trapped radiation zones surrounding those planets that have magnetic fields. The Earth's trapped radiation fields are known as Van Allen belts, named for their discoverer. Electrons trapped in the Earth's field range in energy from a few keV to ≈ 7 MeV, and protons range in energy from a few keV to ≈ 500 MeV. Energies higher than these have been observed, but not in sufficient quantity to concern the solar panel designer. The origin of these trapped particles is most probably the solar wind. Since the particle energies in the solar wind are quite low, they have to be accelerated by some mechanism to reach the energies they have in the Van Allen belts. The primary acceleration mechanism is thought to be local time variations in the Earth's magnetic field. See reference [8.4] for a discussion of the various variations of the Earth's magnetic field.

Charged particles trapped in the Earth's magnetic field undergo three types of motion. First, they are subject to the Lorentz force, $\mathbf{F} = q(\mathbf{v} \times \mathbf{B})$. (Bold letters here denote vector quantities). If we consider the velocity vector, $\mathbf{v}$ to have two components, one along the local magnetic field line and one perpendicular to the field line, then the Lorentz force does not act on the component of $\mathbf{v}$ which is parallel to $\mathbf{B}$, so it undergoes no acceleration along that direction. However, it does act on the perpendicular component of $\mathbf{v}$ and causes the particle to rotate in a circle. The circular component of motion has a radius, r called the Larmor radius, and an oscillation frequency called the cyclotron frequency or Larmor frequency. The particle therefore moves in a helical path and spirals about a magnetic field line. The angle between $\mathbf{v}$ and $\mathbf{B}$ is known as the pitch angle, α.

The fact that the Earth's magnetic field diminishes with radial distance gives rise to a second type of force which is called longitudinal drift [8.6, 8.7]. Figure 8.1 (taken from reference [8.3]) illustrates a positively charged particle moving in a magnetic field with a strong component and a weak component. As the particle moves from a region of high field strength to a region of lower field strength, the Larmor radius increases, so that it returns to the region of high field strength farther to the right in the figure. In the Earth's magnetic field, this action can be seen to cause the charged particles to undergo a longitudinal drift. Positively charged particles will drift to the west and electrons will drift to the east. This drift is reinforced by the action of the Earth's electric field. This field is radially directed near the Earth and points from dawn to dusk at greater distances, thus adding to drift force induced by the magnetic field gradient.

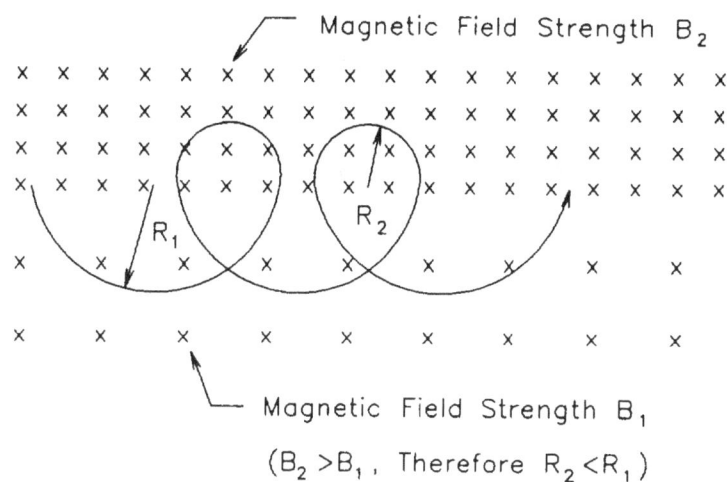

Figure 8.1.   Motion of a Positive Charged Particle in a Magnetic Field of Two Intensities Giving Rise to a Gradient

Another motion experienced by charged particles in the Earth's magnetic field arises because of the gradients along the magnetic field line and is responsible for the trapping of the particles. If one were to pick up a magnetic field line near the equator and follow it poleward, the magnetic field will curve in towards the Earth and increase to a maximum at the pole. Figure 8.2, taken from reference [8.3] illustrates the resulting force on a particle moving in a converging magnetic field. Due to the convergence, there is a component of the magnetic field which is perpendicular to the circular motion of the particle and therefore

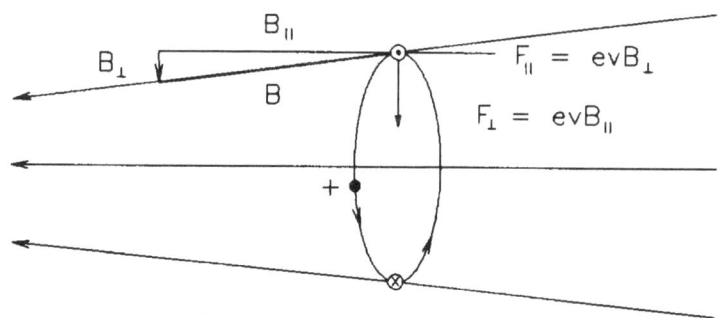

Figure 8.2.   Forces on a Charged Particle in a Converging Magnetic Field. $F_{\parallel}$ Acts Along the Magnetic Field and is Responsible for "Mirroring"

exerts a force away from the converging region. This force becomes stronger as the particle moves poleward, decreasing the pitch angle of the particle until at some point the pitch angle becomes 90°. At this point, called the mirror point, the force on the particle away from the pole is at its strongest and forces the particle to change direction and spiral back in the opposite direction. The particle goes back into the opposite hemisphere until it reaches another mirror point and again reverses itself. This bouncing back and forth between mirror points is the third type of motion characteristic of trapped particles. If the mirror points happen to occur at low altitudes where they can be scattered by the Earth's atmosphere, the particle density will soon diminish to a very low value. This is the reason that there are essentially no trapped particles occurring at equatorial altitudes below ≈ 300 km. Table 8-1, taken from reference [8.4] gives an idea of the scale of these phenomena for 1 MeV electrons and 1 MeV protons at an altitude of 2000 km near the equator.

Table 8-1.   Characteristic Orbital Parameters for Trapped 1 MeV Electrons and 1 MeV Protons at 2000 km Altitude Near the Equator [8.4]

Type	Radius (cm)	Larmor Period (sec)	Bounce Period (sec)	Drift Period (min)
1 MeV e$^-$	3×10^4	7×10^{-6}	0.10	53
1 MeV p$^+$	1×10^6	4×10^{-3}	2.2	32

Models for describing the Van Allen radiation belts in quantitative terms have been developed over the years and extensively documented [8.8 - 8.20]. The models all use the B,L coordinate system introduced by McIlwain in 1961 [8.21]. This coordinate system consists of the magnetic field, B, and the integral invariant, I, which can adequately relate measurements made at different geographic locations. The quantity I is the length of the field line between mirror points weighed by a function of the magnetic field along the line and is an adiabatic invariant of the motion. McIlwain introduced the magnetic shell parameter, L = f(B,I), analogous to a physical distance in a dipole field. In a dipole field, L is equivalent to the distance from the center of the Earth in the magnetic equatorial plane. The L parameter is measured in units of Earth radii (mean radius = 6371.2 km; equatorial radius = 6378.2 km). The beauty of the (B,L) coordinate system is that, with some exceptions, the trapped radiation is identical at all the geographical points which have the same B,L value. The use of the B and L parameters therefore allows the mapping of the three-dimensional trapped radiation environment onto a two-dimensional coordinate system. B and L can be transformed to polar coordinates by the following relations:

$$B = \frac{M}{R^3} \sqrt{4 - \frac{3R}{L}} \quad ; \quad R = L\cos^2\lambda \tag{8-1}$$

where M is the magnetic dipole moment of the Earth, and λ is the magnetic latitude. In order to apply this concept to the Earth's field, which is not a simple dipole, McIlwain expanded the parameter L into a polynomial function of a variable which is a function of I, B, and M, so that the trapped particles can be represented in terms of two dimensions instead of three.

The responsibility for assembling the available data on the Earth's trapped radiation has fallen on the National Space Science Data Center (NSSDC) in Greenbelt, Maryland. Workers at the NSSDC continue to construct and update models of the radiation environment. This data is regarded as the best consolidated source of information available on trapped radiation environments and are used as the single source of data on this subject in this publication. Their work may be consulted in publications listed in references [8.8 through 8.20]. Reference [8.20] is a particularly valuable review of the modeling activity by this group between 1964 and 1991.

8.1.1 Trapped Protons

The AP8 proton model by Sawyer and Vette is the most recent model from the NSSDC describing the Earth's trapped-proton environment [8.18]. The trapped protons are encountered between $\approx 1.2\ R_e$ (Earth radii) and $6.6\ R_e$ (synchronous altitude). The highest energies are found closest to Earth, peaking at about $L = 1.5\ R_e$. The largest proton concentrations at intermediate energies peak at about $L = 2\ R_e$, and the energy spectrum becomes softer (relatively less high energy protons) as the L value increases. At synchronous altitude, the spectrum is so soft that practically no protons with energies greater than 2 MeV exist. Since 2 MeV protons have a range of $\approx 45\ \mu m$ in typical coverslide materials, they are not of concern to the solar panel designer unless there are exposed gaps where the solar cell is not completely covered. The AP8 model was issued in two versions, AP8-MAX and AP8-MIN, which describe the proton distributions at the maxima and at the minima of the 11-year solar cycle. There are good theoretical reasons for changes in the trapped radiation belts induced by the solar cycle. These changes have been verified by experimental observations, but only at low altitudes in the vicinity of the atmospheric cutoff regions. No changes of consequence have been observed in the heart of the proton belts or at synchronous altitude, since the observed variations with time are no larger than the variations between radiation detectors on different satellites [8.22]. The AP8-MIN model was generated from data taken in the 1964 time period, and the AP8-MAX model used data from the 1970 time period. Therefore, when using these models, it is important to use an EPOCH of 1964 when making calculations with AP8-MIN, and an EPOCH of 1970 for calculations for AP8-MAX. Use of the wrong EPOCH in these calculations can induce errors of as much as an order of magnitude in the calculated spectra. The difference in the predicted radiation levels between solar minimum and solar maximum is seen in the models to occur only at altitudes below ≈ 2000 km. The fluxes are higher at solar minimum than at solar maximum for the AP8 proton models.

8.1.2 Trapped Electrons

Trapped electrons with energies of a few hundred keV extend to the outer boundary of the magnetosphere, which fluctuates between altitudes of 8 to 10 Earth radii. There are two intense regions: an inner zone extends between L values of 1.2 and 2.8, with a peak at $\approx 1.4\ R_e$. The outer zone extends between L values of 3 and 11, and peaks at $\approx R_e = 4$ to 5.

The outer zone is a very dynamic region of space where some particles are stably trapped but others are considered to be pseudo-trapped, because their lifetimes are shorter than the drift time around the Earth. However, strong external galactic and solar sources supply electrons to this region of space so that substantial fluxes are always present. In this zone, the flux has large, short-term, temporal variations related to the local time as well as a long-term change in average flux associated with a solar cycle.

The current model for trapped electrons, AE8, was issued in its computer form in 1983 and documented in 1991 [8.19]. It is a merging of the older AE4 model for the outer zone electrons with models AE5P and AE6 for the inner zone electrons. A transition region between the zones was completely reconstructed. The AEI7-HI and AEI7-LO models which had been in interim use for several years had never been verified and were abandoned [8.19, 8.20]. AE8 was issued in two versions, SOLMAX and SOLMIN, describing the trapped electrons at solar maximum and solar minimum, respectively. The AE8 models have been used in the orbital integrations and calculations of GaAs solar panel degradation in Chapter 9. As with the proton models, it is important to use the correct magnetic field EPOCH in making calculations with the AE8 models. The recommendation is to use a magnetic field EPOCH of 1964 for AE8-MIN and an EPOCH of 1970 for AE8-MAX. In contrast to the AP8 models, the AE8-MAX model gives higher fluxes for solar maximum than AE8-MIN during solar minimum.

Since the geomagnetic equator is tilted with respect to the equator defined by the Earth's rotation, there is a longitudinal dependence on the fluence incident on a spacecraft at synchronous orbit [8.21]. The practical effect of this dependence on solar panels populated with Si solar cells was calculated in reference [8.1]. It was found that there are two maxima and two minima in the radiation intensity vs. longitude dependence, with the maxima occurring at 170° West and 20° East, and the minima occurring at 70° West and 110° East. It was found that the highest fluence was experienced at a longitude of 160° West, and the least fluence at 70° West. Chapter 9 includes a summary of orbital integration results for satellites in circular orbits at various altitudes and inclinations. A special calculation is presented in the last four tables of the chapter, reflecting this longitudinal dependence of radiation exposure as a function of longitude. The data in these tables show that the difference in 1 MeV equivalent fluences affecting panel degradation is approximately a factor of two, depending on the coverglass thickness used. Hughes Aircraft Company has built a large number of satellites that are operating in synchronous orbit. Lee Goldhammer of Hughes has verified that there is indeed a longitudinal dependence, and has verified that satellites in orbit near 160° West have experienced more solar panel degradation than those near 70° West [8.23].

One of the basic difficulties with the AE/AP models is the changing of the Earth's magnetic field. If one considers a given point in the geocentric coordinate system, this point experienced a certain magnetic field in 1964 when the minimum models were generated and on a long-term time averaged basis, it also experienced given values for the electron and proton fluence-energy spectra. Today, the magnetic field at that point has changed and, consequently, the fluence-energy spectra have changed also. But we still calculate values for the fluence-energy spectra as they were at that point in 1964. For this reason alone it is apparent that there are uncertainties in the use of the models.

Other models have been developed, more recently, based on the results of the radiation spectrometers flown on the CRRES spacecraft. This satellite was launched on July 25, 1990 and failed on October 12, 1991. It was in a highly elliptical orbit, 350 km by 33,000 km at an inclination of 18.2°.

It carried a complete set of radiation detectors and was in an ideal orbit for mapping the trapped radiation belts. The premature failure of the spacecraft only allowed the accumulation of data during solar maximum, but it generated sufficient data to allow a fresh look at modeling the trapped radiation and to allow a comparison with the AE8 and AP8 models. The results of this modeling may be found in references [8.24 - 8.26] and a brief summary of the modeling activity is given in reference [8.2]. The CRRES data was divided into two parts. The first part, called the "quiet period," covered the time period from July 27, 1990 to March 19, 1991. The second part, called the "active period," covered the time from March 31, 1991 to October 8, 1991. The separation of the quiet and active periods is based on an intense solar particle event that occurred on March 24, 1991, and the subsequent solar wind shock which rearranged the inner magnetosphere radiation populations. These experimenters have found differences between the CRRES models and the AE8 and AP8 models to be as much as 3 orders of magnitude at times near the "slot" region ($L \approx 2.5$). The "slot" region is the region between the inner (mostly proton) belt and the outer electron belt. There appear to be two major differences that occur at low L values. The first is attributable to a second, highly variable proton belt in the region $L = 1.8$ and $L = 4$ that is present in the CRRES active model, but not in the AP8 models. The second difference is due to a lack of high-energy electrons in the AE8 models, that are present in the CRRES models. At high L values, the CRRES experimenters found that the AE8 model is typically higher than the CRRES measurements at all L values above ≈ 3.4.

8.2 Trapped Radiation at Other Planets

Jupiter, Saturn, and Uranus have magnetic fields of some consequence and also have trapped radiation belts. The magnetic field at Jupiter is the strongest magnetic field in the solar system (≈ 8 Gauss at the poles as compared to ≈ 0.6 Gauss at the Earth's poles). The trapped radiation belts near Jupiter are therefore much more intense than they are at Earth. The Jovian magnetic field and particle distributions were measured by Pioneers 10 and 11 and by Voyagers 1 and 2 during their encounters. These measurements, along with theoretical considerations, have given rise to models of the trapped radiation belts at Jupiter [8.29]. A model of the fields and trapped radiation at Saturn has been developed, and some information about the fields and trapped radiation of Uranus has recently become available, but the information has not been published [8.30]. Enough information should be available to calculate the effect of the Jovian radiation belts on spacecraft solar panels as it flies by or goes into orbit around Jupiter. The intense belts at Jupiter would require such solar panels to be very well shielded if they are to survive an extended mission at this planet.

8.3   Solar Flare Protons

The Sun produces events on its surface known as sunspots. Sunspots are produced in a periodic fashion, with approximately an 11-year cycle between maxima. The sunspot cycle consists of an active time of about 7 years, during which solar flare events are probable, and a quiescent period of 4 years when solar flare events are rare. The active period usually begins about 2 years prior to the year of solar maximum, and lasts through the fourth year after the maximum. The maxima for the last 3 solar cycles as reported by reference [8.31] are shown in Table 8-2. We note that the time between peaks for cycles 21 and 22 was 10 years rather than the average 11 years. Such variances between the peak times are common.

Table 8-2.   Solar Cycle Maxima [8.31]

Cycle Number	Peak Occurrence
20	1968.9
21	1979.9
22	1989.9
23	2001. (?)

The occurrence of solar flare proton events is of a statistical nature. A spacecraft flying during the active part of the cycle may cruise along intercepting protons from numerous small flares, then be hit with a large event that totally overwhelms all other radiation it may receive during the mission. The statistics of the events have been studied thoroughly, especially during the past 3 solar cycles when we have had spacecraft available to measure the radiation in space. Each cycle adds to the data available and each cycle spawns a new and improved solar proton model. King made a probabilistic study of the solar proton fluence levels based on the 1966-1972 data [8.32]. He allowed for anomalously large (AL) events, assuming they had fluence-energy spectra given by the very large solar proton event of August 1972. The smaller, or ordinary (OR) events were assumed to obey a log normal distribution. A computer code, SOLPRO [8.33] was developed to calculate the expected fluence during a mission. The result depended on the time and length of the mission (i.e., whether there would be a dominating AL event) and on a probability factor. The probability factor is assumed by users of the code. If a spacecraft designer wishes to design a spacecraft for a fluence such that there is a 99% probability that the computed radiation level will not be exceeded during the mission, he would input a 99% confidence limit to the computer program. This will naturally result in a much higher fluence than if he were to select an 80% confidence limit. The adjustment of these confidence limits is therefore based on how much risk a flight project wishes to assume.

King's model was revised by Feynman et al. [8.34] based on the proton data available from an additional solar cycle (up until 1985). This model is in the NSSDC data base and has been widely used. A more recent model has been developed by Feynman and her coworkers [8.31] which adds the data from one more solar cycle, and in the author's words "has been able to use a data set collected by a single set of experiments over such a long period of time that the population of major events is probably well sampled." The data base comes from a series of closely related instruments on the IMP 1, 2, and 3, OGO 1, and IMP 5, 6, 7, and 8 series of spacecraft. Average fluxes above threshold energies of 1, 4, 10, 30, and 60 MeV were measured. The model produced by Feynman et al., known as the JPL 91 model, is distilled into Figures 8.3 through 8.7. (Note that the corresponding figures from reference [8.31] had an error in the labeling of abscissa. In the paper the fluence was labeled with units of cm^{-2} sr^{-1}. The correct units are cm^{-2}, as depicted in Figures 8.3 through 8.7). The plots may be used to predict the integral fluence as a function of confidence level and exposure time. To use the figures, locate the probability desired on the ordinate, using probability = (1 - confidence level), then read the integral fluence for the desired exposure time. For example, a confidence level of 0.99 (99%) translates to a probability of 0.01 (1%) that the fluence read from the figure will be exceeded during the mission. The exposure time must be correlated to the time of solar maximum. As an example, if a 7-year mission is to take place which would include the 4 years during solar minimum, then the exposure time used should only be 3 years. The calculated integral fluence spectra are for a distance of 1 AU from the Sun. For other radial distances from the Sun, an interplanetary charged-particle working group [8.35] recommended that for distances less than 1 AU, the fluence should be modified by an inverse cube ($1/R^3$) dependence, and for radial distances greater than 1 AU, the fluence should be modified by an inverse square ($1/R^2$) dependence. That is, for distances less than 1 AU, compute the average $1/R^3$ for the mission and multiply the results from the figures by that number. Similarly, for distances greater than 1 AU, perform the calculation for the average inverse square dependence. The $1/R^3$ dependence is intended to calculate a worst-case scenario. In most cases it will probably overestimate the radiation environment, particularly for missions that go very near the Sun. In their 1993 paper, Feynman et al.

[8.31] suggested using an inverse square dependence for all radial distances, stating that this should result in a "most probable" estimate rather than a "worst-case" estimate.

A Fortran program, "Solar Flare Estimator (SPE)" was developed by Feynman and Spitale [8.36] to perform the above steps by machine. It estimates proton integral fluences during the period from 1988 through 2010. It will perform these calculations either at 1 AU or in interplanetary space. The starting time and mission length must be input to the program. The program will calculate the radiation exposure for the most widely used confidence levels of 50%, 75%, 90%, 95%, and 99%. At the present time, there is no formal mechanism in place for distributing this program, but it may be available by contacting the authors.

There are other approaches to estimating the solar flare proton environment for spacecraft missions. One such approach is to design the solar panels for the total proton fluence observed during one of the recent solar cycles. This approach will usually result in a more benign environment than would be calculated by any of the above probabilistic models. Such an approach has worked well for several of the spacecraft designed by Hughes Aircraft Company [8.37]. Table 8-3 lists the total proton fluence observed from major solar proton events between September 1986 and July 1995. The data covers most of cycle 22, but must be considered preliminary, since cycle 22 had not ended in July 1995 [8.38]. For comparison, Table 8-3 also lists the fluence-energy spectra given by the SPE program as a function of various confidence levels. For the purposes of this calculation the mission was assumed to start in July 1988 with a duration of 7 years so that it was exposed to the full time period of solar maximum activity. As the table shows, the program's 75% confidence level calculates fluence values that are quite close to measured fluences. Calculations based on higher confidence levels overestimated the measured fluences, and the calculation at the 50% confidence level underestimated the measured values.

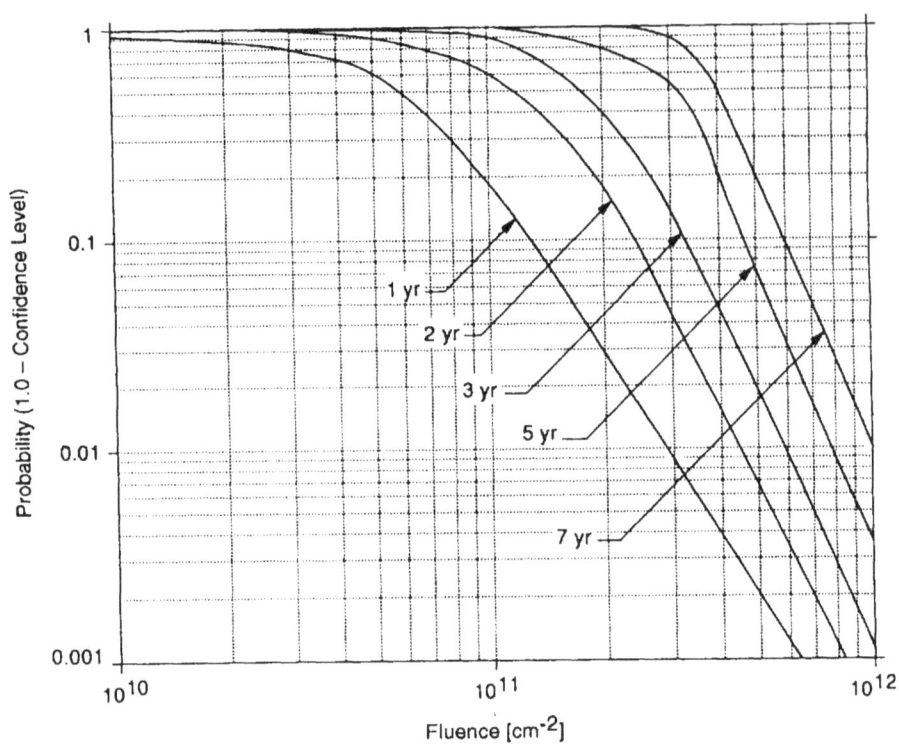

Figure 8.3.   Fluence Probability Curves for Protons of Energy Greater than 1 MeV for Various Exposure Times

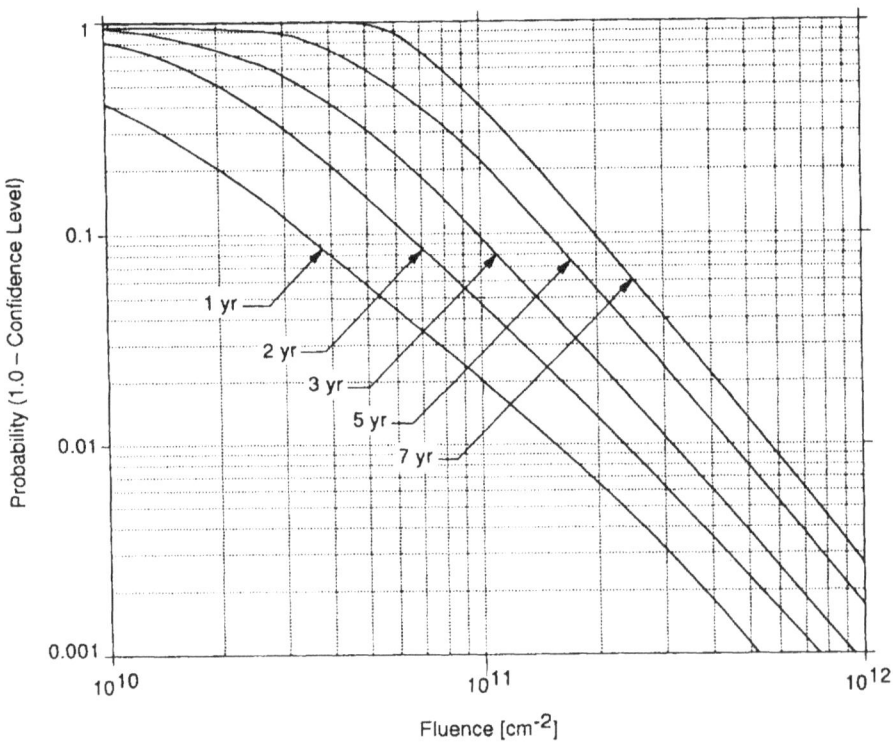

Figure 8.4.   Fluence Probability Curves for Protons of Energy Greater than 4 MeV for Various Exposure Times

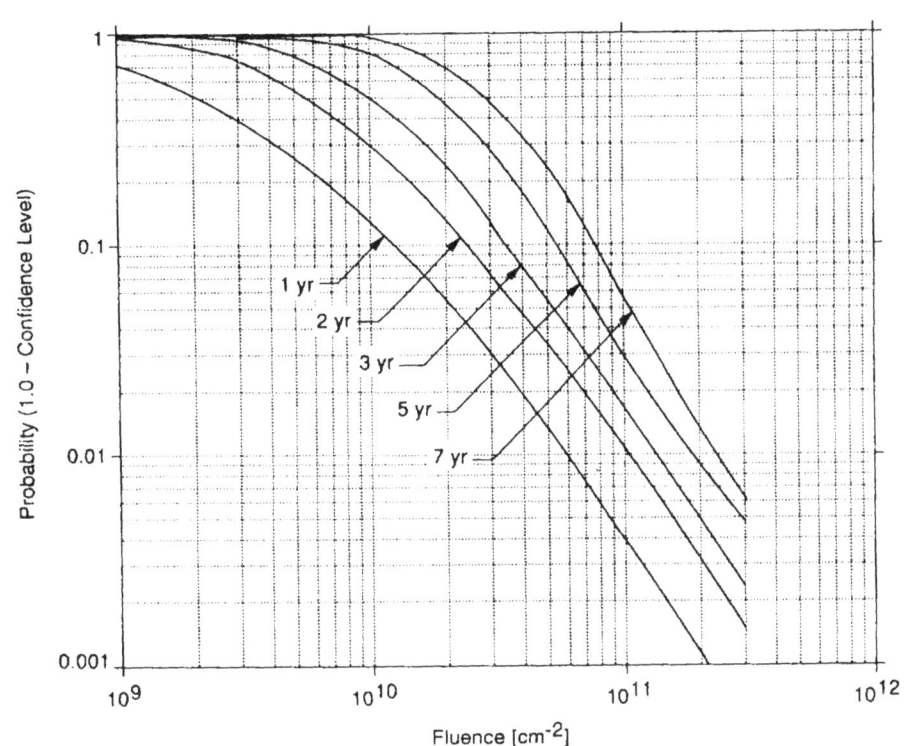

Figure 8.5. Fluence Probability Curves for Protons of Energy Greater than 10 MeV for Various Exposure Times

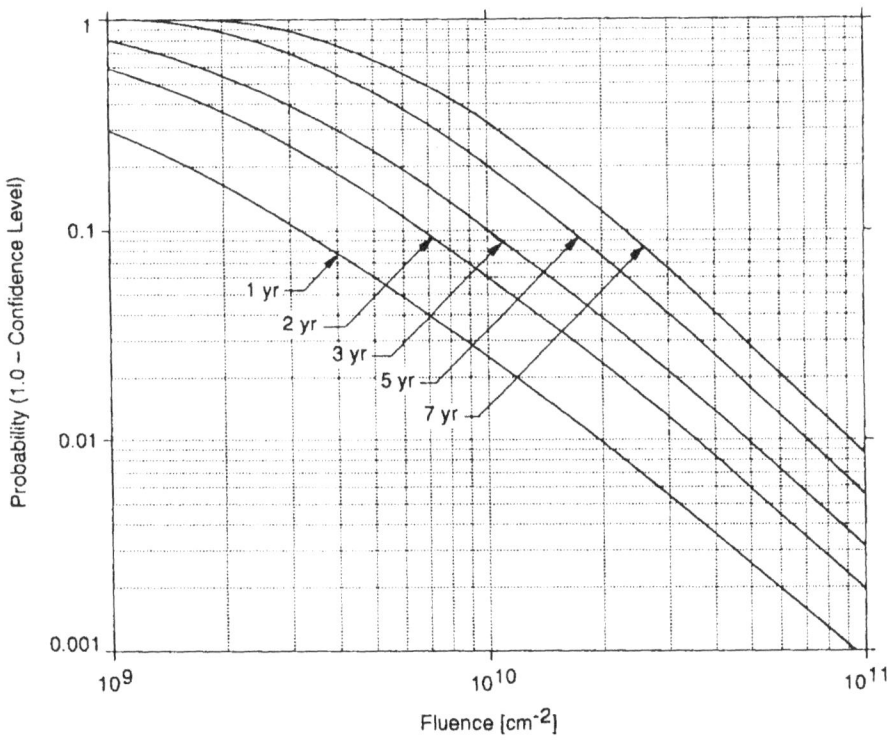

Figure 8.6. Fluence Probability Curves for Protons of Energy Greater than 30 MeV for Various Exposure Times

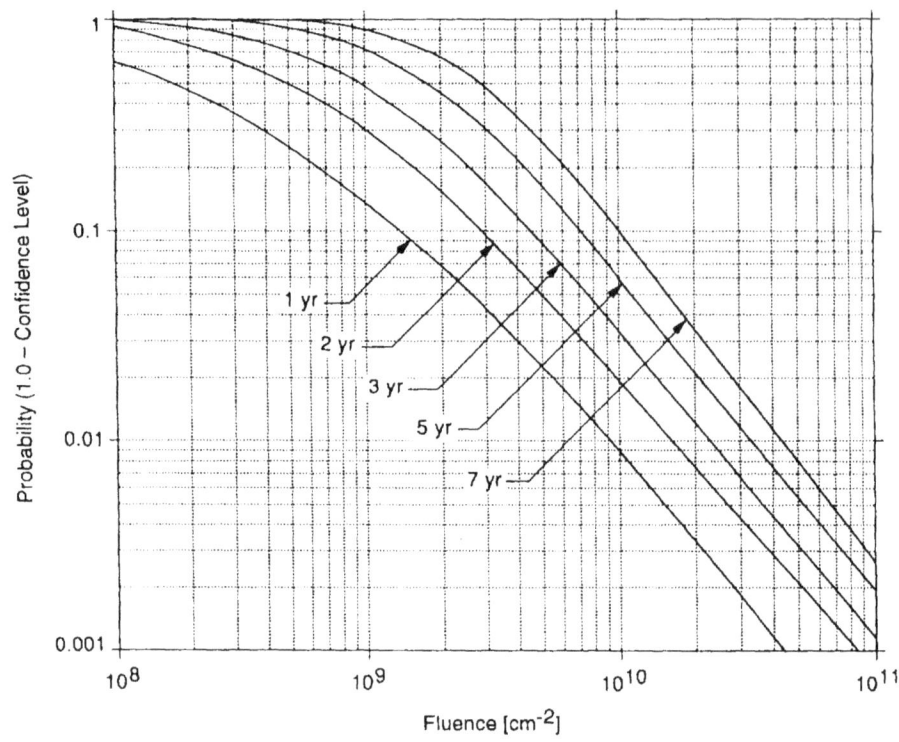

Figure 8.7.   Fluence Probability Curves for Protons of Energy Greater than 60 MeV for Various Exposure Times

Table 8-3.   Cumulative Fluence-Energy Spectra for the Major Flares during Solar Cycle 22

Date of Event	$\Phi(> 1\ MeV)$	$\Phi(> 5\ MeV)$	$\Phi(>10\ MeV)$	$\Phi(> 30\ MeV)$	$\Phi(> 60\ MeV)$
11/08/87	6.93 E08	1.17 E08	3.08 E07	1.20 E06	2.64 E05
3/08/89	2.14 E10	3.54 E09	1.11 E09	4.91 E07	4.03 E06
4/11/89	4.79 E09	7.33 E08	2.03 E08	4.69 E06	6.26 E05
5/01/89	2.51 E09	1.94 E08	4.18 E07	1.28 E06	4.24 E05
8/12/89	3.14 E10	1.30 E10	7.85 E09	1.52 E09	2.11 E08
9/29/89	9.91 E09	5.43 E09	3.78 E09	1.39 E09	4.79 E08
10/19/89	1.03 E11	3.90 E10	1.92 E10	4.24 E09	1.21 E09
11/27/89	1.89 E10	5.94 E09	2.21 E09	1.31 E08	6.30 E06
3/19/90	5.98 E09	1.88 E09	7.19 E08	2.17 E07	9.87 E05
4/28/90	4.32 E08	1.69 E08	7.34 E07	4.80 E06	3.32 E05
5/21/90	2.46 E09	6.32 E08	3.57 E08	1.38 E08	5.98 E07
8/01/90	3.10 E09	5.40 E08	1.71 E08	7.41 E06	8.61 E05
1/31/91	3.00 E09	3.96 E08	8.45 E07	1.28 E06	1.72 E05
3/23/91	4.24 E10	1.63 E10	9.51 E09	1.78 E09	1.64 E08
5/13/91	1.39 E09	3.21 E08	1.38 E08	1.82 E07	3.72 E06
6/04/91	3.10 E10	6.32 E09	3.20 E09	7.85 E08	2.00 E08
6/29/91	2.93 E10	5.04 E09	1.20 E09	3.00 E07	4.02 E06
8/25/91	5.16 E09	5.84 E08	1.27 E08	3.41 E06	4.99 E05
5/09/92	1.95 E10	2.78 E09	6.60 E08	1.32 E07	1.08 E06
6/25/92	1.60 E09	5.35 E08	2.87 E08	4.77 E07	9.53 E06
2/20/94	1.63 E10	3.82 E09	9.92 E08	4.13 E06	5.06 E05
Total	3.54 E11	1.37 E11	5.20 E10	1.02 E10	2.36 E09

Data from H. Sauer, NOAA Boulder, Events with Peak 10-MeV Proton Integral Fluxes > 100 Protons/(cm^2-s-sr)

SPE Confidence Level	$\Phi(> 1\ MeV)$	$\Phi(> 4\ MeV)$	$\Phi(> 10\ MeV)$	$\Phi(> 30\ MeV)$	$\Phi(> 60\ MeV)$
0.99	9.86 E11	5.55 E11	2.33 E11	8.92 E10	4.19 E10
0.95	6.85 E11	2.66 E11	1.04 E11	3.47 E10	1.51 E10
0.90	5.81 E11	1.93 E11	7.45 E10	2.21 E10	9.97 E09
0.75	4.70 E11	1.25 E11	4.64 E10	1.15 E10	5.19 E09
0.50	3.93 E11	8.72 E10	2.74 E10	6.52 E09	2.81 E09

Solar flare proton events are associated with phenomena on the Sun known as coronal mass ejections. These ejections occur at fairly localized places on the Sun, and as the Sun rotates, the sites of the ejections rotate with it. When one of these ejections occurs, the associated high-energy protons are ejected into space and they follow the field lines of the interplanetary magnetic field. The field lines emanate from the Sun and any particular field line may be described as an Archimedean spiral. The field line coming from the Sun that intersects the Earth originates from a so-called "foot point" on the Sun which is ≈ 57° to the east of the Earth-Sun line (west of central meridian [the receding limb as seen from Earth]). If the ejection occurs at this point on the Sun, the protons will immediately propagate to the Earth. The proton intensity seen at Earth will rise very rapidly, with the first protons arriving an hour or so after the flare. These protons are likely to be highly anisotropic since they are travelling along the magnetic field line. The protons released by ejections occurring at other sites on the Sun will propagate along magnetic field lines that miss the Earth. For this reason, variations as large as 100 in the particle fluxes from the same flare at different points around the Earth's orbit have been observed. However, the protons released at other than the foot point site can still reach the Earth, but they must first diffuse through the solar corona to the foot point before they propagate toward the Earth. When this occurs, the proton intensity at the Earth will rise at a less rapid rate, and their arrival may be delayed as much as

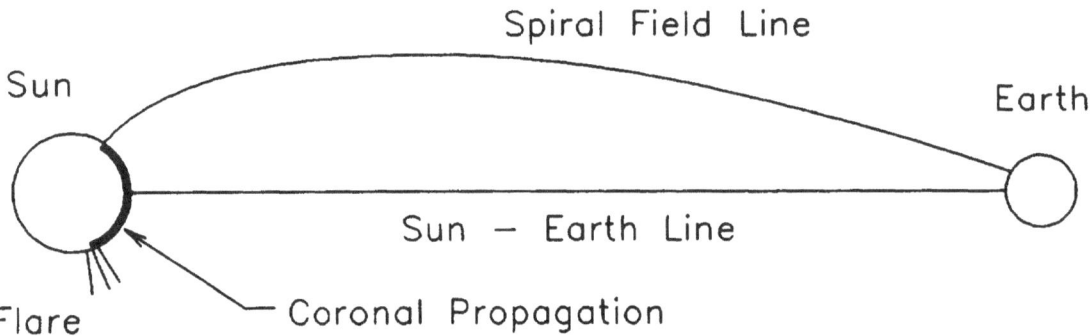

Figure 8.8.   Propagation of Solar Flare Protons from Sun to Earth

10 hours or so. In either event, the solar proton intensity rises to a maximum after a time determined by the solar longitude of the ejection and by the propagation time along the spiral, which is an inverse function of the particle velocity. After the maximum intensity occurs, the flux tends to become isotropic, and decays with an approximate 1/e relationship. There is, however, a tendency for a second peak in the flux to occur some 3 or 4 days later. The geometry of the solar flare proton propagation is illustrated in Figure 8.8 [8.39].

This chapter has given a broad-brush treatment of the important aspects of the space radiation environment, but references have been included that may be consulted for additional, more detailed study. In the next chapter, we will discuss the procedures that may be used with these models to calculate the effect of the space radiation environment on GaAs solar panels.

References for Chapter 8

8.1 H.Y. Tada, J.R. Carter, Jr., B.E. Anspaugh, and R.G. Downing, "Solar Cell Radiation Handbook," JPL Publication 82-69, Jet Propulsion Laboratory, California Institute of Technology, Pasadena, Calif., 1982.

8.2 H.B. Garrett and D. Hastings, "The Space Radiation Models," Paper No. 94-0590, *32nd AIAA Aerospace Sciences Meeting & Exhibit*, Reno, Nevada, January 1994.

8.3 H.B. Garrett, P. Rustan, S.P. Worden, and S. Nozette, "Radiation Environments within Satellites," *IEEE Nuclear Space Radiation Effects Conference Short Course*, Snowbird, Utah, Chapter 2, 1993.

8.4 R.C. Haymes, *Introduction to Space Science*, John Wiley & Sons, New York, 1971.

8.5 D. Bilitza, "Private Communication," National Space Science Data Center (NSSDC), Greenbelt Maryland, June 1995.

8.6 D.P. Le Galley and A. Rosen, *Space Physics*, John Wiley & Sons, New York, 1964.

8.7 J.S. Lew, "Drift Rate in a Dipole Field," *Journal of Geophysical Research*, 66, 2681, 1961.

8.8 J.I. Vette, "Models of the Trapped Radiation Environment, Vol. I: Inner Zone Protons and Electrons," NASA SP-3024, 1966.

8.9 J.I. Vette, A.B. Lucero, and J.A. Wright, "Models of the Trapped Radiation Environment, Vol. II: Inner and Outer Zone Electrons," NASA SP-3024, 1966.

8.10 J.I. Vette and A.B. Lucero, "Models of Trapped Radiation Environment, Vol. III: Electrons at Synchronous Altitudes," NASA SP-3024, 1967.

8.11 J.H. King, "Models of Trapped Radiation Environment, Vol. IV: Low Energy Protons," NASA SP-3024, 1967.

8.12 J.P. Lavine and J.I. Vette, "Models of the Trapped Radiation Environment, Vol. V: Inner Belt Protons," NASA SP-3024, 1969.

8.13 J.P. Lavine and J.I. Vette, "Model of the Trapped Radiation Environment, Vol. VI: High Energy Protons," NASA SP-3024, 1970.

8.14 G.W. Singley and J.I. Vette, "The AE-4 Model of the Outer Radiation Zone Electron Environment," NASA, NSSDC 72-06, 1972.

8.15 M.J. Teague and J.I. Vette, "The Inner Zone Electron Model AE-5," NASA, NSSDC 72-10, 1972.

8.16 M.J. Teague, K.W. Chan, and J.I. Vette, "AE6: A Model Environment of Trapped Electrons for Solar Maximum," NASA, NSSDC/WDC-A-R&S 76-04, 1976.

8.17 G.W. Singley and J.I. Vette, "A Model Environment for Outer Zone Electrons," NASA, NSSDC 72-13, 1972.

8.18 D.M. Sawyer and J.I. Vette, "AP8 Trapped Proton Environment for Solar Maximum and Solar Minimum," NSSDC/WDC-A-R&S 76-06, 1976.

8.19 J.I. Vette, "The AE8 Trapped Electron Model Environment," NSSDC/WDC-A-R&S 91-24, November 1991.

8.20 J.I. Vette, "The NASA/National Space Science Data Center Trapped Radiation Environment Model Program (1964-1991)," NSSDC/WDC-A-R&S 91-29, November 1991.

8.21 C.W. McIlwain, "Coordinates for Mapping the Distribution of Magnetically Trapped Particles," *Journal of Geophysical Research*, 66, 3681, 1961.

8.22 E.G. Stassinopoulos, "The Geostationary Radiation Environment," *Journal of Spacecraft and Rockets*, 17, 145, 1980.

8.23 L.J. Goldhammer, Private Communication, January 1995.

8.24 K.J. Kerns and M.S. Gussenhoven, "CRRESRAD Documentation," Air Force Phillips Laboratory, Directorate of Geophysics: PL-TR-92-2201, August 1992.

8.25 D.H. Brautigam, M.S. Gussenhoven, and E.G. Mullen, "Quasi-Static Model of Outer Zone Electrons," *IEEE Transactions on Nuclear Science*, 39, 1797, 1992.

8.26 M.S. Gussenhoven, E.G. Mullen, and D.H. Brautigam, "Near-Earth Radiation Model Deficiencies as Seen on CRRES." COSPAR, 1993.

8.27 E.G. Mullen, M.S. Gussenhoven, K. Ray, and M. Violet, "A Double-Peaked Inner Radiation Belt: Cause and Effect as Seen on CRRES," *IEEE Transactions on Nuclear Science*, 38, 1713, 1991.

8.28 M.S. Gussenhoven, E.G. Mullen, M.D. Violet, C. Hein, J. Bass, and D. Madden, "CRRES High Energy Proton Flux Maps," *IEEE Transactions on Nuclear Science*, 40, 1450, December 1993.

8.29 N. Divine and H.B. Garrett, "Charged Particle Distributions in Jupiter's Magnetosphere," *Journal of Geophysical Research*, 88, No. A9, 6889, September 1983.

8.30 G. Spitale and M. Ratliff, Private Communication, June 1995.

8.31 J. Feynman, G. Spitale, J. Wang, and S. Gabriel, "Interplanetary Proton Fluence Model: JPL 1991," *Journal of Geophysical Research*, 98, No. A8, 13281, August 1993.

8.32 J.H. King, "Solar Proton Fluences for 1977-1983 Space Missions," *Journal of Spacecraft and Rockets*, 11, No. 6, 401, June 1974.

8.33   E.G. Stassinopoulos, "SOLPRO: A Computer Code to Calculate Probabilistic Energetic Solar Proton Fluences," NASA, NSSDC 75-11, 1975.

8.34   J. Feynman, T.P. Armstrong, L. Dao-Gibner, and S.M. Silverman, "A New Interplanetary Proton Fluence Model," *Journal of Spacecraft and Rockets*, 27, 403, 1990.

8.35   J. Feynman, S.B. Gabriel (Ed.), *Interplanetary Particle Environment: Proceedings of a Conference*, JPL Publication 88-23, Jet Propulsion Laboratory, California Institute of Technology, Pasadena, Calif., 1988.

8.36   G. Spitale and J. Feynman, "Program SPE," Jet Propulsion Laboratory, California Institute of Technology, Pasadena, Calif., 1992.

8.37   L.J. Goldhammer, "Recent Solar Flare Activity and its Effect on In-Orbit Solar Arrays," *Proceedings of the 21st IEEE Photovoltaic Specialists Conference*, 1241, 1990.

8.38   H.H. Sauer, NOAA, Private Communication, July 1995.

8.39   D.F. Smart, "Predicting the Arrival Times of Solar Particles," *Interplanetary Particle Environment: Proceedings of a Conference*, JPL Publication 88-23, Jet Propulsion Laboratory, California Institute of Technology, Pasadena, Calif., 1988.

Chapter 9

GaAs Solar Array Degradation Calculations

9.1   General Procedure

In the previous chapters, the three basic elements required to perform solar array degradation calculations were developed. The first of these elements is degradation data for GaAs solar cells after irradiation with 1 MeV electrons at normal incidence. The second element is the calculated tables of relative damage coefficients for omnidirectional electron and proton exposure. The third element is the definition of the space radiation environment for the orbit of interest. This chapter will cover the use of this data to perform a solar array degradation calculation. The discussion in this chapter closely follows the treatment given in the *Solar Cell Radiation Handbook* [9.1].

The relative damage coefficients allow the conversion of various energy spectra of omnidirectional space electrons and protons into equivalent fluences. The equivalent fluences are based on normal incidence, monoenergetic irradiations for which the degradations of the solar cells are characterized in laboratory measurements. The process of weighting an integral energy spectrum of electrons for a given orbit can be described by the following equation:

$$\phi_{1\,MeV\,e} = \sum_{E=0}^{\infty} \left[\phi(>E) - \phi(>E + \Delta E) \right] \cdot D(E,t) \qquad (9\text{-}1)$$

where $\Phi_{1\,MeV\,e}$ =   the damage equivalent 1 MeV electron fluence (e/cm^2-year)

$\Phi(>E) - \Phi(>E + \Delta E)$ =   the isotropic particle fluence having energies in a small energy increment greater than energy E (e/cm^2-year)

$D(E,t)$ =   the relative damage coefficient for isotropic fluences of energy E incident on solar cells shielded by coverglasses of thickness t (dimensionless)

The quantities $\Phi(>E) - \Phi(>E + \Delta E)$ for a range of energies are also known as the difference spectrum. This spectrum can be generated from an integral fluence-energy spectrum (the type most commonly quoted) for any energy increments desired.

Equation (9-1) is also applicable to isotropic proton space environments, but the damage coefficients as defined here convert the proton spectra to equivalent 10 MeV fluences of normal incidence. Once the 10 MeV equivalent proton fluences have been calculated, they may be converted to 1 MeV equivalent electron fluences by multiplying by the factors listed in Table 9-1.

Table 9-1. Conversion Factors for Protons to Electrons for GaAs Solar Cells
(Converting from 10 MeV Protons to 1 MeV Electrons)

Parameter	Factor
P_{max}	1000
V_{oc}	1400
I_{sc}	400

The above values are approximations which must be made for the purpose of combining electron and proton damage. As we observed in the discussion associated with Figure 5.1, the degradation curves, as a function of various energies of both protons and electrons, would have all had to be parallel to each other for the above factors to hold with great precision.

An additional complication arises in calculating equivalent fluences for proton environments. The damage coefficients describing solar cell degradation are quite different for I_{sc} than they are for P_{max} and V_{oc}, and for this reason two sets of damage coefficients were derived for proton degradation in Chapter 5. This differs with the calculations of electron damage coefficients, where we found that only one set was necessary for all the cell electrical parameters. This requirement for one set of damage coefficients for electron irradiation and two sets for proton irradiation was also found to be true for Si cells. But with GaAs cells, after the two sets of equivalent, normal 10 MeV proton fluences have been calculated, we must apply the factors listed in Table 9-1 to compute equivalent 1 MeV electron fluences, and we end up with a set of three 1 MeV equivalent fluences for the proton degradation calculation. The calculation for the 1 MeV equivalent fluence for omnidirectional electrons, computed separately, must then be added to each of the three 1 MeV equivalent fluences computed for protons. The state of the cell degradation may then be assessed by consulting the curves in Chapter 6, which show degradation vs. 1 MeV electron fluence.

9.2   Rear-Incidence Radiation

The omnidirectional damage coefficients, $D(E,t)$, have been calculated with the assumption that an infinitely thick shield protects the rear surface of the solar panel. This condition is approximately met for body-mounted solar arrays of spinning spacecraft, but is not generally true for spacecraft with "wing-type" solar panels. With the ever increasing popularity of making high power-to-weight ratio panels, the protective substrates of the panels become thinner and thinner, and there may be as much radiation incident on the cells from their rear surfaces as from their front surfaces. The calculation of the effect of rear-surface radiation is not well established, but some work at JPL on Si solar cells (reference [9.2]) indicated that certain equivalent rear-shielding thicknesses could be assumed, depending on the type of solar cell in use. Since GaAs/Ge solar cells are grown on relatively thick Ge substrates and GaAs/GaAs cells are likewise grown on relatively thick GaAs substrates, while the active cell area is only a few microns thick, it should be accurate to assume that whatever structure is beneath the active GaAs cell should be considered as shielding. In order to account for the rear-incidence radiation, the shielding on the rear surfaces must be computed. This is done by adding up the thicknesses of the panels, the panel-to-cell adhesive, the solar cell rear contact (silver will probably be the most important ingredient), and the thickness of the Ge or GaAs cell substrates. Multiply the thickness (in cm) of each item by its density (g/cm^3) to convert each thickness to units of g/cm^2. Add the thicknesses in g/cm^2 units, then convert this thickness to an equivalent coverglass thickness by dividing by the coverglass density of 2.2 g/cm^3, which was used to calculate the orbital damage coefficients and degradation tables in this chapter. The equivalent fluence incident on the solar cell due to a coverglass of this thickness (interpolation between the tabulated values will probably be necessary) should then be added to the equivalent fluence incident on the front surface of the cell. An example of this procedure will be given in Section 9.5.

9.3   Rough Degradation Calculations

For circular orbits, Vette et al. [9.3, 9.4] have performed time integrations for both electrons and protons. Their calculations include convenient energy ranges, various altitudes, and inclinations of $0°$, $30°$, $60°$, and $90°$. They have performed the computations for both solar maximum and solar minimum conditions. The average daily omnidirectional integral fluences are presented in the form of carpet plots in reference [9.3] and in tables in reference [9.4]. A rough determination of the equivalent fluence can be made by using the procedure described by eq. (9-1) and the factors given in Table 9-1. An example of such a calculation for a spacecraft flying in a 3000 nmi circular orbit at an inclination of $0°$ is presented in Table 9-2, where the equivalent 1 MeV electron fluence due to the AP-8 proton

Table 9-2.   Rough Degradation Calculation

AP8 MIN Solar Protons, 3000 nmi Circular Orbit, 0 Deg. Inclination Calc. for Voc and Pmax									
Energy Increment		Integral Energy Spectrum	Difference Spectrum	Shielding Thickness (6 mils)		Shielding Thickness (12 mils)		Shielding Thickness (30 mils)	
E1	E2	F(>E1)	F(>E1)-F(>E2)	D(E,6)	Equiv Fluence	D(E,12)	Equiv Fluence	D(E,6)	Equiv Fluence
(MeV)	(MeV)	(p/cm2-day)	(p/cm2-day)	Table 5.3	(p/cm2-day)	Table 5.3	(p/cm2-day)	Table 5.3	(p/cm2-day)
2	4	2.80E+11	1.50E+11	0.100	1.50E+10	0.000	0.00E+00	0.000	0.00E+00
4	6	1.30E+11	5.80E+10	0.930	5.39E+10	0.000	0.00E+00	0.000	0.00E+00
6	8	7.20E+10	3.40E+10	0.800	2.72E+10	0.590	2.01E+10	0.000	0.00E+00
8	10	3.80E+10	1.80E+10	0.640	1.15E+10	0.660	1.19E+10	0.000	0.00E+00
10	15	2.00E+10	1.30E+10	0.460	5.98E+09	0.500	6.50E+09	0.300	3.90E+09
15	20	7.00E+09	4.90E+09	0.380	1.86E+09	0.380	1.86E+09	0.370	1.81E+09
20	30	2.10E+09	1.10E+09	0.330	3.63E+08	0.340	3.74E+08	0.330	3.63E+08
30	40	1.00E+09	5.70E+08	0.310	1.77E+08	0.310	1.77E+08	0.305	1.74E+08
40	60	4.30E+08	2.30E+08	0.300	6.90E+07	0.295	6.79E+07	0.295	6.79E+07
60	80	2.00E+08	9.00E+07	0.280	2.52E+07	0.281	2.53E+07	0.284	2.56E+07
80	100	1.10E+08	4.00E+07	0.270	1.08E+07	0.280	1.12E+07	0.280	1.12E+07
100	150	7.00E+07	3.70E+07	0.265	9.81E+06	0.266	9.84E+06	0.266	9.84E+06
150		3.30E+07							
Equivalent 10 MeV protons/cm2-day					1.16E+11		4.10E+10		6.36E+09
Equivalent 10 MeV protons/cm2-year (x 365.25)					4.24E+13		1.50E+13		2.32E+12
Equivalent 1 MeV electrons for Pmax (x 1000)					4.24E+16		1.50E+16		2.32E+15
Machine Calculation:					4.17E+16		1.56E+16		2.04E+15
Equivalent 1 MeV electrons/cm2-year for Voc (x1400)					5.94E+16		2.09E+16		3.25E+15
Machine Calculation:					5.83E+16		2.18E+16		2.85E+15

environment is calculated for three coverglass thicknesses. The fluence-energy spectrum for this orbit at Solar Minimum was read from the carpet plots in reference [9.3]. The proton damage coefficients for the 3 coverglass thicknesses were taken from Table 5-3. We have used the proton damage coefficients appropriate for either P_{max} or V_{oc}. The calculation for 1 MeV equivalent fluence is completed by multiplying by a factor of 1000 for the P_{max} degradation, and by a factor of 1400 for the V_{oc} degradation. A similar procedure would be used for computing the I_{sc} performance for these cells, with the exception that we would use damage coefficients appropriate for I_{sc} degradation (see Table 5-2) and multiply the 10 MeV equivalent proton fluence by a factor of 400 to compute the equivalent 1 MeV electron fluence.

Several observations can be made about the calculation shown in Table 9-2. The largest contribution to the equivalent fluence for the 6 mil cover calculation occurs in the 4 to 6 MeV energy range. The damage coefficient is zero for energies below 4 MeV and is changing rapidly between 4 and 6 Mev, so large errors can be made in using coarse energy steps in the energy regions where the damage coefficients change rapidly. A similar observation may be made for the 12-mil coverglass calculation, where the largest contribution to the total appears between 6 and 8 MeV, and the damage coefficients are changing rapidly in that region. Also in the 30-mil coverglass calculation, the damage coefficients become nonzero at ≈ 11 MeV and the largest contribution to the total fluence is between 10 and 15 MeV. Table 9-2 also shows that the contributions to the equivalent fluence become increasingly less important as the energy increases. The computer calculations, to be discussed later, use the same calculational procedure as illustrated in the spreadsheet, with two important exceptions. The energy increments taken are much smaller, and the interpolation methods used are different. In our spreadsheet calculation, we used eyeball interpolation to estimate the appropriate damage coefficients, whereas the computer uses a formal interpolation procedure. The results of the machine calculation for this orbit are also shown in Table 9-2. The agreement, considering the differences in the two procedures, and also that entirely different orbit generating programs were used, is quite satisfactory.

9.4   Computer-Calculated Equivalent Fluence

The rough calculations discussed above can be improved in accuracy and speed with the aid of a computer. Although the quantity computed is exactly the same as before, the selection of difference fluence and the matching damage coefficient can be programmed to achieve higher accuracy and more consistent results. The increased accuracy is achieved mainly by the use of finer energy increments, and by using interpolation schemes matched to the interpolated functions. For example, when finding the integral fluence vs. energy at energy levels between the tabulated values given by the orbital integration routine, it has been found to be more accurate to use a linear interpolation of log Φ vs. linear E for the electron spectra and to use a linear interpolation of log Φ vs. log E for the integral proton spectra.

Tables 9-4 through 9-57 have been calculated by computing orbital coordinates for various circular orbits, converting the spatial coordinates into B,L coordinates, calculating the contributions to both the electron and proton fluence spectra at each position, and then converting these fluence-energy spectra into 1 MeV equivalent electron fluences appropriate for 8 different coverglass thicknesses. The orbital integrations are performed using equations developed in reference [9.5] to compute appropriate trajectories for Earth-orbiting spacecraft. These orbital calculations are estimated to give accuracies of generally a few tens of kilometers in altitude, $\pm 1°$ in latitude or longitude, and ± 1 minute in time. The conversion to B,L coordinates uses routines developed at the National Space Science Data Center [9.6, 9.7]. As discussed in Chapter 8, the Earth's magnetic field is changing fairly rapidly with time and, therefore, the detailed distribution of the trapped radiation belts also changes. It is important to use a magnetic field model appropriate for the time around which the models were constructed [9.8]. The radiation models used in these calculations were AE8MIN, AE8MAX, AP8MIN, and AP8MAX, depending on whether the calculation was for the minimum or maximum of the solar cycle. The appropriate times to use for these models are 1964 for AE8MIN and AP8MIN and 1970 for AE8MAX and AP8MAX [9.9]. Calculation of the proton and electron fluence at each point in B,L space was performed by the NSSDC program SOFIP [9.10]. The program EQGAFLUX, listed as Table A-10 in the Appendix, is used for converting the fluence-energy spectra into 1 MeV equivalent fluence using the tabulated D(E,t) values.

A comparison of the values computed by AE8MIN and AE8MAX shows that the AE8MIN values are slightly lower than the values computed using AE8MAX. This is generally true at all altitudes. The situation is reversed for the AP8 models, however, where the values calculated using AP8MIN are higher than those given by AP8MAX. The differences in the AP8 models are only significant at altitudes below ≈ 1000 nmi. For these reasons, we have included complete tables (all altitudes) for electron calculations at both solar minimum and solar maximum. Values for the calculation using AP8MIN for altitudes below 1000 nmi are appended to the complete tables for AP8MAX.

The final four tables in this chapter are the results of computing the equivalent fluences at synchronous altitude as a function of longitude. There is a variation in radiation exposure with longitude, because the magnetic equator is tilted with respect to the geocentric equator, with the two maxima and two minima in radiation levels. The maxima occur near 160° west longitude and 20° east longitude, and the minima occur near 70° west longitude and 110° east longitude. This longitude dependence is not important for spacecraft orbiting at lower altitudes because they will sweep through all longitudes and will experience a time-averaged exposure.

9.5   Example of Calculation for Rear-Incidence Radiation on a Thin Solar Panel

Suppose the spacecraft flying in the previously considered 3000 nmi circular orbit had solar panels with flexible substrates. In such a configuration there will be a significant amount of radiation reaching the solar cells through the rear surfaces. We proceed by cataloging the materials on the rear surface of the solar cells, along with their densities, and convert all the thicknesses into units of g/cm^2. Suppose the panel substrate under consideration is made up of the materials shown in Table 9-3:

Table 9-3.   Catalog of Materials and Thicknesses for an Example Substrate

Material	Thickness (cm)	Density (g/cm^3)	Thickness (g/cm^2)
Kapton	50. E-4	1.47	7.35 E-3
Adhesive	50. E-4	1.80	9.00 E-3
Silver	10. E-4	10.50	1.05 E-2
Germanium	100. E-4	5.36	5.36 E-2
Total			8.05 E-2 g/cm^2
Equivalent Coverglass Thickness (Divide by 2.2 g/cm^3)			3.66 E-2 cm (14.4 mils)

The procedure outlined in the table demonstrates how the various material thicknesses are converted to an equivalent coverglass thickness of 14.4 mils (366 μm). The density of 2.2 g/cm^3, appropriate for 7940 fused silica, was used because that material density was used in the calculation of the damage coefficient tables and for the orbital integration results tabulated later in this chapter. Suppose that the cells on our panel are protected with 12-mil-thick CMG coverglasses. This material has a density of 2.554 g/cm^3 (see Table A-5 in the Appendix). They are therefore equivalent to 7940 coverglasses that are (2.554/2.2) times as thick, or 13.9 mils thick. We now refer to Table 9-7, where the results of orbital integrations for various circular orbits at 0° inclination are tabulated for the proton degradation of P$_{max}$. According to this table, a panel with infinite backshielding and protected with 12-mil-thick coverglasses (7940) will experience an equivalent 1 MeV electron fluence of 1.56 E16 e/cm^2, and if the coverglass thickness is 20 mils, the equivalent fluence will be 6.26 E15 e/cm^2. The equivalent thickness protecting the front surface of our example panel is 13.9 mils, and a linear interpolation shows that the equivalent fluence incident on the cell would be 1.34 E16 e/cm^2. A similar consideration for the 14.4-mil-thick equivalent thickness protecting the rear cell surface yields an equivalent fluence of 1.28 E16 e/cm^2. We add these two values together to find a total equivalent 1 MeV electron fluence of 2.62 E16 e/cm^2. We should now repeat this procedure with Table 9-4 or Table 9-5 (depending on which part of the solar cycle is of concern) to calculate the additional equivalent fluence contributed by the trapped electrons, however, in this case, the electron contribution is so totally overwhelmed by the protons that this step can be ignored. We now consult Figure 6.10 to see how much power will remain after 1 year in this orbit, and we find that this value is off scale. An estimate of the approximate degradation can still be made, however, by consulting Figure 6.30. Although the degradation curves illustrated in this figure were not derived from modern cells, we can use the shape of the curve beyond the 1.0 E16 e/cm^2 electron fluence point to estimate the degradation. Figure 6.10 shows that the P$_{max}$ degradation at a fluence of 1.0 E16 e/cm^2 gives a P$_{max}$/P$_{max0}$ of $\approx$0.4. If we translate the bottom curve of Figure 6.30 upward, we estimate that the P$_{max}$/P$_{max0}$ for a fluence of 2.6 E16 e/cm^2 would be $\approx$0.30.

Table 9-4. Annual Equivalent 1 MeV Electron Fluence for GaAs Solar Cells from Trapped Electrons During Solar Max, 0° Inclination (Infinite Backshielding)

ELECTRONS - I_{sc}, V_{oc}, AND P_{max} INCLINATION = 0 DEGREES

EQUIV. 1 MEV ELECTRON FLUENCE FOR I_{sc} - CIRCULAR ORBIT
DUE TO GEOMAGNETICALLY TRAPPED ELECTRONS - MODEL AE8MAX

ALTITUDE (nm)	(km)	SHIELD THICKNESS, cm (mils)							
		0 (0)	2.54E-3 (1)	7.64E-3 (3)	1.52E-2 (6)	3.05E-2 (12)	5.09E-2 (20)	7.64E-2 (30)	1.52E-1 (60)
150	277	0.00E+00	0.00E+00	0.00E+00	0.00E+00	0.00E+00	0.00E+00	0.00E+00	0.00E+00
250	463	5.78E+06	5.54E+06	5.25E+06	4.89E+06	4.38E+06	3.88E+06	3.39E+06	2.39E+06
300	555	2.60E+07	2.40E+07	2.17E+07	1.93E+07	1.62E+07	1.35E+07	1.12E+07	7.16E+06
450	833	4.55E+09	3.48E+09	2.48E+09	1.70E+09	9.87E+08	6.00E+08	3.83E+08	1.54E+08
600	1111	3.54E+11	2.59E+11	1.74E+11	1.10E+11	5.62E+10	3.02E+10	1.74E+10	5.74E+09
800	1481	6.46E+12	4.80E+12	3.30E+12	2.16E+12	1.17E+12	6.75E+11	4.11E+11	1.48E+11
1000	1852	2.37E+13	1.79E+13	1.26E+13	8.57E+12	4.96E+12	3.03E+12	1.94E+12	7.44E+11
1250	2315	5.72E+13	4.40E+13	3.19E+13	2.24E+13	1.37E+13	8.77E+12	5.83E+12	2.36E+12
1500	2778	9.29E+13	7.21E+13	5.28E+13	3.76E+13	2.35E+13	1.54E+13	1.04E+13	4.33E+12
1750	3241	1.21E+14	9.27E+13	6.70E+13	4.66E+13	2.80E+13	1.75E+13	1.15E+13	4.69E+12
2000	3704	1.43E+14	1.08E+14	7.64E+13	5.13E+13	2.86E+13	1.64E+13	9.96E+12	3.70E+12
2250	4167	1.62E+14	1.21E+14	8.36E+13	5.44E+13	2.82E+13	1.46E+13	8.07E+12	2.65E+12
2500	4630	1.73E+14	1.28E+14	8.74E+13	5.54E+13	2.70E+13	1.26E+13	6.18E+12	1.71E+12
2750	5093	1.72E+14	1.27E+14	8.58E+13	5.36E+13	2.49E+13	1.06E+13	4.55E+12	1.00E+12
3000	5556	1.55E+14	1.14E+14	7.71E+13	4.77E+13	2.14E+13	8.45E+12	3.24E+12	5.51E+11
3500	6482	1.12E+14	8.14E+13	5.41E+13	3.28E+13	1.41E+13	5.17E+12	1.77E+12	2.17E+11
4000	7408	8.07E+13	5.85E+13	3.86E+13	2.32E+13	9.90E+12	3.57E+12	1.20E+12	1.42E+11
4500	8334	6.89E+13	5.09E+13	3.45E+13	2.15E+13	9.82E+12	3.91E+12	1.47E+12	2.05E+11
5000	9260	5.59E+13	4.26E+13	3.03E+13	2.02E+13	1.05E+13	5.11E+12	2.49E+12	6.39E+11
5500	10186	4.47E+13	3.60E+13	2.76E+13	2.04E+13	1.28E+13	7.85E+12	4.89E+12	1.86E+12
6000	11112	4.18E+13	3.61E+13	3.01E+13	2.44E+13	1.77E+13	1.27E+13	8.97E+12	4.03E+12
7000	12964	6.44E+13	5.81E+13	5.10E+13	4.38E+13	3.45E+13	2.67E+13	2.03E+13	1.02E+13
8000	14816	9.63E+13	8.75E+13	7.72E+13	6.68E+13	5.31E+13	4.14E+13	3.18E+13	1.64E+13
9000	16668	1.30E+14	1.18E+14	1.04E+14	9.00E+13	7.15E+13	5.57E+13	4.28E+13	2.21E+13
10000	18520	1.61E+14	1.45E+14	1.28E+14	1.10E+14	8.63E+13	6.65E+13	5.03E+13	2.50E+13
11000	20372	1.82E+14	1.64E+14	1.43E+14	1.22E+14	9.50E+13	7.21E+13	5.36E+13	2.49E+13
12000	22224	1.85E+14	1.66E+14	1.44E+14	1.22E+14	9.29E+13	6.91E+13	5.01E+13	2.19E+13
13000	24076	1.69E+14	1.50E+14	1.29E+14	1.07E+14	8.02E+13	5.82E+13	4.13E+13	1.72E+13
14000	25928	1.43E+14	1.26E+14	1.07E+14	8.76E+13	6.40E+13	4.54E+13	3.15E+13	1.28E+13
15000	27780	1.20E+14	1.05E+14	8.80E+13	7.15E+13	5.15E+13	3.58E+13	2.45E+13	9.64E+12
16000	29632	9.46E+13	8.17E+13	6.75E+13	5.41E+13	3.80E+13	2.59E+13	1.74E+13	6.61E+12
17000	31484	7.30E+13	6.24E+13	5.10E+13	4.03E+13	2.78E+13	1.86E+13	1.23E+13	4.47E+12
18000	33336	5.87E+13	4.98E+13	4.04E+13	3.17E+13	2.16E+13	1.43E+13	9.28E+12	3.24E+12
19327	35793	4.49E+13	3.78E+13	3.03E+13	2.35E+13	1.58E+13	1.03E+13	6.57E+12	2.17E+12

Table 9-5.   Annual Equivalent 1 MeV Electron Fluence for GaAs Solar Cells from Trapped Electrons During Solar Min, 0° Inclination (Infinite Backshielding)

ELECTRONS   - I_{sc}, V_{oc}, AND P_{max} INCLINATION = 0 DEGREES

EQUIV. 1 MEV ELECTRON FLUENCE FOR I_{sc} - CIRCULAR ORBIT
DUE TO GEOMAGNETICALLY TRAPPED ELECTRONS - MODEL AE8MIN

| ALTITUDE | | SHIELD THICKNESS, cm (mils) | | | | | | | |
(nm)	(km)	0 (0)	2.54E-3 (1)	7.64E-3 (3)	1.52E-2 (6)	3.05E-2 (12)	5.09E-2 (20)	7.64E-2 (30)	1.52E-1 (60)
150	277	0.00E+00	0.00E+00	0.00E+00	0.00E+00	0.00E+00	0.00E+00	0.00E+00	0.00E+00
250	463	4.65E+06	4.48E+06	4.27E+06	4.01E+06	3.62E+06	3.23E+06	2.84E+06	2.02E+06
300	555	2.10E+07	1.96E+07	1.80E+07	1.62E+07	1.38E+07	1.17E+07	9.81E+06	6.37E+06
450	833	2.54E+09	2.05E+09	1.56E+09	1.14E+09	7.26E+08	4.66E+08	3.08E+08	1.31E+08
600	1111	1.70E+11	1.32E+11	9.62E+10	6.68E+10	3.87E+10	2.25E+10	1.36E+10	4.83E+09
800	1481	3.37E+12	2.68E+12	2.02E+12	1.46E+12	9.04E+11	5.61E+11	3.57E+11	1.36E+11
1000	1852	1.36E+13	1.10E+13	8.54E+12	6.39E+12	4.15E+12	2.70E+12	1.79E+12	7.18E+11
1250	2315	3.46E+13	2.86E+13	2.27E+13	1.75E+13	1.19E+13	8.05E+12	5.50E+12	2.31E+12
1500	2778	5.46E+13	4.58E+13	3.70E+13	2.91E+13	2.03E+13	1.42E+13	9.90E+12	4.30E+12
1750	3241	6.18E+13	5.17E+13	4.16E+13	3.25E+13	2.26E+13	1.56E+13	1.08E+13	4.69E+12
2000	3704	6.06E+13	4.99E+13	3.94E+13	3.01E+13	2.02E+13	1.35E+13	9.09E+12	3.71E+12
2250	4167	5.56E+13	4.49E+13	3.45E+13	2.57E+13	1.65E+13	1.06E+13	6.91E+12	2.65E+12
2500	4630	4.73E+13	3.73E+13	2.79E+13	2.01E+13	1.23E+13	7.59E+12	4.76E+12	1.71E+12
2750	5093	3.72E+13	2.87E+13	2.08E+13	1.45E+13	8.49E+12	5.01E+12	3.02E+12	9.94E+11
3000	5556	2.82E+13	2.13E+13	1.51E+13	1.02E+13	5.69E+12	3.21E+12	1.85E+12	5.39E+11
3500	6482	1.82E+13	1.35E+13	9.36E+12	6.15E+12	3.26E+12	1.72E+12	9.14E+11	2.09E+11
4000	7408	1.46E+13	1.10E+13	7.69E+12	5.10E+12	2.69E+12	1.37E+12	6.86E+11	1.34E+11
4500	8334	1.32E+13	1.01E+13	7.22E+12	4.93E+12	2.71E+12	1.43E+12	7.40E+11	1.54E+11
5000	9260	1.21E+13	9.55E+12	7.15E+12	5.15E+12	3.11E+12	1.84E+12	1.07E+12	2.89E+11
5500	10186	1.19E+13	9.79E+12	7.73E+12	5.95E+12	4.02E+12	2.69E+12	1.78E+12	6.26E+11
6000	11112	1.24E+13	1.05E+13	8.65E+12	6.98E+12	5.08E+12	3.68E+12	2.64E+12	1.15E+12
7000	12964	1.71E+13	1.53E+13	1.33E+13	1.14E+13	8.96E+12	6.96E+12	5.32E+12	2.73E+12
8000	14816	2.63E+13	2.40E+13	2.13E+13	1.86E+13	1.50E+13	1.19E+13	9.31E+12	5.01E+12
9000	16668	4.01E+13	3.68E+13	3.29E+13	2.88E+13	2.34E+13	1.87E+13	1.47E+13	7.89E+12
10000	18520	6.34E+13	5.82E+13	5.21E+13	4.56E+13	3.69E+13	2.92E+13	2.25E+13	1.14E+13
11000	20372	9.20E+13	8.41E+13	7.48E+13	6.51E+13	5.21E+13	4.07E+13	3.10E+13	1.52E+13
12000	22224	1.30E+14	1.17E+14	1.03E+14	8.83E+13	6.89E+13	5.24E+13	3.89E+13	1.79E+13
13000	24076	1.47E+14	1.31E+14	1.14E+14	9.55E+13	7.24E+13	5.34E+13	3.84E+13	1.64E+13
14000	25928	1.37E+14	1.21E+14	1.03E+14	8.51E+13	6.28E+13	4.50E+13	3.15E+13	1.29E+13
15005	27789	1.19E+14	1.04E+14	8.76E+13	7.13E+13	5.13E+13	3.59E+13	2.46E+13	9.77E+12
16000	29632	9.49E+13	8.20E+13	6.79E+13	5.44E+13	3.83E+13	2.61E+13	1.75E+13	6.69E+12
17000	31484	7.33E+13	6.27E+13	5.13E+13	4.06E+13	2.80E+13	1.88E+13	1.23E+13	4.50E+12
18000	33336	5.91E+13	5.02E+13	4.07E+13	3.19E+13	2.18E+13	1.44E+13	9.36E+12	3.27E+12
19327	35793	4.51E+13	3.80E+13	3.05E+13	2.37E+13	1.59E+13	1.04E+13	6.62E+12	2.19E+12

Table 9-6. Annual Equivalent 1 MeV Electron Fluence for GaAs Solar Cells from Trapped Protons During Solar Max and Min, 0° Inclination (Infinite Backshielding)

EQUIV. 1 MEV ELECTRON FLUENCE FOR V_{oc} CIRCULAR ORBIT   INCLINATION = 0 DEGREES
DUE TO GEOMAGNETICALLY TRAPPED PROTONS, MODEL AP8MAX

PROTONS - V_{oc}

SHIELD THICKNESS, cm (mils)

ALTITUDE (nm)	(km)	0 (0)	2.54E-3 (1)	7.64E-3 (3)	1.52E-2 (6)	3.05E-2 (12)	5.09E-2 (20)	7.64E-2 (30)	1.52E-1 (60)
150	277	0.00E+00	0.00E+00	0.00E+00	0.00E+00	0.00E+00	0.00E+00	0.00E+00	0.00E+00
250	463	0.00E+00	0.00E+00	0.00E+00	0.00E+00	0.00E+00	0.00E+00	0.00E+00	0.00E+00
300	555	0.00E+00	0.00E+00	0.00E+00	0.00E+00	0.00E+00	0.00E+00	0.00E+00	0.00E+00
450	833	2.30E+12	8.35E+11	7.61E+11	7.08E+11	6.51E+11	6.17E+11	5.80E+11	5.36E+11
600	1111	2.65E+13	1.25E+13	1.15E+13	1.08E+13	9.84E+12	9.04E+12	8.15E+12	7.09E+12
800	1481	2.07E+14	1.11E+14	9.83E+13	8.89E+13	7.79E+13	6.86E+13	5.82E+13	4.67E+13
1000	1852	8.42E+14	4.99E+14	4.24E+14	3.70E+14	3.08E+14	2.53E+14	1.99E+14	1.47E+14
1250	2315	3.45E+15	2.12E+15	1.72E+15	1.43E+15	1.10E+15	8.06E+14	5.57E+14	3.63E+14
1500	2778	1.09E+16	6.68E+15	5.20E+15	4.14E+15	2.98E+15	1.99E+15	1.20E+15	6.54E+14
1750	3241	2.73E+16	1.62E+16	1.22E+16	9.38E+15	6.39E+15	3.94E+15	2.10E+15	9.39E+14
2000	3704	6.20E+16	3.36E+16	2.39E+16	1.72E+16	1.08E+16	6.10E+15	2.89E+15	1.09E+15
2250	4167	1.45E+17	6.79E+16	4.35E+16	2.81E+16	1.53E+16	7.82E+15	3.29E+15	1.11E+15
2500	4630	3.29E+17	1.21E+17	6.98E+16	4.05E+16	1.91E+16	8.84E+15	3.34E+15	1.03E+15
2750	5093	7.27E+17	1.83E+17	9.52E+16	5.09E+16	2.14E+16	9.18E+15	3.20E+15	9.23E+14
3000	5556	1.59E+18	2.57E+17	1.20E+17	5.86E+16	2.18E+16	8.76E+15	2.84E+15	7.83E+14
3500	6482	5.19E+18	4.80E+17	1.65E+17	6.49E+16	1.80E+16	6.82E+15	2.07E+15	5.28E+14
4000	7408	1.39E+19	7.45E+17	1.94E+17	6.30E+16	1.34E+16	4.71E+15	1.33E+15	3.30E+14
4500	8334	2.93E+19	1.01E+18	2.00E+17	5.45E+16	8.99E+15	2.89E+15	7.45E+14	1.75E+14
5000	9260	5.21E+19	1.23E+18	1.89E+17	4.36E+16	5.67E+15	1.69E+15	4.05E+14	9.02E+13
5500	10186	7.63E+19	1.39E+18	1.68E+17	3.19E+16	3.14E+15	9.08E+14	2.06E+14	4.20E+13
6000	11112	1.03E+20	1.40E+18	1.34E+17	2.15E+16	1.65E+15	4.29E+14	8.61E+13	1.54E+13
7000	12964	1.55E+20	1.01E+18	6.33E+16	7.19E+15	3.34E+14	6.78E+13	1.02E+13	1.39E+12
8000	14816	1.86E+20	5.14E+17	1.99E+16	1.62E+15	4.58E+13	7.03E+12	7.81E+11	8.40E+10
9000	16668	1.28E+20	1.68E+17	3.67E+15	1.95E+14	2.95E+12	3.62E+11	3.96E+10	7.69E+09
10000	18520	8.74E+19	4.49E+16	5.62E+14	2.08E+13	1.38E+11	0.00E+00	0.00E+00	0.00E+00
11000	20372	6.45E+19	1.03E+16	3.48E+13	6.11E+11	2.93E+09	0.00E+00	0.00E+00	0.00E+00
12000	22224	5.29E+19	1.98E+15	8.44E+11	2.93E+09	0.00E+00	0.00E+00	0.00E+00	0.00E+00
13000	24076	4.42E+19	2.53E+14	7.95E+07	0.00E+00	0.00E+00	0.00E+00	0.00E+00	0.00E+00
14000	25928	3.71E+19	4.48E+13	3.62E+07	0.00E+00	0.00E+00	0.00E+00	0.00E+00	0.00E+00
15000	27780	3.23E+19	1.36E+13	2.07E+07	0.00E+00	0.00E+00	0.00E+00	0.00E+00	0.00E+00
16000	29632	2.63E+19	7.19E+10	0.00E+00	0.00E+00	0.00E+00	0.00E+00	0.00E+00	0.00E+00
17000	31484	2.13E+19	5.24E+10	0.00E+00	0.00E+00	0.00E+00	0.00E+00	0.00E+00	0.00E+00
18000	33336	1.80E+19	4.38E+10	0.00E+00	0.00E+00	0.00E+00	0.00E+00	0.00E+00	0.00E+00
19327	35793	1.32E+19	3.56E+10	0.00E+00	0.00E+00	0.00E+00	0.00E+00	0.00E+00	0.00E+00

EQUIV. 1 MEV ELECTRON FLUENCE FOR V_{oc} CIRCULAR ORBIT   INCLINATION = 0 DEGREES
DUE TO GEOMAGNETICALLY TRAPPED PROTONS, MODEL AP8MIN

ALTITUDE (nm)	(km)	0 (0)	2.54E-3 (1)	7.64E-3 (3)	1.52E-2 (6)	3.05E-2 (12)	5.09E-2 (20)	7.64E-2 (30)	1.52E-1 (60)
150	277	0.00E+00	0.00E+00	0.00E+00	0.00E+00	0.00E+00	0.00E+00	0.00E+00	0.00E+00
250	463	0.00E+00	0.00E+00	0.00E+00	0.00E+00	0.00E+00	0.00E+00	0.00E+00	0.00E+00
300	555	9.58E+09	2.15E+09	4.68E+06	0.00E+00	0.00E+00	0.00E+00	0.00E+00	0.00E+00
450	833	6.76E+12	2.24E+12	2.03E+12	1.89E+12	1.73E+12	1.63E+12	1.50E+12	1.31E+12
600	1111	3.84E+13	1.80E+13	1.66E+13	1.55E+13	1.42E+13	1.32E+13	1.20E+13	1.05E+13
800	1481	2.27E+14	1.27E+14	1.14E+14	1.03E+14	9.01E+13	8.04E+13	6.91E+13	5.59E+13
1000	1852	8.63E+14	5.29E+14	4.52E+14	3.93E+14	3.27E+14	2.73E+14	2.16E+14	1.60E+14

Table 9-7. Annual Equivalent 1 MeV Electron Fluence for GaAs Solar Cells from Trapped Protons During Solar Max and Min, 0° Inclination (Infinite Backshielding)

PROTONS - P$_{max}$

EQUIV. 1 MEV ELECTRON FLUENCE FOR P$_{max}$ CIRCULAR ORBIT
DUE TO GEOMAGNETICALLY TRAPPED PROTONS, MODEL AP8MAX
INCLINATION = 0 DEGREES

ALTITUDE (nm)	(km)	SHIELD THICKNESS, cm (mils)							
		0 (0)	2.54E-3 (1)	7.64E-3 (3)	1.52E-2 (6)	3.05E-2 (12)	5.09E-2 (20)	7.64E-2 (30)	1.52E-1 (60)
150	277	0.00E+00	0.00E+00	0.00E+00	0.00E+00	0.00E+00	0.00E+00	0.00E+00	0.00E+00
250	463	0.00E+00	0.00E+00	0.00E+00	0.00E+00	0.00E+00	0.00E+00	0.00E+00	0.00E+00
300	555	0.00E+00	0.00E+00	0.00E+00	0.00E+00	0.00E+00	0.00E+00	0.00E+00	0.00E+00
450	833	1.64E+12	5.97E+11	5.44E+11	5.05E+11	4.65E+11	4.41E+11	4.14E+11	3.83E+11
600	1111	1.89E+13	8.94E+12	8.24E+12	7.68E+12	7.03E+12	6.46E+12	5.82E+12	5.06E+12
800	1481	1.48E+14	7.92E+13	7.02E+13	6.35E+13	5.57E+13	4.90E+13	4.16E+13	3.34E+13
1000	1852	6.01E+14	3.56E+14	3.03E+14	2.64E+14	2.20E+14	1.81E+14	1.42E+14	1.05E+14
1250	2315	2.46E+15	1.52E+15	1.23E+15	1.02E+15	7.85E+14	5.76E+14	3.98E+14	2.59E+14
1500	2778	7.76E+15	4.77E+15	3.71E+15	2.96E+15	2.13E+15	1.42E+15	8.59E+14	4.67E+14
1750	3241	1.95E+16	1.15E+16	8.72E+15	6.70E+15	4.57E+15	2.81E+15	1.50E+15	6.70E+14
2000	3704	4.43E+16	2.40E+16	1.71E+16	1.23E+16	7.69E+15	4.36E+15	2.06E+15	7.80E+14
2250	4167	1.03E+17	4.85E+16	3.11E+16	2.00E+16	1.09E+16	5.59E+15	2.35E+15	7.93E+14
2500	4630	2.35E+17	8.67E+16	4.98E+16	2.90E+16	1.36E+16	6.32E+15	2.39E+15	7.39E+14
2750	5093	5.19E+17	1.31E+17	6.80E+16	3.64E+16	1.53E+16	6.56E+15	2.28E+15	6.60E+14
3000	5556	1.14E+18	1.84E+17	8.55E+16	4.18E+16	1.56E+16	6.26E+15	2.03E+15	5.59E+14
3500	6482	3.70E+18	3.43E+17	1.18E+17	4.63E+16	1.28E+16	4.87E+15	1.47E+15	3.77E+14
4000	7408	9.90E+18	5.32E+17	1.38E+17	4.50E+16	9.54E+15	3.36E+15	9.52E+14	2.36E+14
4500	8334	2.09E+19	7.21E+17	1.43E+17	3.89E+16	6.42E+15	2.06E+15	5.32E+14	1.25E+14
5000	9260	3.72E+19	8.76E+17	1.35E+17	3.11E+16	4.05E+15	1.21E+15	2.89E+14	6.45E+13
5500	10186	5.45E+19	9.95E+17	1.20E+17	2.28E+16	2.24E+15	6.49E+14	1.47E+14	3.00E+13
6000	11112	7.35E+19	1.00E+18	9.59E+16	1.54E+16	1.18E+15	3.07E+14	6.15E+13	1.10E+13
7000	12964	1.10E+20	7.20E+17	4.52E+16	5.14E+15	2.38E+14	4.84E+13	7.29E+12	9.91E+11
8000	14816	1.33E+20	3.67E+17	1.42E+16	1.16E+15	3.27E+13	5.02E+12	5.58E+11	6.00E+10
9000	16668	9.15E+19	1.20E+17	2.62E+15	1.39E+14	2.11E+12	2.59E+11	2.83E+10	5.50E+09
10000	18520	6.24E+19	3.21E+16	4.02E+14	1.48E+13	9.86E+10	0.00E+00	0.00E+00	0.00E+00
11000	20372	4.60E+19	7.34E+15	2.48E+13	4.37E+11	2.09E+09	0.00E+00	0.00E+00	0.00E+00
12000	22224	3.78E+19	1.41E+15	6.03E+11	0.00E+00	0.00E+00	0.00E+00	0.00E+00	0.00E+00
13000	24076	3.16E+19	1.81E+14	5.68E+07	0.00E+00	0.00E+00	0.00E+00	0.00E+00	0.00E+00
14000	25928	2.65E+19	3.20E+13	2.59E+07	0.00E+00	0.00E+00	0.00E+00	0.00E+00	0.00E+00
15000	27780	2.30E+19	9.69E+12	1.48E+07	0.00E+00	0.00E+00	0.00E+00	0.00E+00	0.00E+00
16000	29632	1.88E+19	5.13E+11	0.00E+00	0.00E+00	0.00E+00	0.00E+00	0.00E+00	0.00E+00
17000	31484	1.52E+19	3.74E+10	0.00E+00	0.00E+00	0.00E+00	0.00E+00	0.00E+00	0.00E+00
18000	33336	1.28E+19	3.13E+10	0.00E+00	0.00E+00	0.00E+00	0.00E+00	0.00E+00	0.00E+00
19327	35793	9.46E+18	2.54E+10	0.00E+00	0.00E+00	0.00E+00	0.00E+00	0.00E+00	0.00E+00

EQUIV. 1 MEV ELECTRON FLUENCE FOR P$_{max}$ CIRCULAR ORBIT
DUE TO GEOMAGNETICALLY TRAPPED PROTONS, MODEL AP8MIN
INCLINATION = 0 DEGREES

ALTITUDE (nm)	(km)	0 (0)	2.54E-3 (1)	7.64E-3 (3)	1.52E-2 (6)	3.05E-2 (12)	5.09E-2 (20)	7.64E-2 (30)	1.52E-1 (60)
150	277	0.00E+00	0.00E+00	0.00E+00	0.00E+00	0.00E+00	0.00E+00	0.00E+00	0.00E+00
250	463	0.00E+00	0.00E+00	0.00E+00	0.00E+00	0.00E+00	0.00E+00	0.00E+00	0.00E+00
300	555	6.84E+09	1.53E+09	3.34E+06	0.00E+00	0.00E+00	0.00E+00	0.00E+00	0.00E+00
450	833	4.83E+12	1.60E+12	1.45E+12	1.35E+12	1.24E+12	1.17E+12	1.07E+12	9.38E+11
600	1111	2.74E+13	1.28E+13	1.18E+13	1.10E+13	1.02E+13	9.44E+12	8.59E+12	7.52E+12
800	1481	1.62E+14	9.10E+13	8.12E+13	7.32E+13	6.44E+13	5.74E+13	4.93E+13	3.99E+13
1000	1852	6.16E+14	3.78E+14	3.23E+14	2.81E+14	2.34E+14	1.95E+14	1.55E+14	1.14E+14

Table 9-8. Annual Equivalent 1 MeV Electron Fluence for GaAs Solar Cells from Trapped Protons During Solar Max and Min, 0° Inclination (Infinite Backshielding)

PROTONS - I$_{sc}$

INCLINATION = 0 DEGREES

EQUIV. 1 MEV PROTON FLUENCE FOR I$_{sc}$ - CIRCULAR ORBIT
DUE TO GEOMAGNETICALLY TRAPPED PROTONS, MODEL AP8MAX

SHIELD THICKNESS, cm (mils)

ALTITUDE (nm)	(km)	0 (0)	2.54E-3 (1)	7.64E-3 (3)	1.52E-2 (6)	3.05E-2 (12)	5.09E-2 (20)	7.64E-2 (30)	1.52E-1 (60)
150	277	0.00E+00	0.00E+00	0.00E+00	0.00E+00	0.00E+00	0.00E+00	0.00E+00	0.00E+00
250	463	0.00E+00	0.00E+00	0.00E+00	0.00E+00	0.00E+00	0.00E+00	0.00E+00	0.00E+00
300	555	0.00E+00	0.00E+00	0.00E+00	0.00E+00	0.00E+00	0.00E+00	0.00E+00	0.00E+00
450	833	9.85E+11	1.57E+11	1.30E+11	1.12E+11	9.51E+10	8.57E+10	7.60E+10	6.82E+10
600	1111	1.05E+14	2.60E+12	2.26E+12	2.01E+12	1.75E+12	1.53E+12	1.28E+12	1.04E+12
800	1481	8.05E+13	2.70E+13	2.26E+13	1.96E+13	1.63E+13	1.36E+13	1.06E+13	7.65E+12
1000	1852	3.22E+14	1.37E+14	1.11E+14	9.28E+13	7.35E+13	5.69E+13	4.03E+13	2.62E+13
1250	2315	1.33E+15	6.49E+14	5.02E+14	4.05E+14	3.00E+14	2.06E+14	1.28E+14	7.22E+13
1500	2778	4.22E+15	2.16E+15	1.62E+15	1.26E+15	8.88E+14	5.65E+14	3.11E+14	1.47E+14
1750	3241	1.08E+16	5.40E+15	3.96E+15	2.98E+15	2.00E+15	1.19E+15	5.89E+14	2.34E+14
2000	3704	2.52E+16	1.15E+16	7.97E+15	5.59E+15	3.46E+15	1.91E+15	8.48E+14	2.90E+14
2250	4167	6.02E+16	2.39E+16	1.49E+16	9.33E+15	5.00E+15	2.50E+15	9.93E+14	3.08E+14
2500	4630	1.45E+17	4.36E+16	2.43E+16	1.37E+16	6.34E+15	2.86E+15	1.02E+15	2.95E+14
2750	5093	3.50E+17	6.66E+16	3.35E+16	1.73E+16	7.16E+15	3.00E+15	9.89E+14	2.68E+14
3000	5556	8.27E+17	9.48E+16	4.24E+16	2.01E+16	7.35E+15	2.88E+15	8.85E+14	2.30E+14
3500	6482	2.81E+18	1.82E+17	5.95E+16	2.25E+16	6.08E+15	2.25E+15	6.49E+14	1.58E+14
4000	7408	7.97E+18	2.88E+17	7.06E+16	2.20E+16	4.54E+15	1.56E+15	4.21E+14	1.00E+14
4500	8334	1.77E+19	3.97E+17	7.33E+16	1.91E+16	3.07E+15	9.64E+14	2.37E+14	5.36E+13
5000	9260	3.33E+19	4.89E+17	6.97E+16	1.54E+16	1.94E+15	5.67E+14	1.29E+14	2.78E+13
5500	10186	4.81E+19	5.61E+17	6.23E+16	1.13E+16	1.08E+15	3.05E+14	6.61E+13	1.31E+13
6000	11112	6.32E+19	5.70E+17	5.02E+16	7.65E+15	5.67E+14	1.45E+14	2.78E+13	4.84E+12
7000	12964	9.84E+19	4.16E+17	2.38E+16	2.57E+15	1.16E+14	2.31E+13	3.33E+12	4.41E+11
8000	14816	1.30E+20	2.16E+17	7.54E+15	5.81E+14	1.60E+13	2.41E+12	2.56E+11	2.66E+10
9000	16668	9.24E+19	7.18E+16	1.40E+15	7.03E+13	1.04E+12	1.24E+11	1.28E+10	2.46E+09
10000	18520	6.50E+19	1.96E+16	2.15E+14	7.54E+12	5.43E+10	0.00E+00	0.00E+00	0.00E+00
11000	20372	5.04E+19	4.57E+15	1.36E+13	2.22E+11	1.13E+09	0.00E+00	0.00E+00	0.00E+00
12000	22224	4.47E+19	9.00E+14	3.39E+11	1.13E+09	0.00E+00	0.00E+00	0.00E+00	0.00E+00
13000	24076	4.06E+19	1.19E+14	3.52E+07	0.00E+00	0.00E+00	0.00E+00	0.00E+00	0.00E+00
14000	25928	3.69E+19	2.16E+13	1.57E+07	0.00E+00	0.00E+00	0.00E+00	0.00E+00	0.00E+00
15000	27780	3.43E+19	6.67E+12	8.89E+06	0.00E+00	0.00E+00	0.00E+00	0.00E+00	0.00E+00
16000	29632	2.98E+19	4.40E+10	0.00E+00	0.00E+00	0.00E+00	0.00E+00	0.00E+00	0.00E+00
17000	31484	2.54E+19	3.17E+10	0.00E+00	0.00E+00	0.00E+00	0.00E+00	0.00E+00	0.00E+00
18000	33336	2.22E+19	2.64E+10	0.00E+00	0.00E+00	0.00E+00	0.00E+00	0.00E+00	0.00E+00
19327	35793	1.69E+19	2.13E+10	0.00E+00	0.00E+00	0.00E+00	0.00E+00	0.00E+00	0.00E+00

INCLINATION = 0 DEGREES

EQUIV. 1 MEV PROTON FLUENCE FOR I$_{sc}$ - CIRCULAR ORBIT
DUE TO GEOMAGNETICALLY TRAPPED PROTONS, MODEL AP8MIN

ALTITUDE (nm)	(km)	0 (0)	2.54E-3 (1)	7.64E-3 (3)	1.52E-2 (6)	3.05E-2 (12)	5.09E-2 (20)	7.64E-2 (30)	1.52E-1 (60)
150	277	0.00E+00	0.00E+00	0.00E+00	0.00E+00	0.00E+00	0.00E+00	0.00E+00	0.00E+00
250	463	0.00E+00	0.00E+00	0.00E+00	0.00E+00	0.00E+00	0.00E+00	0.00E+00	0.00E+00
300	555	3.60E+09	8.47E+08	1.92E+06	0.00E+00	0.00E+00	0.00E+00	0.00E+00	0.00E+00
450	833	3.14E+12	4.54E+11	3.81E+11	3.35E+11	2.90E+11	2.64E+11	2.28E+11	1.88E+11
600	1111	1.52E+13	3.64E+12	3.15E+12	2.80E+12	2.44E+12	2.17E+12	1.84E+12	1.51E+12
800	1481	8.48E+13	3.05E+13	2.56E+13	2.20E+13	1.83E+13	1.55E+13	1.22E+13	8.90E+12
1000	1852	3.18E+14	1.44E+14	1.17E+14	9.76E+13	7.71E+13	6.07E+13	4.35E+13	2.82E+13

Table 9-9. Annual Equivalent 1 MeV Electron Fluence for GaAs Solar Cells from Trapped Electrons During Solar Max, 10° Inclination (Infinite Backshielding)

ELECTRONS - I_{sc}, V_{oc}, AND P_{max}

INCLINATION = 10 DEGREES

EQUIV. 1 MEV ELECTRON FLUENCE FOR I_{sc} - CIRCULAR ORBIT
DUE TO GEOMAGNETICALLY TRAPPED ELECTRONS - MODEL AE8MAX

ALTITUDE (nm)	(km)	SHIELD THICKNESS, cm (mils)							
		0 (0)	2.54E-3 (1)	7.64E-3 (3)	1.52E-2 (6)	3.05E-2 (12)	5.09E-2 (20)	7.64E-2 (30)	1.52E-1 (60)
150	277	9.08E+05	8.76E+05	8.35E+05	7.85E+05	7.10E+05	6.34E+05	5.60E+05	4.00E+05
250	463	2.36E+07	2.13E+07	1.87E+07	1.61E+07	1.29E+07	1.04E+07	8.33E+06	4.98E+06
300	555	1.11E+08	9.48E+07	7.81E+07	6.29E+07	4.61E+07	3.42E+07	2.57E+07	1.36E+07
450	833	1.61E+10	1.19E+10	8.13E+09	5.23E+09	2.76E+09	1.52E+09	8.89E+08	3.03E+08
600	1111	5.89E+11	4.30E+11	2.88E+11	1.82E+11	9.37E+10	5.07E+10	2.94E+10	9.86E+09
800	1481	6.65E+12	4.96E+12	3.44E+12	2.28E+12	1.27E+12	7.50E+11	4.66E+11	1.73E+11
1000	1852	2.35E+13	1.78E+13	1.27E+13	8.65E+12	5.08E+12	3.14E+12	2.03E+12	7.93E+11
1250	2315	5.48E+13	4.22E+13	3.06E+13	2.15E+13	1.32E+13	8.50E+12	5.67E+12	2.32E+12
1500	2778	8.47E+13	6.54E+13	4.77E+13	3.37E+13	2.08E+13	1.34E+13	9.00E+12	3.73E+12
1750	3241	1.10E+14	8.44E+13	6.07E+13	4.20E+13	2.48E+13	1.54E+13	9.96E+12	4.02E+12
2000	3704	1.32E+14	9.93E+13	6.99E+13	4.67E+13	2.58E+13	1.47E+13	8.83E+12	3.26E+12
2250	4167	1.48E+14	1.10E+14	7.60E+13	4.93E+13	2.54E+13	1.30E+13	7.12E+12	2.31E+12
2500	4630	1.55E+14	1.15E+14	7.81E+13	4.95E+13	2.40E+13	1.12E+13	5.45E+12	1.50E+12
2750	5093	1.51E+14	1.12E+14	7.56E+13	4.72E+13	2.19E+13	9.28E+12	3.99E+12	8.83E+11
3000	5556	1.38E+14	1.02E+14	6.85E+13	4.23E+13	1.90E+13	7.51E+12	2.90E+12	5.03E+11
3500	6482	1.01E+14	7.40E+13	4.92E+13	2.99E+13	1.29E+13	4.74E+12	1.63E+12	2.04E+11
4000	7408	7.69E+13	5.59E+13	3.71E+13	2.26E+13	9.76E+12	3.62E+12	1.26E+12	1.64E+11
4500	8334	6.51E+13	4.83E+13	3.30E+13	2.09E+13	9.82E+12	4.13E+12	1.71E+12	3.26E+11
5000	9260	5.25E+13	4.05E+13	2.92E+13	2.00E+13	1.09E+13	5.71E+12	3.07E+12	9.60E+11
5500	10186	4.39E+13	3.59E+13	2.81E+13	2.13E+13	1.39E+13	9.00E+12	5.91E+12	2.44E+12
6000	11112	4.39E+13	3.81E+13	3.21E+13	2.64E+13	1.95E+13	1.42E+13	1.02E+13	4.76E+12
7000	12964	6.74E+13	6.09E+13	5.35E+13	4.60E+13	3.63E+13	2.82E+13	2.15E+13	1.09E+13
8000	14816	9.96E+13	9.04E+13	7.98E+13	6.90E+13	5.48E+13	4.27E+13	3.28E+13	1.69E+13
9000	16668	1.32E+14	1.20E+14	1.06E+14	9.14E+13	7.24E+13	5.63E+13	4.31E+13	2.20E+13
10000	18520	1.60E+14	1.45E+14	1.27E+14	1.09E+14	8.54E+13	6.56E+13	4.95E+13	2.42E+13
11000	20372	1.79E+14	1.61E+14	1.41E+14	1.20E+14	9.30E+13	7.03E+13	5.20E+13	2.40E+13
12000	22224	1.78E+14	1.59E+14	1.38E+14	1.16E+14	8.86E+13	6.56E+13	4.75E+13	2.07E+13
13000	24076	1.61E+14	1.43E+14	1.22E+14	1.01E+14	7.56E+13	5.47E+13	3.87E+13	1.61E+13
14000	25928	1.36E+14	1.20E+14	1.01E+14	8.33E+13	6.08E+13	4.31E+13	2.99E+13	1.21E+13
15000	27780	1.11E+14	9.67E+13	8.08E+13	6.54E+13	4.68E+13	3.25E+13	2.21E+13	8.65E+12
16000	29632	8.98E+13	7.75E+13	6.40E+13	5.12E+13	3.60E+13	2.45E+13	1.64E+13	6.22E+12
17000	31484	7.05E+13	6.03E+13	4.93E+13	3.90E+13	2.69E+13	1.80E+13	1.19E+13	4.32E+12
18000	33336	5.30E+13	4.49E+13	3.63E+13	2.84E+13	1.92E+13	1.27E+13	8.19E+12	2.83E+12
19327	35793	3.57E+13	2.98E+13	2.37E+13	1.82E+13	1.20E+13	7.65E+12	4.80E+12	1.56E+12

Table 9-10.   Annual Equivalent 1 MeV Electron Fluence for GaAs Solar Cells from Trapped Electrons During Solar Min, 10° Inclination (Infinite Backshielding)

ELECTRONS - I_{sc}, V_{oc}, AND P_{max} INCLINATION = 10 DEGREES

EQUIV. 1 MEV ELECTRON FLUENCE FOR I_{sc} - CIRCULAR ORBIT
DUE TO GEOMAGNETICALLY TRAPPED ELECTRONS - MODEL AE8MIN

ALTITUDE (nm)	(km)	SHIELD THICKNESS, cm (mils)							
		0 (0)	2.54E-3 (1)	7.64E-3 (3)	1.52E-2 (6)	3.05E-2 (12)	5.09E-2 (20)	7.64E-2 (30)	1.52E-1 (60)
150	277	5.76E+05	5.58E+05	5.35E+05	5.06E+05	4.61E+05	4.16E+05	3.70E+05	2.69E+05
250	463	1.74E+07	1.59E+07	1.42E+07	1.25E+07	1.03E+07	8.42E+06	6.85E+06	4.20E+06
300	555	7.66E+07	6.72E+07	5.70E+07	4.74E+07	3.60E+07	2.73E+07	2.08E+07	1.14E+07
450	833	8.78E+09	6.79E+09	4.92E+09	3.40E+09	1.95E+09	1.13E+09	6.83E+08	2.46E+08
600	1111	2.60E+11	2.03E+11	1.49E+11	1.05E+11	6.19E+10	3.68E+10	2.27E+10	8.28E+09
800	1481	3.57E+12	2.86E+12	2.18E+12	1.59E+12	1.00E+12	6.33E+11	4.09E+11	1.60E+11
1000	1852	1.36E+13	1.10E+13	8.59E+12	6.47E+12	4.25E+12	2.79E+12	1.86E+12	7.59E+11
1250	2315	3.28E+13	2.72E+13	2.16E+13	1.67E+13	1.14E+13	7.80E+12	5.36E+12	2.27E+12
1500	2778	4.82E+13	4.03E+13	3.25E+13	2.55E+13	1.77E+13	1.23E+13	8.57E+12	3.71E+12
1750	3241	5.48E+13	4.57E+13	3.66E+13	2.85E+13	1.97E+13	1.36E+13	9.37E+12	4.02E+12
2000	3704	5.41E+13	4.45E+13	3.50E+13	2.67E+13	1.78E+13	1.19E+13	8.00E+12	3.26E+12
2250	4167	4.92E+13	3.97E+13	3.04E+13	2.26E+13	1.45E+13	9.29E+12	6.04E+12	2.31E+12
2500	4630	4.15E+13	3.28E+13	2.45E+13	1.76E+13	1.08E+13	6.65E+12	4.18E+12	1.50E+12
2750	5093	3.27E+13	2.52E+13	1.83E+13	1.28E+13	7.47E+12	4.41E+12	2.66E+12	8.73E+11
3000	5556	2.53E+13	1.92E+13	1.36E+13	9.20E+12	5.15E+12	2.90E+12	1.68E+12	4.92E+11
3500	6482	1.68E+13	1.25E+13	8.69E+12	5.73E+12	3.05E+12	1.61E+12	8.53E+11	1.96E+11
4000	7408	1.27E+13	1.05E+13	7.42E+12	4.95E+12	2.64E+12	1.36E+12	6.91E+11	1.40E+11
4500	8334	1.18E+13	9.79E+12	7.09E+12	4.90E+12	2.75E+12	1.50E+12	8.08E+11	1.87E+11
5000	9260	1.18E+13	9.36E+12	7.10E+12	5.19E+12	3.23E+12	1.98E+12	1.21E+12	3.68E+11
5500	10186	1.26E+13	9.76E+12	7.79E+12	6.07E+12	4.19E+12	2.87E+12	1.95E+12	7.51E+11
6000	11112	1.40E+13	1.09E+13	9.02E+12	7.36E+12	5.44E+12	4.00E+12	2.91E+12	1.32E+12
7000	12964	1.81E+13	1.62E+13	1.42E+13	1.22E+13	9.61E+12	7.50E+12	5.77E+12	2.99E+12
8000	14816	2.80E+13	2.56E+13	2.28E+13	1.99E+13	1.61E+13	1.28E+13	9.99E+12	5.36E+12
9000	16668	4.35E+13	3.99E+13	3.57E+13	3.13E+13	2.54E+13	2.02E+13	1.58E+13	8.33E+12
10000	18520	6.76E+13	6.20E+13	5.53E+13	4.83E+13	3.89E+13	3.06E+13	2.35E+13	1.19E+13
11000	20372	9.69E+13	8.85E+13	7.84E+13	6.80E+13	5.41E+13	4.20E+13	3.18E+13	1.54E+13
12000	22224	1.28E+14	1.16E+14	1.01E+14	8.65E+13	6.72E+13	5.08E+13	3.75E+13	1.71E+13
13000	24076	1.40E+14	1.25E+14	1.08E+14	9.06E+13	6.84E+13	5.02E+13	3.60E+13	1.54E+13
14000	25928	1.31E+14	1.16E+14	9.81E+13	8.10E+13	5.96E+13	4.27E+13	2.98E+13	1.21E+13
15000	27780	1.11E+14	9.64E+13	8.07E+13	6.54E+13	4.69E+13	3.26E+13	2.23E+13	8.77E+12
16000	29632	9.01E+13	7.78E+13	6.44E+13	5.15E+13	3.62E+13	2.47E+13	1.66E+13	6.30E+12
17000	31484	7.07E+13	6.05E+13	4.95E+13	3.92E+13	2.70E+13	1.81E+13	1.19E+13	4.35E+12
18000	33336	5.32E+13	4.51E+13	3.65E+13	2.85E+13	1.94E+13	1.28E+13	8.25E+12	2.85E+12
19327	35793	3.60E+13	3.00E+13	2.39E+13	1.83E+13	1.21E+13	7.73E+12	4.86E+12	1.58E+12

Table 9-11. Annual Equivalent 1 MeV Electron Fluence for GaAs Solar Cells from Trapped Protons During Solar Max and Min, 10° Inclination (Infinite Backshielding)

PROTONS - V_oc

EQUIV. 1 MEV ELECTRON FLUENCE FOR V_oc CIRCULAR ORBIT
DUE TO GEOMAGNETICALLY TRAPPED PROTONS, MODEL AP8MAX

INCLINATION = 10 DEGREES

ALTITUDE (nm)	(km)	SHIELD THICKNESS, cm (mils)							
		0 (0)	2.54E-3 (1)	7.64E-3 (3)	1.52E-2 (6)	3.05E-2 (12)	5.09E-2 (20)	7.64E-2 (30)	1.52E-1 (60)
150	277	0.00E+00	0.00E+00	0.00E+00	0.00E+00	0.00E+00	0.00E+00	0.00E+00	0.00E+00
250	463	1.11E+09	1.55E+08	1.27E+08	1.19E+08	1.10E+08	1.05E+08	9.35E+07	7.17E+07
300	555	6.24E+10	2.11E+10	1.86E+10	1.68E+10	1.50E+10	1.41E+10	1.31E+10	1.17E+10
450	833	3.63E+12	1.42E+12	1.33E+12	1.26E+12	1.17E+12	1.10E+12	1.02E+12	9.27E+11
600	1111	3.01E+13	1.40E+13	1.28E+13	1.19E+13	1.08E+13	9.85E+12	8.77E+12	7.49E+12
800	1481	2.16E+14	1.18E+14	1.04E+14	9.34E+13	8.13E+13	7.11E+13	5.98E+13	4.75E+13
1000	1852	8.56E+14	5.11E+14	4.31E+14	3.73E+14	3.07E+14	2.50E+14	1.95E+14	1.43E+14
1250	2315	3.44E+15	2.13E+15	1.70E+15	1.40E+15	1.06E+15	7.76E+14	5.32E+14	3.42E+14
1500	2778	1.08E+16	6.52E+15	5.02E+15	3.96E+15	2.82E+15	1.86E+15	1.11E+15	5.96E+14
1750	3241	2.69E+16	1.55E+16	1.15E+16	8.74E+15	5.87E+15	3.59E+15	1.90E+15	8.42E+14
2000	3704	6.36E+16	3.27E+16	2.27E+16	1.60E+16	9.79E+15	5.50E+15	2.59E+15	9.80E+14
2250	4167	1.50E+17	6.49E+16	4.06E+16	2.58E+16	1.38E+16	7.04E+15	2.95E+15	9.93E+14
2500	4630	3.35E+17	1.12E+17	6.33E+16	3.65E+16	1.70E+16	7.90E+15	2.99E+15	9.28E+14
2750	5093	7.60E+17	1.72E+17	8.74E+16	4.61E+16	1.91E+16	8.19E+15	2.85E+15	8.25E+14
3000	5556	1.62E+18	2.44E+17	1.10E+17	5.30E+16	1.95E+16	7.84E+15	2.55E+15	7.03E+14
3500	6482	5.25E+18	4.44E+17	1.49E+17	5.83E+16	1.61E+16	6.09E+15	1.85E+15	4.72E+14
4000	7408	1.38E+19	6.86E+17	1.74E+17	5.61E+16	1.18E+16	4.17E+15	1.18E+15	2.90E+14
4500	8334	2.81E+19	9.15E+17	1.78E+17	4.85E+16	8.05E+15	2.59E+15	6.71E+14	1.57E+14
5000	9260	5.01E+19	1.12E+18	1.69E+17	3.89E+16	5.08E+15	1.52E+15	3.62E+14	8.03E+13
5500	10186	7.40E+19	1.25E+18	1.49E+17	2.85E+16	2.82E+15	8.12E+14	1.83E+14	3.72E+13
6000	11112	9.85E+19	1.26E+18	1.20E+17	1.93E+16	1.49E+15	3.88E+14	7.80E+13	1.40E+13
7000	12964	1.45E+20	9.06E+17	5.76E+16	6.63E+15	3.14E+14	6.50E+13	9.97E+12	1.38E+12
8000	14816	1.66E+20	4.66E+17	1.86E+16	1.54E+15	4.49E+13	4.49E+12	7.92E+11	8.50E+10
9000	16668	1.17E+20	1.55E+17	3.51E+15	1.92E+14	3.07E+12	3.83E+11	4.49E+10	1.03E+10
10000	18520	7.97E+19	4.26E+16	5.57E+14	2.12E+13	1.48E+11	0.00E+00	0.00E+00	0.00E+00
11000	20372	5.94E+19	8.96E+15	2.92E+13	5.02E+11	2.41E+09	0.00E+00	0.00E+00	0.00E+00
12000	22224	4.92E+19	2.01E+15	9.71E+11	0.00E+00	0.00E+00	0.00E+00	0.00E+00	0.00E+00
13000	24076	4.06E+19	2.75E+14	8.42E+07	0.00E+00	0.00E+00	0.00E+00	0.00E+00	0.00E+00
14000	25928	3.43E+19	4.79E+13	3.77E+07	0.00E+00	0.00E+00	0.00E+00	0.00E+00	0.00E+00
15000	27780	2.86E+19	1.09E+13	1.92E+07	0.00E+00	0.00E+00	0.00E+00	0.00E+00	0.00E+00
16000	29632	2.40E+19	6.90E+10	0.00E+00	0.00E+00	0.00E+00	0.00E+00	0.00E+00	0.00E+00
17000	31484	1.98E+19	5.12E+10	0.00E+00	0.00E+00	0.00E+00	0.00E+00	0.00E+00	0.00E+00
18000	33336	1.51E+19	3.90E+10	0.00E+00	0.00E+00	0.00E+00	0.00E+00	0.00E+00	0.00E+00
19327	35793	5.53E+18	2.71E+10	0.00E+00	0.00E+00	0.00E+00	0.00E+00	0.00E+00	0.00E+00

EQUIV. 1 MEV ELECTRON FLUENCE FOR V_oc CIRCULAR ORBIT
DUE TO GEOMAGNETICALLY TRAPPED PROTONS, MODEL AP8MIN

INCLINATION = 10 DEGREES

ALTITUDE (nm)	(km)	0 (0)	2.54E-3 (1)	7.64E-3 (3)	1.52E-2 (6)	3.05E-2 (12)	5.09E-2 (20)	7.64E-2 (30)	1.52E-1 (60)
150	277	0.00E+00	0.00E+00	0.00E+00	0.00E+00	0.00E+00	0.00E+00	0.00E+00	0.00E+00
250	463	2.82E+10	9.25E+09	6.91E+09	5.85E+09	4.79E+09	4.02E+09	3.13E+09	2.16E+09
300	555	4.23E+11	1.26E+11	1.11E+11	1.03E+11	9.34E+10	8.31E+10	6.98E+10	5.33E+10
450	833	7.01E+12	2.66E+12	2.45E+12	2.16E+12	2.05E+12	2.05E+12	1.90E+12	1.72E+12
600	1111	4.26E+13	1.99E+13	1.82E+13	1.68E+13	1.53E+13	1.42E+13	1.27E+13	1.10E+13
800	1481	2.37E+14	1.34E+14	1.19E+14	1.07E+14	9.30E+13	8.24E+13	7.02E+13	5.62E+13
1000	1852	8.69E+14	5.36E+14	4.54E+14	3.93E+14	3.25E+14	2.69E+14	2.12E+14	1.56E+14

Table 9-12. Annual Equivalent 1 MeV Electron Fluence for GaAs Solar Cells from Trapped Protons During Solar Max and Min, 10° Inclination (Infinite Backshielding)

PROTONS – P_{max}

EQUIV. 1 MEV ELECTRON FLUENCE FOR P_{max} CIRCULAR ORBIT   INCLINATION = 10 DEGREES
DUE TO GEOMAGNETICALLY TRAPPED PROTONS, MODEL AP8MAX

SHIELD THICKNESS, cm (mils)

ALTITUDE (nm)	(km)	0 (0)	2.54E-3 (1)	7.64E-3 (3)	1.52E-2 (6)	3.05E-2 (12)	5.09E-2 (20)	7.64E-2 (30)	1.52E-1 (60)
150	277	0.0E+00	0.00E+00	0.00E+00	0.00E+00	0.00E+00	0.00E+00	0.00E+00	0.00E+00
250	463	7.96E+08	1.11E+08	9.08E+07	8.47E+07	7.85E+07	7.50E+07	6.68E+07	5.12E+07
300	555	4.46E+10	1.51E+10	1.33E+10	1.20E+10	1.07E+10	1.01E+10	9.34E+09	8.39E+09
450	833	2.59E+12	1.02E+12	9.47E+11	8.97E+11	8.37E+11	7.84E+11	7.28E+11	6.62E+11
600	1111	2.15E+13	1.00E+13	9.16E+12	8.50E+12	7.73E+12	7.04E+12	6.26E+12	5.35E+12
800	1481	1.55E+14	8.40E+13	7.41E+13	6.67E+13	5.81E+13	5.08E+13	4.27E+13	3.39E+13
1000	1852	6.11E+14	3.65E+14	3.08E+14	2.67E+14	2.19E+14	1.79E+14	1.39E+14	1.02E+14
1250	2315	2.46E+15	1.52E+15	1.21E+15	1.00E+15	7.60E+14	5.54E+14	3.80E+14	2.44E+14
1500	2778	7.70E+15	4.66E+15	3.59E+15	2.83E+15	2.01E+15	1.33E+15	7.93E+14	4.25E+14
1750	3241	1.92E+16	1.10E+16	8.22E+15	6.24E+15	4.20E+15	2.57E+15	1.36E+15	6.02E+14
2000	3704	4.55E+16	2.34E+16	1.62E+16	1.14E+16	7.00E+15	3.93E+15	1.85E+15	7.00E+14
2250	4167	1.07E+17	4.64E+16	2.90E+16	1.85E+16	9.89E+15	5.03E+15	2.11E+15	7.09E+14
2500	4630	2.39E+17	8.02E+16	4.52E+16	2.60E+16	1.22E+16	5.64E+15	2.14E+15	6.63E+14
2750	5093	5.43E+17	1.23E+17	6.25E+16	3.29E+16	1.39E+16	5.85E+15	2.03E+15	5.89E+14
3000	5556	1.16E+18	1.74E+17	7.85E+16	3.78E+16	1.15E+16	5.60E+15	1.82E+15	5.02E+14
3500	6482	3.75E+18	3.17E+17	1.07E+17	4.16E+16	8.44E+15	4.35E+15	1.32E+15	3.37E+14
4000	7408	9.86E+18	4.90E+17	1.24E+17	4.01E+16	5.75E+15	2.98E+15	8.42E+14	2.07E+14
4500	8334	2.01E+19	6.53E+17	1.27E+17	3.46E+16	3.63E+15	1.85E+15	4.80E+14	1.12E+14
5000	9260	3.58E+19	7.97E+17	1.21E+17	2.78E+16	2.01E+15	1.08E+15	2.59E+14	5.74E+13
5500	10186	5.29E+19	8.93E+17	1.07E+17	2.03E+16	1.06E+15	5.80E+14	1.31E+14	2.66E+13
6000	11112	7.04E+19	8.96E+17	8.57E+16	1.38E+16	2.25E+14	2.77E+14	5.57E+13	9.84E+11
7000	12964	1.04E+20	6.47E+17	4.11E+16	4.73E+15	3.21E+13	4.64E+13	7.12E+12	6.07E+10
8000	14816	1.18E+20	3.33E+17	1.33E+16	1.10E+15	2.20E+12	5.01E+12	5.65E+11	7.37E+09
9000	16668	8.39E+19	1.11E+17	2.51E+15	1.37E+14	1.06E+11	2.74E+11	3.21E+10	0.00E+00
10000	18520	5.69E+19	3.04E+16	3.98E+14	1.52E+13	1.72E+09	0.00E+00	0.00E+00	0.00E+00
11000	20372	4.24E+19	6.40E+15	2.09E+13	3.58E+11	0.00E+00	0.00E+00	0.00E+00	0.00E+00
12000	22224	3.51E+19	1.44E+15	6.93E+11	1.06E+11	0.00E+00	0.00E+00	0.00E+00	0.00E+00
13000	24076	2.90E+19	1.97E+14	6.01E+07	1.72E+09	0.00E+00	0.00E+00	0.00E+00	0.00E+00
14000	25928	2.45E+19	3.42E+13	2.70E+07	0.00E+00	0.00E+00	0.00E+00	0.00E+00	0.00E+00
15000	27780	2.04E+19	7.81E+12	1.37E+07	0.00E+00	0.00E+00	0.00E+00	0.00E+00	0.00E+00
16000	29632	1.71E+19	4.93E+10	0.00E+00	0.00E+00	0.00E+00	0.00E+00	0.00E+00	0.00E+00
17000	31484	1.41E+19	3.66E+10	0.00E+00	0.00E+00	0.00E+00	0.00E+00	0.00E+00	0.00E+00
18000	33336	1.08E+19	2.78E+10	0.00E+00	0.00E+00	0.00E+00	0.00E+00	0.00E+00	0.00E+00
19327	35793	3.95E+18	1.94E+10	0.00E+00	0.00E+00	0.00E+00	0.00E+00	0.00E+00	0.00E+00

EQUIV. 1 MEV ELECTRON FLUENCE FOR P_{max} CIRCULAR ORBIT   INCLINATION = 10 DEGREES
DUE TO GEOMAGNETICALLY TRAPPED PROTONS, MODEL AP8MIN

ALTITUDE (nm)	(km)	0 (0)	2.54E-3 (1)	7.64E-3 (3)	1.52E-2 (6)	3.05E-2 (12)	5.09E-2 (20)	7.64E-2 (30)	1.52E-1 (60)
150	277	0.00E+00	0.00E+00	0.00E+00	0.00E+00	0.00E+00	0.00E+00	0.00E+00	0.00E+00
250	463	2.02E+10	6.61E+09	4.94E+09	4.18E+09	3.42E+09	2.87E+09	2.24E+09	1.54E+09
300	555	3.02E+11	8.99E+10	7.96E+10	7.39E+10	6.67E+10	5.93E+10	4.99E+10	3.81E+10
450	833	5.01E+12	1.90E+12	1.75E+12	1.65E+12	1.54E+12	1.46E+12	1.36E+12	1.23E+12
600	1111	3.04E+13	1.42E+13	1.30E+13	1.20E+13	1.10E+13	1.01E+13	9.10E+12	7.83E+12
800	1481	1.69E+14	9.55E+13	8.48E+13	7.61E+13	6.64E+13	5.89E+13	5.01E+13	4.02E+13
1000	1852	6.21E+14	3.83E+14	3.24E+14	2.81E+14	2.32E+14	1.92E+14	1.51E+14	1.11E+14

Table 9-13. Annual Equivalent 1 MeV Electron Fluence for GaAs Solar Cells from Trapped Protons During Solar Max and Min, 10° Inclination (Infinite Backshielding)

PROTONS - I$_{sc}$ INCLINATION = 10 DEGREES

EQUIV. 1 MEV PROTON FLUENCE FOR I$_{sc}$ - CIRCULAR ORBIT
DUE TO GEOMAGNETICALLY TRAPPED PROTONS, MODEL AP8MAX

ALTITUDE (nm)	(km)	SHIELD THICKNESS, cm (mils)							
		0 (0)	2.54E-3 (1)	7.64E-3 (3)	1.52E-2 (6)	3.05E-2 (12)	5.09E-2 (20)	7.64E-2 (30)	1.52E-1 (60)
150	277	0.00E+00	0.00E+00	0.00E+00	0.00E+00	0.00E+00	0.00E+00	0.00E+00	0.00E+00
250	463	5.38E+08	3.80E+07	2.73E+07	2.48E+07	2.25E+07	2.18E+07	1.88E+07	1.34E+07
300	555	2.90E+10	4.54E+09	3.62E+09	3.00E+09	2.45E+09	2.22E+09	1.96E+09	1.71E+09
450	833	1.51E+12	2.64E+11	2.30E+11	2.08E+11	1.84E+11	1.64E+11	1.43E+11	1.26E+11
600	1111	1.21E+13	3.02E+12	2.60E+12	2.31E+12	1.99E+12	1.73E+12	1.42E+12	1.12E+12
800	1481	8.42E+13	2.92E+13	2.43E+13	2.09E+13	1.73E+13	1.43E+13	1.10E+13	7.84E+12
1000	1852	3.27E+14	1.43E+14	1.14E+14	9.46E+13	7.40E+13	5.66E+13	3.97E+13	2.57E+13
1250	2315	1.33E+15	6.56E+14	5.02E+14	4.00E+14	2.93E+14	2.00E+14	1.23E+14	6.86E+13
1500	2778	4.23E+15	2.12E+15	1.58E+15	1.21E+15	8.43E+14	5.30E+14	2.88E+14	1.35E+14
1750	3241	1.07E+16	5.19E+15	3.75E+15	2.78E+15	1.84E+15	1.09E+15	5.33E+14	2.10E+14
2000	3704	2.62E+16	1.13E+16	7.61E+15	5.22E+15	3.15E+15	1.72E+15	7.60E+14	2.61E+14
2250	4167	6.38E+16	2.29E+16	1.39E+16	8.60E+15	4.54E+15	2.25E+15	8.90E+14	2.75E+14
2500	4630	1.51E+17	4.04E+16	2.21E+16	1.23E+16	5.66E+15	2.56E+15	9.16E+14	2.64E+14
2750	5093	3.73E+17	6.27E+16	3.08E+16	1.57E+16	6.40E+15	2.67E+15	8.81E+14	2.39E+14
3000	5556	8.49E+17	9.02E+16	3.91E+16	1.81E+16	6.55E+15	2.58E+15	7.95E+14	2.07E+14
3500	6482	2.90E+18	1.69E+17	5.39E+16	2.02E+16	5.44E+15	2.01E+15	5.80E+14	1.42E+14
4000	7408	8.03E+18	2.66E+17	6.34E+16	1.96E+16	4.01E+15	1.38E+15	3.72E+14	8.81E+13
4500	8334	1.71E+19	3.60E+17	6.52E+16	1.70E+16	2.75E+15	8.66E+14	2.13E+14	4.82E+13
5000	9260	3.20E+19	4.45E+17	6.23E+16	1.37E+16	1.74E+15	5.08E+14	1.15E+14	2.48E+13
5500	10186	4.69E+19	5.04E+17	5.55E+16	1.01E+16	9.67E+14	2.73E+14	5.88E+13	1.16E+13
6000	11112	6.12E+19	5.10E+17	4.48E+16	6.85E+15	5.12E+14	1.31E+14	2.52E+13	4.40E+12
7000	12964	9.35E+19	3.74E+17	2.17E+16	2.37E+15	1.09E+14	2.21E+13	3.25E+12	4.38E+11
8000	14816	1.16E+20	1.96E+17	7.03E+15	5.52E+14	1.57E+13	2.41E+12	2.60E+11	2.68E+10
9000	16668	8.50E+19	6.62E+16	1.34E+15	6.94E+13	1.08E+12	1.31E+11	1.44E+10	3.31E+09
10000	18520	5.96E+19	1.86E+16	2.13E+14	7.69E+12	5.83E+10	0.00E+00	0.00E+00	0.00E+00
11000	20372	4.69E+19	3.98E+15	1.14E+13	1.82E+11	9.28E+08	0.00E+00	0.00E+00	0.00E+00
12000	22224	4.18E+19	9.12E+14	3.89E+11	0.00E+00	0.00E+00	0.00E+00	0.00E+00	0.00E+00
13000	24076	3.74E+19	1.29E+14	3.73E+07	0.00E+00	0.00E+00	0.00E+00	0.00E+00	0.00E+00
14000	25928	3.40E+19	2.31E+13	1.64E+07	0.00E+00	0.00E+00	0.00E+00	0.00E+00	0.00E+00
15000	27780	3.07E+19	5.38E+12	8.20E+06	0.00E+00	0.00E+00	0.00E+00	0.00E+00	0.00E+00
16000	29632	2.72E+19	4.22E+10	0.00E+00	0.00E+00	0.00E+00	0.00E+00	0.00E+00	0.00E+00
17000	31484	2.35E+19	3.10E+10	0.00E+00	0.00E+00	0.00E+00	0.00E+00	0.00E+00	0.00E+00
18000	33336	1.88E+19	2.34E+10	0.00E+00	0.00E+00	0.00E+00	0.00E+00	0.00E+00	0.00E+00
19327	35793	6.91E+18	1.61E+10	0.00E+00	0.00E+00	0.00E+00	0.00E+00	0.00E+00	0.00E+00

EQUIV. 1 MEV PROTON FLUENCE FOR I$_{sc}$ - CIRCULAR ORBIT   INCLINATION = 10 DEGREES
DUE TO GEOMAGNETICALLY TRAPPED PROTONS, MODEL AP8MIN

ALTITUDE (nm)	(km)	0 (0)	2.54E-3 (1)	7.64E-3 (3)	1.52E-2 (6)	3.05E-2 (12)	5.09E-2 (20)	7.64E-2 (30)	1.52E-1 (60)
150	277	0.00E+00	0.00E+00	0.00E+00	0.00E+00	0.00E+00	0.00E+00	0.00E+00	0.00E+00
250	463	1.17E+10	2.78E+09	1.90E+09	1.54E+09	1.21E+09	9.94E+08	7.28E+08	4.72E+08
300	555	2.09E+11	3.11E+10	2.59E+10	2.35E+10	2.08E+10	1.80E+10	1.42E+10	1.01E+10
450	833	3.03E+12	4.98E+11	4.22E+11	3.80E+11	3.37E+11	3.07E+11	2.69E+11	2.32E+11
600	1111	1.71E+13	4.20E+12	3.60E+12	3.18E+12	2.74E+12	2.42E+12	2.02E+12	1.61E+12
800	1481	8.87E+13	3.25E+13	2.73E+13	2.34E+13	1.93E+13	1.62E+13	1.26E+13	9.09E+12
1000	1852	3.20E+14	1.48E+14	1.19E+14	9.85E+13	7.72E+13	6.01E+13	4.27E+13	2.75E+13

9-18

Table 9-14. Annual Equivalent 1 MeV Electron Fluence for GaAs Solar Cells from Trapped Electrons During Solar Max, 20° Inclination (Infinite Backshielding)

ELECTRONS - I_{sc}, V_{oc}, AND P_{max}

INCLINATION = 20 DEGREES

EQUIV. 1 MEV ELECTRON FLUENCE FOR I_{sc} - CIRCULAR ORBIT
DUE TO GEOMAGNETICALLY TRAPPED ELECTRONS - MODEL AE8MAX

ALTITUDE (nm)	(km)	SHIELD THICKNESS, cm (mils)							
		0 (0)	2.54E-3 (1)	7.64E-3 (3)	1.52E-2 (6)	3.05E-2 (12)	5.09E-2 (20)	7.64E-2 (30)	1.52E-1 (60)
150	277	8.44E+06	7.71E+06	6.89E+06	6.07E+06	5.04E+06	4.19E+06	3.48E+06	2.24E+06
250	463	2.72E+08	2.13E+08	1.58E+08	1.13E+08	7.17E+07	4.77E+07	3.32E+07	1.59E+07
300	555	2.24E+09	1.69E+09	1.19E+09	8.08E+08	4.66E+08	2.83E+08	1.81E+08	7.22E+07
450	833	1.48E+11	1.11E+11	7.80E+10	5.25E+10	3.01E+10	1.83E+10	1.17E+10	4.53E+09
600	1111	1.26E+12	9.40E+11	6.55E+11	4.37E+11	2.47E+11	1.48E+11	9.36E+10	3.58E+10
800	1481	6.86E+12	5.16E+12	3.63E+12	2.45E+12	1.41E+12	8.54E+11	5.45E+11	2.10E+11
1000	1852	2.14E+13	1.63E+13	1.16E+13	7.98E+12	4.72E+12	2.93E+12	1.91E+12	7.52E+11
1250	2315	4.70E+13	3.61E+13	2.61E+13	1.83E+13	1.11E+13	7.12E+12	4.72E+12	1.92E+12
1500	2778	7.02E+13	5.39E+13	3.91E+13	2.73E+13	1.66E+13	1.05E+13	6.95E+12	2.86E+12
1750	3241	8.91E+13	6.79E+13	4.84E+13	3.31E+13	1.92E+13	1.16E+13	7.35E+12	2.91E+12
2000	3704	1.04E+14	7.83E+13	5.47E+13	3.62E+13	1.96E+13	1.09E+13	6.38E+12	2.30E+12
2250	4167	1.15E+14	8.55E+13	5.88E+13	3.78E+13	1.92E+13	9.61E+12	5.12E+12	1.61E+12
2500	4630	1.18E+14	8.73E+13	5.93E+13	3.74E+13	1.79E+13	8.18E+12	3.89E+12	1.04E+12
2750	5093	1.14E+14	8.39E+13	5.66E+13	3.53E+13	1.62E+13	6.76E+12	2.85E+12	6.04E+11
3000	5556	1.03E+14	7.61E+13	5.11E+13	3.15E+13	1.41E+13	5.53E+12	2.11E+12	3.54E+11
3500	6482	7.84E+13	5.73E+13	3.82E+13	2.34E+13	1.02E+13	3.88E+12	1.41E+12	2.17E+11
4000	7408	6.22E+13	4.58E+13	3.09E+13	1.93E+13	8.94E+12	3.78E+12	1.64E+12	4.03E+11
4500	8334	5.48E+13	4.16E+13	2.94E+13	1.96E+13	1.04E+13	5.33E+12	2.90E+12	9.97E+11
5000	9260	4.79E+13	3.82E+13	2.90E+13	2.12E+13	1.31E+13	8.13E+12	5.22E+12	2.16E+12
5500	10186	4.55E+13	3.85E+13	3.14E+13	2.50E+13	1.77E+13	1.24E+13	8.79E+12	4.05E+12
6000	11112	4.99E+13	4.41E+13	3.79E+13	3.18E+13	2.42E+13	1.82E+13	1.35E+13	6.57E+12
7000	12964	7.60E+13	6.88E+13	6.05E+13	5.21E+13	4.12E+13	3.20E+13	2.44E+13	1.24E+13
8000	14816	1.06E+14	9.61E+13	8.47E+13	7.30E+13	5.78E+13	4.48E+13	3.42E+13	1.73E+13
9000	16668	1.34E+14	1.21E+14	1.06E+14	9.12E+13	7.18E+13	5.53E+13	4.19E+13	2.09E+13
10000	18520	1.52E+14	1.37E+14	1.20E+14	1.02E+14	7.96E+13	6.05E+13	4.52E+13	2.16E+13
11000	20372	1.63E+14	1.47E+14	1.27E+14	1.08E+14	8.29E+13	6.21E+13	4.55E+13	2.05E+13
12000	22224	1.56E+14	1.39E+14	1.19E+14	1.00E+14	7.56E+13	5.56E+13	3.99E+13	1.71E+13
13000	24076	1.36E+14	1.20E+14	1.02E+14	8.48E+13	6.27E+13	4.50E+13	3.16E+13	1.30E+13
14000	25928	1.14E+14	9.99E+13	8.42E+13	6.88E+13	4.99E+13	3.51E+13	2.42E+13	9.70E+12
15000	27780	9.03E+13	7.83E+13	6.51E+13	5.25E+13	3.73E+13	2.57E+13	1.74E+13	6.72E+12
16000	29632	7.18E+13	6.17E+13	5.08E+13	4.05E+13	2.82E+13	1.91E+13	1.27E+13	4.76E+12
17000	31484	5.57E+13	4.74E+13	3.86E+13	3.04E+13	2.09E+13	1.39E+13	9.10E+12	3.28E+12
18000	33336	4.02E+13	3.39E+13	2.72E+13	2.12E+13	1.42E+13	9.32E+12	5.98E+12	2.04E+12
19327	35793	2.50E+13	2.07E+13	1.63E+13	1.23E+13	7.99E+12	5.03E+12	3.12E+12	9.94E+11

Table 9-15. Annual Equivalent 1 MeV Electron Fluence for GaAs Solar Cells from Trapped Electrons During Solar Min, 20° Inclination (Infinite Backshielding)

ELECTRONS - I_{sc}, V_{oc}, AND P_{max} INCLINATION = 20 DEGREES

EQUIV. 1 MEV ELECTRON FLUENCE FOR I_{sc} - CIRCULAR ORBIT
DUE TO GEOMAGNETICALLY TRAPPED ELECTRONS - MODEL AE8MIN

| ALTITUDE | | SHIELD THICKNESS, cm (mils) | | | | | | | |
(nm)	(km)	0 (0)	2.54E-3 (1)	7.64E-3 (3)	1.52E-2 (6)	3.05E-2 (12)	5.09E-2 (20)	7.64E-2 (30)	1.52E-1 (60)
150	277	5.88E+06	5.46E+06	4.98E+06	4.48E+06	3.81E+06	3.22E+06	2.71E+06	1.78E+06
250	463	1.28E+08	1.07E+08	8.52E+07	6.65E+07	4.65E+07	3.30E+07	2.41E+07	1.24E+07
300	555	8.77E+08	7.05E+08	5.37E+08	3.96E+08	2.52E+08	1.63E+08	1.08E+08	4.61E+07
450	833	6.90E+10	5.55E+10	4.24E+10	3.14E+10	2.01E+10	1.30E+10	8.62E+09	3.50E+09
600	1111	6.20E+11	4.99E+11	3.82E+11	2.83E+11	1.81E+11	1.17E+11	7.73E+10	3.12E+10
800	1481	3.71E+12	3.00E+12	2.32E+12	1.73E+12	1.12E+12	7.31E+11	4.85E+11	1.96E+11
1000	1852	1.22E+13	9.97E+12	7.80E+12	5.92E+12	3.93E+12	2.61E+12	1.76E+12	7.24E+11
1250	2315	2.71E+13	2.25E+13	1.79E+13	1.39E+13	9.49E+12	6.49E+12	4.46E+12	1.89E+12
1500	2778	3.78E+13	3.15E+13	2.53E+13	1.98E+13	1.37E+13	9.50E+12	6.60E+12	2.85E+12
1750	3241	4.15E+13	3.44E+13	2.74E+13	2.13E+13	1.46E+13	9.96E+12	6.85E+12	2.91E+12
2000	3704	4.02E+13	3.29E+13	2.57E+13	1.95E+13	1.29E+13	8.54E+12	5.70E+12	2.30E+12
2250	4167	3.63E+13	2.91E+13	2.21E+13	1.63E+13	1.04E+13	6.60E+12	4.26E+12	1.61E+12
2500	4630	3.03E+13	2.37E+13	1.76E+13	1.26E+13	7.66E+12	4.68E+12	2.92E+12	1.03E+12
2750	5093	2.38E+13	1.83E+13	1.32E+13	9.14E+12	5.29E+12	3.09E+12	1.85E+12	5.95E+11
3000	5556	1.87E+13	1.41E+13	9.98E+12	6.73E+12	3.74E+12	2.09E+12	1.19E+12	3.41E+11
3500	6482	1.33E+13	9.97E+12	6.97E+12	4.64E+12	2.50E+12	1.33E+12	7.14E+11	1.68E+11
4000	7408	1.18E+13	9.01E+12	6.46E+12	4.42E+12	2.47E+12	1.36E+12	7.46E+11	1.92E+11
4500	8334	1.14E+13	8.95E+12	6.68E+12	4.80E+12	2.92E+12	1.76E+12	1.07E+12	3.51E+11
5000	9260	1.14E+13	9.31E+12	7.32E+12	5.60E+12	3.77E+12	2.53E+12	1.70E+12	6.72E+11
5500	10186	1.23E+13	1.05E+13	8.67E+12	7.01E+12	5.11E+12	3.71E+12	2.67E+12	1.20E+12
6000	11112	1.42E+13	1.25E+13	1.06E+13	8.90E+12	6.81E+12	5.17E+12	3.88E+12	1.90E+12
7000	12964	2.19E+13	1.98E+13	1.75E+13	1.51E+13	1.21E+13	9.52E+12	7.36E+12	3.84E+12
8000	14816	3.50E+13	3.20E+13	2.85E+13	2.49E+13	2.00E+13	1.58E+13	1.23E+13	6.43E+12
9000	16668	5.38E+13	4.91E+13	4.37E+13	3.80E+13	3.05E+13	2.39E+13	1.84E+13	9.35E+12
10000	18520	7.67E+13	6.98E+13	6.17E+13	5.34E+13	4.23E+13	3.27E+13	2.47E+13	1.20E+13
11000	20372	1.01E+14	9.12E+13	8.01E+13	6.87E+13	5.38E+13	4.11E+13	3.07E+13	1.44E+13
12000	22224	1.19E+14	1.07E+14	9.31E+13	7.88E+13	6.04E+13	4.51E+13	3.29E+13	1.47E+13
13000	24076	1.22E+14	1.09E+14	9.31E+13	7.76E+13	5.80E+13	4.22E+13	3.00E+13	1.26E+13
14000	25928	1.10E+14	9.70E+13	8.20E+13	6.73E+13	4.92E+13	3.49E+13	2.42E+13	9.76E+12
15000	27780	9.00E+13	7.82E+13	6.51E+13	5.25E+13	3.74E+13	2.58E+13	1.75E+13	6.81E+12
16000	29632	7.20E+13	6.20E+13	5.10E+13	4.07E+13	2.84E+13	1.93E+13	1.28E+13	4.82E+12
17000	31484	5.59E+13	4.76E+13	3.88E+13	3.06E+13	2.10E+13	1.40E+13	9.15E+12	3.30E+12
18000	33336	4.04E+13	3.40E+13	2.74E+13	2.13E+13	1.43E+13	9.38E+12	6.02E+12	2.05E+12
19327	35793	2.52E+13	2.09E+13	1.64E+13	1.25E+13	8.08E+12	5.10E+12	3.16E+12	1.01E+12

Table 9-16. Annual Equivalent 1 MeV Electron Fluence for GaAs Solar Cells from Trapped Protons During Solar Max and Min, 20° Inclination (Infinite Backshielding)

PROTONS – V_{oc}

EQUIV. 1 MEV ELECTRON FLUENCE FOR V_{oc} CIRCULAR ORBIT
DUE TO GEOMAGNETICALLY TRAPPED PROTONS, MODEL AP8MAX   INCLINATION = 20 DEGREES

ALTITUDE (nm)	(km)	SHIELD THICKNESS, cm (mils)							
		0 (0)	2.54E-3 (1)	7.64E-3 (3)	1.52E-2 (6)	3.05E-2 (12)	5.09E-2 (20)	7.64E-2 (30)	1.52E-1 (60)
150	277	2.06E+07	6.94E+06	6.10E+06	5.46E+06	4.82E+06	4.47E+06	4.09E+06	3.46E+06
250	463	1.51E+11	4.47E+11	3.97E+11	3.64E+10	3.29E+10	3.07E+10	2.86E+10	2.61E+10
300	555	6.11E+11	2.08E+11	1.88E+11	1.75E+11	1.60E+11	1.50E+11	1.40E+11	1.28E+11
450	833	7.55E+12	3.47E+12	3.19E+12	2.98E+12	2.73E+12	2.51E+12	2.27E+12	1.99E+12
600	1111	4.11E+13	2.10E+13	1.87E+13	1.70E+13	1.51E+13	1.34E+13	1.16E+13	9.64E+12
800	1481	2.38E+14	1.34E+14	1.14E+14	1.00E+14	8.43E+13	7.14E+13	5.83E+13	4.52E+13
1000	1852	9.26E+14	5.44E+14	4.44E+14	3.73E+14	2.96E+14	2.32E+14	1.75E+14	1.25E+14
1250	2315	3.68E+15	2.15E+15	1.66E+15	1.32E+15	9.62E+14	6.77E+14	4.48E+14	2.80E+14
1500	2778	1.17E+16	6.30E+15	4.64E+15	3.51E+15	2.39E+15	1.53E+15	8.85E+14	4.64E+14
1750	3241	3.12E+16	1.47E+16	1.02E+16	7.40E+15	4.73E+15	2.81E+15	1.45E+15	6.33E+14
2000	3704	7.94E+16	3.04E+16	1.96E+16	1.31E+16	7.60E+15	4.15E+15	1.91E+15	7.21E+14
2250	4167	1.94E+17	5.85E+16	3.40E+16	2.06E+16	1.05E+16	5.23E+15	2.16E+15	7.26E+14
2500	4630	4.41E+17	9.75E+16	5.09E+16	2.82E+16	1.26E+16	5.77E+15	2.16E+15	6.71E+14
2750	5093	9.63E+17	1.47E+17	6.88E+16	3.50E+16	1.40E+16	5.91E+15	2.04E+15	5.93E+14
3000	5556	1.92E+18	2.07E+17	8.55E+16	3.97E+16	1.40E+16	5.61E+15	1.82E+15	5.04E+14
3500	6482	5.77E+18	3.66E+17	1.13E+17	4.28E+16	1.14E+16	4.31E+15	1.30E+15	3.36E+14
4000	7408	1.41E+19	5.50E+17	1.29E+17	4.06E+16	8.30E+15	2.91E+15	8.22E+14	2.04E+14
4500	8334	2.75E+19	7.11E+17	1.29E+17	3.45E+16	5.57E+15	1.79E+15	4.62E+14	1.09E+14
5000	9260	4.72E+19	8.41E+17	1.20E+17	2.70E+16	3.42E+15	1.02E+15	2.44E+14	5.41E+13
5500	10186	6.79E+19	9.27E+17	1.06E+17	1.97E+16	1.90E+15	5.43E+14	1.22E+14	2.46E+13
6000	11112	8.79E+19	8.99E+17	8.21E+16	1.30E+16	9.79E+14	2.54E+14	5.06E+13	9.04E+12
7000	12964	1.20E+20	6.24E+17	3.81E+16	4.33E+15	2.02E+14	4.14E+13	6.30E+12	8.64E+11
8000	14816	1.31E+20	3.22E+17	1.23E+16	1.01E+15	2.88E+13	4.47E+12	4.96E+11	5.17E+10
9000	16668	9.15E+19	1.04E+17	2.29E+15	1.23E+14	1.92E+12	2.36E+11	2.64E+10	5.23E+09
10000	18520	6.30E+19	2.80E+16	3.55E+14	1.33E+13	9.20E+10	0.00E+00	0.00E+00	0.00E+00
11000	20372	4.80E+19	5.78E+15	5.32E+11	3.65E+11	1.78E+09	0.00E+00	0.00E+00	0.00E+00
12000	22224	3.95E+19	1.22E+15	6.90E+07	0.00E+00	0.00E+00	0.00E+00	0.00E+00	0.00E+00
13000	24076	3.25E+19	1.62E+14	3.13E+07	0.00E+00	0.00E+00	0.00E+00	0.00E+00	0.00E+00
14000	25928	2.72E+19	2.89E+13	1.67E+07	0.00E+00	0.00E+00	0.00E+00	0.00E+00	0.00E+00
15000	27780	2.19E+19	7.36E+12	0.00E+00	0.00E+00	0.00E+00	0.00E+00	0.00E+00	0.00E+00
16000	29632	1.78E+19	5.41E+10	0.00E+00	0.00E+00	0.00E+00	0.00E+00	0.00E+00	0.00E+00
17000	31484	1.43E+19	4.06E+10	0.00E+00	0.00E+00	0.00E+00	0.00E+00	0.00E+00	0.00E+00
18000	33336	9.99E+18	2.98E+10	0.00E+00	0.00E+00	0.00E+00	0.00E+00	0.00E+00	0.00E+00
19327	35793	2.70E+18	1.89E+10	0.00E+00	0.00E+00	0.00E+00	0.00E+00	0.00E+00	0.00E+00

EQUIV. 1 MEV ELECTRON FLUENCE FOR V_{oc} CIRCULAR ORBIT
DUE TO GEOMAGNETICALLY TRAPPED PROTONS, MODEL AP8MIN   INCLINATION = 20 DEGREES

ALTITUDE (nm)	(km)	0 (0)	2.54E-3 (1)	7.64E-3 (3)	1.52E-2 (6)	3.05E-2 (12)	5.09E-2 (20)	7.64E-2 (30)	1.52E-1 (60)
150	277	2.30E+09	6.44E+08	4.68E+08	3.69E+08	3.10E+08	2.46E+08	1.92E+08	1.34E+08
250	463	5.86E+11	1.77E+11	1.57E+11	1.45E+11	1.32E+11	1.21E+11	1.07E+11	8.92E+10
300	555	1.77E+12	5.77E+11	5.20E+11	4.85E+11	4.46E+11	4.13E+11	3.73E+11	3.22E+11
450	833	1.09E+13	5.00E+12	4.60E+12	4.30E+12	3.96E+12	3.68E+12	3.37E+12	2.97E+12
600	1111	5.00E+13	2.59E+13	2.32E+13	2.12E+13	1.88E+13	1.70E+13	1.50E+13	1.26E+13
800	1481	2.53E+14	1.46E+14	1.26E+14	1.11E+14	9.39E+13	8.08E+13	6.69E+13	5.24E+13
1000	1852	9.33E+14	5.60E+14	4.61E+14	3.88E+14	3.09E+14	2.47E+14	1.88E+14	1.36E+14

Table 9-17. Annual Equivalent 1 MeV Electron Fluence for GaAs Solar Cells from Trapped Protons During Solar Max and Min, 20° Inclination (Infinite Backshielding)

PROTONS - P_{max} INCLINATION = 20 DEGREES

EQUIV. 1 MEV ELECTRON FLUENCE FOR P_{max} CIRCULAR ORBIT
DUE TO GEOMAGNETICALLY TRAPPED PROTONS, MODEL AP8MAX

ALTITUDE		SHIELD THICKNESS, cm (mils)							
(nm)	(km)	0 (0)	2.54E-3 (1)	7.64E-3 (3)	1.52E-2 (6)	3.05E-2 (12)	5.09E-2 (20)	7.64E-2 (30)	1.52E-1 (60)
150	277	1.47E+07	4.96E+06	4.36E+06	3.90E+06	3.44E+06	3.19E+06	2.92E+06	2.47E+06
250	463	1.08E+11	3.19E+10	2.84E+10	2.60E+10	2.35E+10	2.20E+10	2.04E+10	1.87E+10
300	555	4.37E+11	1.48E+11	1.34E+11	1.25E+11	1.14E+11	1.07E+11	1.00E+11	9.16E+10
450	833	5.39E+12	2.48E+12	2.28E+12	2.13E+12	1.95E+12	1.79E+12	1.62E+12	1.42E+12
600	1111	2.94E+13	1.50E+13	1.34E+13	1.22E+13	1.08E+13	9.59E+12	8.30E+12	6.88E+12
800	1481	1.70E+14	9.56E+13	8.17E+13	7.16E+13	6.02E+13	5.10E+13	4.17E+13	3.23E+13
1000	1852	6.61E+14	3.89E+14	3.17E+14	2.67E+14	2.11E+14	1.66E+14	1.25E+14	8.93E+13
1250	2315	2.63E+15	1.53E+15	1.18E+15	9.42E+14	6.87E+14	4.83E+14	3.20E+14	2.00E+14
1500	2778	8.34E+15	4.50E+15	3.31E+15	2.51E+15	1.71E+15	1.09E+15	6.32E+14	3.31E+14
1750	3241	2.23E+16	1.05E+16	7.31E+15	5.28E+15	3.38E+15	2.01E+15	1.03E+15	4.52E+14
2000	3704	5.67E+16	2.17E+16	1.40E+16	9.35E+15	5.43E+15	2.97E+15	1.37E+15	5.15E+14
2250	4167	1.39E+17	4.18E+16	2.42E+16	1.47E+16	7.51E+15	3.74E+15	1.54E+15	5.18E+14
2500	4630	3.15E+17	6.97E+16	3.63E+16	2.01E+16	9.97E+15	4.12E+15	1.54E+15	4.80E+14
2750	5093	6.88E+17	1.05E+17	4.91E+16	2.50E+16	1.00E+16	4.22E+15	1.46E+15	4.24E+14
3000	5556	1.38E+18	1.48E+17	6.10E+16	2.83E+16	8.14E+15	4.01E+15	1.30E+15	3.60E+14
3500	6482	4.12E+18	2.61E+17	8.10E+16	3.05E+16	5.93E+15	3.08E+15	9.32E+14	2.40E+14
4000	7408	1.01E+19	3.93E+17	9.24E+16	2.90E+16	3.98E+15	2.08E+15	5.87E+14	1.45E+14
4500	8334	1.96E+19	5.08E+17	9.24E+16	2.46E+16	2.44E+15	1.28E+15	3.30E+14	7.77E+13
5000	9260	3.37E+19	6.01E+17	8.57E+16	1.93E+16	1.36E+15	7.29E+14	1.74E+14	3.86E+13
5500	10186	4.85E+19	6.62E+17	7.54E+16	1.41E+16	6.99E+14	3.88E+14	8.69E+13	1.76E+13
6000	11112	6.28E+19	6.42E+17	5.86E+16	9.25E+15	1.44E+14	1.81E+14	3.62E+13	6.46E+12
7000	12964	8.57E+19	4.46E+17	2.72E+16	3.09E+15	2.06E+13	2.96E+13	4.50E+12	6.17E+11
8000	14816	9.35E+19	2.30E+17	8.82E+15	7.18E+14	1.37E+12	3.19E+12	3.54E+11	3.69E+10
9000	16668	6.53E+19	7.44E+16	1.63E+15	8.79E+13	6.57E+10	1.69E+11	1.88E+10	3.73E+09
10000	18520	4.50E+19	2.00E+16	2.53E+14	9.53E+12	1.27E+09	0.00E+00	0.00E+00	0.00E+00
11000	20372	3.43E+19	4.13E+15	1.41E+14	2.61E+11	0.00E+00	0.00E+00	0.00E+00	0.00E+00
12000	22224	2.82E+19	8.74E+14	3.80E+11	1.27E+09	0.00E+00	0.00E+00	0.00E+00	0.00E+00
13000	24076	2.32E+19	1.16E+14	4.93E+07	0.00E+00	0.00E+00	0.00E+00	0.00E+00	0.00E+00
14000	25928	1.94E+19	2.06E+13	2.24E+07	0.00E+00	0.00E+00	0.00E+00	0.00E+00	0.00E+00
15000	27780	1.56E+19	5.26E+12	1.19E+07	0.00E+00	0.00E+00	0.00E+00	0.00E+00	0.00E+00
16000	29632	1.27E+19	3.86E+10	0.00E+00	0.00E+00	0.00E+00	0.00E+00	0.00E+00	0.00E+00
17000	31484	1.02E+19	2.90E+10	0.00E+00	0.00E+00	0.00E+00	0.00E+00	0.00E+00	0.00E+00
18000	33336	7.13E+18	2.13E+10	0.00E+00	0.00E+00	0.00E+00	0.00E+00	0.00E+00	0.00E+00
19327	35793	1.93E+18	1.35E+10	0.00E+00	0.00E+00	0.00E+00	0.00E+00	0.00E+00	0.00E+00

EQUIV. 1 MEV ELECTRON FLUENCE FOR P_{max} CIRCULAR ORBIT   INCLINATION = 20 DEGREES
DUE TO GEOMAGNETICALLY TRAPPED PROTONS, MODEL AP8MIN

ALTITUDE									
(nm)	(km)	0 (0)	2.54E-3 (1)	7.64E-3 (3)	1.52E-2 (6)	3.05E-2 (12)	5.09E-2 (20)	7.64E-2 (30)	1.52E-1 (60)
150	277	1.65E+09	4.60E+08	3.34E+08	2.63E+08	2.21E+08	1.75E+08	1.37E+08	9.58E+07
250	463	4.19E+11	1.27E+11	1.12E+11	1.04E+11	9.45E+10	8.62E+10	7.61E+10	6.37E+10
300	555	1.26E+12	4.12E+11	3.71E+11	3.47E+11	3.18E+11	2.95E+11	2.66E+11	2.30E+11
450	833	7.77E+12	3.57E+12	3.28E+12	3.07E+12	2.83E+12	2.63E+12	2.40E+12	2.12E+12
600	1111	3.57E+13	1.85E+13	1.66E+13	1.51E+13	1.35E+13	1.22E+13	1.07E+13	8.98E+12
800	1481	1.80E+14	1.04E+14	9.01E+13	7.91E+13	6.71E+13	5.77E+13	4.78E+13	3.74E+13
1000	1852	6.66E+14	4.00E+14	3.29E+14	2.77E+14	2.21E+14	1.76E+14	1.34E+14	9.68E+13

9-22

Table 9-18. Annual Equivalent 1 MeV Electron Fluence for GaAs Solar Cells from Trapped Protons During Solar Max and Min, 20° Inclination (Infinite Backshielding)

PROTONS - I_{sc}

INCLINATION = 20 DEGREES

EQUIV. 1 MEV PROTON FLUENCE FOR I_{sc} - CIRCULAR ORBIT
DUE TO GEOMAGNETICALLY TRAPPED PROTONS, MODEL AP8MAX

ALTITUDE		SHIELD THICKNESS, cm (mils)							
(nm)	(km)	0 (0)	2.54E-3 (1)	7.64E-3 (3)	1.52E-2 (6)	3.05E-2 (12)	5.09E-2 (20)	7.64E-2 (30)	1.52E-1 (60)
150	277	9.92E+06	1.69E+06	1.40E+06	1.19E+06	1.02E+06	9.44E+05	8.70E+05	8.01E+05
250	463	6.96E+10	9.09E+09	7.28E+09	6.17E+09	5.07E+09	4.47E+09	3.88E+09	3.42E+09
300	555	2.69E+11	4.04E+10	3.34E+10	2.89E+10	2.44E+10	2.17E+10	1.90E+10	1.68E+10
450	833	3.03E+12	7.14E+11	6.15E+11	5.48E+11	4.75E+11	4.13E+11	3.46E+11	2.87E+11
600	1111	1.62E+13	4.91E+12	4.13E+12	3.58E+12	2.99E+12	2.51E+12	1.99E+12	1.51E+12
800	1481	9.21E+13	3.55E+13	2.85E+13	2.38E+13	1.89E+13	1.50E+13	1.11E+13	7.64E+12
1000	1852	3.58E+14	1.59E+14	1.23E+14	9.89E+13	7.42E+13	5.42E+13	3.64E+13	2.28E+13
1250	2315	1.45E+15	6.81E+14	5.04E+14	3.86E+14	2.71E+14	1.79E+14	1.06E+14	5.71E+13
1500	2778	4.72E+15	2.09E+15	1.49E+15	1.09E+15	7.23E+14	4.40E+14	2.32E+14	1.06E+14
1750	3241	1.32E+16	5.01E+15	3.38E+15	2.37E+15	1.49E+15	8.54E+14	4.08E+14	1.59E+14
2000	3704	3.53E+16	1.06E+16	6.62E+15	4.30E+15	2.46E+15	1.30E+15	5.63E+14	1.92E+14
2250	4167	9.06E+16	2.09E+16	1.17E+16	6.89E+15	3.46E+15	1.68E+15	6.51E+14	2.01E+14
2500	4630	2.19E+17	3.55E+16	1.78E+16	9.52E+15	4.20E+15	1.87E+15	6.62E+14	1.91E+14
2750	5093	5.07E+17	5.42E+16	2.43E+16	1.19E+16	4.68E+15	1.93E+15	6.31E+14	1.72E+14
3000	5556	1.06E+18	7.73E+16	3.04E+16	1.36E+16	4.72E+15	1.84E+15	5.67E+14	1.48E+14
3500	6482	3.31E+18	1.40E+17	4.10E+16	1.48E+16	3.85E+15	1.42E+15	4.10E+14	1.01E+14
4000	7408	8.45E+18	2.14E+17	4.72E+16	1.42E+16	2.82E+15	9.68E+14	2.60E+14	6.17E+13
4500	8334	1.72E+19	2.81E+17	4.75E+16	1.21E+16	1.90E+15	5.97E+14	1.47E+14	3.33E+13
5000	9260	3.08E+19	3.37E+17	4.44E+16	9.53E+15	1.17E+15	3.42E+14	7.77E+13	1.67E+13
5500	10186	4.40E+19	3.75E+17	3.93E+16	6.98E+15	6.53E+14	1.83E+14	3.90E+13	7.65E+12
6000	11112	5.62E+19	3.66E+17	3.07E+16	4.61E+15	3.37E+14	8.57E+13	1.63E+13	2.84E+12
7000	12964	7.93E+19	2.58E+17	1.44E+16	1.55E+15	7.02E+13	1.41E+13	2.05E+12	2.75E+11
8000	14816	9.27E+19	1.35E+17	4.68E+15	3.61E+14	1.01E+13	1.53E+12	1.63E+11	1.63E+10
9000	16668	6.73E+19	4.46E+16	8.72E+14	4.44E+13	6.73E+11	8.11E+10	8.51E+09	1.68E+09
10000	18520	4.81E+19	1.22E+16	1.36E+14	4.83E+12	3.61E+10	0.00E+00	0.00E+00	0.00E+00
11000	20372	3.89E+19	2.57E+15	7.69E+12	1.33E+11	6.84E+08	0.00E+00	0.00E+00	0.00E+00
12000	22224	3.45E+19	5.57E+14	2.13E+11	0.00E+00	0.00E+00	0.00E+00	0.00E+00	0.00E+00
13000	24076	3.08E+19	7.58E+13	3.04E+07	0.00E+00	0.00E+00	0.00E+00	0.00E+00	0.00E+00
14000	25928	2.75E+19	1.39E+13	1.36E+07	0.00E+00	0.00E+00	0.00E+00	0.00E+00	0.00E+00
15000	27780	2.38E+19	3.62E+12	7.11E+06	0.00E+00	0.00E+00	0.00E+00	0.00E+00	0.00E+00
16000	29632	2.04E+19	3.28E+10	0.00E+00	0.00E+00	0.00E+00	0.00E+00	0.00E+00	0.00E+00
17000	31484	1.72E+19	2.44E+10	0.00E+00	0.00E+00	0.00E+00	0.00E+00	0.00E+00	0.00E+00
18000	33336	1.25E+19	1.77E+10	0.00E+00	0.00E+00	0.00E+00	0.00E+00	0.00E+00	0.00E+00
19327	35793	3.35E+18	1.11E+10	0.00E+00	0.00E+00	0.00E+00	0.00E+00	0.00E+00	0.00E+00

EQUIV. 1 MEV PROTON FLUENCE FOR I_{sc} - CIRCULAR ORBIT
DUE TO GEOMAGNETICALLY TRAPPED PROTONS, MODEL AP8MIN

INCLINATION = 20 DEGREES

ALTITUDE									
(nm)	(km)	0 (0)	2.54E-3 (1)	7.64E-3 (3)	1.52E-2 (6)	3.05E-2 (12)	5.09E-2 (20)	7.64E-2 (30)	1.52E-1 (60)
150	277	9.94E+08	1.99E+08	1.32E+08	9.80E+07	8.03E+07	6.09E+07	4.51E+07	3.01E+07
250	463	2.77E+11	4.01E+10	3.26E+10	2.91E+10	2.53E+10	2.21E+10	1.82E+10	1.41E+10
300	555	8.19E+11	1.22E+11	1.01E+11	9.06E+10	7.92E+10	7.06E+10	5.96E+10	4.81E+10
450	833	4.37E+12	1.00E+12	8.61E+11	7.65E+11	6.65E+11	5.91E+11	5.04E+11	4.20E+11
600	1111	1.94E+13	5.88E+12	4.96E+12	4.29E+12	3.59E+12	3.08E+12	2.49E+12	1.91E+12
800	1481	9.42E+13	3.79E+13	3.08E+13	2.57E+13	2.05E+13	1.66E+13	1.25E+13	8.68E+12
1000	1852	3.51E+14	1.62E+14	1.26E+14	1.01E+14	7.64E+13	5.69E+13	3.88E+13	2.44E+13

Table 9-19.   Annual Equivalent 1 MeV Electron Fluence for GaAs Solar Cells from Trapped Electrons During Solar Max, 30° Inclination (Infinite Backshielding)

ELECTRONS - I_{sc}, V_{oc}, AND P_{max}

INCLINATION = 30 DEGREES

EQUIV. 1 MEV ELECTRON FLUENCE FOR I_{sc} - CIRCULAR ORBIT
DUE TO GEOMAGNETICALLY TRAPPED ELECTRONS - MODEL AE8MAX

| ALTITUDE | | SHIELD THICKNESS, cm (mils) | | | | | | | |
(nm)	(km)	0 (0)	2.54E-3 (1)	7.64E-3 (3)	1.52E-2 (6)	3.05E-2 (12)	5.09E-2 (20)	7.64E-2 (30)	1.52E-1 (60)
150	277	2.08E+08	1.60E+08	1.16E+08	8.11E+07	4.98E+07	3.23E+07	2.18E+07	9.29E+06
250	463	1.81E+10	1.37E+10	9.78E+09	6.73E+09	3.98E+09	2.48E+09	1.62E+09	6.54E+08
300	555	5.00E+10	3.78E+10	2.68E+10	1.83E+10	1.07E+10	6.63E+09	4.29E+09	1.71E+09
450	833	3.82E+11	2.88E+11	2.03E+11	1.37E+11	7.91E+10	4.80E+10	3.07E+10	1.19E+10
600	1111	1.63E+12	1.23E+12	8.64E+11	5.83E+11	3.34E+11	2.02E+11	1.28E+11	4.97E+10
800	1481	6.53E+12	4.94E+12	3.50E+12	2.39E+12	1.39E+12	8.54E+11	5.50E+11	2.16E+11
1000	1852	1.76E+13	1.34E+13	9.59E+12	6.61E+12	3.93E+12	2.46E+12	1.60E+12	6.40E+11
1250	2315	3.59E+13	2.75E+13	1.98E+13	1.38E+13	8.31E+12	5.26E+12	3.46E+12	1.40E+12
1500	2778	5.18E+13	3.96E+13	2.84E+13	1.97E+13	1.17E+13	7.27E+12	4.73E+12	1.91E+12
1750	3241	6.52E+13	4.94E+13	3.50E+13	2.36E+13	1.34E+13	7.88E+12	4.91E+12	1.90E+12
2000	3704	7.51E+13	5.62E+13	3.91E+13	2.57E+13	1.36E+13	7.34E+12	4.22E+12	1.48E+12
2250	4167	8.17E+13	6.07E+13	4.16E+13	2.66E+13	1.33E+13	6.53E+12	3.40E+12	1.04E+12
2500	4630	8.24E+13	6.09E+13	4.13E+13	2.60E+13	1.24E+13	5.61E+12	2.64E+12	6.98E+11
2750	5093	7.87E+13	5.81E+13	3.92E+13	2.45E+13	1.13E+13	4.79E+12	2.07E+12	4.81E+11
3000	5556	7.23E+13	5.33E+13	3.60E+13	2.24E+13	1.03E+13	4.28E+12	1.82E+12	4.31E+11
3500	6482	5.84E+13	4.33E+13	2.96E+13	1.88E+13	9.08E+12	4.17E+12	2.08E+12	6.85E+11
4000	7408	5.17E+13	3.92E+13	2.78E+13	1.86E+13	1.01E+13	5.48E+12	3.24E+12	1.30E+12
4500	8334	4.90E+13	3.88E+13	2.91E+13	2.10E+13	1.30E+13	8.11E+12	5.33E+12	2.34E+12
5000	9260	4.69E+13	3.91E+13	3.13E+13	2.45E+13	1.69E+13	1.17E+13	8.24E+12	3.81E+12
5500	10186	4.91E+13	4.28E+13	3.61E+13	2.97E+13	2.22E+13	1.63E+13	1.20E+13	5.74E+12
6000	11112	5.63E+13	5.03E+13	4.36E+13	3.70E+13	2.86E+13	2.17E+13	1.63E+13	7.98E+12
7000	12964	8.00E+13	7.23E+13	6.35E+13	5.46E+13	4.29E+13	3.31E+13	2.51E+13	1.25E+13
8000	14816	1.02E+14	9.26E+13	8.12E+13	6.97E+13	5.48E+13	4.21E+13	3.19E+13	1.59E+13
9000	16668	1.21E+14	1.09E+14	9.55E+13	8.17E+13	6.37E+13	4.87E+13	3.66E+13	1.78E+13
10000	18520	1.31E+14	1.17E+14	1.02E+14	8.68E+13	6.70E+13	5.05E+13	3.74E+13	1.75E+13
11000	20372	1.35E+14	1.20E+14	1.04E+14	8.76E+13	6.67E+13	4.95E+13	3.60E+13	1.59E+13
12000	22224	1.23E+14	1.10E+14	9.39E+13	7.83E+13	5.86E+13	4.27E+13	3.04E+13	1.29E+13
13000	24076	1.05E+14	9.24E+13	7.83E+13	6.44E+13	4.73E+13	3.37E+13	2.35E+13	9.59E+12
14000	25928	8.68E+13	7.58E+13	6.35E+13	5.17E+13	3.72E+13	2.60E+13	1.78E+13	7.05E+12
15000	27780	6.72E+13	5.81E+13	4.81E+13	3.86E+13	2.73E+13	1.87E+13	1.26E+13	4.83E+12
16000	29632	5.21E+13	4.46E+13	3.66E+13	2.90E+13	2.01E+13	1.36E+13	8.98E+12	3.31E+12
17000	31484	4.03E+13	3.42E+13	2.78E+13	2.18E+13	1.49E+13	9.86E+12	6.43E+12	2.29E+12
18000	33336	2.76E+13	2.31E+13	1.85E+13	1.43E+13	9.59E+12	6.24E+12	3.98E+12	1.34E+12
19327	35793	1.59E+13	1.31E+13	1.03E+13	7.79E+12	5.03E+12	3.16E+12	1.96E+12	6.20E+11

Table 9-20. Annual Equivalent 1 MeV Electron Fluence for GaAs Solar Cells from Trapped Electrons During Solar Min, 30° Inclination (Infinite Backshielding)

ELECTRONS - I_{sc}, V_{oc}, AND P_{max} INCLINATION = 30 DEGREES

EQUIV. 1 MEV ELECTRON FLUENCE FOR I_{sc} - CIRCULAR ORBIT
DUE TO GEOMAGNETICALLY TRAPPED ELECTRONS - MODEL AE8MIN

| ALTITUDE | | SHIELD THICKNESS, cm (mils) | | | | | | | |
(nm)	(km)	0 (0)	2.54E-3 (1)	7.64E-3 (3)	1.52E-2 (6)	3.05E-2 (12)	5.09E-2 (20)	7.64E-2 (30)	1.52E-1 (60)
150	277	1.72E+07	1.41E+07	1.12E+07	8.58E+06	5.89E+06	4.12E+06	2.97E+06	1.51E+06
250	463	6.10E+09	5.02E+09	3.97E+09	3.05E+09	2.07E+09	1.42E+09	9.76E+08	4.18E+08
300	555	1.88E+10	1.54E+10	1.21E+10	9.26E+09	6.24E+09	4.22E+09	2.89E+09	1.22E+09
450	833	1.72E+11	1.40E+11	1.09E+11	8.19E+10	5.39E+10	3.57E+10	2.40E+10	9.89E+09
600	1111	8.02E+11	6.50E+11	5.03E+11	3.77E+11	2.46E+11	1.61E+11	1.08E+11	4.39E+10
800	1481	3.51E+12	2.86E+12	2.23E+12	1.68E+12	1.11E+12	7.36E+11	4.94E+11	2.03E+11
1000	1852	9.82E+12	8.07E+12	6.35E+12	4.85E+12	3.26E+12	2.19E+12	1.49E+12	6.22E+11
1250	2315	1.98E+13	1.64E+13	1.31E+13	1.01E+13	6.94E+12	4.75E+12	3.27E+12	1.39E+12
1500	2778	2.63E+13	2.19E+13	1.75E+13	1.36E+13	9.36E+12	6.45E+12	4.46E+12	1.91E+12
1750	3241	2.85E+13	2.36E+13	1.86E+13	1.44E+13	9.74E+12	6.62E+12	4.52E+12	1.90E+12
2000	3704	2.72E+13	2.21E+13	1.72E+13	1.29E+13	8.48E+12	5.58E+12	3.70E+12	1.48E+12
2250	4167	2.44E+13	1.94E+13	1.47E+13	1.08E+13	6.79E+12	4.29E+12	2.76E+12	1.03E+12
2500	4630	2.03E+13	1.58E+13	1.17E+13	8.33E+12	5.02E+12	3.05E+12	1.89E+12	6.62E+11
2750	5093	1.61E+13	1.23E+13	8.90E+12	6.16E+12	3.57E+12	2.09E+12	1.25E+12	4.03E+11
3000	5556	1.31E+13	9.95E+12	7.07E+12	4.81E+12	2.72E+12	1.55E+12	9.06E+11	2.77E+11
3500	6482	1.05E+13	8.00E+12	5.76E+12	3.99E+12	2.31E+12	1.36E+12	8.16E+11	2.73E+11
4000	7408	1.06E+13	8.32E+12	6.25E+12	4.55E+12	2.85E+12	1.80E+12	1.16E+12	4.53E+11
4500	8334	1.13E+13	9.27E+12	7.32E+12	5.64E+12	3.85E+12	2.64E+12	1.83E+12	7.93E+11
5000	9260	1.26E+13	1.08E+13	8.91E+12	7.22E+12	5.29E+12	3.86E+12	2.81E+12	1.30E+12
5500	10186	1.51E+13	1.32E+13	1.13E+13	9.47E+12	7.25E+12	5.49E+12	4.11E+12	1.99E+12
6000	11112	1.86E+13	1.67E+13	1.45E+13	1.23E+13	9.65E+12	7.44E+12	5.65E+12	2.80E+12
7000	12964	2.88E+13	2.61E+13	2.30E+13	1.99E+13	1.58E+13	1.23E+13	9.40E+12	4.75E+12
8000	14816	4.14E+13	3.76E+13	3.32E+13	2.87E+13	2.28E+13	1.77E+13	1.35E+13	6.79E+12
9000	16668	5.67E+13	5.14E+13	4.53E+13	3.91E+13	3.09E+13	2.39E+13	1.81E+13	8.89E+12
10000	18520	7.34E+13	6.64E+13	5.82E+13	4.99E+13	3.90E+13	2.98E+13	2.23E+13	1.06E+13
11000	20372	9.16E+13	8.25E+13	7.18E+13	6.11E+13	4.72E+13	3.55E+13	2.62E+13	1.20E+13
12000	22224	1.00E+14	8.97E+13	7.73E+13	6.49E+13	4.92E+13	3.63E+13	2.62E+13	1.14E+13
13000	24076	9.65E+13	8.53E+13	7.26E+13	6.01E+13	4.45E+13	3.20E+13	2.26E+13	9.35E+12
14000	25928	8.46E+13	7.41E+13	6.23E+13	5.08E+13	3.68E+13	2.59E+13	1.78E+13	7.10E+12
15000	27780	6.70E+13	5.80E+13	4.82E+13	3.87E+13	2.74E+13	1.88E+13	1.27E+13	4.89E+12
16000	29632	5.22E+13	4.48E+13	3.67E+13	2.92E+13	2.02E+13	1.36E+13	9.04E+12	3.35E+12
17000	31484	4.04E+13	3.44E+13	2.79E+13	2.19E+13	1.49E+13	9.91E+12	6.46E+12	2.30E+12
18000	33336	2.77E+13	2.33E+13	1.86E+13	1.44E+13	9.65E+12	6.27E+12	4.01E+12	1.35E+12
19327	35793	1.61E+13	1.33E+13	1.04E+13	7.88E+12	5.09E+12	3.20E+12	1.98E+12	6.30E+11

Table 9-21. Annual Equivalent 1 MeV Electron Fluence for GaAs Solar Cells from Trapped Protons During Solar Max and Min, 30° Inclination (Infinite Backshielding)

PROTONS - V_{oc}

INCLINATION = 30 DEGREES

EQUIV. 1 MEV ELECTRON FLUENCE FOR V_{oc} CIRCULAR ORBIT
DUE TO GEOMAGNETICALLY TRAPPED PROTONS, MODEL AP8MAX

ALTITUDE (nm)	(km)	SHIELD THICKNESS, cm (mils)							
		0 (0)	2.54E-3 (1)	7.64E-3 (3)	1.52E-2 (6)	3.05E-2 (12)	5.09E-2 (20)	7.64E-2 (30)	1.52E-1 (60)
150	277	2.28E+10	3.30E+09	2.62E+09	2.22E+09	1.78E+09	1.49E+09	1.24E+09	1.06E+09
250	463	7.46E+11	2.64E+11	2.35E+11	2.17E+11	1.97E+11	1.81E+11	1.65E+11	1.49E+11
300	555	1.87E+12	7.99E+11	7.12E+11	6.53E+11	5.87E+11	5.35E+11	4.84E+11	4.31E+11
450	833	1.41E+13	7.06E+12	6.04E+12	5.33E+12	4.56E+12	3.99E+12	3.43E+12	2.89E+12
600	1111	6.70E+13	3.31E+13	2.71E+13	2.29E+13	1.86E+13	1.56E+13	1.28E+13	1.02E+13
800	1481	3.51E+14	1.70E+14	1.30E+14	1.05E+14	8.12E+13	6.48E+13	5.05E+13	3.80E+13
1000	1852	2.01E+15	6.69E+14	4.72E+14	3.61E+14	2.62E+14	1.95E+14	1.41E+14	9.84E+13
1250	2315	1.07E+16	2.53E+15	1.64E+15	1.18E+15	7.98E+14	5.39E+14	3.45E+14	2.09E+14
1500	2778	3.67E+16	6.82E+15	4.18E+15	2.91E+15	1.85E+15	1.15E+15	6.47E+14	3.32E+14
1750	3241	9.99E+16	1.49E+16	8.70E+15	5.83E+15	3.50E+15	2.02E+15	1.02E+15	4.40E+14
2000	3704	2.31E+17	2.87E+16	1.57E+16	9.86E+15	5.41E+15	2.89E+15	1.31E+15	4.90E+14
2250	4167	4.88E+17	5.16E+16	2.61E+16	1.51E+16	7.35E+15	3.59E+15	1.46E+15	4.90E+14
2500	4630	9.51E+17	8.20E+16	3.78E+16	2.01E+16	8.67E+15	3.91E+15	1.45E+15	4.51E+14
2750	5093	1.76E+18	1.20E+17	5.01E+16	2.44E+16	9.39E+15	3.95E+15	1.36E+15	3.95E+14
3000	5556	3.08E+18	1.66E+17	6.14E+16	2.74E+16	9.34E+15	3.73E+15	1.21E+15	3.35E+14
3500	6482	7.63E+18	2.78E+17	7.89E+16	2.89E+16	7.44E+15	2.80E+15	8.48E+14	2.19E+14
4000	7408	1.55E+19	4.04E+17	8.85E+16	2.71E+16	5.38E+15	1.88E+15	5.30E+14	1.32E+14
4500	8334	2.69E+19	5.08E+17	8.71E+16	2.27E+16	3.58E+15	1.15E+15	2.96E+14	6.96E+13
5000	9260	4.20E+19	5.83E+17	7.88E+16	1.74E+16	2.19E+15	6.40E+14	1.52E+14	3.37E+13
5500	10186	5.73E+19	6.26E+17	6.81E+16	1.25E+16	1.19E+15	3.39E+14	7.57E+13	1.52E+13
6000	11112	7.16E+19	5.96E+17	5.25E+16	8.18E+15	6.09E+14	1.57E+14	3.11E+13	5.53E+12
7000	12964	9.20E+19	4.06E+17	2.41E+16	2.70E+15	1.25E+14	2.54E+13	3.84E+12	5.24E+11
8000	14816	9.41E+19	2.00E+17	7.45E+15	5.98E+14	1.69E+13	2.61E+12	2.87E+11	2.94E+10
9000	16668	6.58E+19	6.43E+16	1.38E+15	7.35E+13	1.13E+12	1.38E+11	1.55E+10	3.21E+09
10000	18520	4.60E+19	1.72E+16	2.13E+14	7.98E+12	5.51E+10	7.98E+09	0.00E+00	0.00E+00
11000	20372	3.54E+19	3.64E+15	1.40E+13	2.82E+11	1.37E+09	0.00E+00	0.00E+00	0.00E+00
12000	22224	2.88E+19	7.05E+14	3.00E+11	0.00E+00	0.00E+00	0.00E+00	0.00E+00	0.00E+00
13000	24076	2.34E+19	9.80E+13	5.84E+07	0.00E+00	0.00E+00	0.00E+00	0.00E+00	0.00E+00
14000	25928	1.95E+19	1.76E+13	2.64E+07	0.00E+00	0.00E+00	0.00E+00	0.00E+00	0.00E+00
15000	27780	1.54E+19	4.55E+12	1.40E+07	0.00E+00	0.00E+00	0.00E+00	0.00E+00	0.00E+00
16000	29632	1.21E+19	4.11E+10	0.00E+00	0.00E+00	0.00E+00	0.00E+00	0.00E+00	0.00E+00
17000	31484	9.68E+18	3.13E+10	0.00E+00	0.00E+00	0.00E+00	0.00E+00	0.00E+00	0.00E+00
18000	33336	6.16E+18	2.26E+10	0.00E+00	0.00E+00	0.00E+00	0.00E+00	0.00E+00	0.00E+00
19327	35793	1.75E+18	1.46E+10	0.00E+00	0.00E+00	0.00E+00	0.00E+00	0.00E+00	0.00E+00

INCLINATION = 30 DEGREES

EQUIV. 1 MEV ELECTRON FLUENCE FOR V_{oc} CIRCULAR ORBIT
DUE TO GEOMAGNETICALLY TRAPPED PROTONS, MODEL AP8MIN

ALTITUDE (nm)	(km)	0 (0)	2.54E-3 (1)	7.64E-3 (3)	1.52E-2 (6)	3.05E-2 (12)	5.09E-2 (20)	7.64E-2 (30)	1.52E-1 (60)
150	277	8.14E+10	2.07E+10	1.79E+10	1.64E+10	1.48E+10	1.37E+10	1.26E+10	1.14E+10
250	463	1.34E+12	5.05E+11	4.55E+11	4.25E+11	3.90E+11	3.60E+11	3.28E+11	2.93E+11
300	555	3.00E+12	1.27E+12	1.14E+12	1.06E+12	9.60E+11	8.81E+11	7.97E+11	7.02E+11
450	833	1.72E+13	8.63E+12	7.53E+12	6.71E+12	5.84E+12	5.19E+12	4.53E+12	3.84E+12
600	1111	7.40E+13	3.72E+13	3.08E+13	2.62E+13	2.17E+13	1.85E+13	1.55E+13	1.25E+13
800	1481	4.79E+14	1.97E+14	1.47E+14	1.17E+14	9.02E+13	7.32E+13	5.80E+13	4.41E+13
1000	1852	2.47E+15	7.39E+14	5.05E+14	3.80E+14	2.75E+14	2.08E+14	1.52E+14	1.07E+14

Table 9-22. Annual Equivalent 1 MeV Electron Fluence for GaAs Solar Cells from Trapped Protons During Solar Max and Min, 30° Inclination (Infinite Backshielding)

PROTONS - P_{max}

INCLINATION = 30 DEGREES

EQUIV. 1 MEV ELECTRON FLUENCE FOR P_{max} CIRCULAR ORBIT
DUE TO GEOMAGNETICALLY TRAPPED PROTONS, MODEL AP8MAX

ALTITUDE (nm)	(km)	SHIELD THICKNESS, cm (mils)							
		0 (0)	2.54E-3 (1)	7.64E-3 (3)	1.52E-2 (6)	3.05E-2 (12)	5.09E-2 (20)	7.64E-2 (30)	1.52E-1 (60)
150	277	1.63E+10	2.36E+09	1.87E+09	1.59E+09	1.27E+09	1.06E+09	8.87E+08	7.58E+08
250	463	5.33E+11	1.88E+11	1.68E+11	1.55E+11	1.41E+11	1.29E+11	1.18E+11	1.06E+11
300	555	1.34E+12	5.70E+11	5.09E+11	4.66E+11	4.19E+11	3.82E+11	3.46E+11	3.08E+11
450	833	1.01E+13	5.04E+12	4.31E+12	3.80E+12	3.26E+12	2.85E+12	2.45E+12	2.06E+12
600	1111	4.78E+13	2.37E+13	1.93E+13	1.50E+13	1.33E+13	1.11E+13	9.12E+12	7.26E+12
800	1481	2.51E+14	1.21E+14	9.31E+13	7.50E+13	5.80E+13	4.63E+13	3.60E+13	2.71E+13
1000	1852	1.43E+15	4.78E+14	3.37E+14	2.58E+14	1.87E+14	1.39E+14	1.01E+14	7.03E+13
1250	2315	7.65E+15	1.81E+15	1.17E+15	8.46E+14	5.70E+14	3.85E+14	2.46E+14	1.50E+14
1500	2778	2.62E+16	4.87E+15	2.99E+15	2.08E+15	1.32E+15	8.21E+14	4.62E+14	2.37E+14
1750	3241	7.14E+16	1.06E+16	6.22E+15	4.16E+15	2.50E+15	1.44E+15	7.26E+14	3.14E+14
2000	3704	1.65E+17	2.05E+16	1.12E+16	7.04E+15	3.86E+15	2.06E+15	9.33E+14	3.50E+14
2250	4167	3.49E+17	3.68E+16	1.86E+16	1.08E+16	5.25E+15	2.56E+15	1.04E+15	3.50E+14
2500	4630	6.79E+17	5.86E+16	2.70E+16	1.43E+16	6.19E+15	2.79E+15	9.68E+14	3.22E+14
2750	5093	1.26E+18	8.60E+16	3.58E+16	1.75E+16	6.70E+15	2.82E+15	8.63E+14	2.82E+14
3000	5556	2.20E+18	1.19E+17	4.39E+16	1.96E+16	6.67E+15	2.66E+15	6.05E+14	2.39E+14
3500	6482	5.45E+18	1.99E+17	5.64E+16	2.06E+16	5.32E+15	2.00E+15	3.79E+14	1.57E+14
4000	7408	1.11E+19	2.89E+17	6.32E+16	1.93E+16	3.84E+15	1.35E+15	2.11E+14	9.39E+13
4500	8334	1.92E+19	3.63E+17	6.22E+16	1.62E+16	2.56E+15	8.19E+14	1.09E+14	4.97E+13
5000	9260	3.00E+19	4.16E+17	5.63E+16	1.24E+16	1.54E+15	4.57E+14	5.40E+13	2.41E+13
5500	10186	4.10E+19	4.47E+17	4.87E+16	8.95E+15	8.51E+14	2.42E+14	2.22E+13	1.09E+13
6000	11112	5.11E+19	4.26E+17	3.75E+16	5.84E+15	4.35E+14	1.12E+14	2.74E+12	3.95E+12
7000	12964	6.57E+19	2.90E+17	1.72E+16	1.93E+15	8.90E+13	1.81E+13	2.05E+11	3.74E+11
8000	14816	6.72E+19	1.43E+17	5.32E+15	4.27E+14	1.20E+13	1.86E+12	1.11E+10	2.10E+10
9000	16668	4.70E+19	4.59E+16	9.85E+14	5.25E+13	8.08E+11	9.83E+10	0.00E+00	2.29E+09
10000	18520	3.29E+19	1.23E+16	1.52E+14	5.70E+12	3.94E+10	0.00E+00	0.00E+00	0.00E+00
11000	20372	2.53E+19	2.60E+15	1.00E+13	2.01E+11	9.81E+08	0.00E+00	0.00E+00	0.00E+00
12000	22224	2.05E+19	5.04E+14	2.15E+11	0.00E+00	0.00E+00	0.00E+00	0.00E+00	0.00E+00
13000	24076	1.67E+19	7.00E+13	4.17E+07	0.00E+00	0.00E+00	0.00E+00	0.00E+00	0.00E+00
14000	25928	1.39E+19	1.26E+13	1.89E+07	0.00E+00	0.00E+00	0.00E+00	0.00E+00	0.00E+00
15000	27780	1.10E+19	3.25E+12	1.00E+07	0.00E+00	0.00E+00	0.00E+00	0.00E+00	0.00E+00
16000	29632	8.66E+18	2.93E+10	0.00E+00	0.00E+00	0.00E+00	0.00E+00	0.00E+00	0.00E+00
17000	31484	6.91E+18	2.24E+10	0.00E+00	0.00E+00	0.00E+00	0.00E+00	0.00E+00	0.00E+00
18000	33336	4.40E+18	1.62E+10	0.00E+00	0.00E+00	0.00E+00	0.00E+00	0.00E+00	0.00E+00
19327	35793	1.25E+18	1.05E+10	0.00E+00	0.00E+00	0.00E+00	0.00E+00	0.00E+00	0.00E+00

PROTONS - P_{max}

INCLINATION = 30 DEGREES

EQUIV. 1 MEV ELECTRON FLUENCE FOR P_{max} CIRCULAR ORBIT
DUE TO GEOMAGNETICALLY TRAPPED PROTONS, MODEL AP8MIN

ALTITUDE (nm)	(km)	0 (0)	2.54E-3 (1)	7.64E-3 (3)	1.52E-2 (6)	3.05E-2 (12)	5.09E-2 (20)	7.64E-2 (30)	1.52E-1 (60)
150	277	5.81E+10	1.48E+10	1.28E+10	1.17E+10	1.06E+10	9.77E+09	8.98E+09	8.15E+09
250	463	9.60E+11	3.61E+11	3.25E+11	3.04E+11	2.79E+11	2.57E+11	2.35E+11	2.09E+11
300	555	2.14E+12	9.05E+11	8.16E+11	7.55E+11	6.86E+11	6.30E+11	5.70E+11	5.01E+11
450	833	1.23E+13	6.16E+12	5.38E+12	4.80E+12	4.17E+12	3.71E+12	3.24E+12	2.74E+12
600	1111	5.29E+13	2.65E+13	2.20E+13	1.87E+13	1.55E+13	1.32E+13	1.11E+13	8.90E+12
800	1481	3.42E+14	1.40E+14	1.05E+14	8.34E+13	6.44E+13	5.23E+13	4.14E+13	3.15E+13
1000	1852	1.77E+15	5.28E+14	3.61E+14	2.72E+14	1.97E+14	1.49E+14	1.09E+14	7.65E+13

Table 9-23. Annual Equivalent 1 MeV Electron Fluence for GaAs Solar Cells from Trapped Protons During Solar Max and Min, 30° Inclination (Infinite Backshielding)

PROTONS - I$_{sc}$ INCLINATION = 30 DEGREES

EQUIV. 1 MEV PROTON FLUENCE FOR I$_{sc}$ - CIRCULAR ORBIT
DUE TO GEOMAGNETICALLY TRAPPED PROTONS, MODEL AP8MAX

ALTITUDE (nm)	(km)	0 (0)	2.54E-3 (1)	7.64E-3 (3)	1.52E-2 (6)	3.05E-2 (12)	5.09E-2 (20)	7.64E-2 (30)	1.52E-1 (60)
					SHIELD THICKNESS, cm (mils)				
150	277	1.18E+10	8.89E+08	6.37E+08	5.01E+08	3.55E+08	2.62E+08	1.84E+08	1.38E+08
250	463	3.28E+11	5.50E+10	4.45E+10	3.86E+10	3.24E+10	2.77E+10	2.33E+10	2.00E+10
300	555	7.71E+11	1.72E+11	1.41E+11	1.21E+11	1.01E+11	8.58E+10	7.15E+10	6.04E+10
450	833	5.53E+12	1.75E+12	1.39E+12	1.15E+12	9.08E+11	7.36E+11	5.73E+11	4.43E+11
600	1111	2.69E+13	9.11E+12	6.91E+12	5.47E+12	4.12E+12	3.19E+12	2.34E+12	1.67E+12
800	1481	1.43E+14	5.04E+13	3.59E+13	2.72E+13	1.95E+13	1.44E+13	9.92E+12	6.59E+12
1000	1852	9.41E+14	2.13E+14	1.40E+14	1.01E+14	6.85E+13	4.72E+13	3.01E+13	1.83E+13
1250	2315	5.36E+15	8.56E+14	5.17E+14	3.57E+14	2.30E+14	1.45E+14	8.27E+13	4.34E+13
1500	2778	1.90E+16	2.37E+15	1.37E+15	9.20E+14	5.67E+14	3.35E+14	1.71E+14	7.67E+13
1750	3241	5.32E+16	5.26E+15	2.92E+15	1.89E+15	1.11E+15	6.17E+14	2.88E+14	1.11E+14
2000	3704	1.25E+17	1.03E+16	5.37E+15	3.26E+15	1.75E+15	9.09E+14	3.85E+14	1.31E+14
2250	4167	2.70E+17	1.88E+16	9.06E+15	5.05E+15	2.42E+15	1.15E+15	4.41E+14	1.36E+14
2500	4630	5.42E+17	3.03E+16	1.33E+16	6.81E+15	2.88E+15	1.27E+15	4.44E+14	1.28E+14
2750	5093	1.04E+18	4.49E+16	1.78E+16	8.35E+15	3.14E+15	1.29E+15	4.19E+14	1.15E+14
3000	5556	1.87E+18	6.26E+16	2.19E+16	9.42E+15	3.14E+15	1.26E+15	3.76E+14	9.85E+13
3500	6482	4.80E+18	1.07E+17	2.86E+16	1.00E+16	2.52E+15	9.26E+14	2.66E+14	6.57E+13
4000	7408	9.92E+18	1.58E+17	3.23E+16	9.46E+15	1.83E+15	6.25E+14	1.67E+14	3.99E+13
4500	8334	1.76E+19	2.02E+17	3.20E+16	7.98E+15	1.22E+15	3.83E+14	9.38E+13	2.13E+13
5000	9260	2.81E+19	2.34E+17	2.92E+16	6.13E+15	7.37E+14	2.15E+14	4.86E+13	1.04E+13
5500	10186	3.81E+19	2.53E+17	2.53E+16	4.44E+15	4.09E+14	1.14E+14	2.43E+13	4.74E+12
6000	11112	4.72E+19	2.43E+17	1.96E+16	2.91E+15	2.10E+14	5.30E+13	1.00E+13	1.74E+12
7000	12964	6.79E+19	1.68E+17	9.06E+15	9.66E+14	4.33E+13	8.64E+12	1.25E+12	1.67E+12
8000	14816	6.79E+19	8.42E+16	2.82E+15	2.15E+14	5.90E+12	8.95E+11	9.42E+10	9.30E+09
9000	16668	4.93E+19	2.75E+16	5.26E+14	2.65E+13	3.97E+11	4.72E+10	5.01E+09	1.03E+09
10000	18520	3.59E+19	7.52E+15	8.18E+13	2.89E+12	2.16E+10	0.00E+00	0.00E+00	0.00E+00
11000	20372	2.94E+19	1.62E+15	5.45E+12	1.03E+11	5.26E+08	0.00E+00	0.00E+00	0.00E+00
12000	22224	2.58E+19	3.21E+14	1.21E+11	0.00E+00	0.00E+00	0.00E+00	0.00E+00	0.00E+00
13000	24076	2.27E+19	4.60E+13	2.57E+07	0.00E+00	0.00E+00	0.00E+00	0.00E+00	0.00E+00
14000	25928	2.02E+19	8.48E+12	1.14E+07	0.00E+00	0.00E+00	0.00E+00	0.00E+00	0.00E+00
15000	27780	1.69E+19	2.24E+12	5.94E+06	0.00E+00	0.00E+00	0.00E+00	0.00E+00	0.00E+00
16000	29632	1.40E+19	2.47E+10	0.00E+00	0.00E+00	0.00E+00	0.00E+00	0.00E+00	0.00E+00
17000	31484	1.17E+19	1.87E+10	0.00E+00	0.00E+00	0.00E+00	0.00E+00	0.00E+00	0.00E+00
18000	33336	7.70E+18	1.34E+10	0.00E+00	0.00E+00	0.00E+00	0.00E+00	0.00E+00	0.00E+00
19327	35793	2.18E+18	8.51E+09						

EQUIV. 1 MEV PROTON FLUENCE FOR I$_{sc}$ - CIRCULAR ORBIT   INCLINATION = 30 DEGREES
DUE TO GEOMAGNETICALLY TRAPPED PROTONS, MODEL AP8MIN

ALTITUDE (nm)	(km)	0 (0)	2.54E-3 (1)	7.64E-3 (3)	1.52E-2 (6)	3.05E-2 (12)	5.09E-2 (20)	7.64E-2 (30)	1.52E-1 (60)
150	277	3.84E+10	4.45E+09	3.38E+09	2.89E+09	2.40E+09	2.07E+09	1.77E+09	1.55E+09
250	463	5.64E+11	1.04E+11	8.56E+10	7.60E+10	6.57E+10	5.71E+10	4.83E+10	4.11E+10
300	555	1.23E+12	2.68E+11	2.23E+11	1.96E+11	1.67E+11	1.45E+11	1.21E+11	1.01E+11
450	833	6.73E+12	2.06E+12	1.67E+12	1.40E+12	1.13E+12	9.34E+11	7.43E+11	5.79E+11
600	1111	2.94E+13	9.93E+12	7.63E+12	6.08E+12	4.64E+12	3.68E+12	2.76E+12	2.00E+12
800	1481	2.08E+14	5.86E+13	4.03E+13	2.97E+13	2.11E+13	1.59E+13	1.12E+13	7.50E+12
1000	1852	1.18E+15	2.37E+14	1.50E+14	1.05E+14	7.11E+13	4.97E+13	3.22E+13	1.96E+13

9-28

Table 9-24.   Annual Equivalent 1 MeV Electron Fluence for GaAs Solar Cells from Trapped Electrons During Solar Max, 40° Inclination (Infinite Backshielding)

ELECTRONS - I_{sc}, V_{oc}, AND P_{max} INCLINATION = 40 DEGREES

EQUIV. 1 MEV ELECTRON FLUENCE FOR I_{sc} - CIRCULAR ORBIT
DUE TO GEOMAGNETICALLY TRAPPED ELECTRONS - MODEL AE8MAX

ALTITUDE		SHIELD THICKNESS, cm (mils)							
(nm)	(km)	0 (0)	2.54E-3 (1)	7.64E-3 (3)	1.52E-2 (6)	3.05E-2 (12)	5.09E-2 (20)	7.64E-2 (30)	1.52E-1 (60)
150	277	6.80E+09	5.06E+09	3.52E+09	2.33E+09	1.28E+09	7.26E+08	4.42E+08	1.69E+08
250	463	4.66E+10	3.49E+10	2.45E+10	1.65E+10	9.31E+09	5.53E+09	3.49E+09	1.37E+09
300	555	9.01E+10	6.77E+10	4.78E+10	3.24E+10	1.86E+10	1.12E+10	7.16E+09	2.83E+09
450	833	4.11E+11	3.09E+11	2.19E+11	1.48E+11	8.56E+10	5.20E+10	3.33E+10	1.31E+10
600	1111	1.42E+12	1.07E+12	7.54E+11	5.11E+11	2.95E+11	1.80E+11	1.15E+11	4.54E+10
800	1481	5.06E+12	3.82E+12	2.71E+12	1.84E+12	1.07E+12	6.52E+11	4.19E+11	1.65E+11
1000	1852	1.32E+13	1.00E+13	7.14E+12	4.91E+12	2.89E+12	1.79E+12	1.16E+12	4.61E+11
1250	2315	2.68E+13	2.05E+13	1.48E+13	1.03E+13	6.15E+12	3.87E+12	2.54E+12	1.03E+12
1500	2778	3.85E+13	2.94E+13	2.12E+13	1.47E+13	8.77E+12	5.47E+12	3.58E+12	1.46E+12
1750	3241	4.84E+13	3.68E+13	2.62E+13	1.78E+13	1.02E+13	6.10E+12	3.85E+12	1.53E+12
2000	3704	5.57E+13	4.19E+13	2.94E+13	1.95E+13	1.06E+13	5.91E+12	3.51E+12	1.30E+12
2250	4167	6.12E+13	4.58E+13	3.17E+13	2.07E+13	1.07E+13	5.57E+12	3.11E+12	1.08E+12
2500	4630	6.28E+13	4.69E+13	3.23E+13	2.08E+13	1.05E+13	5.22E+12	2.78E+12	9.35E+11
2750	5093	6.10E+13	4.56E+13	3.15E+13	2.03E+13	1.02E+13	4.95E+12	2.60E+12	8.86E+11
3000	5556	5.76E+13	4.33E+13	3.00E+13	1.95E+13	9.88E+12	4.89E+12	2.64E+12	9.46E+11
3500	6482	4.90E+13	3.73E+13	2.65E+13	1.79E+13	9.82E+12	5.44E+12	3.28E+12	1.33E+12
4000	7408	4.53E+13	3.55E+13	2.63E+13	1.88E+13	1.15E+13	7.14E+12	4.71E+12	2.07E+12
4500	8334	4.54E+13	3.70E+13	2.89E+13	2.20E+13	1.47E+13	9.89E+12	6.87E+12	3.14E+12
5000	9260	4.59E+13	3.91E+13	3.21E+13	2.57E+13	1.85E+13	1.33E+13	9.54E+12	4.49E+12
5500	10186	4.88E+13	4.29E+13	3.64E+13	3.03E+13	2.28E+13	1.70E+13	1.25E+13	6.00E+12
6000	11112	5.50E+13	4.92E+13	4.27E+13	3.63E+13	2.80E+13	2.13E+13	1.59E+13	7.78E+12
7005	12973	7.26E+13	6.54E+13	5.73E+13	4.90E+13	3.83E+13	2.94E+13	2.21E+13	1.09E+13
8000	14816	8.88E+13	8.01E+13	7.01E+13	5.99E+13	4.68E+13	3.58E+13	2.69E+13	1.31E+13
9000	16668	9.97E+13	8.97E+13	7.82E+13	6.66E+13	5.16E+13	3.91E+13	2.92E+13	1.40E+13
10000	18520	1.04E+14	9.28E+13	8.04E+13	6.81E+13	5.22E+13	3.92E+13	2.88E+13	1.33E+13
11000	20372	1.03E+14	9.18E+13	7.91E+13	6.64E+13	5.03E+13	3.71E+13	2.68E+13	1.18E+13
12000	22224	9.22E+13	8.16E+13	6.97E+13	5.79E+13	4.32E+13	3.13E+13	2.22E+13	9.34E+12
13000	24076	7.77E+13	6.82E+13	5.77E+13	4.73E+13	3.46E+13	2.46E+13	1.71E+13	6.93E+12
14000	25928	6.36E+13	5.54E+13	4.64E+13	3.76E+13	2.70E+13	1.88E+13	1.29E+13	5.07E+12
15000	27780	4.99E+13	4.31E+13	3.57E+13	2.86E+13	2.02E+13	1.38E+13	9.29E+12	3.55E+12
16000	29632	3.74E+13	3.20E+13	2.62E+13	2.08E+13	1.44E+13	9.67E+12	6.40E+12	2.36E+12
17000	31484	2.92E+13	2.48E+13	2.01E+13	1.58E+13	1.07E+13	7.11E+12	4.63E+12	1.65E+12
18000	33336	1.96E+13	1.65E+13	1.32E+13	1.02E+13	6.83E+12	4.45E+12	2.84E+12	9.61E+11
19327	35793	1.17E+13	9.65E+12	7.59E+12	5.75E+12	3.72E+12	2.35E+12	1.46E+12	4.64E+11

Table 9-25. Annual Equivalent 1 MeV Electron Fluence for GaAs Solar Cells from Trapped Electrons During Solar Min, 40° Inclination (Infinite Backshielding)

ELECTRONS - I_{sc}, V_{oc}, AND P_{max} — CIRCULAR ORBIT

INCLINATION = 40 DEGREES

EQUIV. 1 MEV ELECTRON FLUENCE FOR I_{sc} - CIRCULAR ORBIT
DUE TO GEOMAGNETICALLY TRAPPED ELECTRONS - MODEL AE8MIN

ALTITUDE (nm)	(km)	SHIELD THICKNESS, cm (mils)							
		0 (0)	2.54E-3 (1)	7.64E-3 (3)	1.52E-2 (6)	3.05E-2 (12)	5.09E-2 (20)	7.64E-2 (30)	1.52E-1 (60)
150	277	1.07E+09	8.67E+08	6.73E+08	5.09E+08	3.40E+08	2.29E+08	1.56E+08	6.59E+07
250	463	1.53E+10	1.25E+10	9.82E+09	7.52E+09	5.08E+09	3.46E+09	2.38E+09	1.01E+09
300	555	3.29E+10	2.69E+10	2.12E+10	1.62E+10	1.09E+10	7.44E+09	5.10E+09	2.15E+09
450	833	1.80E+11	1.47E+11	1.15E+11	8.73E+10	5.83E+10	3.91E+10	2.66E+10	1.11E+10
600	1111	6.82E+11	5.56E+11	4.33E+11	3.27E+11	2.17E+11	1.44E+11	9.75E+10	4.05E+10
800	1481	2.61E+12	2.13E+12	1.66E+12	1.25E+12	8.28E+11	5.49E+11	3.69E+11	1.52E+11
1000	1852	7.05E+12	5.78E+12	4.53E+12	3.45E+12	2.31E+12	1.55E+12	1.05E+12	4.36E+11
1250	2315	1.43E+13	1.19E+13	9.42E+12	7.28E+12	4.97E+12	3.39E+12	2.33E+12	9.91E+11
1500	2778	1.91E+13	1.59E+13	1.27E+13	9.87E+12	6.80E+12	4.69E+12	3.25E+12	1.40E+12
1750	3241	2.09E+13	1.73E+13	1.37E+13	1.06E+13	7.20E+12	4.91E+12	3.37E+12	1.43E+12
2000	3704	2.02E+13	1.64E+13	1.28E+13	9.69E+12	6.41E+12	4.25E+12	2.85E+12	1.16E+12
2250	4167	1.84E+13	1.48E+13	1.13E+13	8.35E+12	5.34E+12	3.44E+12	2.25E+12	8.74E+11
2500	4630	1.59E+13	1.25E+13	9.39E+12	6.81E+12	4.24E+12	2.68E+12	1.72E+12	6.57E+11
2750	5093	1.32E+13	1.03E+13	7.64E+12	5.47E+12	3.37E+12	2.12E+12	1.36E+12	5.21E+11
3000	5556	1.15E+13	8.97E+12	6.65E+12	4.79E+12	2.98E+12	1.89E+12	1.24E+12	4.85E+11
3500	6482	1.05E+13	8.39E+12	6.44E+12	4.82E+12	3.18E+12	2.13E+12	1.45E+12	6.07E+11
4000	7408	1.14E+13	9.43E+12	7.51E+12	5.85E+12	4.06E+12	2.83E+12	1.98E+12	8.76E+11
4500	8334	1.30E+13	1.11E+13	9.09E+12	7.32E+12	5.31E+12	3.84E+12	2.77E+12	1.27E+12
5000	9260	1.51E+13	1.31E+13	1.11E+13	9.19E+12	6.91E+12	5.14E+12	3.78E+12	1.78E+12
5500	10186	1.78E+13	1.58E+13	1.36E+13	1.15E+13	8.80E+12	6.66E+12	4.96E+12	2.38E+12
6000	11112	2.09E+13	1.87E+13	1.63E+13	1.38E+13	1.08E+13	8.27E+12	6.24E+12	3.06E+12
7000	12964	2.99E+13	2.70E+13	2.37E+13	2.03E+13	1.60E+13	1.23E+13	9.34E+12	4.61E+12
8000	14816	4.04E+13	3.65E+13	3.21E+13	2.76E+13	2.17E+13	1.67E+13	1.27E+13	6.22E+12
9000	16668	5.24E+13	4.73E+13	4.14E+13	3.55E+13	2.77E+13	2.12E+13	1.59E+13	7.61E+12
10000	18520	6.25E+13	5.63E+13	4.91E+13	4.18E+13	3.24E+13	2.45E+13	1.81E+13	8.43E+12
11000	20372	7.29E+13	6.53E+13	5.65E+13	4.78E+13	3.65E+13	2.73E+13	1.99E+13	8.99E+12
12000	22224	7.60E+13	6.76E+13	5.80E+13	4.85E+13	3.65E+13	2.68E+13	1.92E+13	8.31E+12
13000	24076	7.16E+13	6.31E+13	5.35E+13	4.42E+13	3.26E+13	2.33E+13	1.64E+13	6.75E+12
14000	25928	6.20E+13	5.42E+13	4.55E+13	3.70E+13	2.67E+13	1.88E+13	1.29E+13	5.11E+12
15000	27780	4.98E+13	4.31E+13	3.57E+13	2.86E+13	2.02E+13	1.39E+13	9.36E+12	3.59E+12
16000	29632	3.75E+13	3.22E+13	2.63E+13	2.09E+13	1.45E+13	9.74E+12	6.45E+12	2.39E+12
17000	31484	2.93E+13	2.49E+13	2.02E+13	1.58E+13	1.08E+13	7.15E+12	4.66E+12	1.66E+12
18000	33336	1.97E+13	1.65E+13	1.33E+13	1.03E+13	6.87E+12	4.47E+12	2.86E+12	9.68E+11
19327	35793	1.18E+13	9.76E+12	7.68E+12	5.82E+12	3.77E+12	2.38E+12	1.48E+12	4.71E+11

Table 9-26. Annual Equivalent 1 MeV Electron Fluence for GaAs Solar Cells from Trapped Protons During Solar Max and Min, 40° Inclination (Infinite Backshielding)

PROTONS - V_{oc}

INCLINATION = 40 DEGREES

EQUIV. 1 MEV ELECTRON FLUENCE FOR V_{oc} CIRCULAR ORBIT
DUE TO GEOMAGNETICALLY TRAPPED PROTONS, MODEL AP8MAX

ALTITUDE		SHIELD THICKNESS, cm (mils)							
(nm)	(km)	0 (0)	2.54E-3 (1)	7.64E-3 (3)	1.52E-2 (6)	3.05E-2 (12)	5.09E-2 (20)	7.64E-2 (30)	1.52E-1 (60)
150	277	8.09E+10	2.34E+10	1.94E+10	1.69E+10	1.44E+10	1.30E+10	1.17E+10	1.06E+10
250	463	1.97E+12	6.33E+11	4.93E+11	4.18E+11	3.48E+11	3.06E+11	2.67E+11	2.32E+11
300	555	7.64E+12	1.92E+12	1.37E+12	1.10E+12	8.69E+11	7.45E+11	6.37E+11	5.41E+11
450	833	1.57E+14	1.95E+13	1.10E+13	7.70E+12	5.38E+12	4.30E+12	3.44E+12	2.76E+12
600	1111	7.53E+14	7.97E+13	4.17E+13	2.79E+13	1.88E+13	1.45E+13	1.12E+13	8.60E+12
800	1481	5.74E+15	2.93E+14	1.51E+14	1.03E+14	7.07E+13	5.36E+13	4.01E+13	2.94E+13
1000	1852	3.87E+16	9.32E+14	4.71E+14	3.20E+14	2.14E+14	1.53E+14	1.08E+14	7.34E+13
1250	2315	1.49E+17	2.96E+15	1.47E+15	9.83E+14	6.23E+14	4.10E+14	2.57E+14	1.54E+14
1500	2778	3.35E+17	7.05E+15	3.52E+15	2.30E+15	1.40E+15	8.52E+14	4.75E+14	2.42E+14
1750	3241	6.05E+17	1.41E+16	7.01E+15	4.48E+15	2.60E+15	1.49E+15	7.46E+14	3.22E+14
2000	3704	9.46E+17	2.56E+16	1.23E+16	7.42E+15	3.98E+15	2.12E+15	9.58E+14	3.59E+14
2250	4167	1.41E+18	4.37E+16	1.99E+16	1.12E+16	5.37E+15	2.63E+15	1.07E+15	3.59E+14
2500	4630	2.12E+18	6.74E+16	2.85E+16	1.48E+16	6.34E+15	2.86E+15	1.06E+15	3.30E+14
2750	5093	3.07E+18	9.65E+16	3.73E+16	1.79E+16	6.84E+15	2.88E+15	9.89E+14	2.88E+14
3000	5556	4.36E+18	1.30E+17	4.54E+16	2.00E+16	6.80E+15	2.71E+15	8.79E+14	2.44E+14
3500	6482	8.39E+18	2.12E+17	5.76E+16	2.09E+16	5.40E+15	2.04E+15	6.15E+14	1.59E+14
4000	7408	1.50E+19	3.03E+17	6.47E+16	1.98E+16	3.94E+15	1.38E+15	3.88E+14	9.61E+13
4500	8334	2.42E+19	3.76E+17	6.38E+16	1.66E+16	2.63E+15	8.42E+14	2.17E+14	5.10E+13
5000	9260	3.59E+19	4.25E+17	5.73E+16	1.27E+16	1.58E+15	4.70E+14	1.12E+14	2.48E+13
5500	10186	4.71E+19	4.55E+17	4.96E+16	9.15E+15	8.76E+14	2.50E+14	5.59E+13	1.13E+13
6000	11112	5.63E+19	4.35E+17	3.85E+16	6.02E+15	4.51E+14	1.16E+14	2.31E+13	4.13E+12
7005	12973	6.90E+19	2.98E+17	1.78E+16	2.00E+15	9.20E+13	1.87E+13	2.83E+12	3.87E+11
8000	14816	6.96E+19	1.45E+17	5.43E+15	4.37E+14	1.24E+13	1.91E+12	2.11E+11	2.18E+10
9000	16668	4.89E+19	4.68E+16	1.01E+15	5.38E+13	8.30E+11	1.01E+11	1.14E+10	2.36E+09
10000	18520	3.47E+19	1.28E+16	1.59E+14	5.96E+12	4.12E+10	0.00E+00	0.00E+00	0.00E+00
11000	20372	2.63E+19	2.74E+15	1.13E+13	2.34E+11	1.16E+09	0.00E+00	0.00E+00	0.00E+00
12000	22224	2.10E+19	5.21E+14	2.25E+11	0.00E+00	0.00E+00	0.00E+00	0.00E+00	0.00E+00
13000	24076	1.70E+19	7.35E+13	5.32E+07	0.00E+00	0.00E+00	0.00E+00	0.00E+00	0.00E+00
14000	25928	1.31E+19	1.31E+13	4.40E+07	0.00E+00	0.00E+00	0.00E+00	0.00E+00	0.00E+00
15000	27780	1.13E+19	3.40E+12	1.27E+07	0.00E+00	0.00E+00	0.00E+00	0.00E+00	0.00E+00
16000	29632	8.55E+18	3.44E+10	0.00E+00	0.00E+00	0.00E+00	0.00E+00	0.00E+00	0.00E+00
17000	31484	6.96E+18	2.64E+10	0.00E+00	0.00E+00	0.00E+00	0.00E+00	0.00E+00	0.00E+00
18000	33336	4.53E+18	1.90E+10	0.00E+00	0.00E+00	0.00E+00	0.00E+00	0.00E+00	0.00E+00
19327	35793	1.35E+18	1.25E+10	0.00E+00	0.00E+00	0.00E+00	0.00E+00	0.00E+00	0.00E+00

EQUIV. 1 MEV ELECTRON FLUENCE FOR V_{oc} CIRCULAR ORBIT
DUE TO GEOMAGNETICALLY TRAPPED PROTONS, MODEL AP8MIN

INCLINATION = 40 DEGREES

ALTITUDE		SHIELD THICKNESS, cm (mils)							
(nm)	(km)	0 (0)	2.54E-3 (1)	7.64E-3 (3)	1.52E-2 (6)	3.05E-2 (12)	5.09E-2 (20)	7.64E-2 (30)	1.52E-1 (60)
150	277	1.71E+12	4.20E+11	2.01E+11	1.11E+11	6.09E+10	4.67E+10	3.76E+10	3.14E+10
250	463	2.34E+13	3.96E+12	2.02E+12	1.21E+12	7.48E+11	5.94E+11	4.87E+11	4.05E+11
300	555	4.98E+13	7.66E+12	3.89E+12	2.41E+12	1.54E+12	1.23E+12	1.00E+12	8.30E+11
450	833	2.96E+14	3.35E+13	1.68E+13	1.06E+13	6.94E+12	5.50E+12	4.43E+12	3.58E+12
600	1111	1.09E+15	9.90E+13	4.93E+13	3.21E+13	2.15E+13	1.69E+13	1.34E+13	1.04E+13
800	1481	7.71E+15	3.50E+14	1.72E+14	1.15E+14	7.79E+13	6.00E+13	4.57E+13	3.39E+13
1000	1852	4.25E+16	1.05E+15	5.08E+14	3.39E+14	2.24E+14	1.63E+14	1.16E+14	7.97E+13

Table 9-27.   Annual Equivalent 1 MeV Electron Fluence for GaAs Solar Cells from Trapped Protons During Solar Max and Min, 40° Inclination (Infinite Backshielding)

PROTONS - P_max

EQUIV. 1 MEV ELECTRON FLUENCE FOR P_max CIRCULAR ORBIT
DUE TO GEOMAGNETICALLY TRAPPED PROTONS, MODEL AP8MAX

INCLINATION = 40 DEGREES

SHIELD THICKNESS, cm (mils)

ALTITUDE (nm)	(km)	0 (0)	2.54E-3 (1)	7.64E-3 (3)	1.52E-2 (6)	3.05E-2 (12)	5.09E-2 (20)	7.64E-2 (30)	1.52E-1 (60)
150	277	5.78E+10	1.67E+10	1.38E+10	1.21E+10	1.03E+10	9.26E+09	8.35E+09	7.55E+09
250	463	1.40E+12	4.52E+11	3.52E+11	2.99E+11	2.49E+11	2.18E+11	1.91E+11	1.66E+11
300	555	5.46E+12	1.37E+12	9.77E+11	7.85E+11	6.21E+11	5.32E+11	4.55E+11	3.87E+11
450	833	1.12E+14	1.39E+13	7.88E+12	5.50E+12	3.85E+12	3.07E+12	2.46E+12	1.97E+12
600	1111	5.38E+14	5.69E+13	2.98E+13	1.99E+13	1.34E+13	1.04E+13	8.02E+12	6.14E+12
800	1481	4.10E+15	2.09E+14	1.08E+14	7.39E+13	5.05E+13	3.83E+13	2.87E+13	2.10E+13
1000	1852	2.76E+16	6.66E+14	3.36E+14	2.29E+14	1.53E+14	1.10E+14	7.68E+13	5.24E+13
1250	2315	1.06E+17	2.12E+15	1.05E+15	7.02E+14	4.45E+14	2.93E+14	1.83E+14	1.10E+14
1500	2778	2.40E+17	5.04E+15	2.51E+15	1.64E+15	9.97E+14	6.09E+14	3.39E+14	1.73E+14
1750	3241	4.32E+17	1.01E+16	5.01E+15	3.06E+15	1.86E+15	1.06E+15	5.33E+14	2.30E+14
2000	3704	6.76E+17	1.83E+16	8.76E+15	5.30E+15	2.84E+15	1.51E+15	6.84E+14	2.56E+14
2250	4167	1.01E+18	3.12E+16	1.42E+16	8.01E+15	3.84E+15	1.88E+15	7.63E+14	2.56E+14
2500	4630	1.51E+18	4.81E+16	2.04E+16	1.06E+16	4.53E+15	2.04E+15	7.59E+14	2.36E+14
2750	5093	2.20E+18	6.89E+16	2.66E+16	1.28E+16	4.88E+15	2.06E+15	7.07E+14	2.06E+14
3000	5556	3.12E+18	9.31E+16	3.24E+16	1.43E+16	4.86E+15	1.94E+15	6.28E+14	1.74E+14
3500	6482	5.99E+18	1.51E+17	4.11E+16	1.50E+16	3.86E+15	1.45E+15	4.39E+14	1.13E+14
4000	7408	1.07E+19	2.17E+17	4.62E+16	1.41E+16	2.81E+15	9.84E+14	2.77E+14	6.86E+13
4500	8334	1.73E+19	2.69E+17	4.56E+16	1.19E+16	1.88E+15	6.01E+14	1.55E+14	3.64E+13
5000	9260	2.57E+19	3.04E+17	4.10E+16	9.05E+15	1.13E+15	3.36E+14	8.00E+13	1.77E+13
5500	10186	3.36E+19	3.25E+17	3.54E+16	6.53E+15	6.26E+14	1.78E+14	4.00E+13	8.06E+12
6000	11112	4.02E+19	3.11E+17	2.75E+16	4.30E+15	3.22E+14	8.31E+13	1.65E+13	2.95E+12
7005	12973	4.93E+19	2.13E+17	1.27E+16	1.43E+15	6.57E+13	1.34E+13	2.02E+12	2.76E+11
8000	14816	4.97E+19	1.03E+17	3.88E+15	3.12E+14	8.83E+12	1.37E+12	1.51E+11	1.56E+10
9000	16668	3.49E+19	3.34E+16	7.20E+14	3.84E+13	5.93E+11	7.24E+10	8.17E+09	1.69E+09
10000	18520	2.48E+19	9.14E+15	1.14E+14	4.26E+12	2.95E+10	0.00E+00	0.00E+00	0.00E+00
11000	20372	1.88E+19	1.96E+15	8.05E+12	1.67E+11	8.29E+08	0.00E+00	0.00E+00	0.00E+00
12000	22224	1.50E+19	3.72E+14	1.61E+11	0.00E+00	0.00E+00	0.00E+00	0.00E+00	0.00E+00
13000	24076	1.00E+19	5.25E+13	3.80E+07	0.00E+00	0.00E+00	0.00E+00	0.00E+00	0.00E+00
14000	25928	8.07E+18	9.34E+12	1.71E+07	0.00E+00	0.00E+00	0.00E+00	0.00E+00	0.00E+00
15000	27780	6.11E+18	2.43E+12	9.11E+06	0.00E+00	0.00E+00	0.00E+00	0.00E+00	0.00E+00
16000	29632	4.97E+18	2.46E+10	0.00E+00	0.00E+00	0.00E+00	0.00E+00	0.00E+00	0.00E+00
17000	31484	3.24E+18	1.88E+10	0.00E+00	0.00E+00	0.00E+00	0.00E+00	0.00E+00	0.00E+00
18000	33336	9.63E+17	1.36E+10	0.00E+00	0.00E+00	0.00E+00	0.00E+00	0.00E+00	0.00E+00
19327	35793		8.94E+09	0.00E+00	0.00E+00	0.00E+00	0.00E+00	0.00E+00	0.00E+00

EQUIV. 1 MEV ELECTRON FLUENCE FOR P_max CIRCULAR ORBIT
DUE TO GEOMAGNETICALLY TRAPPED PROTONS, MODEL AP8MIN

INCLINATION = 40 DEGREES

ALTITUDE (nm)	(km)	0 (0)	2.54E-3 (1)	7.64E-3 (3)	1.52E-2 (6)	3.05E-2 (12)	5.09E-2 (20)	7.64E-2 (30)	1.52E-1 (60)
150	277	1.22E+12	3.00E+11	1.44E+11	7.91E+10	4.35E+10	3.34E+10	2.68E+10	2.24E+10
250	463	1.67E+13	2.83E+12	1.44E+12	8.67E+11	5.34E+11	4.24E+11	3.48E+11	2.89E+11
300	555	3.56E+13	5.47E+12	2.78E+12	1.72E+12	1.10E+12	8.76E+11	7.16E+11	5.93E+11
450	833	2.11E+14	2.39E+13	1.20E+13	7.60E+12	4.96E+12	3.93E+12	3.17E+12	2.55E+12
600	1111	7.76E+14	7.07E+13	3.52E+13	2.29E+13	1.54E+13	1.21E+13	9.55E+12	7.45E+12
800	1481	5.50E+15	2.50E+14	1.23E+14	8.20E+13	5.56E+13	4.28E+13	3.26E+13	2.42E+13
1000	1852	3.03E+16	7.48E+14	3.63E+14	2.42E+14	1.60E+14	1.17E+14	8.28E+13	5.69E+13

Table 9-28. Annual Equivalent 1 MeV Electron Fluence for GaAs Solar Cells from Trapped Protons During Solar Max and Min, 40° Inclination (Infinite Backshielding)

PROTONS - I$_{sc}$

INCLINATION = 40 DEGREES

EQUIV. 1 MEV PROTON FLUENCE FOR I$_{sc}$ - CIRCULAR ORBIT
DUE TO GEOMAGNETICALLY TRAPPED PROTONS, MODEL AP8MAX

ALTITUDE		SHIELD THICKNESS, cm (mils)							
(nm)	(km)	0 (0)	2.54E-3 (1)	7.64E-3 (3)	1.52E-2 (6)	3.05E-2 (12)	5.09E-2 (20)	7.64E-2 (30)	1.52E-1 (60)
150	277	3.63E+10	5.57E+09	4.10E+09	3.23E+09	2.43E+09	1.99E+09	1.62E+09	1.39E+09
250	463	8.84E+11	1.67E+11	1.14E+11	8.85E+10	6.58E+10	5.29E+10	4.16E+10	3.36E+10
300	555	3.70E+12	5.53E+11	3.44E+11	2.49E+11	1.74E+11	1.36E+11	1.04E+11	8.10E+10
450	833	8.76E+13	6.44E+12	3.20E+12	2.00E+12	1.23E+12	8.80E+11	6.17E+11	4.44E+11
600	1111	4.22E+14	2.72E+13	1.26E+13	7.58E+12	4.54E+12	3.17E+12	2.14E+12	1.45E+12
800	1481	3.66E+15	1.01E+14	4.60E+13	2.87E+13	1.78E+13	1.23E+13	8.12E+12	5.20E+12
1000	1852	2.77E+16	3.26E+14	1.47E+14	9.29E+13	5.74E+13	3.80E+13	2.34E+13	1.38E+13
1250	2315	1.10E+17	1.06E+15	4.78E+14	3.01E+14	1.81E+14	1.12E+14	6.22E+13	3.22E+13
1500	2778	2.45E+17	2.54E+15	1.17E+15	7.32E+14	4.29E+14	2.49E+14	1.26E+14	5.62E+13
1750	3241	4.37E+17	5.12E+15	2.37E+15	1.46E+15	8.25E+14	4.56E+14	2.12E+14	8.13E+13
2000	3704	6.66E+17	9.35E+15	4.21E+15	2.46E+15	1.29E+15	6.67E+14	2.82E+14	9.58E+13
2250	4167	9.72E+17	1.61E+16	6.94E+15	3.76E+15	1.77E+15	8.42E+14	3.23E+14	9.97E+13
2500	4630	1.45E+18	2.51E+16	1.00E+16	5.03E+15	2.11E+15	9.28E+14	3.26E+14	9.41E+13
2750	5093	2.09E+18	3.63E+16	1.32E+16	6.12E+15	2.29E+15	9.40E+14	3.06E+14	8.35E+13
3000	5556	2.93E+18	4.95E+16	1.62E+16	6.89E+15	2.29E+15	8.92E+14	2.74E+14	7.17E+13
3500	6482	5.56E+18	8.18E+16	2.08E+16	7.27E+15	1.83E+15	6.73E+14	1.93E+14	4.76E+13
4000	7408	1.00E+19	1.19E+17	2.37E+16	6.90E+15	1.34E+15	4.58E+14	1.23E+14	2.91E+13
4500	8334	1.63E+19	1.49E+17	2.35E+16	5.84E+15	8.97E+14	2.81E+14	6.88E+13	1.56E+13
5000	9260	2.47E+19	1.71E+17	2.12E+16	4.47E+15	5.41E+14	1.58E+14	3.57E+13	7.64E+12
5500	10186	3.21E+19	1.84E+17	1.84E+16	3.24E+15	3.01E+14	8.40E+13	1.79E+13	3.51E+12
6000	11112	3.78E+19	1.78E+17	1.44E+16	2.14E+15	1.55E+14	3.93E+13	7.47E+12	1.29E+12
7005	12973	4.72E+19	1.23E+17	6.69E+15	7.14E+14	3.20E+13	6.38E+12	9.24E+11	1.23E+11
8000	14816	5.05E+19	6.10E+16	2.06E+15	1.57E+14	4.32E+12	6.56E+11	6.92E+10	6.89E+09
9000	16668	3.69E+19	2.00E+16	3.84E+14	1.94E+13	2.91E+11	3.48E+10	3.69E+09	7.58E+08
10000	18520	2.73E+19	5.58E+15	6.10E+13	2.16E+12	1.61E+10	0.00E+00	0.00E+00	0.00E+00
11000	20372	2.21E+19	1.22E+15	4.37E+12	8.50E+10	4.44E+08	0.00E+00	0.00E+00	0.00E+00
12000	22224	1.90E+19	2.37E+14	9.05E+10	0.00E+00	0.00E+00	0.00E+00	0.00E+00	0.00E+00
13000	24076	1.65E+19	3.44E+13	2.33E+07	0.00E+00	0.00E+00	0.00E+00	0.00E+00	0.00E+00
14000	25928	1.45E+19	6.31E+12	1.03E+07	0.00E+00	0.00E+00	0.00E+00	0.00E+00	0.00E+00
15000	27780	1.24E+19	1.68E+12	5.39E+06	0.00E+00	0.00E+00	0.00E+00	0.00E+00	0.00E+00
16000	29632	9.84E+18	2.06E+10	0.00E+00	0.00E+00	0.00E+00	0.00E+00	0.00E+00	0.00E+00
17000	31484	8.37E+18	1.56E+10	0.00E+00	0.00E+00	0.00E+00	0.00E+00	0.00E+00	0.00E+00
18000	33336	5.66E+18	1.12E+10	0.00E+00	0.00E+00	0.00E+00	0.00E+00	0.00E+00	0.00E+00
19327	35793	1.68E+18	7.23E+09	0.00E+00	0.00E+00	0.00E+00	0.00E+00	0.00E+00	0.00E+00

EQUIV. 1 MEV PROTON FLUENCE FOR I$_{sc}$ - CIRCULAR ORBIT
DUE TO GEOMAGNETICALLY TRAPPED PROTONS, MODEL AP8MIN

INCLINATION = 40 DEGREES

ALTITUDE		0 (0)	2.54E-3 (1)	7.64E-3 (3)	1.52E-2 (6)	3.05E-2 (12)	5.09E-2 (20)	7.64E-2 (30)	1.52E-1 (60)
150	277	8.25E+11	1.48E+11	6.40E+10	3.09E+10	1.37E+10	9.13E+09	6.27E+09	4.76E+09
250	463	1.26E+13	1.36E+12	6.17E+11	3.23E+11	1.64E+11	1.15E+11	8.16E+10	6.17E+10
300	555	2.71E+13	2.62E+12	1.18E+12	6.35E+11	3.38E+11	2.38E+11	1.70E+11	1.27E+11
450	833	1.65E+14	1.15E+13	5.06E+12	2.82E+12	1.57E+12	1.11E+12	7.81E+11	5.63E+11
600	1111	6.16E+14	3.40E+13	1.48E+13	8.61E+12	5.04E+12	3.59E+12	2.48E+12	1.71E+12
800	1481	4.92E+15	1.21E+14	5.23E+13	3.16E+13	1.92E+13	1.35E+13	9.06E+12	5.87E+12
1000	1852	3.01E+16	3.69E+14	1.59E+14	9.79E+13	5.96E+13	4.00E+13	2.50E+13	1.48E+13

Table 9-29. Annual Equivalent 1 MeV Electron Fluence for GaAs Solar Cells from Trapped Electrons During Solar Max, 50° Inclination (Infinite Backshielding)

ELECTRONS - I_{sc}, V_{oc}, AND P_{max} INCLINATION = 50 DEGREES

EQUIV. 1 MEV ELECTRON FLUENCE FOR I_{sc} - CIRCULAR ORBIT
DUE TO GEOMAGNETICALLY TRAPPED ELECTRONS - MODEL AE8MAX

| ALTITUDE | | SHIELD THICKNESS, cm (mils) | | | | | | | |
(nm)	(km)	0 (0)	2.54E-3 (1)	7.64E-3 (3)	1.52E-2 (6)	3.05E-2 (12)	5.09E-2 (20)	7.64E-2 (30)	1.52E-1 (60)
150	277	1.85E+10	1.59E+10	1.32E+10	1.08E+10	8.12E+09	6.09E+09	4.56E+09	2.29E+09
250	463	7.33E+10	6.07E+10	4.85E+10	3.81E+10	2.71E+10	1.96E+10	1.43E+10	6.92E+09
300	555	1.26E+11	1.03E+11	8.12E+10	6.30E+10	4.42E+10	3.16E+10	2.29E+10	1.09E+10
450	833	4.57E+11	3.66E+11	2.81E+11	2.11E+11	1.42E+11	9.82E+10	6.95E+10	3.18E+10
600	1111	1.35E+12	1.05E+12	7.81E+11	5.65E+11	3.60E+11	2.40E+11	1.65E+11	7.22E+10
800	1481	4.36E+12	3.36E+12	2.44E+12	1.71E+12	1.05E+12	6.73E+11	4.49E+11	1.88E+11
1000	1852	1.09E+13	8.39E+12	6.08E+12	4.25E+12	2.58E+12	1.64E+12	1.09E+12	4.50E+11
1250	2315	2.20E+13	1.70E+13	1.24E+13	8.73E+12	5.35E+12	3.43E+12	2.29E+12	9.51E+11
1500	2778	3.17E+13	2.44E+13	1.78E+13	1.25E+13	7.61E+12	4.84E+12	3.21E+12	1.34E+12
1750	3241	4.00E+13	3.06E+13	2.20E+13	1.52E+13	8.93E+12	5.46E+12	3.51E+12	1.42E+12
2000	3704	4.63E+13	3.51E+13	2.49E+13	1.68E+13	9.40E+12	5.41E+12	3.31E+12	1.28E+12
2250	4167	5.12E+13	3.86E+13	2.70E+13	1.79E+13	9.65E+12	5.26E+12	3.07E+12	1.13E+12
2500	4630	5.30E+13	3.99E+13	2.79E+13	1.83E+13	9.66E+12	5.09E+12	2.89E+12	1.05E+12
2750	5093	5.19E+13	3.92E+13	2.75E+13	1.81E+13	9.54E+12	5.00E+12	2.84E+12	1.06E+12
3000	5556	4.94E+13	3.75E+13	2.65E+13	1.77E+13	9.54E+12	5.14E+12	3.02E+12	1.19E+12
3500	6482	4.33E+13	3.35E+13	2.44E+13	1.71E+13	1.00E+13	6.00E+12	3.87E+12	1.69E+12
4000	7408	4.08E+13	3.25E+13	2.46E+13	1.81E+13	1.15E+13	7.49E+12	5.10E+12	2.32E+12
4500	8334	4.10E+13	3.38E+13	2.67E+13	2.06E+13	1.40E+13	9.63E+12	6.76E+12	3.12E+12
5000	9260	4.14E+13	3.54E+13	2.92E+13	2.36E+13	1.70E+13	1.23E+13	8.84E+12	4.15E+12
5500	10186	4.31E+13	3.79E+13	3.22E+13	2.68E+13	2.02E+13	1.50E+13	1.10E+13	5.22E+12
6005	11121	4.73E+13	4.23E+13	3.66E+13	3.10E+13	2.38E+13	1.80E+13	1.34E+13	6.41E+12
7005	12973	5.89E+13	5.30E+13	4.62E+13	3.95E+13	3.07E+13	2.34E+13	1.76E+13	8.55E+12
8000	14816	7.02E+13	6.32E+13	5.51E+13	4.70E+13	3.65E+13	2.78E+13	2.08E+13	1.01E+13
9000	16668	7.74E+13	6.96E+13	6.05E+13	5.15E+13	3.98E+13	3.02E+13	2.25E+13	1.08E+13
10000	18520	8.12E+13	7.27E+13	6.30E+13	5.33E+13	4.09E+13	3.07E+13	2.26E+13	1.05E+13
11000	20372	7.96E+13	7.09E+13	6.11E+13	5.13E+13	3.89E+13	2.87E+13	2.08E+13	9.18E+12
12000	22224	7.11E+13	6.30E+13	5.38E+13	4.48E+13	3.34E+13	2.43E+13	1.72E+13	7.27E+12
13000	24076	6.09E+13	5.29E+13	4.48E+13	3.68E+13	2.69E+13	1.92E+13	1.34E+13	5.44E+12
14000	25928	4.99E+13	4.35E+13	3.64E+13	2.96E+13	2.13E+13	1.49E+13	1.02E+13	4.03E+12
15000	27780	3.98E+13	3.44E+13	2.85E+13	2.28E+13	1.61E+13	1.11E+13	7.45E+12	2.85E+12
16000	29632	2.96E+13	2.53E+13	2.08E+13	1.65E+13	1.14E+13	7.72E+12	5.12E+12	1.90E+12
17000	31484	2.35E+13	2.00E+13	1.62E+13	1.27E+13	8.68E+12	5.76E+12	3.76E+12	1.34E+12
18000	33336	1.58E+13	1.33E+13	1.07E+13	8.28E+12	5.55E+12	3.62E+12	2.31E+12	7.84E+11
19327	35793	9.65E+12	7.98E+12	6.28E+12	4.76E+12	3.09E+12	1.95E+12	1.21E+12	3.85E+11

Table 9-30. Annual Equivalent 1 MeV Electron Fluence for GaAs Solar Cells from Trapped Electrons During Solar Min, 50° Inclination (Infinite Backshielding)

ELECTRONS - I_{sc}, V_{oc}, AND P_{max} INCLINATION = 50 DEGREES

EQUIV. 1 MEV ELECTRON FLUENCE FOR I_{sc} - CIRCULAR ORBIT
DUE TO GEOMAGNETICALLY TRAPPED ELECTRONS - MODEL AE8MIN

ALTITUDE		SHIELD THICKNESS, cm (mils)							
(nm)	(km)	0 (0)	2.54E-3 (1)	7.64E-3 (3)	1.52E-2 (6)	3.05E-2 (12)	5.09E-2 (20)	7.64E-2 (30)	1.52E-1 (60)
150	277	4.62E+09	4.12E+09	3.58E+09	3.06E+09	2.41E+09	1.88E+09	1.44E+09	7.50E+08
250	463	2.40E+10	2.08E+10	1.75E+10	1.45E+10	1.10E+10	8.24E+09	6.15E+09	3.01E+09
300	555	4.44E+10	3.82E+10	3.19E+10	2.61E+10	1.94E+10	1.44E+10	1.06E+10	5.09E+09
450	833	1.93E+11	1.63E+11	1.33E+11	1.06E+11	7.63E+10	5.49E+10	3.94E+10	1.81E+10
600	1111	6.22E+11	5.16E+11	4.12E+11	3.20E+11	2.21E+11	1.54E+11	1.07E+11	4.72E+10
800	1481	2.21E+12	1.82E+12	1.43E+12	1.09E+12	7.38E+11	5.00E+11	3.42E+11	1.45E+11
1000	1852	5.85E+12	4.81E+12	3.79E+12	2.91E+12	1.97E+12	1.33E+12	9.12E+11	3.85E+11
1250	2315	1.19E+13	9.89E+12	7.89E+12	6.14E+12	4.22E+12	2.90E+12	2.01E+12	8.60E+11
1500	2778	1.60E+13	1.34E+13	1.07E+13	8.40E+12	5.83E+12	4.05E+12	2.82E+12	1.22E+12
1750	3241	1.77E+13	1.47E+13	1.17E+13	9.11E+12	6.25E+12	4.30E+12	2.97E+12	1.27E+12
2000	3704	1.73E+13	1.42E+13	1.12E+13	8.51E+12	5.70E+12	3.83E+12	2.59E+12	1.07E+12
2250	4167	1.61E+13	1.30E+13	1.00E+13	7.53E+12	4.92E+12	3.23E+12	2.14E+12	8.57E+11
2500	4630	1.41E+13	1.13E+13	8.61E+12	6.36E+12	4.08E+12	2.65E+12	1.75E+12	6.92E+11
2750	5093	1.21E+13	9.61E+12	7.27E+12	5.35E+12	3.43E+12	2.23E+12	1.48E+12	5.95E+11
3000	5556	1.08E+13	8.61E+12	6.56E+12	4.87E+12	3.18E+12	2.11E+12	1.43E+12	5.95E+11
3500	6482	1.02E+13	8.36E+12	6.58E+12	5.08E+12	3.50E+12	2.44E+12	1.71E+12	7.61E+11
4000	7408	1.14E+13	9.53E+12	7.74E+12	6.15E+12	4.39E+12	3.14E+12	2.24E+12	1.02E+12
4500	8334	1.30E+13	1.12E+13	9.28E+12	7.57E+12	5.57E+12	4.08E+12	2.96E+12	1.37E+12
5000	9260	1.50E+13	1.31E+13	1.12E+13	9.30E+12	7.03E+12	5.25E+12	3.87E+12	1.82E+12
5500	10186	1.72E+13	1.53E+13	1.32E+13	1.11E+13	8.55E+12	6.46E+12	4.81E+12	2.28E+12
6000	11112	2.04E+13	1.83E+13	1.59E+13	1.35E+13	1.04E+13	7.94E+12	5.93E+12	2.84E+12
7000	12964	2.66E+13	2.40E+13	2.09E+13	1.78E+13	1.39E+13	1.06E+13	7.94E+12	3.84E+12
8000	14816	3.39E+13	3.06E+13	2.67E+13	2.28E+13	1.77E+13	1.35E+13	1.01E+13	4.90E+12
9000	16668	4.09E+13	3.68E+13	3.21E+13	2.74E+13	2.13E+13	1.62E+13	1.21E+13	5.77E+12
10000	18520	4.82E+13	4.34E+13	3.78E+13	3.22E+13	2.49E+13	1.88E+13	1.39E+13	6.50E+12
11000	20372	5.54E+13	4.96E+13	4.29E+13	3.63E+13	2.78E+13	2.07E+13	1.52E+13	6.87E+12
12000	22224	5.80E+13	5.16E+13	4.43E+13	3.71E+13	2.80E+13	2.05E+13	1.48E+13	6.43E+12
13000	24076	5.52E+13	4.87E+13	4.14E+13	3.42E+13	2.53E+13	1.82E+13	1.28E+13	5.28E+12
14000	25928	4.85E+13	4.24E+13	3.57E+13	2.91E+13	2.10E+13	1.48E+13	1.02E+13	4.06E+12
15000	27780	3.97E+13	3.43E+13	2.85E+13	2.29E+13	1.62E+13	1.11E+13	7.51E+12	2.89E+12
16000	29632	2.97E+13	2.55E+13	2.09E+13	1.66E+13	1.15E+13	7.77E+12	5.16E+12	1.92E+12
17000	31484	2.36E+13	2.01E+13	1.63E+13	1.28E+13	8.73E+12	5.79E+12	3.78E+12	1.35E+12
18000	33336	1.59E+13	1.34E+13	1.07E+13	8.32E+12	5.58E+12	3.64E+12	2.33E+12	7.90E+11
19327	35793	9.75E+12	8.07E+12	6.35E+12	4.82E+12	3.12E+12	1.97E+12	1.23E+12	3.91E+11

Table 9-31. Annual Equivalent 1 MeV Electron Fluence for GaAs Solar Cells from Trapped Protons During Solar Max and Min, 50° Inclination (Infinite Backshielding)

PROTONS - V_{oc}

EQUIV. 1 MEV ELECTRON FLUENCE FOR V_{oc} CIRCULAR ORBIT   INCLINATION = 50 DEGREES
DUE TO GEOMAGNETICALLY TRAPPED PROTONS, MODEL AP8MAX

ALTITUDE (nm)	(km)	SHIELD THICKNESS, cm (mils)							
		0 (0)	2.54E-3 (1)	7.64E-3 (3)	1.52E-2 (6)	3.05E-2 (12)	5.09E-2 (20)	7.64E-2 (30)	1.52E-1 (60)
150	277	9.84E+12	6.37E+10	1.95E+10	1.27E+10	9.59E+09	8.19E+09	7.15E+09	6.25E+09
250	463	1.93E+14	1.51E+12	5.33E+11	3.60E+11	2.61E+11	2.21E+11	1.88E+11	1.60E+11
300	555	5.03E+14	4.36E+12	1.57E+12	9.93E+11	6.64E+11	5.41E+11	4.47E+11	3.72E+11
450	833	3.73E+15	3.55E+13	1.24E+13	7.15E+12	4.32E+12	3.30E+12	2.56E+12	2.02E+12
600	1111	1.26E+16	1.16E+14	4.21E+13	2.47E+13	1.51E+13	1.13E+13	8.57E+12	6.49E+12
800	1481	4.70E+16	3.53E+14	1.40E+14	8.77E+13	5.65E+13	4.22E+13	3.13E+13	2.29E+13
1000	1852	1.41E+17	1.02E+15	4.20E+14	2.67E+14	1.69E+14	1.21E+14	8.43E+13	5.76E+13
1250	2315	3.40E+17	2.99E+15	1.26E+15	7.99E+14	4.93E+14	3.25E+14	2.04E+14	1.23E+14
1500	2778	6.09E+17	6.62E+15	2.89E+15	1.84E+15	1.10E+15	6.78E+14	3.80E+14	1.95E+14
1750	3241	9.52E+17	1.25E+16	5.63E+15	3.56E+15	2.07E+15	1.20E+15	6.02E+14	2.60E+14
2000	3704	1.38E+18	2.18E+16	9.74E+15	5.89E+15	3.19E+15	1.71E+15	7.75E+14	2.90E+14
2250	4167	1.91E+18	3.63E+16	1.58E+16	8.94E+15	4.33E+15	2.12E+15	8.66E+14	2.91E+14
2500	4630	2.66E+18	5.50E+16	2.28E+16	1.19E+16	5.14E+15	2.32E+15	8.65E+14	2.68E+14
2750	5093	3.68E+18	7.77E+16	2.98E+16	1.44E+16	5.55E+15	2.34E+15	8.05E+14	2.34E+14
3000	5556	4.88E+18	1.04E+17	3.63E+16	1.61E+16	5.53E+15	2.21E+15	7.16E+14	1.98E+14
3500	6482	8.27E+18	1.68E+17	4.62E+16	1.69E+16	4.41E+15	1.66E+15	5.03E+14	1.30E+14
4000	7408	1.34E+19	2.42E+17	5.24E+16	1.61E+16	3.22E+15	1.13E+15	3.18E+14	7.88E+13
4500	8334	2.01E+19	3.02E+17	5.18E+16	1.36E+16	2.16E+15	6.91E+14	1.78E+14	4.19E+13
5000	9260	2.86E+19	3.42E+17	4.68E+16	1.04E+16	1.30E+15	3.87E+14	9.23E+13	2.04E+13
5500	10186	3.70E+19	3.68E+17	4.05E+16	7.51E+15	7.21E+14	2.06E+14	4.61E+13	9.31E+12
6005	11121	4.43E+19	3.54E+17	3.16E+16	4.94E+15	3.70E+14	9.54E+13	1.90E+13	3.38E+12
7005	12973	5.53E+19	2.45E+17	1.47E+16	1.65E+15	7.60E+13	1.55E+13	2.34E+12	3.20E+11
8000	14816	5.58E+19	1.18E+17	4.43E+15	3.57E+14	1.01E+13	1.56E+12	1.72E+11	1.78E+10
9000	16668	3.91E+19	3.82E+16	8.25E+14	4.40E+13	6.78E+11	8.30E+10	9.39E+09	1.95E+09
10000	18520	2.78E+19	1.05E+16	1.31E+14	4.91E+12	3.40E+10	0.00E+00	0.00E+00	0.00E+00
11000	20372	2.06E+19	2.29E+15	9.76E+12	2.07E+11	1.03E+09	0.00E+00	0.00E+00	0.00E+00
12000	22224	1.64E+19	4.30E+14	1.87E+11	0.00E+00	0.00E+00	0.00E+00	0.00E+00	0.00E+00
13000	24076	1.33E+19	6.09E+13	5.00E+07	0.00E+00	0.00E+00	0.00E+00	0.00E+00	0.00E+00
14000	25928	1.11E+19	1.07E+13	2.25E+07	0.00E+00	0.00E+00	0.00E+00	0.00E+00	0.00E+00
15000	27780	9.13E+18	2.81E+12	1.20E+07	0.00E+00	0.00E+00	0.00E+00	0.00E+00	0.00E+00
16000	29632	6.95E+18	3.07E+10	0.00E+00	0.00E+00	0.00E+00	0.00E+00	0.00E+00	0.00E+00
17000	31484	5.72E+18	2.36E+10	0.00E+00	0.00E+00	0.00E+00	0.00E+00	0.00E+00	0.00E+00
18000	33336	3.74E+18	1.70E+10	0.00E+00	0.00E+00	0.00E+00	0.00E+00	0.00E+00	0.00E+00
19327	35793	1.13E+18	1.13E+10	0.00E+00	0.00E+00	0.00E+00	0.00E+00	0.00E+00	0.00E+00

EQUIV. 1 MEV ELECTRON FLUENCE FOR V_{oc} CIRCULAR ORBIT   INCLINATION = 50 DEGREES
DUE TO GEOMAGNETICALLY TRAPPED PROTONS, MODEL AP8MIN

ALTITUDE (nm)	(km)	0 (0)	2.54E-3 (1)	7.64E-3 (3)	1.52E-2 (6)	3.05E-2 (12)	5.09E-2 (20)	7.64E-2 (30)	1.52E-1 (60)
150	277	4.96E+13	9.95E+11	3.15E+11	1.43E+11	5.99E+10	4.00E+10	2.83E+10	2.15E+10
250	463	4.47E+14	9.28E+12	2.98E+12	1.42E+12	6.67E+11	4.74E+11	3.56E+11	2.79E+11
300	555	8.24E+14	1.59E+13	5.30E+12	2.66E+12	1.35E+12	9.86E+11	7.54E+11	5.96E+11
450	833	3.98E+15	5.77E+13	1.95E+13	1.03E+13	5.76E+12	4.34E+12	3.37E+12	2.66E+12
600	1111	1.24E+16	1.38E+14	5.02E+13	2.89E+13	1.76E+13	1.34E+13	1.03E+13	7.98E+12
800	1481	4.73E+16	4.20E+14	1.63E+14	9.96E+13	6.31E+13	4.77E+13	3.59E+13	2.65E+13
1000	1852	1.42E+17	1.17E+15	4.68E+14	2.88E+14	1.80E+14	1.29E+14	9.12E+13	6.27E+13

Table 9-32. Annual Equivalent 1 MeV Electron Fluence for GaAs Solar Cells from Trapped Protons During Solar Max and Min, 50° Inclination (Infinite Backshielding)

PROTONS - P_{max}

INCLINATION = 50 DEGREES

EQUIV. 1 MEV ELECTRON FLUENCE FOR P_{max} CIRCULAR ORBIT
DUE TO GEOMAGNETICALLY TRAPPED PROTONS, MODEL AP8MAX

ALTITUDE (nm)	(km)	SHIELD THICKNESS, cm (mils)							
		0 (0)	2.54E-3 (1)	7.64E-3 (3)	1.52E-2 (6)	3.05E-2 (12)	5.09E-2 (20)	7.64E-2 (30)	1.52E-1 (60)
150	277	7.03E+12	4.55E+10	1.39E+10	9.06E+09	6.85E+09	5.85E+09	5.11E+09	4.47E+09
250	463	1.38E+14	1.08E+12	3.81E+11	2.57E+11	1.87E+11	1.58E+11	1.35E+11	1.14E+11
300	555	3.60E+14	3.12E+12	1.12E+12	7.09E+11	4.74E+11	3.86E+11	3.19E+11	2.65E+11
450	833	2.66E+15	2.54E+13	8.87E+12	5.11E+12	3.08E+12	2.35E+12	1.83E+12	1.44E+12
600	1111	8.98E+15	8.29E+13	3.01E+13	1.77E+13	1.08E+13	8.09E+12	6.12E+12	4.64E+12
800	1481	3.36E+16	2.52E+14	9.98E+13	6.26E+13	4.03E+13	3.01E+13	2.24E+13	1.63E+13
1000	1852	1.01E+17	7.27E+14	3.00E+14	1.90E+14	1.21E+14	8.63E+13	6.02E+13	4.11E+13
1250	2315	2.43E+17	2.13E+15	9.02E+14	5.71E+14	3.52E+14	2.32E+14	1.46E+14	8.76E+13
1500	2778	4.35E+17	4.73E+15	2.06E+15	1.31E+15	7.89E+14	4.84E+14	2.71E+14	1.39E+14
1750	3241	6.80E+17	8.95E+15	4.02E+15	2.54E+15	1.48E+15	8.54E+14	4.30E+14	1.86E+14
2000	3704	9.87E+17	1.56E+16	6.96E+15	4.21E+15	2.28E+15	1.22E+15	5.54E+14	2.07E+14
2250	4167	1.36E+18	2.60E+16	1.13E+16	6.39E+15	3.09E+15	1.52E+15	6.19E+14	2.08E+14
2500	4630	1.90E+18	3.93E+16	1.63E+16	8.51E+15	3.67E+15	1.66E+15	6.17E+14	1.92E+14
2750	5093	2.63E+18	5.55E+16	2.13E+16	1.03E+16	3.97E+15	1.67E+15	5.75E+14	1.67E+14
3000	5556	3.48E+18	7.44E+16	2.59E+16	1.21E+16	3.95E+15	1.58E+15	5.11E+14	1.42E+14
3500	6482	5.91E+18	1.20E+17	3.30E+16	1.15E+16	3.15E+15	1.19E+15	3.59E+14	9.26E+13
4000	7408	9.59E+18	1.73E+17	3.74E+16	1.15E+16	2.30E+15	8.07E+14	2.27E+14	5.63E+13
4500	8334	1.44E+19	2.16E+17	3.70E+16	9.70E+15	1.54E+15	4.94E+14	1.27E+14	2.99E+13
5000	9260	2.04E+19	2.44E+17	3.34E+16	7.42E+15	9.29E+14	2.77E+14	6.59E+13	1.46E+13
5500	10186	2.64E+19	2.63E+17	2.89E+16	5.36E+15	5.15E+14	1.47E+14	3.30E+13	6.65E+12
6005	11121	3.17E+19	2.53E+17	2.25E+16	3.53E+15	2.64E+14	6.82E+13	1.36E+13	2.41E+12
7005	12973	3.95E+19	1.75E+17	1.05E+16	1.18E+15	5.43E+13	1.11E+13	1.67E+12	2.28E+11
8000	14816	3.99E+19	8.44E+16	3.17E+15	2.55E+14	7.22E+12	1.12E+12	1.23E+11	1.27E+11
9000	16668	2.79E+19	2.73E+16	5.89E+14	3.14E+13	4.85E+11	5.93E+10	6.71E+10	1.39E+09
10000	18520	1.99E+19	7.53E+15	9.38E+13	3.51E+12	2.43E+10	0.00E+00	0.00E+00	0.00E+00
11000	20372	1.47E+19	1.63E+15	6.97E+12	1.48E+11	7.37E+08	0.00E+00	0.00E+00	0.00E+00
12000	22224	1.17E+19	3.07E+14	1.33E+11	0.00E+00	0.00E+00	0.00E+00	0.00E+00	0.00E+00
13000	24076	9.53E+18	4.35E+13	3.57E+07	0.00E+00	0.00E+00	0.00E+00	0.00E+00	0.00E+00
14000	25928	7.96E+18	7.67E+12	1.60E+07	0.00E+00	0.00E+00	0.00E+00	0.00E+00	0.00E+00
15000	27780	6.52E+18	2.00E+12	8.54E+06	0.00E+00	0.00E+00	0.00E+00	0.00E+00	0.00E+00
16000	29632	4.96E+18	2.19E+10	0.00E+00	0.00E+00	0.00E+00	0.00E+00	0.00E+00	0.00E+00
17000	31484	4.09E+18	1.68E+10	0.00E+00	0.00E+00	0.00E+00	0.00E+00	0.00E+00	0.00E+00
18000	33336	2.67E+18	1.22E+10	0.00E+00	0.00E+00	0.00E+00	0.00E+00	0.00E+00	0.00E+00
19327	35793	8.08E+17	8.06E+09	0.00E+00	0.00E+00	0.00E+00	0.00E+00	0.00E+00	0.00E+00

INCLINATION = 50 DEGREES

EQUIV. 1 MEV ELECTRON FLUENCE FOR P_{max} CIRCULAR ORBIT
DUE TO GEOMAGNETICALLY TRAPPED PROTONS, MODEL AP8MIN

ALTITUDE (nm)	(km)	0 (0)	2.54E-3 (1)	7.64E-3 (3)	1.52E-2 (6)	3.05E-2 (12)	5.09E-2 (20)	7.64E-2 (30)	1.52E-1 (60)
150	277	3.54E+13	7.11E+11	2.25E+11	1.02E+11	4.28E+10	2.86E+10	2.02E+10	1.53E+10
250	463	3.19E+14	6.63E+12	2.13E+12	1.01E+12	4.77E+11	3.38E+11	2.54E+11	2.00E+11
300	555	5.89E+14	1.14E+13	3.79E+12	1.90E+12	9.66E+11	7.05E+11	5.38E+11	4.26E+11
450	833	2.84E+15	4.12E+13	1.39E+13	7.37E+12	4.12E+12	3.10E+12	2.41E+12	1.90E+12
600	1111	8.83E+15	9.89E+13	3.59E+13	2.07E+13	1.26E+13	9.57E+12	7.39E+12	5.70E+12
800	1481	3.38E+16	3.00E+14	1.17E+14	7.12E+13	4.51E+13	3.40E+13	2.56E+13	1.89E+13
1000	1852	1.01E+17	8.33E+14	3.34E+14	2.06E+14	1.28E+14	9.24E+13	6.52E+13	4.48E+13

Table 9-33. Annual Equivalent 1 MeV Electron Fluence for GaAs Solar Cells from Trapped Protons During Solar Max and Min, 50° Inclination (Infinite Backshielding)

PROTONS - I_sc

EQUIV. 1 MEV PROTON FLUENCE FOR I_sc - CIRCULAR ORBIT
DUE TO GEOMAGNETICALLY TRAPPED PROTONS, MODEL AP8MAX
INCLINATION = 50 DEGREES

ALTITUDE (nm)	(km)	SHIELD THICKNESS, cm (mils)							
		0 (0)	2.54E-3 (1)	7.64E-3 (3)	1.52E-2 (6)	3.05E-2 (12)	5.09E-2 (20)	7.64E-2 (30)	1.52E-1 (60)
150	277	7.50E+12	2.30E+10	5.35E+09	2.85E+09	1.81E+09	1.37E+09	1.06E+09	8.54E+08
250	463	1.42E+14	5.38E+11	1.49E+11	8.59E+10	5.30E+10	4.05E+10	3.06E+10	2.39E+10
300	555	3.75E+14	1.57E+12	4.65E+11	2.55E+11	1.43E+11	1.04E+11	7.52E+10	5.68E+10
450	833	2.82E+15	1.31E+13	3.92E+12	1.99E+12	1.03E+12	7.00E+11	4.71E+11	3.28E+11
600	1111	9.74E+15	4.26E+13	1.34E+13	7.00E+12	3.74E+12	2.52E+12	1.66E+12	1.10E+12
800	1481	3.80E+16	1.28E+14	4.38E+13	2.49E+13	1.44E+13	9.77E+12	6.36E+12	4.05E+12
1000	1852	1.16E+17	3.69E+14	1.34E+14	7.83E+13	4.57E+13	2.99E+13	1.84E+13	1.08E+13
1250	2315	2.75E+17	1.09E+15	4.14E+14	2.46E+14	1.43E+14	8.82E+13	4.92E+13	2.56E+13
1500	2778	4.79E+17	2.42E+15	9.64E+14	5.84E+14	3.39E+14	1.98E+14	1.01E+14	4.50E+13
1750	3241	7.29E+17	4.60E+15	1.90E+15	1.16E+15	6.57E+14	3.65E+14	1.70E+14	6.56E+13
2000	3704	1.03E+18	8.04E+15	3.34E+15	1.95E+15	1.03E+15	5.37E+14	2.28E+14	7.76E+13
2250	4167	1.39E+18	1.35E+16	5.51E+15	3.00E+15	1.42E+15	6.80E+14	2.61E+14	8.08E+13
2500	4630	1.92E+18	2.05E+16	8.00E+15	4.04E+15	1.71E+15	7.53E+14	2.65E+14	7.65E+13
2750	5093	2.64E+18	2.92E+16	1.06E+16	4.92E+15	1.86E+15	7.65E+14	2.49E+14	6.79E+13
3000	5556	3.45E+18	3.95E+16	1.30E+16	5.55E+15	1.86E+15	7.25E+14	2.23E+14	5.83E+13
3500	6482	5.75E+18	6.47E+16	1.67E+16	5.88E+15	1.49E+15	5.49E+14	1.58E+14	3.88E+13
4000	7408	9.21E+18	9.46E+16	1.91E+16	5.62E+15	1.10E+15	3.75E+14	1.00E+14	2.39E+13
4500	8334	1.38E+19	1.20E+17	1.91E+16	4.77E+15	7.36E+14	2.31E+14	5.65E+13	1.28E+13
5000	9260	1.98E+19	1.37E+17	1.73E+16	3.66E+15	4.46E+14	1.30E+14	2.94E+13	6.30E+12
5500	10186	2.52E+19	1.49E+17	1.51E+16	2.66E+15	2.48E+14	6.92E+13	1.48E+13	2.90E+12
6005	11121	2.97E+19	1.45E+17	1.18E+16	1.75E+15	1.27E+14	3.22E+13	6.12E+12	1.06E+12
7005	12973	3.77E+19	1.01E+17	5.52E+15	5.89E+14	2.64E+13	5.27E+12	7.63E+11	1.02E+11
8000	14816	4.05E+19	4.97E+16	1.68E+15	1.28E+14	3.53E+12	5.36E+11	5.65E+10	5.62E+09
9000	16668	2.95E+19	1.64E+16	3.14E+14	1.59E+13	2.38E+11	2.85E+10	3.02E+09	6.26E+08
10000	18520	2.18E+19	4.60E+15	5.03E+13	1.78E+12	1.33E+10	0.00E+00	0.00E+00	0.00E+00
11000	20372	1.72E+19	1.02E+15	3.79E+12	7.53E+10	3.94E+08	0.00E+00	0.00E+00	0.00E+00
12000	22224	1.47E+19	1.96E+14	7.50E+10	0.00E+00	0.00E+00	0.00E+00	0.00E+00	0.00E+00
13000	24076	1.29E+19	2.85E+13	2.19E+07	0.00E+00	0.00E+00	0.00E+00	0.00E+00	0.00E+00
14000	25928	1.15E+19	5.18E+12	9.65E+06	0.00E+00	0.00E+00	0.00E+00	0.00E+00	0.00E+00
15000	27780	1.00E+19	1.38E+12	5.05E+06	0.00E+00	0.00E+00	0.00E+00	0.00E+00	0.00E+00
16000	29632	7.99E+18	1.83E+10	0.00E+00	0.00E+00	0.00E+00	0.00E+00	0.00E+00	0.00E+00
17000	31484	6.88E+18	1.39E+10	0.00E+00	0.00E+00	0.00E+00	0.00E+00	0.00E+00	0.00E+00
18000	33336	4.67E+18	9.96E+09	0.00E+00	0.00E+00	0.00E+00	0.00E+00	0.00E+00	0.00E+00
19327	35793	1.41E+18	6.50E+09	0.00E+00	0.00E+00	0.00E+00	0.00E+00	0.00E+00	0.00E+00

EQUIV. 1 MEV PROTON FLUENCE FOR I_sc - CIRCULAR ORBIT
DUE TO GEOMAGNETICALLY TRAPPED PROTONS, MODEL AP8MIN
INCLINATION = 50 DEGREES

ALTITUDE (nm)	(km)	0 (0)	2.54E-3 (1)	7.64E-3 (3)	1.52E-2 (6)	3.05E-2 (12)	5.09E-2 (20)	7.64E-2 (30)	1.52E-1 (60)
150	277	3.36E+13	3.76E+11	1.08E+11	4.45E+10	1.57E+10	9.15E+09	5.40E+09	3.55E+09
250	463	3.00E+14	3.48E+12	1.00E+12	4.25E+11	1.66E+11	1.02E+11	6.47E+10	4.47E+10
300	555	5.64E+14	5.94E+12	1.75E+12	7.80E+11	3.29E+11	2.10E+11	1.36E+11	9.53E+10
450	833	2.86E+15	2.14E+13	6.30E+12	2.94E+12	1.38E+12	9.16E+11	6.12E+11	4.26E+11
600	1111	9.37E+15	5.07E+13	1.59E+13	8.10E+12	4.24E+12	2.90E+12	1.95E+12	1.32E+12
800	1481	3.76E+16	1.52E+14	5.13E+13	2.81E+13	1.58E+13	1.08E+13	7.15E+12	4.60E+12
1000	1852	1.15E+17	4.25E+14	1.50E+14	8.46E+13	4.81E+13	3.18E+13	1.97E+13	1.17E+13

Table 9-34. Annual Equivalent 1 MeV Electron Fluence for GaAs Solar Cells from Trapped Electrons During Solar Max, 60° Inclination (Infinite Backshielding)

ELECTRONS - I_{sc}, V_{oc}, AND P_{max} INCLINATION = 60 DEGREES

EQUIV. 1 MEV ELECTRON FLUENCE FOR I_{sc} - CIRCULAR ORBIT
DUE TO GEOMAGNETICALLY TRAPPED ELECTRONS - MODEL AE8MAX

ALTITUDE (nm)	(km)	SHIELD THICKNESS, cm (mils)							
		0 (0)	2.54E-3 (1)	7.64E-3 (3)	1.52E-2 (6)	3.05E-2 (12)	5.09E-2 (20)	7.64E-2 (30)	1.52E-1 (60)
150	277	6.84E+10	6.11E+10	5.31E+10	4.52E+10	3.51E+10	2.68E+10	2.02E+10	9.97E+09
250	463	1.43E+11	1.25E+11	1.06E+11	8.79E+10	6.67E+10	5.01E+10	3.73E+10	1.81E+10
300	555	2.03E+11	1.75E+11	1.46E+11	1.20E+11	9.00E+10	6.70E+10	4.97E+10	2.40E+10
450	833	5.29E+11	4.41E+11	3.55E+11	2.81E+11	2.01E+11	1.46E+11	1.06E+11	5.00E+10
600	1111	1.34E+12	1.08E+12	8.28E+11	6.24E+11	4.22E+11	2.93E+11	2.07E+11	9.43E+10
800	1481	4.03E+12	3.15E+12	2.33E+12	1.68E+12	1.07E+12	7.10E+11	4.87E+11	2.11E+11
1000	1852	9.76E+12	7.56E+12	5.54E+12	3.94E+12	2.45E+12	1.60E+12	1.08E+12	4.54E+11
1250	2315	1.96E+13	1.52E+13	1.12E+13	7.98E+12	4.97E+12	3.24E+12	2.19E+12	9.23E+11
1500	2778	2.83E+13	2.20E+13	1.61E+13	1.14E+13	7.08E+12	4.58E+12	3.07E+12	1.30E+12
1750	3241	3.58E+13	2.76E+13	2.00E+13	1.40E+13	8.35E+12	5.20E+12	3.40E+12	1.41E+12
2000	3704	4.16E+13	3.17E+13	2.27E+13	1.55E+13	8.89E+12	5.26E+12	3.29E+12	1.31E+12
2250	4167	4.59E+13	3.49E+13	2.47E+13	1.66E+13	9.16E+12	5.16E+12	3.11E+12	1.19E+12
2500	4630	4.77E+13	3.61E+13	2.55E+13	1.70E+13	9.19E+12	5.02E+12	2.96E+12	1.12E+12
2750	5093	4.67E+13	3.55E+13	2.51E+13	1.68E+13	9.07E+12	4.92E+12	2.89E+12	1.12E+12
3000	5556	4.43E+13	3.38E+13	2.41E+13	1.63E+13	8.94E+12	4.95E+12	2.97E+12	1.19E+12
3500	6482	3.82E+13	2.97E+13	2.17E+13	1.52E+13	8.93E+12	5.36E+12	3.45E+12	1.47E+12
4000	7408	3.53E+13	2.81E+13	2.12E+13	1.55E+13	9.79E+12	6.30E+12	4.25E+12	1.88E+12
4500	8334	3.46E+13	2.83E+13	2.23E+13	1.70E+13	1.14E+13	7.75E+12	5.37E+12	2.42E+12
5000	9260	3.39E+13	2.88E+13	2.36E+13	1.89E+13	1.35E+13	9.58E+12	6.84E+12	3.15E+12
5500	10186	3.48E+13	3.05E+13	2.58E+13	2.14E+13	1.60E+13	1.18E+13	8.59E+12	4.05E+12
6000	11112	3.75E+13	3.34E+13	2.89E+13	2.44E+13	1.87E+13	1.41E+13	1.04E+13	5.01E+12
7005	12973	4.76E+13	4.28E+13	3.74E+13	3.19E+13	2.49E+13	1.90E+13	1.43E+13	6.98E+12
8000	14816	5.72E+13	5.15E+13	4.50E+13	3.84E+13	2.99E+13	2.29E+13	1.72E+13	8.39E+12
9000	16668	6.42E+13	5.77E+13	5.03E+13	4.28E+13	3.32E+13	2.52E+13	1.88E+13	9.07E+12
10000	18520	6.68E+13	6.17E+13	5.35E+13	4.53E+13	3.48E+13	2.62E+13	1.93E+13	9.01E+12
11000	20372	6.68E+13	5.96E+13	5.14E+13	4.33E+13	3.28E+13	2.43E+13	1.76E+13	7.81E+12
12000	22224	6.02E+13	5.34E+13	4.57E+13	3.80E+13	2.84E+13	2.07E+13	1.47E+13	6.23E+12
13000	24076	5.14E+13	4.52E+13	3.83E+13	3.15E+13	2.31E+13	1.65E+13	1.15E+13	4.68E+12
14000	25928	4.29E+13	3.74E+13	3.14E+13	2.55E+13	1.84E+13	1.28E+13	8.80E+12	3.48E+12
15000	27780	3.44E+13	2.98E+13	2.47E+13	1.98E+13	1.40E+13	9.59E+12	6.47E+12	2.48E+12
16000	29632	2.57E+13	2.20E+13	1.80E+13	1.43E+13	9.95E+12	6.71E+12	4.45E+12	1.65E+12
17000	31484	2.05E+13	1.74E+13	1.41E+13	1.11E+13	7.58E+12	5.03E+12	3.29E+12	1.18E+12
18000	33336	1.38E+13	1.16E+13	9.33E+12	7.23E+12	4.85E+12	3.16E+12	2.02E+12	6.86E+11
19327	35793	8.51E+12	7.04E+12	5.54E+12	4.20E+12	2.72E+12	1.72E+12	1.07E+12	3.41E+11

Table 9-35. Annual Equivalent 1 MeV Electron Fluence for GaAs Solar Cells from Trapped Electrons During Solar Min, 60° Inclination (Infinite Backshielding)

GaAs SOLAR CELLS

ELECTRONS - I_{sc}, V_{oc}, AND P_{max}

INCLINATION = 60 DEGREES

EQUIV. 1 MEV ELECTRON FLUENCE FOR I_{sc} - CIRCULAR ORBIT
DUE TO GEOMAGNETICALLY TRAPPED ELECTRONS - MODEL AE8MIN

ALTITUDE (nm)	(km)	\multicolumn{8}{c}{SHIELD THICKNESS, cm (mils)}							
		0 (0)	2.54E-3 (1)	7.64E-3 (3)	1.52E-2 (6)	3.05E-2 (12)	5.09E-2 (20)	7.64E-2 (30)	1.52E-1 (60)
150	277	2.65E+10	2.40E+10	2.11E+10	1.82E+10	1.44E+10	1.11E+10	8.45E+09	4.21E+09
250	463	5.75E+10	5.13E+10	4.45E+10	3.78E+10	2.93E+10	2.24E+10	1.68E+10	8.21E+09
300	555	8.40E+10	7.43E+10	6.39E+10	5.38E+10	4.13E+10	3.12E+10	2.33E+10	1.12E+10
450	833	2.34E+11	2.03E+11	1.70E+11	1.39E+11	1.03E+11	7.58E+10	5.54E+10	2.60E+10
600	1111	6.30E+11	5.31E+11	4.31E+11	3.42E+11	2.43E+11	1.73E+11	1.23E+11	5.55E+10
800	1481	2.04E+12	1.69E+12	1.34E+12	1.04E+12	7.13E+11	4.90E+11	3.39E+11	1.47E+11
1000	1852	5.21E+12	4.31E+12	3.42E+12	2.64E+12	1.80E+12	1.23E+12	8.45E+11	3.61E+11
1250	2315	1.06E+13	8.86E+12	7.09E+12	5.53E+12	3.82E+12	2.64E+12	1.84E+12	7.92E+11
1500	2778	1.43E+13	1.20E+13	9.67E+12	7.59E+12	5.30E+12	3.70E+12	2.58E+12	1.13E+12
1750	3241	1.59E+13	1.33E+13	1.02E+13	8.29E+12	5.73E+12	3.96E+12	2.75E+12	1.19E+12
2000	3704	1.57E+13	1.30E+13	1.01E+13	7.88E+12	5.33E+12	3.61E+12	2.46E+12	1.03E+12
2250	4167	1.48E+13	1.20E+13	9.36E+12	7.08E+12	4.69E+12	3.12E+12	2.10E+12	8.58E+11
2500	4630	1.32E+13	1.06E+13	8.20E+12	6.14E+12	4.02E+12	2.66E+12	1.79E+12	7.29E+11
2750	5093	1.15E+13	9.27E+12	7.13E+12	5.33E+12	3.51E+12	2.34E+12	1.59E+12	6.57E+11
3000	5556	1.04E+13	8.41E+12	6.50E+12	4.91E+12	3.29E+12	2.23E+12	1.53E+12	6.50E+11
3500	6482	9.94E+12	8.20E+12	6.52E+12	5.08E+12	3.53E+12	2.47E+12	1.74E+12	7.61E+11
4000	7408	1.07E+13	9.03E+12	7.35E+12	5.85E+12	4.17E+12	2.97E+12	2.11E+12	9.37E+11
4500	8334	1.17E+13	1.01E+13	8.35E+12	6.77E+12	4.94E+12	3.58E+12	2.57E+12	1.16E+12
5000	9260	1.30E+13	1.13E+13	9.58E+12	7.92E+12	5.92E+12	4.36E+12	3.18E+12	1.45E+12
5500	10186	1.45E+13	1.28E+13	1.09E+13	9.17E+12	6.98E+12	5.22E+12	3.85E+12	1.79E+12
6000	11112	1.61E+13	1.43E+13	1.24E+13	1.05E+13	8.05E+12	6.08E+12	4.52E+12	2.15E+12
7000	12964	2.06E+13	1.86E+13	1.62E+13	1.38E+13	1.07E+13	8.21E+12	6.17E+12	3.00E+12
8000	14816	2.65E+13	2.39E+13	2.09E+13	1.79E+13	1.40E+13	1.07E+13	8.03E+12	3.92E+12
9000	16668	3.28E+13	2.96E+13	2.59E+13	2.21E+13	1.73E+13	1.32E+13	9.88E+12	4.76E+12
10000	18520	4.01E+13	3.62E+13	3.16E+13	2.69E+13	2.09E+13	1.58E+13	1.18E+13	5.51E+12
11000	20372	4.58E+13	4.11E+13	3.56E+13	3.02E+13	2.32E+13	1.74E+13	1.27E+13	5.80E+12
12000	22224	4.88E+13	4.35E+13	3.74E+13	3.13E+13	2.37E+13	1.74E+13	1.26E+13	5.48E+12
13000	24076	4.69E+13	4.15E+13	3.53E+13	2.92E+13	2.16E+13	1.55E+13	1.10E+13	4.54E+12
14000	25928	4.17E+13	3.65E+13	3.07E+13	2.51E+13	1.81E+13	1.28E+13	8.81E+12	3.51E+12
15000	27780	3.43E+13	2.97E+13	2.47E+13	1.98E+13	1.40E+13	9.65E+12	6.52E+12	2.51E+12
16000	29632	2.58E+13	2.21E+13	1.81E+13	1.44E+13	1.00E+13	6.76E+12	4.49E+12	1.67E+12
17000	31484	2.06E+13	1.75E+13	1.42E+13	1.12E+13	7.62E+12	5.06E+12	3.30E+12	1.18E+12
18000	33336	1.39E+13	1.17E+13	9.38E+12	7.27E+12	4.88E+12	3.18E+12	2.04E+12	6.91E+11
19327	35793	8.59E+12	7.11E+12	5.60E+12	4.25E+12	2.76E+12	1.74E+12	1.08E+12	3.46E+11

Table 9-36. Annual Equivalent 1 MeV Electron Fluence for GaAs Solar Cells from Trapped Protons During Solar Max and Min, 60° Inclination (Infinite Backshielding)

PROTONS - V_{oc}

INCLINATION = 60 DEGREES

EQUIV. 1 MEV ELECTRON FLUENCE FOR V_{oc} CIRCULAR ORBIT
DUE TO GEOMAGNETICALLY TRAPPED PROTONS, MODEL AP8MAX

ALTITUDE (nm)	(km)	SHIELD THICKNESS, cm (mils)							
		0 (0)	2.54E-3 (1)	7.64E-3 (3)	1.52E-2 (6)	3.05E-2 (12)	5.09E-2 (20)	7.64E-2 (30)	1.52E-1 (60)
150	277	6.62E+13	1.19E+11	1.88E+10	9.42E+09	6.81E+09	5.83E+09	5.13E+09	4.51E+09
250	463	7.44E+14	1.58E+12	4.00E+11	2.60E+11	1.92E+11	1.64E+11	1.42E+11	1.22E+11
300	555	1.81E+15	4.43E+12	1.15E+12	7.19E+11	4.99E+11	4.14E+11	3.47E+11	2.92E+11
450	833	9.32E+15	3.05E+13	9.39E+12	5.39E+12	3.31E+12	2.56E+12	2.02E+12	1.60E+12
600	1111	2.61E+16	1.04E+14	3.37E+13	1.95E+13	1.20E+13	9.11E+12	6.97E+12	5.33E+12
800	1481	7.84E+16	3.30E+14	1.16E+14	7.18E+13	4.65E+13	3.51E+13	2.63E+13	1.94E+13
1000	1852	2.14E+17	9.13E+14	3.41E+14	2.17E+14	1.40E+14	1.01E+14	7.12E+13	4.90E+13
1250	2315	4.82E+17	2.53E+15	1.02E+15	6.59E+14	4.15E+14	2.76E+14	1.75E+14	1.06E+14
1500	2778	8.07E+17	5.41E+15	2.37E+15	1.54E+15	9.46E+14	5.84E+14	3.29E+14	1.69E+14
1750	3241	1.16E+18	1.02E+16	4.70E+15	3.03E+15	1.79E+15	1.04E+15	5.23E+14	2.26E+14
2000	3704	1.59E+18	1.78E+16	8.24E+15	5.06E+15	2.77E+15	1.49E+15	6.75E+14	2.53E+14
2250	4167	2.03E+18	2.99E+16	1.35E+16	7.71E+15	3.76E+15	1.85E+15	7.55E+14	2.54E+14
2500	4630	2.54E+18	4.58E+16	1.96E+16	1.03E+16	4.48E+15	2.03E+15	7.04E+14	2.34E+14
2750	5093	3.23E+18	6.51E+16	2.57E+16	1.25E+16	4.85E+15	2.04E+15	6.24E+14	2.05E+14
3000	5556	4.10E+18	8.77E+16	3.13E+16	1.40E+16	4.82E+15	1.93E+15	4.40E+14	1.73E+14
3500	6482	6.62E+18	1.43E+17	4.01E+16	1.48E+16	3.86E+15	1.45E+15	2.78E+14	1.13E+14
4000	7408	1.08E+19	2.08E+17	4.56E+16	1.40E+16	2.82E+15	9.88E+14	1.56E+14	6.90E+13
4500	8334	1.65E+19	2.62E+17	4.52E+16	1.19E+16	1.82E+15	6.05E+14	8.10E+13	3.67E+13
5000	9260	2.37E+19	2.98E+17	4.09E+16	9.10E+15	1.14E+15	3.40E+14	4.05E+13	1.79E+13
5500	10186	3.11E+19	3.21E+17	3.55E+16	6.58E+15	6.32E+14	1.81E+14	1.68E+13	8.17E+12
6000	11112	3.75E+19	3.09E+17	2.77E+16	4.35E+15	3.27E+14	8.44E+13	2.05E+12	3.00E+12
7005	12973	4.75E+19	2.14E+17	1.29E+16	1.45E+15	6.67E+13	1.36E+13	1.51E+11	2.80E+11
8000	14816	4.79E+19	1.03E+17	3.87E+15	3.12E+14	8.83E+12	1.37E+12	8.19E+09	1.56E+10
9000	16668	3.36E+19	3.34E+16	7.20E+14	3.84E+13	5.93E+11	7.24E+10	0.00E+00	1.71E+09
10000	18520	2.41E+19	9.25E+15	1.15E+14	4.30E+12	2.98E+10	0.00E+00	0.00E+00	0.00E+00
11000	20372	1.76E+19	2.02E+15	8.84E+12	1.90E+11	9.45E+08	0.00E+00	0.00E+00	0.00E+00
12000	22224	1.41E+19	3.77E+14	1.64E+11	0.00E+00	0.00E+00	0.00E+00	0.00E+00	0.00E+00
13000	24076	1.15E+19	5.37E+13	4.80E+07	0.00E+00	0.00E+00	0.00E+00	0.00E+00	0.00E+00
14000	25928	9.65E+18	9.39E+12	2.15E+07	0.00E+00	0.00E+00	0.00E+00	0.00E+00	0.00E+00
15000	27780	7.95E+18	2.47E+12	1.15E+07	0.00E+00	0.00E+00	0.00E+00	0.00E+00	0.00E+00
16000	29632	6.07E+18	2.85E+10	0.00E+00	0.00E+00	0.00E+00	0.00E+00	0.00E+00	0.00E+00
17000	31484	5.02E+18	2.19E+10	0.00E+00	0.00E+00	0.00E+00	0.00E+00	0.00E+00	0.00E+00
18000	33336	3.29E+18	1.58E+10	0.00E+00	0.00E+00	0.00E+00	0.00E+00	0.00E+00	0.00E+00
19327	35793	1.00E+18	1.05E+10	0.00E+00	0.00E+00	0.00E+00	0.00E+00	0.00E+00	0.00E+00

EQUIV. 1 MEV ELECTRON FLUENCE FOR V_{oc} CIRCULAR ORBIT
DUE TO GEOMAGNETICALLY TRAPPED PROTONS, MODEL AP8MIN

INCLINATION = 60 DEGREES

ALTITUDE (nm)	(km)	0 (0)	2.54E-3 (1)	7.64E-3 (3)	1.52E-2 (6)	3.05E-2 (12)	5.09E-2 (20)	7.64E-2 (30)	1.52E-1 (60)
150	277	6.84E+13	6.83E+11	2.00E+11	9.12E+10	4.06E+10	2.82E+10	2.09E+10	1.65E+10
250	463	6.91E+14	6.36E+12	1.95E+12	9.52E+11	4.77E+11	3.50E+11	2.70E+11	2.16E+11
300	555	1.63E+15	1.30E+13	3.90E+12	1.93E+12	1.00E+12	7.45E+11	5.81E+11	4.67E+11
450	833	8.64E+15	5.02E+13	1.56E+13	8.15E+12	4.55E+12	3.45E+12	2.70E+12	2.14E+12
600	1111	2.42E+16	1.31E+14	4.29E+13	2.39E+13	1.43E+13	1.10E+13	8.53E+12	6.61E+12
800	1481	7.70E+16	3.96E+14	1.37E+14	8.21E+13	5.21E+13	3.97E+13	3.02E+13	2.25E+13
1000	1852	2.12E+17	1.04E+15	3.78E+14	2.33E+14	1.48E+14	1.08E+14	7.70E+13	5.33E+13

Table 9-37.   Annual Equivalent 1 MeV Electron Fluence for GaAs Solar Cells from Trapped Protons During Solar Max and Min, 60° Inclination (Infinite Backshielding)

PROTONS - P_max

EQUIV. 1 MEV ELECTRON FLUENCE FOR P_max CIRCULAR ORBIT

DUE TO GEOMAGNETICALLY TRAPPED PROTONS, MODEL AP8MAX

INCLINATION = 60 DEGREES

ALTITUDE (nm)	(km)	SHIELD THICKNESS, cm (mils)							
		0 (0)	2.54E-3 (1)	7.64E-3 (3)	1.52E-2 (6)	3.05E-2 (12)	5.09E-2 (20)	7.64E-2 (30)	1.52E-1 (60)
150	277	4.73E+13	8.53E+10	1.34E+10	6.73E+09	4.86E+09	4.17E+09	3.66E+09	3.22E+09
250	463	5.31E+14	1.13E+12	2.85E+11	1.86E+11	1.37E+11	1.17E+11	1.01E+11	8.71E+10
300	555	1.29E+15	3.16E+12	8.24E+11	5.14E+11	3.57E+11	2.96E+11	2.48E+11	2.09E+11
450	833	6.66E+15	2.18E+13	6.71E+12	3.85E+12	2.37E+12	1.83E+12	1.44E+12	1.14E+12
600	1111	1.87E+16	7.41E+13	2.41E+13	1.39E+13	8.58E+12	6.51E+12	4.98E+12	3.80E+12
800	1481	5.60E+16	2.36E+14	8.31E+13	5.13E+13	3.32E+13	2.51E+13	1.88E+13	1.38E+13
1000	1852	1.53E+17	6.52E+14	2.44E+14	1.55E+14	1.00E+14	7.21E+13	5.09E+13	3.50E+13
1250	2315	3.45E+17	1.80E+15	7.30E+14	2.97E+14	2.97E+14	1.97E+14	1.25E+14	7.55E+13
1500	2778	5.77E+17	3.87E+15	1.69E+15	4.71E+14	6.76E+14	4.17E+14	2.35E+14	1.21E+14
1750	3241	8.30E+17	7.29E+15	3.36E+15	1.10E+15	1.28E+15	7.40E+14	3.74E+14	1.62E+14
2000	3704	1.14E+18	1.27E+16	5.88E+15	2.16E+15	1.98E+15	1.06E+15	4.82E+14	1.81E+14
2250	4167	1.45E+18	2.14E+16	9.66E+15	3.62E+15	2.69E+15	1.32E+15	5.39E+14	1.81E+14
2500	4630	1.81E+18	3.27E+16	1.40E+16	5.51E+15	3.20E+15	1.45E+15	5.39E+14	1.67E+14
2750	5093	2.31E+18	4.65E+16	1.84E+16	7.38E+15	3.46E+15	1.46E+15	5.03E+14	1.46E+14
3000	5556	2.93E+18	6.26E+16	2.24E+16	8.95E+15	3.44E+15	1.38E+15	4.46E+14	1.24E+14
3500	6482	4.73E+18	1.02E+17	2.87E+16	1.00E+16	2.75E+15	1.04E+15	3.14E+14	8.10E+13
4000	7408	7.68E+18	1.49E+17	3.26E+16	1.06E+16	2.01E+15	7.06E+14	1.99E+14	4.93E+13
4500	8334	1.18E+19	1.87E+17	3.23E+16	1.00E+16	1.35E+15	4.32E+14	1.12E+14	2.62E+13
5000	9260	1.69E+19	2.13E+17	2.92E+16	8.48E+15	8.15E+14	2.43E+14	5.78E+13	1.28E+13
5500	10186	2.22E+19	2.29E+17	2.53E+16	6.50E+15	4.52E+14	1.29E+14	2.89E+13	5.84E+12
6000	11112	2.68E+19	2.21E+17	1.98E+16	4.70E+15	2.33E+14	6.03E+13	1.20E+13	2.14E+12
7005	12973	3.39E+19	1.53E+17	9.19E+15	3.10E+15	4.77E+13	9.71E+12	1.47E+12	2.00E+11
8000	14816	3.42E+19	7.36E+16	2.77E+15	1.03E+15	6.31E+12	9.75E+11	1.08E+11	1.11E+10
9000	16668	2.40E+19	2.38E+16	5.14E+14	2.23E+14	4.23E+11	5.17E+10	5.85E+09	1.22E+09
10000	18520	1.72E+19	6.61E+15	8.22E+13	2.75E+13	2.13E+10	0.00E+00	0.00E+00	0.00E+00
11000	20372	1.26E+19	1.44E+15	6.31E+12	3.07E+12	6.75E+08	0.00E+00	0.00E+00	0.00E+00
12000	22224	1.01E+19	2.70E+14	1.17E+11	1.36E+11	0.00E+00	0.00E+00	0.00E+00	0.00E+00
13000	24076	8.22E+18	3.83E+13	3.43E+07	0.00E+00	0.00E+00	0.00E+00	0.00E+00	0.00E+00
14000	25928	6.89E+18	6.70E+12	1.53E+07	0.00E+00	0.00E+00	0.00E+00	0.00E+00	0.00E+00
15000	27780	5.68E+18	1.76E+12	8.19E+06	0.00E+00	0.00E+00	0.00E+00	0.00E+00	0.00E+00
16000	29632	4.33E+18	2.03E+10	0.00E+00	0.00E+00	0.00E+00	0.00E+00	0.00E+00	0.00E+00
17000	31484	3.59E+18	1.56E+10	0.00E+00	0.00E+00	0.00E+00	0.00E+00	0.00E+00	0.00E+00
18000	33336	2.35E+18	1.13E+10	0.00E+00	0.00E+00	0.00E+00	0.00E+00	0.00E+00	0.00E+00
19327	35793	7.17E+17	7.52E+09	0.00E+00	0.00E+00	0.00E+00	0.00E+00	0.00E+00	0.00E+00

EQUIV. 1 MEV ELECTRON FLUENCE FOR P_max CIRCULAR ORBIT

DUE TO GEOMAGNETICALLY TRAPPED PROTONS, MODEL AP8MIN

INCLINATION = 60 DEGREES

ALTITUDE (nm)	(km)	0 (0)	2.54E-3 (1)	7.64E-3 (3)	1.52E-2 (6)	3.05E-2 (12)	5.09E-2 (20)	7.64E-2 (30)	1.52E-1 (60)
150	277	4.88E+13	4.88E+11	1.43E+11	6.52E+10	2.90E+10	2.02E+10	1.50E+10	1.18E+10
250	463	4.93E+14	4.54E+12	1.39E+12	6.80E+11	3.41E+11	2.50E+11	1.93E+11	1.54E+11
300	555	1.17E+15	9.26E+12	2.78E+12	1.38E+12	7.15E+11	5.32E+11	4.15E+11	3.34E+11
450	833	6.17E+15	3.58E+13	1.11E+13	5.82E+12	3.25E+12	2.46E+12	1.93E+12	1.53E+12
600	1111	1.73E+16	9.38E+13	3.06E+13	1.71E+13	1.02E+13	7.84E+12	6.09E+12	4.72E+12
800	1481	5.50E+16	2.83E+14	9.82E+13	5.87E+13	3.72E+13	2.84E+13	2.16E+13	1.60E+13
1000	1852	1.52E+17	7.40E+14	2.70E+14	1.67E+14	1.06E+14	7.71E+13	5.50E+13	3.81E+13

Table 9-38. Annual Equivalent 1 MeV Electron Fluence for GaAs Solar Cells from Trapped Protons During Solar Max and Min, 60° Inclination (Infinite Backshielding)

PROTONS - I$_{sc}$

EQUIV. 1 MEV PROTON FLUENCE FOR I$_{sc}$ - CIRCULAR ORBIT
DUE TO GEOMAGNETICALLY TRAPPED PROTONS, MODEL AP8MAX

INCLINATION = 60 DEGREES

ALTITUDE					SHIELD THICKNESS, cm (mils)			
(nm)	(km)	0 (0)	2.54E-3 (1)	7.64E-3 (3)	1.52E-2 (6)	3.05E-2 (12)	5.09E-2 (20)	7.64E-2 (30)	1.52E-1 (60)
150	277	5.65E+13	4.65E+10	5.65E+09	2.16E+09	1.27E+09	9.66E+08	7.54E+08	6.13E+08
250	463	6.34E+14	5.89E+11	1.12E+11	6.05E+10	3.78E+10	2.91E+10	2.23E+10	1.77E+10
300	555	1.51E+15	1.66E+12	3.38E+11	1.79E+11	1.04E+11	7.77E+10	5.74E+10	4.41E+10
450	833	7.99E+15	1.14E+13	2.95E+12	1.48E+12	7.75E+11	5.35E+11	3.65E+11	2.58E+11
600	1111	2.23E+16	3.85E+13	1.07E+13	5.49E+12	2.94E+12	2.01E+12	1.34E+12	8.98E+11
800	1481	6.72E+16	1.21E+14	3.64E+13	2.02E+13	1.17E+13	8.05E+12	5.30E+12	3.41E+12
1000	1852	1.83E+17	3.35E+14	1.08E+14	6.31E+13	3.74E+13	2.48E+13	1.54E+13	9.18E+12
1250	2315	4.07E+17	9.27E+14	3.33E+14	2.02E+14	1.20E+14	7.47E+13	4.20E+13	2.20E+13
1500	2778	6.65E+17	1.98E+15	7.87E+14	4.89E+14	2.89E+14	1.70E+14	8.70E+13	3.90E+13
1750	3241	9.37E+17	3.74E+15	1.59E+15	9.83E+14	5.67E+14	3.16E+14	1.48E+14	5.71E+13
2000	3704	1.26E+18	6.52E+15	2.82E+15	1.67E+15	8.97E+14	4.67E+14	1.99E+14	6.76E+13
2250	4167	1.56E+18	1.11E+16	4.70E+15	2.59E+15	1.24E+15	5.92E+14	2.28E+14	7.04E+13
2500	4630	1.91E+18	1.70E+16	6.88E+15	3.50E+15	1.49E+15	6.58E+14	2.31E+14	6.68E+13
2750	5093	2.38E+18	2.44E+16	9.11E+15	4.28E+15	1.62E+15	6.68E+14	2.18E+14	5.94E+13
3000	5556	2.96E+18	3.32E+16	1.12E+16	4.82E+15	1.62E+15	6.33E+14	1.94E+14	5.08E+13
3500	6482	4.62E+18	5.51E+16	1.45E+16	5.13E+15	1.30E+15	4.81E+14	1.38E+14	3.40E+13
4000	7408	7.33E+18	8.15E+16	1.67E+16	4.91E+15	9.58E+14	3.28E+14	8.79E+13	2.09E+13
4500	8334	1.12E+19	1.19E+17	1.66E+16	4.17E+15	6.44E+14	2.02E+14	4.95E+13	1.12E+13
5000	9260	1.63E+19	1.30E+17	1.51E+16	3.21E+15	3.91E+14	1.14E+14	2.58E+13	5.53E+12
5500	10186	2.11E+19	1.26E+17	1.32E+16	2.33E+15	2.17E+14	6.07E+13	1.30E+13	2.54E+12
6000	11112	2.50E+19	1.04E+17	1.04E+16	1.54E+15	1.13E+14	2.85E+13	5.43E+12	9.41E+11
7005	12973	3.23E+19	8.87E+16	4.84E+15	5.17E+14	2.32E+13	4.63E+12	6.70E+11	8.92E+10
8000	14816	3.46E+19	4.33E+16	1.47E+15	1.12E+14	3.09E+12	4.68E+11	4.94E+10	4.92E+09
9000	16668	2.52E+19	1.43E+16	2.75E+14	1.39E+13	2.08E+11	2.48E+10	2.64E+09	5.47E+08
10000	18520	1.88E+19	4.03E+15	4.41E+13	1.56E+12	1.61E+10	0.00E+00	0.00E+00	0.00E+00
11000	20372	1.46E+19	8.98E+14	3.43E+12	6.91E+10	3.61E+08	0.00E+00	0.00E+00	0.00E+00
12000	22224	1.26E+19	1.72E+14	6.59E+10	0.00E+00	0.00E+00	0.00E+00	0.00E+00	0.00E+00
13000	24076	1.11E+19	2.51E+13	2.10E+07	0.00E+00	0.00E+00	0.00E+00	0.00E+00	0.00E+00
14000	25928	9.91E+18	4.52E+12	9.21E+06	0.00E+00	0.00E+00	0.00E+00	0.00E+00	0.00E+00
15000	27780	8.71E+18	1.22E+12	4.84E+06	0.00E+00	0.00E+00	0.00E+00	0.00E+00	0.00E+00
16000	29632	6.97E+18	1.69E+10	0.00E+00	0.00E+00	0.00E+00	0.00E+00	0.00E+00	0.00E+00
17000	31484	6.03E+18	1.29E+10	0.00E+00	0.00E+00	0.00E+00	0.00E+00	0.00E+00	0.00E+00
18000	33336	4.10E+18	9.22E+09	0.00E+00	0.00E+00	0.00E+00	0.00E+00	0.00E+00	0.00E+00
19327	35793	1.25E+18	6.05E+09	0.00E+00	0.00E+00	0.00E+00	0.00E+00	0.00E+00	0.00E+00

EQUIV. 1 MEV PROTON FLUENCE FOR I$_{sc}$ - CIRCULAR ORBIT
DUE TO GEOMAGNETICALLY TRAPPED PROTONS, MODEL AP8MIN

INCLINATION = 60 DEGREES

ALTITUDE					SHIELD THICKNESS, cm (mils)			
(nm)	(km)	0 (0)	2.54E-3 (1)	7.64E-3 (3)	1.52E-2 (6)	3.05E-2 (12)	5.09E-2 (20)	7.64E-2 (30)	1.52E-1 (60)
150	277	5.08E+13	2.59E+11	6.78E+10	2.78E+10	1.02E+10	6.11E+09	3.79E+09	2.63E+09
250	463	5.25E+14	2.39E+12	6.45E+11	2.79E+11	1.15E+11	7.30E+10	4.78E+10	3.39E+10
300	555	1.25E+15	4.88E+12	1.28E+12	5.59E+11	2.37E+11	1.54E+11	1.03E+11	7.35E+10
450	833	7.07E+15	1.88E+13	5.03E+12	2.31E+12	1.07E+12	7.18E+11	4.84E+11	3.40E+11
600	1111	2.02E+16	4.87E+13	1.36E+13	6.70E+12	3.43E+12	2.36E+12	1.59E+12	1.09E+12
800	1481	6.48E+16	1.46E+14	4.33E+13	2.30E+13	1.29E+13	8.95E+12	5.97E+12	3.88E+12
1000	1852	1.80E+17	3.81E+14	1.21E+14	6.79E+13	3.93E+13	2.63E+13	1.65E+13	9.85E+12

Table 9-39. Annual Equivalent 1 MeV Electron Fluence for GaAs Solar Cells from Trapped Electrons During Solar Max, 70° Inclination (Infinite Backshielding)

ELECTRONS – I_{sc}, V_{oc}, AND P_{max} INCLINATION = 70 DEGREES

EQUIV. 1 MEV ELECTRON FLUENCE FOR I_{sc} - CIRCULAR ORBIT
DUE TO GEOMAGNETICALLY TRAPPED ELECTRONS - MODEL AE8MAX

ALTITUDE (nm)	(km)	0 (0)	2.54E-3 (1)	7.64E-3 (3)	SHIELD THICKNESS, cm (mils) 1.52E-2 (6)	3.05E-2 (12)	5.09E-2 (20)	7.64E-2 (30)	1.52E-1 (60)
150	277	8.33E+10	7.45E+10	6.47E+10	5.49E+10	4.25E+10	3.23E+10	2.42E+10	1.18E+10
250	463	1.72E+11	1.51E+11	1.29E+11	1.08E+11	8.21E+10	6.17E+10	4.59E+10	2.21E+10
300	555	2.42E+11	2.11E+11	1.78E+11	1.48E+11	1.11E+11	8.32E+10	6.16E+10	2.94E+10
450	833	5.42E+11	4.92E+11	4.03E+11	3.23E+11	2.35E+11	1.72E+11	1.25E+11	5.87E+10
600	1111	1.34E+12	1.09E+12	8.54E+11	6.55E+11	4.51E+11	3.17E+11	2.26E+11	1.03E+11
800	1481	3.82E+12	3.00E+12	2.25E+12	1.64E+12	1.05E+12	7.06E+11	4.86E+11	2.11E+11
1000	1852	9.02E+12	7.01E+12	5.16E+12	3.69E+12	2.31E+12	1.51E+12	1.02E+12	4.31E+11
1250	2315	1.80E+13	1.40E+13	1.03E+13	7.35E+12	4.59E+12	2.99E+12	2.02E+12	8.49E+11
1500	2778	2.59E+13	2.01E+13	1.47E+13	1.04E+13	6.46E+12	4.17E+12	2.79E+12	1.17E+12
1750	3241	3.26E+13	2.50E+13	1.81E+13	1.26E+13	7.53E+12	4.67E+12	3.03E+12	1.24E+12
2000	3704	3.76E+13	2.86E+13	2.04E+13	1.39E+13	7.89E+12	4.62E+12	2.87E+12	1.12E+12
2250	4167	4.14E+13	3.13E+13	2.21E+13	1.47E+13	8.04E+12	4.45E+12	2.63E+12	9.84E+11
2500	4630	4.28E+13	3.23E+13	2.26E+13	1.50E+13	7.95E+12	4.24E+12	2.43E+12	8.90E+11
2750	5093	4.17E+13	3.15E+13	2.21E+13	1.46E+13	7.74E+12	4.07E+12	2.32E+12	8.60E+11
3000	5556	3.92E+13	2.97E+13	2.10E+13	1.40E+13	7.50E+12	4.01E+12	2.33E+12	8.97E+11
3500	6482	3.34E+13	2.57E+13	1.85E+13	1.28E+13	7.32E+12	4.27E+12	2.68E+12	1.13E+12
4000	7408	3.05E+13	2.41E+13	1.80E+13	1.30E+13	8.04E+12	5.08E+12	3.38E+12	1.49E+12
4500	8334	2.98E+13	2.43E+13	1.90E+13	1.44E+13	9.56E+12	6.42E+12	4.43E+12	2.00E+12
5000	9260	2.89E+13	2.45E+13	2.00E+13	1.59E+13	1.13E+13	8.02E+12	5.72E+12	2.64E+12
5500	10186	2.99E+13	2.61E+13	2.21E+13	1.83E+13	1.36E+13	1.01E+13	7.35E+12	3.47E+12
6000	11112	3.25E+13	2.90E+13	2.50E+13	2.12E+13	1.62E+13	1.22E+13	9.10E+12	4.38E+12
7000	12964	4.20E+13	3.78E+13	3.30E+13	2.82E+13	2.20E+13	1.69E+13	1.27E+13	6.22E+12
8005	14825	5.09E+13	4.59E+13	4.01E+13	3.43E+13	2.68E+13	2.05E+13	1.54E+13	7.53E+12
9000	16668	5.76E+13	5.18E+13	4.52E+13	3.85E+13	2.99E+13	2.27E+13	1.70E+13	8.20E+12
10000	18520	6.24E+13	5.59E+13	4.86E+13	4.12E+13	3.17E+13	2.38E+13	1.76E+13	8.20E+12
11000	20372	6.02E+13	5.37E+13	4.63E+13	3.90E+13	2.96E+13	2.20E+13	1.59E+13	7.07E+12
12000	22224	5.45E+13	4.83E+13	4.13E+13	3.45E+13	2.58E+13	1.88E+13	1.34E+13	5.66E+12
13000	24076	4.66E+13	4.10E+13	3.47E+13	2.86E+13	2.10E+13	1.50E+13	1.04E+13	4.26E+12
14000	25928	3.90E+13	3.40E+13	2.85E+13	2.32E+13	1.67E+13	1.17E+13	8.01E+12	3.17E+12
15000	27780	3.14E+13	2.71E+13	2.25E+13	1.81E+13	1.28E+13	8.76E+12	5.91E+12	2.26E+12
16000	29632	2.35E+13	2.01E+13	1.65E+13	1.31E+13	9.09E+12	6.14E+12	4.07E+12	1.51E+12
17000	31484	1.88E+13	1.59E+13	1.30E+13	1.02E+13	6.95E+12	4.61E+12	3.01E+12	1.08E+12
18000	33336	1.27E+13	1.06E+13	8.55E+12	6.63E+12	4.44E+12	2.90E+12	1.86E+12	6.29E+11
19327	35793	7.85E+12	6.50E+12	5.12E+12	3.88E+12	2.52E+12	1.59E+12	9.86E+11	3.15E+11

Table 9-40. Annual Equivalent 1 MeV Electron Fluence for GaAs Solar Cells from Trapped Electrons During Solar Min, 70° Inclination (Infinite Backshielding)

ELECTRONS - I_{sc}, V_{oc}, AND P_{max} INCLINATION = 70 DEGREES

EQUIV. 1 MEV ELECTRON FLUENCE FOR I_{sc} - CIRCULAR ORBIT
DUE TO GEOMAGNETICALLY TRAPPED ELECTRONS - MODEL AE8MIN

ALTITUDE (nm)	(km)	SHIELD THICKNESS, cm (mils)							
		0 (0)	2.54E-3 (1)	7.64E-3 (3)	1.52E-2 (6)	3.05E-2 (12)	5.09E-2 (20)	7.64E-2 (30)	1.52E-1 (60)
150	277	3.60E+10	3.24E+10	2.83E+10	2.42E+10	1.89E+10	1.45E+10	1.09E+10	5.30E+09
250	463	7.71E+10	6.89E+10	5.96E+10	5.05E+10	3.90E+10	2.96E+10	2.21E+10	1.06E+10
300	555	1.11E+11	9.87E+10	8.50E+10	7.16E+10	5.49E+10	4.14E+10	3.08E+10	1.46E+10
450	833	2.77E+11	2.41E+11	2.04E+11	1.68E+11	1.26E+11	9.29E+10	6.79E+10	3.15E+10
600	1111	6.57E+11	5.59E+11	4.58E+11	3.67E+11	2.64E+11	1.89E+11	1.35E+11	6.12E+10
800	1481	1.98E+12	1.65E+12	1.32E+12	1.03E+12	7.08E+11	4.90E+11	3.41E+11	1.48E+11
1000	1852	4.88E+12	4.05E+12	3.22E+12	2.49E+12	1.71E+12	1.17E+12	8.06E+11	3.44E+11
1250	2315	9.87E+12	8.23E+12	6.59E+12	5.15E+12	3.57E+12	2.47E+12	1.71E+12	7.38E+11
1500	2778	1.33E+13	1.11E+13	8.97E+12	7.05E+12	4.92E+12	3.43E+12	2.40E+12	1.04E+12
1750	3241	1.47E+13	1.22E+13	9.80E+12	7.65E+12	5.28E+12	3.65E+12	2.53E+12	1.08E+12
2000	3704	1.44E+13	1.19E+13	9.39E+12	7.21E+12	4.86E+12	3.28E+12	2.23E+12	9.26E+11
2250	4167	1.35E+13	1.09E+13	8.49E+12	6.40E+12	4.22E+12	2.79E+12	1.86E+12	7.51E+11
2500	4630	1.19E+13	9.53E+12	7.31E+12	5.44E+12	3.53E+12	2.31E+12	1.53E+12	6.12E+11
2750	5093	1.02E+13	8.13E+12	6.21E+12	4.60E+12	2.98E+12	1.96E+12	1.31E+12	5.26E+11
3000	5556	9.01E+12	7.21E+12	5.51E+12	4.11E+12	2.70E+12	1.80E+12	1.22E+12	5.00E+11
3500	6482	8.25E+12	6.73E+12	5.29E+12	4.06E+12	2.77E+12	1.91E+12	1.33E+12	5.70E+11
4000	7408	8.72E+12	7.29E+12	5.87E+12	4.63E+12	3.26E+12	2.29E+12	1.62E+12	7.12E+11
4500	8334	9.51E+12	8.11E+12	6.69E+12	5.40E+12	3.92E+12	2.82E+12	2.02E+12	9.13E+11
5000	9260	1.05E+13	9.11E+12	7.68E+12	6.33E+12	4.73E+12	3.48E+12	2.54E+12	1.17E+12
5500	10186	1.19E+13	1.05E+13	8.97E+12	7.52E+12	5.72E+12	4.29E+12	3.17E+12	1.49E+12
6000	11112	1.34E+13	1.20E+13	1.04E+13	8.76E+12	6.76E+12	5.13E+12	3.82E+12	1.83E+12
7000	12964	1.78E+13	1.60E+13	1.40E+13	1.19E+13	9.33E+12	7.15E+12	5.38E+12	2.63E+12
8000	14816	2.31E+13	2.09E+13	1.83E+13	1.57E+13	1.23E+13	9.42E+12	7.10E+12	3.48E+12
9000	16668	2.91E+13	2.63E+13	2.30E+13	1.97E+13	1.54E+13	1.18E+13	8.83E+12	4.27E+12
10000	18520	3.61E+13	3.26E+13	2.84E+13	2.43E+13	1.89E+13	1.43E+13	1.06E+13	5.00E+12
11000	20372	4.10E+13	3.67E+13	3.19E+13	2.70E+13	2.08E+13	1.56E+13	1.15E+13	5.22E+12
12000	22224	4.40E+13	3.92E+13	3.38E+13	2.83E+13	2.14E+13	1.58E+13	1.14E+13	4.97E+12
13000	24076	4.25E+13	3.76E+13	3.20E+13	2.65E+13	1.96E+13	1.41E+13	9.96E+12	4.13E+12
14000	25928	3.79E+13	3.32E+13	2.79E+13	2.28E+13	1.65E+13	1.16E+13	8.02E+12	3.19E+12
15000	27780	3.13E+13	2.71E+13	2.25E+13	1.81E+13	1.28E+13	8.81E+12	5.95E+12	2.29E+12
16000	29632	2.35E+13	2.02E+13	1.66E+13	1.32E+13	9.15E+12	6.18E+12	4.11E+12	1.53E+12
17000	31484	1.88E+13	1.60E+13	1.30E+13	1.02E+13	6.98E+12	4.64E+12	3.03E+12	1.08E+12
18000	33336	1.27E+13	1.07E+13	8.59E+12	6.66E+12	4.47E+12	2.92E+12	1.87E+12	6.34E+11
19327	35793	7.93E+12	6.57E+12	5.17E+12	3.92E+12	2.55E+12	1.61E+12	1.00E+12	3.19E+11

Table 9-41. Annual Equivalent 1 MeV Electron Fluence for GaAs Solar Cells from Trapped Protons During Solar Max and Min, 70° Inclination (Infinite Backshielding)

PROTONS - V_{oc}

EQUIV. 1 MEV ELECTRON FLUENCE FOR V_{oc} CIRCULAR ORBIT   INCLINATION = 70 DEGREES
DUE TO GEOMAGNETICALLY TRAPPED PROTONS, MODEL AP8MAX

ALTITUDE (nm)	(km)	SHIELD THICKNESS, cm (mils)							
		0 (0)	2.54E-3 (1)	7.64E-3 (3)	1.52E-2 (6)	3.05E-2 (12)	5.09E-2 (20)	7.64E-2 (30)	1.52E-1 (60)
150	277	4.81E+13	9.69E+10	1.57E+10	8.04E+09	5.90E+09	5.11E+09	4.53E+09	4.03E+09
250	463	5.18E+14	1.26E+12	3.36E+11	2.23E+11	1.66E+11	1.42E+11	1.23E+11	1.06E+11
300	555	1.18E+15	3.07E+12	9.24E+11	6.01E+11	4.25E+11	3.54E+11	2.98E+11	2.51E+11
450	833	8.09E+15	2.39E+13	7.71E+12	4.54E+12	2.86E+12	2.23E+12	1.77E+12	1.42E+12
600	1111	2.64E+16	8.16E+13	2.74E+13	1.64E+13	1.04E+13	8.00E+12	6.17E+12	4.74E+12
800	1481	8.44E+16	2.62E+14	9.65E+13	6.18E+13	4.12E+13	3.14E+13	2.37E+13	1.75E+13
1000	1852	2.07E+17	7.32E+14	2.90E+14	1.89E+14	1.25E+14	9.05E+13	6.42E+13	4.43E+13
1250	2315	4.19E+17	2.08E+15	8.95E+14	5.88E+14	3.75E+14	2.50E+14	1.59E+14	9.63E+13
1500	2778	6.54E+17	4.57E+15	2.11E+15	1.39E+15	8.61E+14	5.33E+14	3.00E+14	1.54E+14
1750	3241	9.04E+17	8.83E+15	4.24E+15	2.75E+15	1.63E+15	9.46E+14	4.78E+14	2.07E+14
2000	3704	1.19E+18	1.56E+16	7.46E+15	4.61E+15	2.53E+15	1.36E+15	6.18E+14	2.32E+14
2250	4167	1.54E+18	2.66E+16	1.23E+16	7.04E+15	3.44E+15	1.69E+15	6.91E+14	2.32E+14
2500	4630	1.95E+18	4.11E+16	1.78E+16	9.45E+15	4.10E+15	1.86E+15	6.92E+14	2.15E+14
2750	5093	2.53E+18	5.87E+16	2.34E+16	1.15E+16	4.44E+15	1.87E+15	6.45E+14	1.87E+14
3000	5556	3.29E+18	7.92E+16	2.85E+16	1.28E+16	4.41E+15	1.76E+15	5.71E+14	1.58E+14
3500	6482	5.50E+18	1.30E+17	3.67E+16	1.35E+16	3.54E+15	1.33E+15	4.03E+14	1.04E+14
4000	7408	9.28E+18	1.90E+17	4.17E+16	1.29E+16	2.59E+15	9.06E+14	2.55E+14	6.32E+13
4500	8334	1.45E+19	2.39E+17	4.14E+16	1.09E+16	1.73E+15	5.55E+14	1.43E+14	3.36E+13
5000	9260	2.11E+19	2.72E+17	3.75E+16	8.35E+15	1.05E+15	3.12E+14	7.43E+13	1.65E+13
5500	10186	2.79E+19	2.94E+17	3.26E+16	6.04E+15	5.80E+14	1.66E+14	3.72E+13	7.50E+12
6000	11112	3.38E+19	2.83E+17	2.54E+16	3.99E+15	3.00E+14	7.75E+13	1.54E+13	2.76E+12
7000	12964	4.31E+19	1.97E+17	1.18E+16	1.33E+15	6.15E+13	1.25E+13	1.90E+12	2.59E+11
8005	14825	4.34E+19	9.38E+16	3.52E+15	2.83E+14	8.00E+12	1.24E+12	1.36E+11	1.41E+10
9000	16668	3.05E+19	3.06E+16	6.60E+14	3.53E+13	5.44E+11	6.65E+10	7.54E+09	1.59E+09
10000	18520	2.20E+19	8.49E+15	1.06E+14	3.94E+12	2.73E+10	0.00E+00	0.00E+00	0.00E+00
11000	20372	1.60E+19	1.87E+15	8.34E+12	1.81E+11	8.98E+08	0.00E+00	0.00E+00	0.00E+00
12000	22224	1.28E+19	3.46E+14	1.51E+11	0.00E+00	0.00E+00	0.00E+00	0.00E+00	0.00E+00
13000	24076	1.05E+19	4.94E+13	4.67E+07	0.00E+00	0.00E+00	0.00E+00	0.00E+00	0.00E+00
14000	25928	8.79E+18	8.57E+12	2.08E+07	0.00E+00	0.00E+00	0.00E+00	0.00E+00	0.00E+00
15000	27780	7.26E+18	2.26E+12	1.11E+07	0.00E+00	0.00E+00	0.00E+00	0.00E+00	0.00E+00
16000	29632	5.56E+18	2.71E+10	0.00E+00	0.00E+00	0.00E+00	0.00E+00	0.00E+00	0.00E+00
17000	31484	4.61E+18	2.08E+10	0.00E+00	0.00E+00	0.00E+00	0.00E+00	0.00E+00	0.00E+00
18000	33336	3.02E+18	1.51E+10	0.00E+00	0.00E+00	0.00E+00	0.00E+00	0.00E+00	0.00E+00
19327	35793	9.30E+17	1.01E+10	0.00E+00	0.00E+00	0.00E+00	0.00E+00	0.00E+00	0.00E+00

EQUIV. 1 MEV ELECTRON FLUENCE FOR V_{oc} CIRCULAR ORBIT   INCLINATION = 70 DEGREES
DUE TO GEOMAGNETICALLY TRAPPED PROTONS, MODEL AP8MIN

ALTITUDE (nm)	(km)	0 (0)	2.54E-3 (1)	7.64E-3 (3)	1.52E-2 (6)	3.05E-2 (12)	5.09E-2 (20)	7.64E-2 (30)	1.52E-1 (60)
150	277	4.54E+13	5.37E+11	1.71E+11	7.96E+10	3.56E+10	2.47E+10	1.82E+10	1.43E+10
250	463	4.73E+14	4.85E+12	1.55E+12	7.77E+11	4.04E+11	3.01E+11	2.36E+11	1.91E+11
300	555	1.05E+15	9.56E+12	3.07E+12	1.58E+12	8.47E+11	6.38E+11	5.02E+11	4.06E+11
450	833	6.97E+15	3.86E+13	1.25E+13	6.73E+12	3.47E+12	2.98E+12	2.36E+12	1.89E+12
600	1111	2.37E+16	1.02E+14	3.44E+13	1.99E+13	1.24E+13	9.63E+12	7.56E+12	5.90E+12
800	1481	8.11E+16	3.11E+14	1.13E+14	7.04E+13	4.60E+13	3.55E+13	2.72E+13	2.03E+13
1000	1852	2.04E+17	8.26E+14	3.20E+14	2.03E+14	1.32E+14	9.68E+13	6.94E+13	4.82E+13

Table 9-42. Annual Equivalent 1 MeV Electron Fluence for GaAs Solar Cells from Trapped Protons During Solar Max and Min, 70° Inclination (Infinite Backshielding)

PROTONS – P$_{max}$

INCLINATION = 70 DEGREES

EQUIV. 1 MEV ELECTRON FLUENCE FOR P$_{max}$ CIRCULAR ORBIT
DUE TO GEOMAGNETICALLY TRAPPED PROTONS, MODEL AP8MAX

ALTITUDE (nm)	(km)	SHIELD THICKNESS, cm (mils)							
		0 (0)	2.54E-3 (1)	7.64E-3 (3)	1.52E-2 (6)	3.05E-2 (12)	5.09E-2 (20)	7.64E-2 (30)	1.52E-1 (60)
150	277	3.44E+13	6.92E+10	1.12E+10	5.74E+09	4.22E+09	3.65E+09	3.23E+09	2.88E+09
250	463	3.70E+14	8.98E+11	2.40E+11	1.59E+11	1.19E+11	1.02E+11	8.77E+10	7.55E+10
300	555	8.44E+14	2.19E+12	6.60E+11	4.29E+11	3.04E+11	2.53E+11	2.13E+11	1.79E+11
450	833	5.78E+15	1.71E+13	5.51E+12	3.24E+12	2.04E+12	1.60E+12	1.27E+12	1.01E+12
600	1111	1.88E+16	5.83E+13	1.96E+13	1.17E+13	7.45E+12	5.71E+12	4.41E+12	3.39E+12
800	1481	6.03E+16	1.87E+14	6.89E+13	4.41E+13	2.94E+13	2.24E+13	1.69E+13	1.25E+13
1000	1852	1.48E+17	5.23E+14	2.07E+14	1.35E+14	8.90E+13	6.46E+13	4.59E+13	3.17E+13
1250	2315	2.99E+17	1.48E+15	6.39E+14	4.20E+14	2.68E+14	1.79E+14	1.13E+14	6.88E+13
1500	2778	4.67E+17	3.27E+15	1.51E+15	9.95E+14	6.15E+14	3.80E+14	2.14E+14	1.10E+14
1750	3241	6.46E+17	6.31E+15	3.03E+15	1.96E+15	1.17E+15	6.76E+14	3.42E+14	1.48E+14
2000	3704	8.53E+17	1.11E+16	5.33E+15	3.29E+15	1.81E+15	9.71E+14	4.42E+14	1.66E+14
2250	4167	1.10E+18	1.90E+16	8.77E+15	5.03E+15	2.46E+15	1.21E+15	4.94E+14	1.66E+14
2500	4630	1.39E+18	2.94E+16	1.27E+16	6.75E+15	2.93E+15	1.33E+15	4.94E+14	1.53E+14
2750	5093	1.81E+18	4.19E+16	1.67E+16	8.18E+15	3.17E+15	1.34E+15	4.61E+14	1.34E+14
3000	5556	2.35E+18	5.66E+16	2.04E+16	9.14E+15	3.15E+15	1.26E+15	4.08E+14	1.13E+14
3500	6482	3.93E+18	9.28E+16	2.62E+16	9.67E+15	2.53E+15	9.53E+14	2.88E+14	7.43E+13
4000	7408	6.63E+18	1.36E+17	2.98E+16	9.18E+15	1.85E+15	6.47E+14	1.82E+14	4.51E+13
4500	8334	1.04E+19	1.71E+17	2.96E+16	7.77E+15	1.24E+15	3.96E+14	1.02E+14	2.40E+13
5000	9260	1.51E+19	1.95E+17	2.68E+16	5.96E+15	7.48E+14	2.23E+14	5.31E+13	1.18E+13
5500	10186	1.99E+19	2.10E+17	2.33E+16	4.31E+15	4.15E+14	1.18E+14	2.65E+13	5.36E+12
6000	11112	2.42E+19	2.02E+17	1.82E+16	2.85E+15	2.14E+14	5.54E+13	1.10E+13	1.97E+12
7000	12964	3.08E+19	1.41E+17	8.45E+15	9.51E+14	4.39E+13	1.35E+13	1.35E+12	1.85E+11
8005	14825	3.10E+19	6.70E+16	2.51E+15	2.02E+14	5.72E+12	8.83E+11	9.73E+10	1.01E+10
9000	16668	2.18E+19	2.18E+16	4.72E+14	2.51E+13	3.88E+11	4.75E+10	5.39E+09	1.13E+09
10000	18520	1.57E+19	6.06E+15	7.54E+13	2.82E+12	1.95E+10	0.00E+00	0.00E+00	0.00E+00
11000	20372	1.14E+19	1.34E+15	5.96E+12	1.29E+11	6.41E+08	0.00E+00	0.00E+00	0.00E+00
12000	22224	9.16E+18	2.47E+14	1.08E+11	0.00E+00	0.00E+00	0.00E+00	0.00E+00	0.00E+00
13000	24076	7.48E+18	3.53E+13	3.33E+07	0.00E+00	0.00E+00	0.00E+00	0.00E+00	0.00E+00
14000	25928	6.28E+18	6.12E+12	1.49E+07	0.00E+00	0.00E+00	0.00E+00	0.00E+00	0.00E+00
15000	27780	5.19E+18	1.61E+12	7.94E+06	0.00E+00	0.00E+00	0.00E+00	0.00E+00	0.00E+00
16000	29632	3.97E+18	1.94E+10	0.00E+00	0.00E+00	0.00E+00	0.00E+00	0.00E+00	0.00E+00
17000	31484	3.29E+18	1.49E+10	0.00E+00	0.00E+00	0.00E+00	0.00E+00	0.00E+00	0.00E+00
18000	33336	2.16E+18	1.08E+10	0.00E+00	0.00E+00	0.00E+00	0.00E+00	0.00E+00	0.00E+00
19327	35793	6.64E+17	7.19E+09	0.00E+00	0.00E+00	0.00E+00	0.00E+00	0.00E+00	0.00E+00

EQUIV. 1 MEV ELECTRON FLUENCE FOR P$_{max}$ CIRCULAR ORBIT
DUE TO GEOMAGNETICALLY TRAPPED PROTONS, MODEL AP8MIN

INCLINATION = 70 DEGREES

ALTITUDE (nm)	(km)	0 (0)	2.54E-3 (1)	7.64E-3 (3)	1.52E-2 (6)	3.05E-2 (12)	5.09E-2 (20)	7.64E-2 (30)	1.52E-1 (60)
150	277	3.24E+13	3.84E+11	1.22E+11	5.69E+10	2.54E+10	1.76E+10	1.30E+10	1.02E+10
250	463	3.38E+14	3.47E+12	1.10E+12	5.55E+11	2.88E+11	2.15E+11	1.69E+11	1.37E+11
300	555	7.51E+14	6.83E+12	2.19E+12	1.13E+12	6.05E+11	4.56E+11	3.59E+11	2.90E+11
450	833	4.98E+15	2.75E+13	8.94E+12	4.80E+12	2.77E+12	2.13E+12	1.68E+12	1.35E+12
600	1111	1.69E+16	7.29E+13	2.45E+13	1.42E+13	8.85E+12	6.88E+12	5.40E+12	4.21E+12
800	1481	5.79E+16	2.22E+14	8.10E+13	5.03E+13	3.29E+13	2.54E+13	1.94E+13	1.45E+13
1000	1852	1.46E+17	5.90E+14	2.29E+14	1.45E+14	9.42E+13	6.91E+13	4.96E+13	3.44E+13

9-47

Table 9-43. Annual Equivalent 1 MeV Electron Fluence for GaAs Solar Cells from Trapped Protons During Solar Max and Min, 70° Inclination (Infinite Backshielding)

PROTONS - I_{sc}

EQUIV. 1 MEV PROTON FLUENCE FOR I_{sc} - CIRCULAR ORBIT
DUE TO GEOMAGNETICALLY TRAPPED PROTONS, MODEL AP8MAX

INCLINATION = 70 DEGREES

ALTITUDE (nm)	(km)	0 (0)	2.54E-3 (1)	7.64E-3 (3)	SHIELD THICKNESS, cm (mils) 1.52E-2 (6)	3.05E-2 (12)	5.09E-2 (20)	7.64E-2 (30)	1.52E-1 (60)
150	277	4.02E+13	3.77E+10	4.64E+09	1.79E+09	1.07E+09	8.25E+08	6.51E+08	5.42E+08
250	463	4.33E+14	4.65E+11	9.28E+10	5.16E+10	3.25E+10	2.52E+10	1.93E+10	1.54E+10
300	555	1.01E+15	1.13E+12	2.65E+11	1.48E+11	8.84E+10	6.62E+10	4.91E+10	3.78E+10
450	833	7.07E+15	8.84E+12	2.39E+12	1.23E+12	6.60E+11	4.61E+11	3.19E+11	2.27E+11
600	1111	2.32E+16	3.01E+13	8.56E+12	4.55E+12	2.52E+12	1.75E+12	1.18E+12	7.96E+11
800	1481	7.51E+16	9.55E+13	2.98E+13	1.72E+13	1.03E+13	7.16E+12	4.76E+12	3.07E+12
1000	1852	1.83E+17	2.66E+14	9.13E+13	5.48E+13	3.32E+13	2.22E+13	1.38E+13	8.28E+12
1250	2315	3.62E+17	7.55E+14	2.90E+14	1.80E+14	1.08E+14	6.77E+13	3.82E+13	2.00E+13
1500	2778	5.50E+17	1.66E+15	7.00E+14	4.41E+14	2.63E+14	1.55E+14	7.94E+13	3.57E+13
1750	3241	7.35E+17	3.22E+15	1.43E+15	8.92E+14	5.17E+14	2.89E+14	1.35E+14	5.22E+13
2000	3704	9.42E+17	5.71E+15	2.55E+15	1.56E+15	8.20E+14	4.28E+14	1.82E+14	6.45E+13
2250	4167	1.17E+18	9.82E+15	4.27E+15	2.36E+15	1.13E+15	5.42E+14	2.09E+14	6.12E+13
2500	4630	1.44E+18	1.53E+16	6.27E+15	3.20E+15	1.36E+15	6.02E+14	2.12E+14	5.44E+13
2750	5093	1.83E+18	2.20E+16	8.31E+15	3.91E+15	1.49E+15	6.12E+14	1.99E+14	4.65E+13
3000	5556	2.34E+18	2.99E+16	1.02E+16	4.40E+15	1.48E+15	5.79E+14	1.78E+14	3.12E+13
3500	6482	3.78E+18	5.00E+16	1.33E+16	4.70E+15	1.20E+15	4.41E+14	1.27E+14	1.92E+13
4000	7408	6.26E+18	7.42E+16	1.52E+16	4.49E+15	8.78E+14	3.01E+14	8.06E+13	1.03E+13
4500	8334	9.82E+18	9.48E+16	1.52E+16	3.82E+15	5.91E+14	1.85E+14	4.54E+13	5.07E+12
5000	9260	1.44E+19	1.09E+17	1.39E+16	2.95E+15	3.59E+14	1.05E+14	2.37E+13	2.33E+12
5500	10186	1.88E+19	1.19E+17	1.21E+16	2.14E+15	1.99E+14	5.57E+13	1.19E+13	8.65E+11
6000	11112	2.25E+19	1.16E+17	9.50E+15	1.42E+15	1.03E+14	2.62E+13	4.98E+12	8.24E+10
7000	12964	2.93E+19	8.15E+16	4.45E+15	4.76E+14	2.14E+13	4.27E+12	6.18E+11	4.45E+09
8005	14825	3.14E+19	3.95E+16	1.33E+15	1.02E+14	2.80E+12	4.24E+11	4.47E+10	5.08E+08
9000	16668	2.29E+19	1.31E+16	2.52E+14	1.27E+13	1.91E+11	2.28E+10	2.43E+09	0.00E+00
10000	18520	1.72E+19	3.70E+15	4.04E+13	1.43E+12	1.07E+10	0.00E+00	0.00E+00	0.00E+00
11000	20372	1.32E+19	8.31E+14	3.23E+12	6.59E+10	3.43E+08	0.00E+00	0.00E+00	0.00E+00
12000	22224	1.14E+19	1.57E+14	6.05E+10	0.00E+00	0.00E+00	0.00E+00	0.00E+00	0.00E+00
13000	24076	1.01E+19	2.31E+13	2.04E+07	0.00E+00	0.00E+00	0.00E+00	0.00E+00	0.00E+00
14000	25928	9.03E+18	4.13E+12	8.93E+06	0.00E+00	0.00E+00	0.00E+00	0.00E+00	0.00E+00
15000	27780	7.96E+18	1.11E+12	4.68E+06	0.00E+00	0.00E+00	0.00E+00	0.00E+00	0.00E+00
16000	29632	6.38E+18	1.61E+10	0.00E+00	0.00E+00	0.00E+00	0.00E+00	0.00E+00	0.00E+00
17000	31484	5.54E+18	1.22E+10	0.00E+00	0.00E+00	0.00E+00	0.00E+00	0.00E+00	0.00E+00
18000	33336	3.77E+18	8.77E+09	0.00E+00	0.00E+00	0.00E+00	0.00E+00	0.00E+00	0.00E+00
19327	35793	1.16E+18	5.78E+09	0.00E+00	0.00E+00	0.00E+00	0.00E+00	0.00E+00	0.00E+00

EQUIV. 1 MEV PROTON FLUENCE FOR I_{sc} - CIRCULAR ORBIT
DUE TO GEOMAGNETICALLY TRAPPED PROTONS, MODEL AP8MIN

INCLINATION = 70 DEGREES

ALTITUDE (nm)	(km)	0 (0)	2.54E-3 (1)	7.64E-3 (3)	1.52E-2 (6)	3.05E-2 (12)	5.09E-2 (20)	7.64E-2 (30)	1.52E-1 (60)
150	277	3.36E+13	2.02E+11	5.78E+10	2.43E+10	8.99E+09	5.37E+09	3.30E+09	2.28E+09
250	463	3.54E+14	1.82E+12	5.07E+11	2.24E+11	9.49E+10	6.17E+10	4.12E+10	2.98E+10
300	555	8.08E+14	3.57E+12	9.98E+11	4.49E+11	1.98E+11	1.31E+11	8.81E+10	6.35E+10
450	833	5.78E+15	1.43E+13	4.00E+12	1.88E+12	9.00E+11	6.12E+11	4.18E+11	2.98E+11
600	1111	2.04E+16	3.76E+13	1.08E+13	5.48E+12	2.93E+12	2.04E+12	1.40E+12	9.65E+11
800	1481	7.10E+16	1.13E+14	3.51E+13	1.95E+13	1.13E+13	7.94E+12	5.35E+12	3.49E+12
1000	1852	1.78E+17	3.01E+14	1.01E+14	5.87E+13	3.48E+13	2.35E+13	1.48E+13	8.89E+12

Table 9-44. Annual Equivalent 1 MeV Electron Fluence for GaAs Solar Cells from Trapped Electrons During Solar Max, 80° Inclination (Infinite Backshielding)

ELECTRONS - I_{sc}, V_{oc}, AND P_{max}

INCLINATION = 80 DEGREES

EQUIV. 1 MEV ELECTRON FLUENCE FOR I_{sc} - CIRCULAR ORBIT
DUE TO GEOMAGNETICALLY TRAPPED ELECTRONS - MODEL AE8MAX

ALTITUDE		SHIELD THICKNESS, cm (mils)							
(nm)	(km)	0 (0)	2.54E-3 (1)	7.64E-3 (3)	1.52E-2 (6)	3.05E-2 (12)	5.09E-2 (20)	7.64E-2 (30)	1.52E-1 (60)
150	277	7.34E+10	6.56E+10	5.68E+10	4.81E+10	3.71E+10	2.81E+10	2.10E+10	1.01E+10
250	463	1.49E+11	1.30E+11	1.11E+11	9.24E+10	6.99E+10	5.23E+10	3.87E+10	1.84E+10
300	555	2.06E+11	1.79E+11	1.51E+11	1.24E+11	9.33E+10	6.93E+10	5.10E+10	2.41E+10
450	833	4.92E+11	4.14E+11	3.36E+11	2.67E+11	1.92E+11	1.39E+11	1.01E+11	4.68E+10
600	1111	1.17E+12	9.45E+11	7.31E+11	5.54E+11	3.76E+11	2.61E+11	1.84E+11	8.29E+10
800	1481	3.45E+12	2.70E+12	2.00E+12	1.44E+12	9.15E+11	6.05E+11	4.13E+11	1.77E+11
1000	1852	8.32E+12	6.44E+12	4.71E+12	3.34E+12	2.07E+12	1.34E+12	9.02E+11	3.77E+11
1250	2315	1.68E+13	1.30E+13	9.54E+12	6.78E+12	4.20E+12	2.73E+12	1.83E+12	7.67E+11
1500	2778	2.42E+13	1.87E+13	1.37E+13	9.69E+12	5.96E+12	3.83E+12	2.56E+12	1.07E+12
1750	3241	3.05E+13	2.34E+13	1.69E+13	1.17E+13	6.96E+12	4.29E+12	2.78E+12	1.13E+12
2000	3704	3.53E+13	2.68E+13	1.91E+13	1.29E+13	7.29E+12	4.23E+12	2.61E+12	1.01E+12
2250	4167	3.89E+13	2.94E+13	2.06E+13	1.37E+13	7.41E+12	4.06E+12	2.38E+12	8.81E+11
2500	4630	4.02E+13	3.03E+13	2.11E+13	1.39E+13	7.32E+12	3.86E+12	2.18E+12	7.89E+11
2750	5093	3.93E+13	2.95E+13	2.07E+13	1.36E+13	7.11E+12	3.69E+12	2.07E+12	7.59E+11
3000	5556	3.67E+13	2.78E+13	1.95E+13	1.30E+13	6.87E+12	3.62E+12	2.08E+12	7.91E+11
3500	6482	3.11E+13	2.39E+13	1.72E+13	1.18E+13	6.68E+12	3.85E+12	2.40E+12	1.00E+12
4000	7408	2.84E+13	2.24E+13	1.67E+13	1.20E+13	7.38E+12	4.63E+12	3.07E+12	1.35E+12
4500	8334	2.78E+13	2.26E+13	1.76E+13	1.33E+13	8.83E+12	5.91E+12	4.07E+12	1.83E+12
5000	9260	2.69E+13	2.28E+13	1.86E+13	1.48E+13	1.05E+13	7.41E+12	5.28E+12	2.44E+12
5500	10186	2.78E+13	2.43E+13	2.06E+13	1.70E+13	1.27E+13	9.33E+12	6.82E+12	3.23E+12
6000	11112	3.03E+13	2.70E+13	2.34E+13	1.97E+13	1.52E+13	1.14E+13	8.50E+12	4.09E+12
7000	12964	3.94E+13	3.55E+13	3.10E+13	2.65E+13	2.07E+13	1.58E+13	1.19E+13	5.86E+12
8000	14816	4.79E+13	4.32E+13	3.78E+13	3.23E+13	2.52E+13	1.93E+13	1.45E+13	7.10E+12
9000	16668	5.44E+13	4.89E+13	4.27E+13	3.64E+13	2.82E+13	2.15E+13	1.61E+13	7.75E+12
10000	18520	5.92E+13	5.31E+13	4.61E+13	3.90E+13	3.00E+13	2.26E+13	1.67E+13	7.79E+12
11000	20372	5.67E+13	5.06E+13	4.37E+13	3.67E+13	2.79E+13	2.07E+13	1.50E+13	6.67E+12
12000	22224	5.15E+13	4.57E+13	3.91E+13	3.26E+13	2.44E+13	1.78E+13	1.27E+13	5.36E+12
13000	24076	4.41E+13	3.89E+13	3.29E+13	2.71E+13	1.99E+13	1.42E+13	9.91E+12	4.04E+12
14000	25928	3.69E+13	3.22E+13	2.70E+13	2.20E+13	1.58E+13	1.11E+13	7.59E+12	3.01E+12
15000	27780	2.98E+13	2.58E+13	2.14E+13	1.72E+13	1.21E+13	8.32E+12	5.61E+12	2.15E+12
16000	29632	2.23E+13	1.91E+13	1.57E+13	1.25E+13	8.65E+12	5.84E+12	3.87E+12	1.44E+12
17000	31484	1.78E+13	1.52E+13	1.23E+13	9.67E+12	6.61E+12	4.39E+12	2.86E+12	1.02E+12
18000	33336	1.21E+13	1.01E+13	8.14E+12	6.31E+12	4.23E+12	2.76E+12	1.77E+12	6.00E+11
19327	35793	7.53E+12	6.24E+12	4.91E+12	3.72E+12	2.42E+12	1.52E+12	9.47E+11	3.02E+11

Table 9-45. Annual Equivalent 1 MeV Electron Fluence for GaAs Solar Cells from Trapped Electrons During Solar Min, 80° Inclination (Infinite Backshielding)

ELECTRONS - I_{sc}, V_{oc}, AND P_{max} INCLINATION = 80 DEGREES

EQUIV. 1 MEV ELECTRON FLUENCE FOR I_{sc} - CIRCULAR ORBIT
DUE TO GEOMAGNETICALLY TRAPPED ELECTRONS - MODEL AE8MIN

ALTITUDE		SHIELD THICKNESS, cm (mils)							
(nm)	(km)	0 (0)	2.54E-3 (1)	7.64E-3 (3)	1.52E-2 (6)	3.05E-2 (12)	5.09E-2 (20)	7.64E-2 (30)	1.52E-1 (60)
150	277	3.30E+10	2.97E+10	2.59E+10	2.21E+10	1.72E+10	1.31E+10	9.82E+09	4.73E+09
250	463	7.00E+10	6.24E+10	5.39E+10	4.55E+10	3.50E+10	2.64E+10	1.96E+10	9.27E+09
300	555	9.81E+10	8.69E+10	7.47E+10	6.27E+10	4.79E+10	3.59E+10	2.66E+10	1.25E+10
450	833	2.40E+11	2.08E+11	1.75E+11	1.44E+11	1.07E+11	7.83E+10	5.70E+10	2.62E+10
600	1111	5.78E+11	4.90E+11	3.99E+11	3.18E+11	2.27E+11	1.61E+11	1.15E+11	5.13E+10
800	1481	1.80E+12	1.49E+12	1.19E+12	9.21E+11	6.31E+11	4.33E+11	2.99E+11	1.29E+11
1000	1852	4.53E+12	3.74E+12	2.97E+12	2.29E+12	1.56E+12	1.06E+12	7.31E+11	3.11E+11
1250	2315	9.21E+12	7.67E+12	6.14E+12	4.78E+12	3.30E+12	2.28E+12	1.58E+12	6.79E+11
1500	2778	1.24E+13	1.04E+13	8.36E+12	6.56E+12	4.57E+12	3.18E+12	2.22E+12	9.62E+11
1750	3241	1.37E+13	1.14E+13	9.13E+12	7.11E+12	4.90E+12	3.38E+12	2.34E+12	1.00E+12
2000	3704	1.34E+13	1.11E+13	8.70E+12	6.67E+12	4.48E+12	3.02E+12	2.05E+12	8.46E+11
2250	4167	1.25E+13	1.01E+13	7.83E+12	5.89E+12	3.86E+12	2.54E+12	1.69E+12	6.79E+11
2500	4630	1.10E+13	8.78E+12	6.70E+12	4.97E+12	3.20E+12	2.08E+12	1.38E+12	5.45E+11
2750	5093	9.34E+12	7.43E+12	5.64E+12	4.16E+12	2.67E+12	1.74E+12	1.16E+12	4.62E+11
3000	5556	8.21E+12	6.54E+12	4.98E+12	3.69E+12	2.40E+12	1.59E+12	1.07E+12	4.37E+11
3500	6482	7.47E+12	6.08E+12	4.75E+12	3.63E+12	2.46E+12	1.69E+12	1.17E+12	5.00E+11
4000	7408	7.93E+12	6.61E+12	5.31E+12	4.17E+12	2.93E+12	2.05E+12	1.45E+12	6.36E+11
4500	8334	8.69E+12	7.41E+12	6.10E+12	4.92E+12	3.56E+12	2.56E+12	1.84E+12	8.29E+11
5000	9260	9.56E+12	8.31E+12	7.00E+12	5.78E+12	4.31E+12	3.18E+12	2.31E+12	1.07E+12
5500	10186	1.09E+13	9.61E+12	8.23E+12	6.90E+12	5.25E+12	3.94E+12	2.91E+12	1.37E+12
6000	11112	1.24E+13	1.11E+13	9.56E+12	8.09E+12	6.25E+12	4.74E+12	3.54E+12	1.70E+12
7000	12964	1.65E+13	1.49E+13	1.30E+13	1.11E+13	8.70E+12	6.67E+12	5.02E+12	2.46E+12
8000	14816	2.16E+13	1.95E+13	1.71E+13	1.47E+13	1.15E+13	8.83E+12	6.66E+12	3.26E+12
9000	16668	2.73E+13	2.47E+13	2.16E+13	1.85E+13	1.45E+13	1.11E+13	8.31E+12	4.02E+12
10000	18520	3.42E+13	3.08E+13	2.69E+13	2.30E+13	1.79E+13	1.36E+13	1.01E+13	4.74E+12
11000	20372	3.85E+13	3.45E+13	3.00E+13	2.54E+13	1.95E+13	1.47E+13	1.08E+13	4.91E+12
12000	22224	4.16E+13	3.71E+13	3.19E+13	2.68E+13	2.03E+13	1.49E+13	1.08E+13	4.71E+12
13000	24076	4.03E+13	3.56E+13	3.03E+13	2.51E+13	1.86E+13	1.34E+13	9.45E+12	3.92E+12
14000	25928	3.59E+13	3.14E+13	2.64E+13	2.16E+13	1.56E+13	1.10E+13	7.60E+12	3.03E+12
15000	27780	2.97E+13	2.57E+13	2.14E+13	1.72E+13	1.22E+13	8.37E+12	5.65E+12	2.18E+12
16000	29632	2.24E+13	1.92E+13	1.58E+13	1.25E+13	8.70E+12	5.88E+12	3.90E+12	1.45E+12
17000	31484	1.79E+13	1.52E+13	1.24E+13	9.72E+12	6.64E+12	4.41E+12	2.88E+12	1.03E+12
18000	33336	1.21E+13	1.02E+13	8.19E+12	6.35E+12	4.26E+12	2.78E+12	1.78E+12	6.04E+11
19327	35793	7.61E+12	6.30E+12	4.96E+12	3.77E+12	2.44E+12	1.54E+12	9.60E+11	3.07E+11

Table 9-46. Annual Equivalent 1 MeV Electron Fluence for GaAs Solar Cells from Trapped Protons During Solar Max and Min, 80° Inclination (Infinite Backshielding)

PROTONS - V_{oc}

EQUIV. 1 MEV ELECTRON FLUENCE FOR V_{oc} CIRCULAR ORBIT   INCLINATION = 80 DEGREES
DUE TO GEOMAGNETICALLY TRAPPED PROTONS, MODEL AP8MAX

ALTITUDE (nm)	(km)	SHIELD THICKNESS, cm (mils)							
		0 (0)	2.54E-3 (1)	7.64E-3 (3)	1.52E-2 (6)	3.05E-2 (12)	5.09E-2 (20)	7.64E-2 (30)	1.52E-1 (60)
150	277	4.70E+13	9.39E+10	1.51E+10	7.84E+09	5.60E+09	4.86E+09	4.28E+09	3.79E+09
250	463	4.15E+14	1.05E+12	2.97E+11	2.01E+11	1.51E+11	1.29E+11	1.11E+11	9.58E+10
300	555	9.79E+14	2.69E+12	8.28E+11	5.45E+11	3.88E+11	3.24E+11	2.73E+11	2.31E+11
450	833	6.25E+15	2.08E+13	6.84E+12	4.11E+12	2.63E+12	2.07E+12	1.66E+12	1.33E+12
600	1111	2.03E+16	7.17E+13	2.50E+13	1.52E+13	9.73E+12	7.49E+12	5.80E+12	4.46E+12
800	1481	6.41E+16	2.32E+14	8.91E+13	5.77E+13	3.88E+13	2.96E+13	2.24E+13	1.66E+13
1000	1852	1.61E+17	6.56E+14	2.68E+14	1.76E+14	1.17E+14	8.54E+13	6.07E+13	4.20E+13
1250	2315	3.33E+17	1.90E+15	8.38E+14	5.54E+14	3.55E+14	2.37E+14	1.51E+14	9.14E+13
1500	2778	5.27E+17	4.24E+15	1.99E+15	1.32E+15	8.17E+14	5.06E+14	2.85E+14	1.47E+14
1750	3241	7.50E+17	8.25E+15	4.01E+15	2.61E+15	1.55E+15	8.99E+14	4.55E+14	1.97E+14
2000	3704	1.01E+18	1.47E+16	7.07E+15	4.38E+15	2.41E+15	1.29E+15	5.88E+14	2.21E+14
2250	4167	1.32E+18	2.51E+16	1.16E+16	6.68E+15	3.27E+15	1.61E+15	6.57E+14	2.21E+14
2500	4630	1.70E+18	3.88E+16	1.69E+16	8.98E+15	3.90E+15	1.77E+15	6.58E+14	2.04E+14
2750	5093	2.25E+18	5.55E+16	2.23E+16	1.09E+16	4.22E+15	1.78E+15	6.13E+14	1.78E+14
3000	5556	2.96E+18	7.48E+16	2.71E+16	1.22E+16	4.19E+15	1.67E+15	5.43E+14	1.50E+14
3500	6482	5.04E+18	1.23E+17	3.49E+16	1.29E+16	3.37E+15	1.27E+15	3.84E+14	9.90E+13
4000	7408	8.64E+18	1.80E+17	3.97E+16	1.04E+16	2.46E+15	8.62E+14	2.43E+14	6.01E+13
4500	8334	1.36E+19	2.27E+17	3.94E+16	1.04E+16	1.65E+15	5.28E+14	1.36E+14	3.20E+13
5000	9260	1.99E+19	2.59E+17	3.57E+16	7.95E+15	9.98E+14	2.97E+14	7.08E+13	1.57E+13
5500	10186	2.64E+19	2.80E+17	3.10E+16	5.75E+15	5.53E+14	1.58E+14	3.54E+13	7.14E+12
6000	11112	3.20E+19	2.70E+17	2.42E+16	3.80E+15	2.86E+14	7.39E+13	1.47E+13	2.63E+12
7000	12964	4.09E+19	1.87E+17	1.13E+16	1.27E+15	5.85E+13	1.19E+13	1.80E+12	2.46E+11
8000	14816	4.13E+19	8.98E+16	3.38E+15	2.73E+14	7.72E+12	1.19E+12	1.32E+11	1.36E+10
9000	16668	2.90E+19	2.91E+16	6.29E+14	3.36E+13	5.18E+11	6.34E+10	7.18E+09	1.51E+09
10000	18520	2.09E+19	2.09E+16	1.00E+14	3.75E+12	2.59E+10	0.00E+00	0.00E+00	0.00E+00
11000	20372	1.51E+19	1.79E+15	8.14E+12	1.78E+11	8.91E+08	0.00E+00	0.00E+00	0.00E+00
12000	22224	1.22E+19	3.29E+14	1.43E+11	0.00E+00	0.00E+00	0.00E+00	0.00E+00	0.00E+00
13000	24076	9.95E+18	4.71E+13	4.59E+07	0.00E+00	0.00E+00	0.00E+00	0.00E+00	0.00E+00
14000	25928	8.34E+18	8.13E+12	2.05E+07	0.00E+00	0.00E+00	0.00E+00	0.00E+00	0.00E+00
15000	27780	6.91E+18	2.15E+12	1.09E+07	0.00E+00	0.00E+00	0.00E+00	0.00E+00	0.00E+00
16000	29632	5.29E+18	2.64E+10	0.00E+00	0.00E+00	0.00E+00	0.00E+00	0.00E+00	0.00E+00
17000	31484	4.39E+18	2.02E+10	0.00E+00	0.00E+00	0.00E+00	0.00E+00	0.00E+00	0.00E+00
18000	33336	2.88E+18	1.47E+10	0.00E+00	0.00E+00	0.00E+00	0.00E+00	0.00E+00	0.00E+00
19327	35793	8.94E+17	9.84E+09	0.00E+00	0.00E+00	0.00E+00	0.00E+00	0.00E+00	0.00E+00

EQUIV. 1 MEV ELECTRON FLUENCE FOR V_{oc} CIRCULAR ORBIT   INCLINATION = 80 DEGREES
DUE TO GEOMAGNETICALLY TRAPPED PROTONS, MODEL AP8MIN

ALTITUDE (nm)	(km)	0 (0)	2.54E-3 (1)	7.64E-3 (3)	1.52E-2 (6)	3.05E-2 (12)	5.09E-2 (20)	7.64E-2 (30)	1.52E-1 (60)
150	277	3.92E+13	5.28E+11	1.73E+11	7.98E+10	3.46E+10	2.34E+10	1.70E+10	1.32E+10
250	463	3.87E+14	4.35E+12	1.43E+12	7.20E+11	3.71E+11	2.76E+11	2.15E+11	1.74E+11
300	555	8.83E+14	8.52E+12	2.78E+12	1.43E+12	7.72E+11	5.82E+11	4.59E+11	3.71E+11
450	833	5.43E+15	3.35E+13	1.11E+13	6.06E+12	3.57E+12	2.77E+12	2.20E+12	1.77E+12
600	1111	1.84E+16	8.95E+13	3.12E+13	1.83E+13	1.15E+13	9.00E+12	7.09E+12	5.55E+12
800	1481	6.18E+16	2.76E+14	1.04E+14	6.56E+13	4.33E+13	3.35E+13	2.57E+13	1.92E+13
1000	1852	1.58E+17	7.41E+14	2.95E+14	1.90E+14	1.24E+14	9.13E+13	6.57E+13	4.57E+13

Table 9-47. Annual Equivalent 1 MeV Electron Fluence for GaAs Solar Cells from Trapped Protons During Solar Max and Min, 80° Inclination (Infinite Backshielding)

PROTONS - P_max

INCLINATION = 80 DEGREES

EQUIV. 1 MEV ELECTRON FLUENCE FOR P_max CIRCULAR ORBIT
DUE TO GEOMAGNETICALLY TRAPPED PROTONS, MODEL AP8MAX

ALTITUDE (nm)	(km)	SHIELD THICKNESS, cm (mils)							
		0 (0)	2.54E-3 (1)	7.64E-3 (3)	1.52E-2 (6)	3.05E-2 (12)	5.09E-2 (20)	7.64E-2 (30)	1.52E-1 (60)
150	277	3.36E+13	6.71E+10	1.08E+10	5.60E+09	4.00E+09	3.47E+09	3.06E+09	2.71E+09
250	463	2.96E+14	7.48E+11	2.12E+11	1.44E+11	1.08E+11	9.23E+10	7.95E+10	6.85E+10
300	555	6.99E+14	1.92E+12	5.91E+11	3.89E+11	2.77E+11	2.32E+11	1.95E+11	1.65E+11
450	833	4.46E+15	4.89E+13	4.89E+12	2.93E+12	1.88E+12	1.48E+12	1.18E+12	9.47E+11
600	1111	1.45E+16	5.12E+13	1.79E+13	1.08E+13	6.95E+12	5.35E+12	4.14E+12	3.19E+12
800	1481	4.58E+16	1.66E+14	6.37E+13	4.12E+13	2.77E+13	2.12E+13	1.60E+13	1.18E+13
1000	1852	1.15E+17	4.68E+14	1.91E+14	1.26E+14	8.37E+13	6.10E+13	4.34E+13	3.00E+13
1250	2315	2.38E+17	1.36E+15	5.98E+14	3.96E+14	2.53E+14	1.69E+14	1.08E+14	6.53E+13
1500	2778	3.76E+17	3.03E+15	1.42E+15	9.42E+14	5.84E+14	3.61E+14	2.04E+14	1.05E+14
1750	3241	5.36E+17	5.89E+15	2.86E+15	1.86E+15	1.11E+15	6.42E+14	3.25E+14	1.41E+14
2000	3704	7.19E+17	1.05E+16	5.05E+15	3.13E+15	1.72E+15	9.24E+14	4.20E+14	1.58E+14
2250	4167	9.43E+17	1.79E+16	8.31E+15	4.77E+15	2.34E+15	1.15E+15	4.70E+14	1.58E+14
2500	4630	1.22E+18	2.77E+16	1.21E+16	6.42E+15	2.79E+15	1.26E+15	4.70E+14	1.46E+14
2750	5093	1.61E+18	3.96E+16	1.59E+16	7.78E+15	3.02E+15	1.27E+15	4.38E+14	1.27E+14
3000	5556	2.11E+18	5.35E+16	1.93E+16	8.68E+15	2.99E+15	1.20E+15	3.88E+14	1.07E+14
3500	6482	3.60E+18	8.81E+16	2.49E+16	9.20E+15	2.40E+15	9.07E+14	2.74E+14	7.07E+13
4000	7408	6.17E+18	1.29E+17	2.83E+16	8.73E+15	1.76E+15	6.16E+14	1.74E+14	4.30E+13
4500	8334	9.74E+18	1.62E+17	2.82E+16	7.39E+15	1.18E+15	3.77E+14	9.73E+13	2.29E+13
5000	9260	1.42E+19	1.85E+17	2.55E+16	5.68E+15	7.13E+14	2.12E+14	5.06E+13	1.12E+13
5500	10186	1.88E+19	2.00E+17	2.21E+16	4.11E+15	3.95E+14	1.13E+14	2.53E+13	5.10E+12
6000	11112	2.29E+19	1.93E+17	1.73E+16	2.71E+15	2.04E+14	5.28E+13	1.05E+13	1.88E+12
7000	12964	2.92E+19	1.34E+17	8.04E+15	9.05E+14	4.18E+13	8.52E+12	1.29E+12	1.76E+11
8000	14816	2.95E+19	6.41E+16	2.41E+15	1.95E+14	5.51E+12	8.52E+11	9.41E+10	9.75E+09
9000	16668	2.07E+19	2.08E+16	4.49E+14	2.40E+13	3.70E+11	4.53E+10	5.13E+09	1.08E+09
10000	18520	1.49E+19	5.77E+15	7.17E+13	2.68E+12	1.85E+10	0.00E+00	0.00E+00	0.00E+00
11000	20372	1.08E+19	1.28E+15	5.82E+12	1.27E+11	6.36E+08	0.00E+00	0.00E+00	0.00E+00
12000	22224	8.68E+18	2.35E+14	3.28E+11	0.00E+00	0.00E+00	0.00E+00	0.00E+00	0.00E+00
13000	24076	7.10E+18	3.37E+13	3.28E+07	0.00E+00	0.00E+00	0.00E+00	0.00E+00	0.00E+00
14000	25928	5.95E+18	5.81E+12	1.46E+07	0.00E+00	0.00E+00	0.00E+00	0.00E+00	0.00E+00
15000	27780	4.93E+18	1.54E+12	7.82E+06	0.00E+00	0.00E+00	0.00E+00	0.00E+00	0.00E+00
16000	29632	3.78E+18	1.88E+10	0.00E+00	0.00E+00	0.00E+00	0.00E+00	0.00E+00	0.00E+00
17000	31484	3.13E+18	1.44E+10	0.00E+00	0.00E+00	0.00E+00	0.00E+00	0.00E+00	0.00E+00
18000	33336	2.06E+18	1.05E+10	0.00E+00	0.00E+00	0.00E+00	0.00E+00	0.00E+00	0.00E+00
19327	35793	6.39E+17	7.03E+09	0.00E+00	0.00E+00	0.00E+00	0.00E+00	0.00E+00	0.00E+00

INCLINATION = 80 DEGREES

EQUIV. 1 MEV ELECTRON FLUENCE FOR P_max CIRCULAR ORBIT
DUE TO GEOMAGNETICALLY TRAPPED PROTONS, MODEL AP8MIN

ALTITUDE (nm)	(km)	0 (0)	2.54E-3 (1)	7.64E-3 (3)	1.52E-2 (6)	3.05E-2 (12)	5.09E-2 (20)	7.64E-2 (30)	1.52E-1 (60)
150	277	2.80E+13	3.77E+11	1.23E+11	5.70E+10	2.47E+10	1.67E+10	1.21E+10	9.43E+09
250	463	2.76E+14	3.10E+12	1.02E+12	5.14E+11	2.65E+11	1.97E+11	1.54E+11	1.24E+11
300	555	6.31E+14	6.09E+12	1.90E+12	1.02E+12	5.51E+11	4.16E+11	3.28E+11	2.65E+11
450	833	3.88E+15	2.39E+13	7.90E+12	4.33E+12	2.55E+12	1.98E+12	1.57E+12	1.27E+12
600	1111	1.32E+16	6.39E+13	2.23E+13	1.31E+13	8.23E+12	6.43E+12	5.07E+12	3.96E+12
800	1481	4.41E+16	1.97E+14	7.46E+13	4.68E+13	3.09E+13	2.39E+13	1.84E+13	1.37E+13
1000	1852	1.13E+17	5.29E+14	2.11E+14	1.35E+14	8.86E+13	6.52E+13	4.69E+13	3.26E+13

9-52

Table 9-48. Annual Equivalent 1 MeV Electron Fluence for GaAs Solar Cells from Trapped Protons During Solar Max and Min, 80° Inclination (Infinite Backshielding)

PROTONS - I_{sc}

EQUIV. 1 MEV PROTON FLUENCE FOR I_{sc} - CIRCULAR ORBIT DUE TO GEOMAGNETICALLY TRAPPED PROTONS, MODEL AP8MAX   INCLINATION = 80 DEGREES

| ALTITUDE | | SHIELD THICKNESS, cm (mils) | | | | | | | |
(nm)	(km)	0 (0)	2.54E-3 (1)	7.64E-3 (3)	1.52E-2 (6)	3.05E-2 (12)	5.09E-2 (20)	7.64E-2 (30)	1.52E-1 (60)
150	277	3.85E+13	3.66E+10	4.50E+09	1.78E+09	1.02E+09	7.90E+08	6.18E+08	5.07E+08
250	463	3.48E+14	3.84E+11	8.13E+10	4.64E+10	2.96E+10	2.29E+10	1.76E+10	1.40E+10
300	555	8.31E+14	9.82E+11	2.36E+11	1.33E+11	8.04E+10	6.03E+10	4.49E+10	3.48E+10
450	833	5.41E+15	7.65E+12	2.10E+12	1.10E+12	6.04E+11	4.26E+11	2.96E+11	2.12E+11
600	1111	1.78E+16	2.63E+13	7.78E+12	4.19E+12	2.34E+12	1.63E+12	1.10E+12	7.48E+11
800	1481	5.67E+16	8.39E+13	2.74E+13	1.60E+13	9.65E+12	6.74E+12	4.49E+12	2.91E+12
1000	1852	1.41E+17	2.37E+14	8.39E+13	5.09E+13	3.11E+13	2.09E+13	1.31E+13	7.84E+12
1250	2315	2.84E+17	6.88E+14	2.71E+14	1.69E+14	1.02E+14	6.40E+13	3.62E+13	1.90E+13
1500	2778	4.36E+17	1.54E+15	6.60E+14	4.17E+14	2.50E+14	1.47E+14	7.55E+13	3.39E+13
1750	3241	6.01E+17	3.00E+15	1.35E+15	8.45E+14	4.91E+14	2.75E+14	1.29E+14	4.96E+13
2000	3704	7.83E+17	5.35E+15	2.42E+15	1.45E+15	7.79E+14	4.07E+14	1.73E+14	5.89E+13
2250	4167	9.95E+17	9.23E+15	4.04E+15	2.24E+15	1.08E+15	5.15E+14	1.98E+14	6.13E+13
2500	4630	1.25E+18	1.44E+16	5.95E+15	3.04E+15	1.30E+15	5.73E+14	2.02E+14	5.82E+13
2750	5093	1.61E+18	2.08E+16	7.89E+15	3.72E+15	1.41E+15	5.82E+14	1.90E+14	5.17E+13
3000	5556	2.09E+18	2.83E+16	9.66E+15	4.17E+15	1.41E+15	5.50E+14	1.69E+14	4.42E+13
3500	6482	3.44E+18	4.75E+16	1.26E+16	4.47E+15	1.14E+15	4.20E+14	1.21E+14	2.97E+13
4000	7408	5.80E+18	7.04E+16	1.45E+16	4.27E+15	8.35E+14	2.86E+14	7.67E+13	1.82E+13
4500	8334	9.20E+18	9.01E+16	1.45E+16	3.64E+15	5.62E+14	1.76E+14	4.32E+13	9.80E+12
5000	9260	1.36E+19	1.04E+17	1.32E+16	2.80E+15	3.42E+14	9.96E+13	2.26E+13	4.83E+12
5500	10186	1.78E+19	1.13E+17	1.15E+16	2.04E+15	1.90E+14	5.31E+13	1.13E+13	2.22E+12
6000	11112	2.12E+19	1.10E+17	9.04E+15	1.35E+15	9.84E+13	2.49E+13	4.75E+12	8.24E+11
7000	12964	2.77E+19	7.75E+16	4.24E+15	4.53E+14	2.03E+13	4.06E+12	5.88E+11	7.83E+10
8000	14816	2.98E+19	3.78E+16	1.28E+15	9.79E+13	2.70E+12	4.09E+11	4.32E+10	4.32E+09
9000	16668	2.17E+19	1.25E+16	2.40E+14	1.21E+13	1.82E+11	2.17E+10	2.31E+09	4.84E+08
10000	18520	1.63E+19	3.52E+15	3.84E+13	1.36E+12	1.01E+10	0.00E+00	0.00E+00	0.00E+00
11000	20372	1.25E+19	7.97E+14	3.16E+12	6.49E+10	3.40E+08	0.00E+00	0.00E+00	0.00E+00
12000	22224	1.08E+19	1.50E+14	5.74E+10	0.00E+00	0.00E+00	0.00E+00	0.00E+00	0.00E+00
13000	24076	9.54E+18	2.21E+13	2.01E+07	0.00E+00	0.00E+00	0.00E+00	0.00E+00	0.00E+00
14000	25928	8.56E+18	3.92E+12	8.76E+06	0.00E+00	0.00E+00	0.00E+00	0.00E+00	0.00E+00
15000	27780	7.57E+18	1.06E+12	4.61E+06	0.00E+00	0.00E+00	0.00E+00	0.00E+00	0.00E+00
16000	29632	6.08E+18	1.56E+10	0.00E+00	0.00E+00	0.00E+00	0.00E+00	0.00E+00	0.00E+00
17000	31484	5.27E+18	1.19E+10	0.00E+00	0.00E+00	0.00E+00	0.00E+00	0.00E+00	0.00E+00
18000	33336	3.60E+18	8.53E+09	0.00E+00	0.00E+00	0.00E+00	0.00E+00	0.00E+00	0.00E+00
19327	35793	1.11E+18	5.64E+09	0.00E+00	0.00E+00	0.00E+00	0.00E+00	0.00E+00	0.00E+00

EQUIV. 1 MEV PROTON FLUENCE FOR I_{sc} - CIRCULAR ORBIT DUE TO GEOMAGNETICALLY TRAPPED PROTONS, MODEL AP8MIN   INCLINATION = 80 DEGREES

| ALTITUDE | | 0 | 2.54E-3 | 7.64E-3 | 1.52E-2 | 3.05E-2 | 5.09E-2 | 7.64E-2 | 1.52E-1 |
(nm)	(km)	(0)	(1)	(3)	(6)	(12)	(20)	(30)	(60)
150	277	2.88E+13	1.99E+11	5.88E+10	2.46E+10	8.89E+09	5.19E+09	3.13E+09	2.13E+09
250	463	2.90E+14	1.62E+12	4.70E+11	2.08E+11	8.78E+10	5.68E+10	3.78E+10	2.71E+10
300	555	6.78E+14	3.17E+12	9.03E+11	4.08E+11	1.81E+11	1.19E+11	8.04E+10	5.80E+10
450	833	4.45E+15	1.24E+13	3.51E+12	1.67E+12	8.22E+11	5.64E+11	3.89E+11	2.79E+11
600	1111	1.57E+16	3.28E+13	9.71E+12	5.01E+12	2.71E+12	1.90E+12	1.31E+12	9.06E+11
800	1481	5.37E+16	1.00E+14	3.22E+13	1.81E+13	1.06E+13	7.47E+12	5.05E+12	3.30E+12
1000	1852	1.37E+17	2.69E+14	9.29E+13	5.45E+13	3.26E+13	2.21E+13	1.40E+13	8.42E+12

9-53

Table 9-49.   Annual Equivalent 1 MeV Electron Fluence for GaAs Solar Cells from Trapped Electrons During Solar Max, 90° Inclination (Infinite Backshielding)

ELECTRONS - I$_{sc}$, V$_{oc}$, AND P$_{max}$ INCLINATION = 90 DEGREES

EQUIV. 1 MEV ELECTRON FLUENCE FOR I$_{sc}$ - CIRCULAR ORBIT
DUE TO GEOMAGNETICALLY TRAPPED ELECTRONS - MODEL AE8MAX

| ALTITUDE | | SHIELD THICKNESS, cm (mils) | | | | | | | |
(nm)	(km)	0 (0)	2.54E-3 (1)	7.64E-3 (3)	1.52E-2 (6)	3.05E-2 (12)	5.09E-2 (20)	7.64E-2 (30)	1.52E-1 (60)
150	277	6.45E+10	5.77E+10	5.00E+10	4.24E+10	3.27E+10	2.48E+10	1.85E+10	8.97E+09
250	463	1.32E+11	1.16E+11	9.83E+10	8.18E+10	6.19E+10	4.64E+10	3.43E+10	1.64E+10
300	555	1.85E+11	1.61E+11	1.35E+11	1.11E+11	8.33E+10	6.19E+10	4.56E+10	2.16E+10
450	833	4.58E+11	3.84E+11	3.10E+11	2.46E+11	1.77E+11	1.28E+11	9.25E+10	4.30E+10
600	1111	1.11E+12	8.92E+11	6.88E+11	5.19E+11	3.50E+11	2.43E+11	1.71E+11	7.70E+10
800	1481	3.34E+12	2.61E+12	1.93E+12	1.39E+12	8.77E+11	5.78E+11	3.93E+11	1.68E+11
1000	1852	8.13E+12	6.28E+12	4.59E+12	3.26E+12	2.01E+12	1.30E+12	8.74E+11	3.65E+11
1250	2315	1.64E+13	1.28E+13	9.33E+12	6.62E+12	4.10E+12	2.66E+12	1.79E+12	7.47E+11
1500	2778	2.37E+13	1.84E+13	1.34E+13	9.49E+12	5.83E+12	3.74E+12	2.50E+12	1.04E+12
1750	3241	3.00E+13	2.30E+13	1.66E+13	1.15E+13	6.81E+12	4.19E+12	2.71E+12	1.11E+12
2000	3704	3.46E+13	2.63E+13	1.87E+13	1.27E+13	7.13E+12	4.13E+12	2.54E+12	9.84E+11
2250	4167	3.82E+13	2.88E+13	2.02E+13	1.34E+13	7.25E+12	3.97E+12	2.32E+12	8.56E+11
2500	4630	3.94E+13	2.97E+13	2.07E+13	1.36E+13	7.16E+12	3.76E+12	2.12E+12	7.65E+11
2750	5093	3.84E+13	2.90E+13	2.02E+13	1.33E+13	6.94E+12	3.59E+12	2.01E+12	7.33E+11
3000	5556	3.60E+13	2.72E+13	1.91E+13	1.27E+13	6.70E+12	3.52E+12	2.01E+12	7.64E+11
3500	6482	3.05E+13	2.34E+13	1.68E+13	1.15E+13	6.50E+12	3.73E+12	2.32E+12	9.65E+11
4000	7408	2.78E+13	2.19E+13	1.63E+13	1.17E+13	7.19E+12	4.51E+12	2.99E+12	1.31E+12
4500	8334	2.72E+13	2.22E+13	1.72E+13	1.30E+13	8.63E+12	5.77E+12	3.97E+12	1.79E+12
5000	9260	2.63E+13	2.23E+13	1.81E+13	1.44E+13	1.02E+13	7.23E+12	5.15E+12	2.38E+12
5500	10186	2.72E+13	2.38E+13	2.01E+13	1.66E+13	1.24E+13	9.12E+12	6.67E+12	3.15E+12
6000	11112	2.97E+13	2.64E+13	2.29E+13	1.93E+13	1.48E+13	1.12E+13	8.32E+12	4.01E+12
7000	12964	3.86E+13	3.48E+13	3.04E+13	2.60E+13	2.03E+13	1.55E+13	1.17E+13	5.74E+12
8000	14816	4.70E+13	4.24E+13	3.70E+13	3.17E+13	2.47E+13	1.89E+13	1.42E+13	6.97E+12
9000	16668	5.34E+13	4.80E+13	4.19E+13	3.57E+13	2.77E+13	2.11E+13	1.58E+13	7.62E+12
10000	18520	5.82E+13	5.22E+13	4.54E+13	3.84E+13	2.96E+13	2.23E+13	1.64E+13	7.68E+12
11000	20372	5.55E+13	4.95E+13	4.27E+13	3.60E+13	2.73E+13	2.03E+13	1.47E+13	6.54E+12
12000	22224	5.06E+13	4.49E+13	3.84E+13	3.20E+13	2.39E+13	1.74E+13	1.24E+13	5.26E+12
13000	24076	4.34E+13	3.82E+13	3.24E+13	2.67E+13	1.96E+13	1.40E+13	9.75E+12	3.98E+12
14000	25928	3.62E+13	3.16E+13	2.65E+13	2.16E+13	1.55E+13	1.09E+13	7.45E+12	2.95E+12
15000	27780	2.93E+13	2.53E+13	2.10E+13	1.69E+13	1.19E+13	8.18E+12	5.52E+12	2.11E+12
16000	29632	2.19E+13	1.88E+13	1.54E+13	1.23E+13	8.51E+12	5.75E+12	3.81E+12	1.42E+12
17000	31484	1.76E+13	1.49E+13	1.21E+13	9.51E+12	6.50E+12	4.31E+12	2.82E+12	1.01E+12
18000	33336	1.19E+13	1.00E+13	8.03E+12	6.23E+12	4.18E+12	2.72E+12	1.74E+12	5.92E+11
19327	35793	7.49E+12	6.20E+12	4.88E+12	3.70E+12	2.40E+12	1.52E+12	9.42E+11	3.01E+11

Table 9-50. Annual Equivalent 1 MeV Electron Fluence for GaAs Solar Cells from Trapped Electrons During Solar Min, 90° Inclination (Infinite Backshielding)

ELECTRONS - I_{sc}, V_{oc}, AND P_{max} INCLINATION = 90 DEGREES

EQUIV. 1 MEV ELECTRON FLUENCE FOR I_{sc} - CIRCULAR ORBIT
DUE TO GEOMAGNETICALLY TRAPPED ELECTRONS - MODEL AE8MIN

ALTITUDE		SHIELD THICKNESS, cm (mils)							
(nm)	(km)	0 (0)	2.54E-3 (1)	7.64E-3 (3)	1.52E-2 (6)	3.05E-2 (12)	5.09E-2 (20)	7.64E-2 (30)	1.52E-1 (60)
150	277	2.85E+10	2.57E+10	2.24E+10	1.92E+10	1.49E+10	1.14E+10	8.57E+09	4.15E+09
250	463	6.06E+10	5.40E+10	4.67E+10	3.95E+10	3.04E+10	2.30E+10	1.71E+10	8.12E+09
300	555	8.60E+10	7.62E+10	6.55E+10	5.50E+10	4.20E+10	3.16E+10	2.34E+10	1.11E+10
450	833	2.19E+11	1.90E+11	1.59E+11	1.31E+11	9.68E+10	7.11E+10	5.18E+10	2.39E+10
600	1111	5.45E+11	4.61E+11	3.75E+11	2.98E+11	2.12E+11	1.51E+11	1.07E+11	4.78E+10
800	1481	1.74E+12	1.44E+12	1.15E+12	8.88E+11	6.07E+11	4.16E+11	2.87E+11	1.24E+11
1000	1852	4.42E+12	3.65E+12	2.89E+12	2.23E+12	1.52E+12	1.03E+12	7.10E+11	3.02E+11
1250	2315	9.02E+12	7.51E+12	6.01E+12	4.68E+12	3.23E+12	2.23E+12	1.54E+12	6.63E+11
1500	2778	1.22E+13	1.02E+13	8.20E+12	6.43E+12	4.48E+12	3.11E+12	2.17E+12	9.42E+11
1750	3241	1.35E+13	1.12E+13	8.95E+12	6.97E+12	4.80E+12	3.31E+12	2.29E+12	9.79E+11
2000	3704	1.32E+13	1.08E+13	8.52E+12	6.53E+12	4.39E+12	2.95E+12	2.00E+12	8.26E+11
2250	4167	1.22E+13	9.90E+12	7.66E+12	5.76E+12	3.77E+12	2.48E+12	1.65E+12	6.61E+11
2500	4630	1.07E+13	8.58E+12	6.55E+12	4.85E+12	3.12E+12	2.03E+12	1.34E+12	5.29E+11
2750	5093	9.12E+12	7.25E+12	5.50E+12	4.05E+12	2.60E+12	1.69E+12	1.12E+12	4.47E+11
3000	5556	8.00E+12	6.37E+12	4.84E+12	3.59E+12	2.33E+12	1.54E+12	1.03E+12	4.22E+11
3500	6482	7.27E+12	5.91E+12	4.61E+12	3.52E+12	2.38E+12	1.63E+12	1.13E+12	4.82E+11
4000	7408	7.72E+12	6.43E+12	5.16E+12	4.05E+12	2.84E+12	1.99E+12	1.40E+12	6.17E+11
4500	8334	8.48E+12	7.22E+12	5.95E+12	4.79E+12	3.47E+12	2.50E+12	1.79E+12	8.07E+11
5000	9260	9.31E+12	8.10E+12	6.82E+12	5.63E+12	4.19E+12	3.09E+12	2.25E+12	1.04E+12
5500	10186	1.06E+13	9.37E+12	8.02E+12	6.72E+12	5.12E+12	3.84E+12	2.84E+12	1.33E+12
6000	11112	1.21E+13	1.08E+13	9.33E+12	7.90E+12	6.10E+12	4.63E+12	3.46E+12	1.66E+12
7000	12964	1.62E+13	1.46E+13	1.27E+13	1.09E+13	8.51E+12	6.53E+12	4.92E+12	2.41E+12
8000	14816	2.12E+13	1.91E+13	1.68E+13	1.44E+13	1.13E+13	8.65E+12	6.53E+12	3.20E+12
9000	16668	2.68E+13	2.42E+13	2.12E+13	1.82E+13	1.42E+13	1.09E+13	8.16E+12	3.95E+12
10000	18520	3.36E+13	3.03E+13	2.65E+13	2.26E+13	1.76E+13	1.34E+13	9.93E+12	4.67E+12
11000	20372	3.76E+13	3.37E+13	2.93E+13	2.48E+13	1.91E+13	1.43E+13	1.05E+13	4.80E+12
12000	22224	4.08E+13	3.64E+13	3.13E+13	2.63E+13	1.99E+13	1.47E+13	1.06E+13	4.63E+12
13000	24076	3.96E+13	3.50E+13	2.98E+13	2.47E+13	1.83E+13	1.32E+13	9.29E+12	3.86E+12
14000	25928	3.52E+13	3.08E+13	2.59E+13	2.12E+13	1.53E+13	1.08E+13	7.46E+12	2.97E+12
15000	27780	2.92E+13	2.53E+13	2.10E+13	1.69E+13	1.20E+13	8.23E+12	5.56E+12	2.14E+12
16000	29632	2.20E+13	1.89E+13	1.55E+13	1.23E+13	8.56E+12	5.79E+12	3.84E+12	1.43E+12
17000	31484	1.76E+13	1.50E+13	1.22E+13	9.56E+12	6.53E+12	4.34E+12	2.83E+12	1.01E+12
18000	33336	1.19E+13	1.01E+13	8.08E+12	6.26E+12	4.20E+12	2.74E+12	1.76E+12	5.96E+11
19327	35793	7.56E+12	6.26E+12	4.93E+12	3.74E+12	2.43E+12	1.54E+12	9.54E+11	3.05E+11

Table 9-51. Annual Equivalent 1 MeV Electron Fluence for GaAs Solar Cells from Trapped Protons During Solar Max and Min, 90° Inclination (Infinite Backshielding)

PROTONS - V_{oc}

EQUIV. 1 MEV ELECTRON FLUENCE FOR V_{oc} CIRCULAR ORBIT DUE TO GEOMAGNETICALLY TRAPPED PROTONS, MODEL AP8MAX

INCLINATION = 90 DEGREES

ALTITUDE (nm)	(km)	SHIELD THICKNESS, cm (mils)							
		0 (0)	2.54E-3 (1)	7.64E-3 (3)	1.52E-2 (6)	3.05E-2 (12)	5.09E-2 (20)	7.64E-2 (30)	1.52E-1 (60)
150	277	3.84E+13	8.30E+10	1.43E+10	7.38E+09	5.18E+09	4.40E+09	3.86E+09	3.40E+09
250	463	3.71E+14	9.89E+11	2.87E+11	1.94E+11	1.45E+11	1.24E+11	1.07E+11	9.21E+10
300	555	9.18E+14	2.59E+12	8.14E+11	5.35E+11	3.80E+11	3.17E+11	2.68E+11	2.26E+11
450	833	5.62E+15	1.95E+13	6.57E+12	3.99E+12	2.58E+12	2.03E+12	1.63E+12	1.31E+12
600	1111	1.91E+16	6.92E+13	2.43E+13	1.48E+13	9.51E+12	7.33E+12	5.68E+12	4.37E+12
800	1481	5.96E+16	2.24E+14	8.67E+13	5.64E+13	3.80E+13	2.91E+13	2.20E+13	1.63E+13
1000	1852	1.50E+17	6.35E+14	2.62E+14	1.73E+14	1.15E+14	8.39E+13	5.97E+13	4.13E+13
1250	2315	3.16E+17	1.85E+15	8.20E+14	5.43E+14	3.48E+14	2.33E+14	1.48E+14	8.98E+13
1500	2778	5.03E+17	4.15E+15	1.96E+15	1.30E+15	8.04E+14	4.98E+14	2.81E+14	1.44E+14
1750	3241	7.18E+17	8.08E+15	3.94E+15	2.56E+15	1.53E+15	8.85E+14	4.47E+14	1.94E+14
2000	3704	9.65E+17	1.44E+16	6.95E+15	4.31E+15	2.37E+15	1.27E+15	5.79E+14	2.17E+14
2250	4167	1.27E+18	2.46E+16	1.14E+16	6.57E+15	3.22E+15	1.58E+15	6.47E+14	2.17E+14
2500	4630	1.65E+18	3.81E+16	1.67E+16	8.84E+15	3.84E+15	1.74E+15	6.47E+14	2.01E+14
2750	5093	2.28E+18	5.45E+16	2.19E+16	1.07E+16	4.12E+15	1.75E+15	6.04E+14	1.75E+14
3000	5556	2.87E+18	7.35E+16	2.66E+16	1.19E+16	4.15E+15	1.65E+15	5.34E+14	1.48E+14
3500	6482	4.90E+18	1.21E+17	3.44E+16	1.27E+16	3.31E+15	1.25E+15	3.78E+14	9.74E+13
4000	7408	8.46E+18	1.77E+17	3.90E+16	1.20E+16	2.42E+15	8.48E+14	2.39E+14	5.92E+13
4500	8334	1.34E+19	2.24E+17	3.88E+16	1.02E+16	1.62E+15	5.20E+14	1.34E+14	3.15E+13
5000	9260	1.95E+19	2.55E+17	3.52E+16	7.82E+15	9.82E+14	2.92E+14	6.97E+13	1.54E+13
5500	10186	2.59E+19	2.76E+17	3.05E+16	5.66E+15	5.44E+14	1.55E+14	3.48E+13	7.03E+12
6000	11112	3.14E+19	2.65E+17	2.38E+16	3.74E+15	2.81E+14	7.27E+13	1.45E+13	2.59E+12
7000	12964	4.02E+19	1.84E+17	1.11E+16	1.25E+15	5.76E+13	1.17E+13	1.77E+12	2.42E+11
8000	14816	4.06E+19	8.85E+16	3.33E+15	2.69E+14	7.61E+12	1.18E+12	1.30E+11	1.35E+10
9000	16668	2.85E+19	2.87E+16	6.20E+14	3.31E+13	5.11E+11	6.25E+10	7.10E+09	1.50E+09
10000	18520	2.06E+19	7.95E+15	9.85E+13	3.68E+12	2.54E+10	0.00E+00	0.00E+00	0.00E+00
11000	20372	1.48E+19	1.78E+15	8.19E+12	1.81E+11	9.09E+08	0.00E+00	0.00E+00	0.00E+00
12000	22224	1.19E+19	3.24E+14	1.41E+11	9.09E+08	0.00E+00	0.00E+00	0.00E+00	0.00E+00
13000	24076	9.79E+18	4.65E+13	4.57E+07	0.00E+00	0.00E+00	0.00E+00	0.00E+00	0.00E+00
14000	25928	8.18E+18	7.98E+12	2.03E+07	0.00E+00	0.00E+00	0.00E+00	0.00E+00	0.00E+00
15000	27780	6.79E+18	2.12E+12	1.09E+07	0.00E+00	0.00E+00	0.00E+00	0.00E+00	0.00E+00
16000	29632	5.21E+18	2.61E+10	0.00E+00	0.00E+00	0.00E+00	0.00E+00	0.00E+00	0.00E+00
17000	31484	4.31E+18	2.00E+10	0.00E+00	0.00E+00	0.00E+00	0.00E+00	0.00E+00	0.00E+00
18000	33336	2.84E+18	1.46E+10	0.00E+00	0.00E+00	0.00E+00	0.00E+00	0.00E+00	0.00E+00
19327	35793	8.90E+17	9.81E+09	0.00E+00	0.00E+00	0.00E+00	0.00E+00	0.00E+00	0.00E+00

EQUIV. 1 MEV ELECTRON FLUENCE FOR V_{oc} CIRCULAR ORBIT DUE TO GEOMAGNETICALLY TRAPPED PROTONS, MODEL AP8MIN

INCLINATION = 90 DEGREES

ALTITUDE (nm)	(km)	0 (0)	2.54E-3 (1)	7.64E-3 (3)	1.52E-2 (6)	3.05E-2 (12)	5.09E-2 (20)	7.64E-2 (30)	1.52E-1 (60)
150	277	3.70E+13	4.07E+11	1.24E+11	5.84E+10	2.71E+10	1.93E+10	1.46E+10	1.17E+10
250	463	3.64E+14	4.13E+12	1.35E+12	6.84E+11	3.56E+11	2.65E+11	2.07E+11	1.67E+11
300	555	8.42E+14	8.27E+12	2.70E+12	1.39E+12	7.49E+11	5.65E+11	4.45E+11	3.60E+11
450	833	4.95E+15	3.22E+13	1.08E+13	5.91E+12	3.49E+12	2.71E+12	2.16E+12	1.74E+12
600	1111	1.72E+16	8.62E+13	3.02E+13	1.78E+13	1.13E+13	8.81E+12	6.95E+12	5.44E+12
800	1481	5.78E+16	2.67E+14	1.02E+14	6.41E+13	4.25E+13	3.29E+13	2.53E+13	1.89E+13
1000	1852	1.48E+17	7.18E+14	2.88E+14	1.86E+14	1.22E+14	8.97E+13	6.46E+13	4.49E+13

Table 9-52. Annual Equivalent 1 MeV Electron Fluence for GaAs Solar Cells from Trapped Protons During Solar Max and Min, 90° Inclination (Infinite Backshielding)

PROTONS - P_max

INCLINATION = 90 DEGREES

EQUIV. 1 MEV ELECTRON FLUENCE FOR P_max CIRCULAR ORBIT
DUE TO GEOMAGNETICALLY TRAPPED PROTONS, MODEL AP8MAX

ALTITUDE		SHIELD THICKNESS, cm (mils)							
(nm)	(km)	0 (0)	2.54E-3 (1)	7.64E-3 (3)	1.52E-2 (6)	3.05E-2 (12)	5.09E-2 (20)	7.64E-2 (30)	1.52E-1 (60)
150	277	2.74E+13	5.93E+10	1.02E+10	5.27E+09	3.70E+09	3.14E+09	2.76E+09	2.43E+09
250	463	2.65E+14	7.06E+11	2.05E+11	1.39E+11	1.04E+11	8.86E+10	7.64E+10	6.58E+10
300	555	6.49E+14	1.85E+12	5.81E+11	3.82E+11	2.72E+11	2.27E+11	1.91E+11	1.61E+11
450	833	4.02E+15	1.39E+13	4.70E+12	2.85E+12	1.84E+12	1.45E+12	1.16E+12	9.34E+11
600	1111	1.36E+16	4.94E+13	1.74E+13	1.06E+13	6.79E+12	5.24E+12	4.06E+12	3.12E+12
800	1481	4.25E+16	1.60E+14	6.19E+13	4.03E+13	2.72E+13	2.08E+13	1.57E+13	1.16E+13
1000	1852	1.07E+17	4.54E+14	1.87E+14	1.24E+14	8.22E+13	5.99E+13	4.26E+13	2.95E+13
1250	2315	2.26E+17	1.32E+15	5.86E+14	3.88E+14	2.49E+14	1.66E+14	1.06E+14	6.42E+13
1500	2778	3.59E+17	2.96E+15	1.40E+15	9.26E+14	5.74E+14	3.56E+14	2.01E+14	1.03E+14
1750	3241	5.13E+17	5.77E+15	2.81E+15	1.83E+15	1.09E+15	6.32E+14	3.20E+14	1.38E+14
2000	3704	6.89E+17	1.03E+16	4.96E+15	3.08E+15	1.69E+15	9.09E+14	4.13E+14	1.55E+14
2250	4167	9.07E+17	1.76E+16	8.17E+15	4.69E+15	2.30E+15	1.13E+15	4.62E+14	1.55E+14
2500	4630	1.18E+18	2.72E+16	1.19E+16	6.31E+15	2.74E+15	1.24E+15	4.62E+14	1.43E+14
2750	5093	1.56E+18	3.89E+16	1.56E+16	7.65E+15	2.97E+15	1.25E+15	4.31E+14	1.25E+14
3000	5556	2.05E+18	5.25E+16	1.90E+16	8.53E+15	2.94E+15	1.18E+15	3.81E+14	1.06E+14
3500	6482	3.50E+18	8.66E+16	2.46E+16	9.06E+15	2.37E+15	8.93E+14	2.70E+14	6.96E+13
4000	7408	6.04E+18	1.26E+17	2.79E+16	8.59E+15	1.73E+15	6.06E+14	1.71E+14	4.23E+13
4500	8334	9.55E+18	1.60E+17	2.77E+16	7.28E+15	1.16E+15	3.71E+14	9.57E+13	2.25E+13
5000	9260	1.39E+19	1.82E+17	2.51E+16	5.59E+15	7.02E+14	2.09E+14	4.98E+13	1.10E+13
5500	10186	1.85E+19	1.97E+17	2.18E+16	4.04E+15	3.89E+14	1.11E+14	2.49E+13	5.02E+12
6000	11112	2.24E+19	1.90E+17	1.70E+16	2.67E+15	2.01E+14	5.19E+13	1.03E+13	1.85E+12
7000	12964	2.87E+19	1.32E+17	7.91E+15	8.90E+14	4.11E+13	8.37E+12	1.27E+12	1.73E+11
8000	14816	2.90E+19	6.32E+16	2.38E+15	1.92E+14	5.44E+12	8.41E+11	9.28E+10	9.62E+09
9000	16668	2.04E+19	2.05E+16	4.43E+14	2.36E+13	3.65E+11	4.47E+10	5.07E+09	1.07E+09
10000	18520	1.47E+19	5.68E+15	7.04E+13	2.63E+12	1.82E+10	0.00E+00	0.00E+00	0.00E+00
11000	20372	1.06E+19	1.27E+15	5.85E+12	1.29E+11	6.49E+08	0.00E+00	0.00E+00	0.00E+00
12000	22224	8.53E+18	2.31E+14	1.01E+11	0.00E+00	0.00E+00	0.00E+00	0.00E+00	0.00E+00
13000	24076	6.99E+18	3.32E+13	3.27E+07	0.00E+00	0.00E+00	0.00E+00	0.00E+00	0.00E+00
14000	25928	5.84E+18	5.70E+12	1.45E+07	0.00E+00	0.00E+00	0.00E+00	0.00E+00	0.00E+00
15000	27780	4.85E+18	1.51E+12	7.77E+06	0.00E+00	0.00E+00	0.00E+00	0.00E+00	0.00E+00
16000	29632	3.72E+18	1.87E+10	0.00E+00	0.00E+00	0.00E+00	0.00E+00	0.00E+00	0.00E+00
17000	31484	3.08E+18	1.43E+10	0.00E+00	0.00E+00	0.00E+00	0.00E+00	0.00E+00	0.00E+00
18000	33336	2.03E+18	1.04E+10	0.00E+00	0.00E+00	0.00E+00	0.00E+00	0.00E+00	0.00E+00
19327	35793	6.36E+17	7.01E+09	0.00E+00	0.00E+00	0.00E+00	0.00E+00	0.00E+00	0.00E+00

INCLINATION = 90 DEGREES

EQUIV. 1 MEV ELECTRON FLUENCE FOR P_max CIRCULAR ORBIT
DUE TO GEOMAGNETICALLY TRAPPED PROTONS, MODEL AP8MIN

(nm)	(km)	0	2.54E-3 (1)	7.64E-3 (3)	1.52E-2 (6)	3.05E-2 (12)	5.09E-2 (20)	7.64E-2 (30)	1.52E-1 (60)
150	277	2.64E+13	2.91E+11	8.88E+10	4.17E+10	1.94E+10	1.38E+10	1.04E+10	8.32E+09
250	463	2.60E+14	2.95E+12	9.67E+11	4.89E+11	2.54E+11	1.90E+11	1.48E+11	1.19E+11
300	555	6.01E+14	5.90E+12	1.93E+12	9.92E+11	5.35E+11	4.04E+11	3.18E+11	2.57E+11
450	833	3.54E+15	2.30E+13	7.70E+12	4.22E+12	2.49E+12	1.94E+12	1.54E+12	1.24E+12
600	1111	1.23E+16	6.16E+13	2.16E+13	1.27E+13	8.05E+12	6.29E+12	4.97E+12	3.89E+12
800	1481	4.13E+16	1.91E+14	7.26E+13	4.58E+13	3.03E+13	2.35E+13	1.80E+13	1.35E+13
1000	1852	1.05E+17	5.13E+14	2.06E+14	1.33E+14	8.69E+13	6.41E+13	4.61E+13	3.21E+13

Table 9-53. Annual Equivalent 1 MeV Electron Fluence for GaAs Solar Cells from Trapped Protons During Solar Max and Min, 90° Inclination (Infinite Backshielding)

PROTONS - I$_{sc}$

EQUIV. 1 MEV PROTON FLUENCE FOR I$_{sc}$ - CIRCULAR ORBIT   INCLINATION = 90 DEGREES
DUE TO GEOMAGNETICALLY TRAPPED PROTONS, MODEL AP8MAX

ALTITUDE (nm)	(km)	SHIELD THICKNESS, cm (mils) 0 (0)	2.54E-3 (1)	7.64E-3 (3)	1.52E-2 (6)	3.05E-2 (12)	5.09E-2 (20)	7.64E-2 (30)	1.52E-1 (60)
150	277	3.20E+13	3.22E+10	4.30E+09	1.73E+09	9.82E+08	7.35E+08	5.76E+08	4.75E+08
250	463	3.11E+14	3.61E+11	7.88E+10	4.49E+10	2.85E+10	2.20E+10	1.69E+10	1.35E+10
300	555	7.70E+14	9.45E+11	2.32E+11	1.31E+11	7.88E+10	5.91E+10	4.39E+10	3.40E+10
450	833	4.85E+15	7.15E+12	2.01E+12	1.07E+12	5.89E+11	4.16E+11	2.90E+11	2.09E+11
600	1111	1.67E+16	2.53E+13	7.55E+12	4.08E+12	2.29E+12	1.59E+12	1.08E+12	7.32E+11
800	1481	5.24E+16	8.09E+13	2.66E+13	1.56E+13	9.46E+12	6.61E+12	4.41E+12	2.85E+12
1000	1852	1.31E+17	2.29E+14	8.19E+13	4.98E+13	3.05E+13	2.05E+13	1.29E+13	7.71E+12
1250	2315	2.69E+17	6.70E+14	2.65E+14	1.66E+14	1.01E+14	6.29E+13	3.56E+13	1.86E+13
1500	2778	4.15E+17	1.50E+15	6.48E+14	4.10E+14	2.46E+14	1.45E+14	7.43E+13	3.34E+13
1750	3241	5.73E+17	2.93E+15	1.32E+15	8.31E+14	4.83E+14	2.70E+14	1.27E+14	4.88E+13
2000	3704	7.48E+17	5.25E+15	2.38E+15	1.42E+15	7.67E+14	4.00E+14	1.71E+14	5.79E+13
2250	4167	9.54E+17	9.05E+15	3.97E+15	2.20E+15	1.06E+15	5.07E+14	1.95E+14	6.03E+13
2500	4630	1.20E+18	1.42E+16	5.85E+15	2.99E+15	1.28E+15	5.64E+14	1.98E+14	5.73E+13
2750	5093	1.56E+18	2.04E+16	7.75E+15	3.66E+15	1.39E+15	5.73E+14	1.87E+14	5.09E+13
3000	5556	2.02E+18	2.78E+16	9.50E+15	4.10E+15	1.12E+15	5.41E+14	1.66E+14	4.35E+13
3500	6482	3.34E+18	4.67E+16	1.24E+16	4.40E+15	8.22E+14	4.13E+14	1.19E+14	2.92E+13
4000	7408	5.68E+18	6.92E+16	1.42E+16	4.20E+15	5.53E+14	2.82E+14	7.55E+13	1.79E+13
4500	8334	9.01E+18	8.86E+16	1.43E+16	3.58E+15	3.36E+14	1.73E+14	4.25E+13	9.64E+12
5000	9260	1.33E+19	1.02E+17	1.30E+16	2.76E+15	1.87E+14	9.80E+13	2.22E+13	4.76E+12
5500	10186	1.75E+19	1.12E+17	1.14E+16	2.01E+15	9.69E+13	5.22E+13	1.12E+13	2.19E+12
6000	11112	2.08E+19	1.08E+17	8.90E+15	1.33E+15	2.00E+13	2.46E+13	4.68E+12	8.11E+11
7000	12964	2.72E+19	7.62E+16	4.17E+15	4.46E+14	2.66E+12	3.99E+12	5.78E+11	7.70E+10
8000	14816	2.93E+19	3.72E+16	1.26E+15	9.65E+13	1.79E+11	4.04E+11	4.26E+10	4.26E+09
9000	16668	2.14E+19	1.23E+16	2.36E+14	1.20E+13	9.92E+09	2.14E+10	2.29E+09	4.80E+08
10000	18520	1.61E+19	3.47E+15	3.78E+13	1.33E+12	3.47E+08	0.00E+00	0.00E+00	0.00E+00
11000	20372	1.22E+19	7.91E+14	3.18E+12	6.58E+10	0.00E+00	0.00E+00	0.00E+00	0.00E+00
12000	22224	1.06E+19	1.47E+14	5.65E+10	0.00E+00	0.00E+00	0.00E+00	0.00E+00	0.00E+00
13000	24076	9.39E+18	2.18E+13	2.00E+07	0.00E+00	0.00E+00	0.00E+00	0.00E+00	0.00E+00
14000	25928	8.40E+18	3.85E+12	8.71E+06	0.00E+00	0.00E+00	0.00E+00	0.00E+00	0.00E+00
15000	27780	7.44E+18	1.04E+12	4.58E+06	0.00E+00	0.00E+00	0.00E+00	0.00E+00	0.00E+00
16000	29632	5.98E+18	1.55E+11	0.00E+00	0.00E+00	0.00E+00	0.00E+00	0.00E+00	0.00E+00
17000	31484	5.18E+18	1.18E+10	0.00E+00	0.00E+00	0.00E+00	0.00E+00	0.00E+00	0.00E+00
18000	33336	3.55E+18	8.46E+09	0.00E+00	0.00E+00	0.00E+00	0.00E+00	0.00E+00	0.00E+00
19327	35793	1.11E+18	5.62E+09	0.00E+00	0.00E+00	0.00E+00	0.00E+00	0.00E+00	0.00E+00

EQUIV. 1 MEV PROTON FLUENCE FOR I$_{sc}$ - CIRCULAR ORBIT   INCLINATION = 90 DEGREES
DUE TO GEOMAGNETICALLY TRAPPED PROTONS, MODEL AP8MIN

ALTITUDE (nm)	(km)	SHIELD THICKNESS, cm (mils) 0 (0)	2.54E-3 (1)	7.64E-3 (3)	1.52E-2 (6)	3.05E-2 (12)	5.09E-2 (20)	7.64E-2 (30)	1.52E-1 (60)
150	277	2.73E+13	1.54E+11	4.18E+10	1.75E+10	6.67E+09	4.09E+09	2.61E+09	1.84E+09
250	463	2.72E+14	1.54E+12	4.44E+11	1.97E+11	8.42E+10	5.46E+10	3.63E+10	2.61E+10
300	555	6.43E+14	3.08E+12	8.76E+11	3.96E+11	1.75E+11	1.15E+11	7.80E+10	5.63E+10
450	833	4.05E+15	1.19E+13	3.42E+12	1.63E+12	8.03E+11	5.52E+11	3.81E+11	2.73E+11
600	1111	1.46E+16	3.16E+13	9.39E+12	4.87E+12	2.64E+12	1.86E+12	1.28E+12	8.88E+11
800	1481	4.99E+16	9.67E+13	3.13E+13	1.77E+13	1.04E+13	7.33E+12	4.96E+12	3.25E+12
1000	1852	1.28E+17	2.60E+14	9.06E+13	5.33E+13	3.20E+13	2.17E+13	1.38E+13	8.28E+12

Table 9-54. Annual Equivalent 1 MeV Electron Fluence for GaAs Solar Cells from Trapped Electrons During Solar Min, 0° Inclination at Synchronous Altitude vs. Longitude (Infinite Backshielding)

ELECTRONS - I$_{sc}$, V$_{oc}$, AND P$_{max}$ INCLINATION = 0 DEGREES

EQUIV. 1 MEV ELECTRON FLUENCE FOR I$_{sc}$ - CIRCULAR ORBIT
DUE TO GEOMAGNETICALLY TRAPPED ELECTRONS - MODEL AE8MIN

ALTITUDE (nm)	LONGITUDE (Deg)	SHIELD THICKNESS, cm (mils)							
		0 (0)	2.54E-3 (1)	7.64E-3 (3)	1.52E-2 (6)	3.05E-2 (12)	5.09E-2 (20)	7.64E-2 (30)	1.52E-1 (60)
19327	180 W	4.07E+13	3.41E+13	2.72E+13	2.10E+13	1.40E+13	9.02E+12	5.71E+12	1.87E+12
19327	170 W	4.19E+13	3.52E+13	2.82E+13	2.17E+13	1.45E+13	9.39E+12	5.95E+12	1.96E+12
19327	160 W	4.25E+13	3.56E+13	2.85E+13	2.21E+13	1.47E+13	9.55E+12	6.06E+12	1.99E+12
19327	150 W	4.23E+13	3.55E+13	2.84E+13	2.20E+13	1.47E+13	9.50E+12	6.03E+12	1.98E+12
19327	140 W	4.15E+13	3.48E+13	2.79E+13	2.15E+13	1.44E+13	9.29E+12	5.89E+12	1.93E+12
19327	130 W	4.03E+13	3.38E+13	2.70E+13	2.08E+13	1.38E+13	8.94E+12	5.66E+12	1.85E+12
19327	120 W	3.88E+13	3.25E+13	2.60E+13	2.00E+13	1.32E+13	8.53E+12	5.38E+12	1.76E+12
19327	110 W	3.75E+13	3.14E+13	2.50E+13	1.92E+13	1.27E+13	8.15E+12	5.14E+12	1.68E+12
19327	100 W	3.62E+13	3.02E+13	2.40E+13	1.84E+13	1.22E+13	7.78E+12	4.89E+12	1.59E+12
19327	90 W	3.52E+13	2.94E+13	2.34E+13	1.79E+13	1.18E+13	7.52E+12	4.72E+12	1.53E+12
19327	80 W	3.45E+13	2.88E+13	2.29E+13	1.75E+13	1.15E+13	7.34E+12	4.60E+12	1.49E+12
19327	70 W	3.44E+13	2.87E+13	2.28E+13	1.74E+13	1.14E+13	7.30E+12	4.58E+12	1.49E+12
19327	60 W	3.47E+13	2.90E+13	2.30E+13	1.76E+13	1.16E+13	7.40E+12	4.64E+12	1.51E+12
19327	50 W	3.56E+13	2.98E+13	2.37E+13	1.81E+13	1.20E+13	7.65E+12	4.81E+12	1.56E+12
19327	40 W	3.70E+13	3.10E+13	2.47E+13	1.90E+13	1.25E+13	8.04E+12	5.07E+12	1.65E+12
19327	30 W	3.89E+13	3.26E+13	2.60E+13	2.00E+13	1.33E+13	8.57E+12	5.42E+12	1.77E+12
19327	20 W	4.10E+13	3.44E+13	2.75E+13	2.12E+13	1.42E+13	9.17E+12	5.81E+12	1.91E+12
19327	10 W	4.32E+13	3.63E+13	2.91E+13	2.25E+13	1.51E+13	9.80E+12	6.23E+12	2.05E+12
19327	0	4.51E+13	3.80E+13	3.05E+13	2.37E+13	1.59E+13	1.04E+13	6.62E+12	2.19E+12
19327	10 E	4.64E+13	3.91E+13	3.15E+13	2.44E+13	1.65E+13	1.08E+13	6.87E+12	2.27E+12
19327	20 E	4.68E+13	3.94E+13	3.17E+13	2.47E+13	1.66E+13	1.09E+13	6.94E+12	2.30E+12
19327	30 E	4.63E+13	3.90E+13	3.13E+13	2.43E+13	1.64E+13	1.07E+13	6.83E+12	2.26E+12
19327	40 E	4.46E+13	3.75E+13	3.01E+13	2.34E+13	1.57E+13	1.02E+13	6.50E+12	2.14E+12
19327	50 E	4.20E+13	3.52E+13	2.82E+13	2.18E+13	1.46E+13	9.44E+12	5.99E+12	1.97E+12
19327	60 E	3.90E+13	3.26E+13	2.60E+13	2.00E+13	1.33E+13	8.56E+12	5.40E+12	1.77E+12
19327	70 E	3.58E+13	2.99E+13	2.38E+13	1.82E+13	1.20E+13	7.67E+12	4.82E+12	1.57E+12
19327	80 E	3.29E+13	2.74E+13	2.17E+13	1.66E+13	1.08E+13	6.87E+12	4.29E+12	1.39E+12
19327	90 E	3.06E+13	2.54E+13	2.01E+13	1.53E+13	9.91E+12	6.25E+12	3.89E+12	1.25E+12
19327	100 E	2.91E+13	2.41E+13	1.90E+13	1.44E+13	9.30E+12	5.84E+12	3.62E+12	1.16E+12
19327	110 E	2.84E+13	2.35E+13	1.85E+13	1.40E+13	9.04E+12	5.66E+12	3.50E+12	1.12E+12
19327	120 E	2.87E+13	2.38E+13	1.87E+13	1.42E+13	9.14E+12	5.73E+12	3.55E+12	1.14E+12
19327	130 E	2.98E+13	2.47E+13	1.95E+13	1.48E+13	9.57E+12	6.02E+12	3.73E+12	1.20E+12
19327	140 E	3.16E+13	2.62E+13	2.07E+13	1.58E+13	1.03E+13	6.49E+12	4.04E+12	1.30E+12
19327	150 E	3.39E+13	2.82E+13	2.24E+13	1.71E+13	1.12E+13	7.12E+12	4.45E+12	1.44E+12
19327	160 E	3.64E+13	3.04E+13	2.41E+13	1.85E+13	1.22E+13	7.80E+12	4.90E+12	1.59E+12
19327	170 E	3.87E+13	3.24E+13	2.58E+13	1.99E+13	1.31E+13	8.46E+12	5.33E+12	1.74E+12

Table 9-55. Annual Equivalent 1 MeV Electron Fluence for GaAs Solar Cells from Trapped Protons During Solar Min, 0° Inclination at Synchronous Altitude vs. Longitude (Infinite Backshielding)

PROTONS - V_{oc} INCLINATION = 0 DEGREES

EQUIV. 1 MEV ELECTRON FLUENCE FOR V_{oc} CIRCULAR ORBIT
DUE TO GEOMAGNETICALLY TRAPPED PROTONS, MODEL AP8MIN

ALTITUDE (nm)	LONGITUDE (Deg)	SHIELD THICKNESS, cm (mils)							
		0 (0)	2.54E-3 (1)	7.64E-3 (3)	1.52E-2 (6)	3.05E-2 (12)	5.09E-2 (20)	7.64E-2 (30)	1.52E-1 (60)
19327	180 W	8.64E+18	3.30E+10	0.00E+00	0.00E+00	0.00E+00	0.00E+00	0.00E+00	0.00E+00
19327	170 W	9.87E+18	3.47E+10	0.00E+00	0.00E+00	0.00E+00	0.00E+00	0.00E+00	0.00E+00
19327	160 W	1.04E+19	3.52E+10	0.00E+00	0.00E+00	0.00E+00	0.00E+00	0.00E+00	0.00E+00
19327	150 W	1.02E+19	3.46E+10	0.00E+00	0.00E+00	0.00E+00	0.00E+00	0.00E+00	0.00E+00
19327	140 W	9.48E+18	3.30E+10	0.00E+00	0.00E+00	0.00E+00	0.00E+00	0.00E+00	0.00E+00
19327	130 W	8.31E+18	3.08E+10	0.00E+00	0.00E+00	0.00E+00	0.00E+00	0.00E+00	0.00E+00
19327	120 W	7.09E+18	2.83E+10	0.00E+00	0.00E+00	0.00E+00	0.00E+00	0.00E+00	0.00E+00
19327	110 W	6.08E+18	2.60E+10	0.00E+00	0.00E+00	0.00E+00	0.00E+00	0.00E+00	0.00E+00
19327	100 W	5.18E+18	2.40E+10	0.00E+00	0.00E+00	0.00E+00	0.00E+00	0.00E+00	0.00E+00
19327	90 W	4.60E+18	2.24E+10	0.00E+00	0.00E+00	0.00E+00	0.00E+00	0.00E+00	0.00E+00
19327	80 W	4.23E+18	2.15E+10	0.00E+00	0.00E+00	0.00E+00	0.00E+00	0.00E+00	0.00E+00
19327	70 W	4.15E+18	2.12E+10	0.00E+00	0.00E+00	0.00E+00	0.00E+00	0.00E+00	0.00E+00
19327	60 W	4.34E+18	2.15E+10	0.00E+00	0.00E+00	0.00E+00	0.00E+00	0.00E+00	0.00E+00
19327	50 W	4.85E+18	2.24E+10	0.00E+00	0.00E+00	0.00E+00	0.00E+00	0.00E+00	0.00E+00
19327	40 W	5.75E+18	2.41E+10	0.00E+00	0.00E+00	0.00E+00	0.00E+00	0.00E+00	0.00E+00
19327	30 W	7.13E+18	2.65E+10	0.00E+00	0.00E+00	0.00E+00	0.00E+00	0.00E+00	0.00E+00
19327	20 W	8.92E+18	2.94E+10	0.00E+00	0.00E+00	0.00E+00	0.00E+00	0.00E+00	0.00E+00
19327	10 W	1.12E+19	3.26E+10	0.00E+00	0.00E+00	0.00E+00	0.00E+00	0.00E+00	0.00E+00
19327	0	1.36E+19	3.57E+10	0.00E+00	0.00E+00	0.00E+00	0.00E+00	0.00E+00	0.00E+00
19327	10 E	1.53E+19	3.79E+10	0.00E+00	0.00E+00	0.00E+00	0.00E+00	0.00E+00	0.00E+00
19327	20 E	1.55E+19	3.88E+10	0.00E+00	0.00E+00	0.00E+00	0.00E+00	0.00E+00	0.00E+00
19327	30 E	1.51E+19	3.83E+10	0.00E+00	0.00E+00	0.00E+00	0.00E+00	0.00E+00	0.00E+00
19327	40 E	1.29E+19	3.61E+10	0.00E+00	0.00E+00	0.00E+00	0.00E+00	0.00E+00	0.00E+00
19327	50 E	9.92E+18	3.25E+10	0.00E+00	0.00E+00	0.00E+00	0.00E+00	0.00E+00	0.00E+00
19327	60 E	7.16E+18	2.85E+10	0.00E+00	0.00E+00	0.00E+00	0.00E+00	0.00E+00	0.00E+00
19327	70 E	4.96E+18	2.45E+10	0.00E+00	0.00E+00	0.00E+00	0.00E+00	0.00E+00	0.00E+00
19327	80 E	3.43E+18	2.11E+10	0.00E+00	0.00E+00	0.00E+00	0.00E+00	0.00E+00	0.00E+00
19327	90 E	2.50E+18	1.87E+10	0.00E+00	0.00E+00	0.00E+00	0.00E+00	0.00E+00	0.00E+00
19327	100 E	1.98E+18	1.73E+10	0.00E+00	0.00E+00	0.00E+00	0.00E+00	0.00E+00	0.00E+00
19327	110 E	1.80E+18	1.68E+10	0.00E+00	0.00E+00	0.00E+00	0.00E+00	0.00E+00	0.00E+00
19327	120 E	1.87E+18	1.71E+10	0.00E+00	0.00E+00	0.00E+00	0.00E+00	0.00E+00	0.00E+00
19327	130 E	2.21E+18	1.84E+10	0.00E+00	0.00E+00	0.00E+00	0.00E+00	0.00E+00	0.00E+00
19327	140 E	2.86E+18	2.05E+10	0.00E+00	0.00E+00	0.00E+00	0.00E+00	0.00E+00	0.00E+00
19327	150 E	3.89E+18	2.35E+10	0.00E+00	0.00E+00	0.00E+00	0.00E+00	0.00E+00	0.00E+00
19327	160 E	5.30E+18	2.69E+10	0.00E+00	0.00E+00	0.00E+00	0.00E+00	0.00E+00	0.00E+00
19327	170 E	6.96E+18	3.02E+10	0.00E+00	0.00E+00	0.00E+00	0.00E+00	0.00E+00	0.00E+00

Table 9-56. Annual Equivalent 1 MeV Electron Fluence for GaAs Solar Cells from Trapped Protons During Solar Min, 0° Inclination at Synchronous Altitude vs. Longitude (Infinite Backshielding)

PROTONS - P_{max} INCLINATION = 0 DEGREES

EQUIV. 1 MEV ELECTRON FLUENCE FOR P_{max} CIRCULAR ORBIT
DUE TO GEOMAGNETICALLY TRAPPED PROTONS, MODEL AP8MIN

ALTITUDE (nm)	LONGITUDE (Deg)	SHIELD THICKNESS, cm (mils)							
		0 (0)	2.54E-3 (1)	7.64E-3 (3)	1.52E-2 (6)	3.05E-2 (12)	5.09E-2 (20)	7.64E-2 (30)	1.52E-1 (60)
19327	180 W	6.17E+18	2.36E+10	0.00E+00	0.00E+00	0.00E+00	0.00E+00	0.00E+00	0.00E+00
19327	170 W	7.05E+18	2.48E+10	0.00E+00	0.00E+00	0.00E+00	0.00E+00	0.00E+00	0.00E+00
19327	160 W	7.45E+18	2.52E+10	0.00E+00	0.00E+00	0.00E+00	0.00E+00	0.00E+00	0.00E+00
19327	150 W	7.31E+18	2.47E+10	0.00E+00	0.00E+00	0.00E+00	0.00E+00	0.00E+00	0.00E+00
19327	140 W	6.77E+18	2.36E+10	0.00E+00	0.00E+00	0.00E+00	0.00E+00	0.00E+00	0.00E+00
19327	130 W	5.94E+18	2.20E+10	0.00E+00	0.00E+00	0.00E+00	0.00E+00	0.00E+00	0.00E+00
19327	120 W	5.06E+18	2.02E+10	0.00E+00	0.00E+00	0.00E+00	0.00E+00	0.00E+00	0.00E+00
19327	110 W	4.34E+18	1.86E+10	0.00E+00	0.00E+00	0.00E+00	0.00E+00	0.00E+00	0.00E+00
19327	100 W	3.70E+18	1.71E+10	0.00E+00	0.00E+00	0.00E+00	0.00E+00	0.00E+00	0.00E+00
19327	90 W	3.29E+18	1.60E+10	0.00E+00	0.00E+00	0.00E+00	0.00E+00	0.00E+00	0.00E+00
19327	80 W	3.02E+18	1.53E+10	0.00E+00	0.00E+00	0.00E+00	0.00E+00	0.00E+00	0.00E+00
19327	70 W	2.96E+18	1.51E+10	0.00E+00	0.00E+00	0.00E+00	0.00E+00	0.00E+00	0.00E+00
19327	60 W	3.10E+18	1.53E+10	0.00E+00	0.00E+00	0.00E+00	0.00E+00	0.00E+00	0.00E+00
19327	50 W	3.47E+18	1.60E+10	0.00E+00	0.00E+00	0.00E+00	0.00E+00	0.00E+00	0.00E+00
19327	40 W	4.11E+18	1.72E+10	0.00E+00	0.00E+00	0.00E+00	0.00E+00	0.00E+00	0.00E+00
19327	30 W	5.09E+18	1.89E+10	0.00E+00	0.00E+00	0.00E+00	0.00E+00	0.00E+00	0.00E+00
19327	20 W	6.37E+18	2.10E+10	0.00E+00	0.00E+00	0.00E+00	0.00E+00	0.00E+00	0.00E+00
19327	10 W	7.98E+18	2.33E+10	0.00E+00	0.00E+00	0.00E+00	0.00E+00	0.00E+00	0.00E+00
19327	0	9.71E+18	2.55E+10	0.00E+00	0.00E+00	0.00E+00	0.00E+00	0.00E+00	0.00E+00
19327	10 E	1.09E+19	2.71E+10	0.00E+00	0.00E+00	0.00E+00	0.00E+00	0.00E+00	0.00E+00
19327	20 E	1.11E+19	2.77E+10	0.00E+00	0.00E+00	0.00E+00	0.00E+00	0.00E+00	0.00E+00
19327	30 E	1.08E+19	2.73E+10	0.00E+00	0.00E+00	0.00E+00	0.00E+00	0.00E+00	0.00E+00
19327	40 E	9.21E+18	2.58E+10	0.00E+00	0.00E+00	0.00E+00	0.00E+00	0.00E+00	0.00E+00
19327	50 E	7.09E+18	2.32E+10	0.00E+00	0.00E+00	0.00E+00	0.00E+00	0.00E+00	0.00E+00
19327	60 E	5.12E+18	2.04E+10	0.00E+00	0.00E+00	0.00E+00	0.00E+00	0.00E+00	0.00E+00
19327	70 E	3.54E+18	1.75E+10	0.00E+00	0.00E+00	0.00E+00	0.00E+00	0.00E+00	0.00E+00
19327	80 E	2.45E+18	1.51E+10	0.00E+00	0.00E+00	0.00E+00	0.00E+00	0.00E+00	0.00E+00
19327	90 E	1.78E+18	1.34E+10	0.00E+00	0.00E+00	0.00E+00	0.00E+00	0.00E+00	0.00E+00
19327	100 E	1.42E+18	1.24E+10	0.00E+00	0.00E+00	0.00E+00	0.00E+00	0.00E+00	0.00E+00
19327	110 E	1.28E+18	1.20E+10	0.00E+00	0.00E+00	0.00E+00	0.00E+00	0.00E+00	0.00E+00
19327	120 E	1.33E+18	1.22E+10	0.00E+00	0.00E+00	0.00E+00	0.00E+00	0.00E+00	0.00E+00
19327	130 E	1.58E+18	1.31E+10	0.00E+00	0.00E+00	0.00E+00	0.00E+00	0.00E+00	0.00E+00
19327	140 E	2.04E+18	1.47E+10	0.00E+00	0.00E+00	0.00E+00	0.00E+00	0.00E+00	0.00E+00
19327	150 E	2.78E+18	1.68E+10	0.00E+00	0.00E+00	0.00E+00	0.00E+00	0.00E+00	0.00E+00
19327	160 E	3.79E+18	1.92E+10	0.00E+00	0.00E+00	0.00E+00	0.00E+00	0.00E+00	0.00E+00
19327	170 E	4.97E+18	2.16E+10	0.00E+00	0.00E+00	0.00E+00	0.00E+00	0.00E+00	0.00E+00

Table 9-57. Annual Equivalent 1 MeV Electron Fluence for GaAs Solar Cells from Trapped Protons During Solar Min, 0° Inclination at Synchronous Altitude vs. Longitude (Infinite Backshielding)

PROTONS - I_{sc} INCLINATION = 0 DEGREES

EQUIV. 1 MEV PROTON FLUENCE FOR I_{sc} - CIRCULAR ORBIT
DUE TO GEOMAGNETICALLY TRAPPED PROTONS, MODEL AP8MIN

ALTITUDE (nm)	LONGITUDE (Deg)	SHIELD THICKNESS, cm (mils)							
		0 (0)	2.54E-3 (1)	7.64E-3 (3)	1.52E-2 (6)	3.05E-2 (12)	5.09E-2 (20)	7.64E-2 (30)	1.52E-1 (60)
19327	180 W	1.09E+19	1.97E+10	0.00E+00	0.00E+00	0.00E+00	0.00E+00	0.00E+00	0.00E+00
19327	170 W	1.24E+19	2.08E+10	0.00E+00	0.00E+00	0.00E+00	0.00E+00	0.00E+00	0.00E+00
19327	160 W	1.32E+19	2.11E+10	0.00E+00	0.00E+00	0.00E+00	0.00E+00	0.00E+00	0.00E+00
19327	150 W	1.29E+19	2.07E+10	0.00E+00	0.00E+00	0.00E+00	0.00E+00	0.00E+00	0.00E+00
19327	140 W	1.20E+19	1.97E+10	0.00E+00	0.00E+00	0.00E+00	0.00E+00	0.00E+00	0.00E+00
19327	130 W	1.05E+19	1.83E+10	0.00E+00	0.00E+00	0.00E+00	0.00E+00	0.00E+00	0.00E+00
19327	120 W	8.93E+18	1.68E+10	0.00E+00	0.00E+00	0.00E+00	0.00E+00	0.00E+00	0.00E+00
19327	110 W	7.65E+18	1.54E+10	0.00E+00	0.00E+00	0.00E+00	0.00E+00	0.00E+00	0.00E+00
19327	100 W	6.51E+18	1.42E+10	0.00E+00	0.00E+00	0.00E+00	0.00E+00	0.00E+00	0.00E+00
19327	90 W	5.78E+18	1.32E+10	0.00E+00	0.00E+00	0.00E+00	0.00E+00	0.00E+00	0.00E+00
19327	80 W	5.31E+18	1.26E+10	0.00E+00	0.00E+00	0.00E+00	0.00E+00	0.00E+00	0.00E+00
19327	70 W	5.21E+18	1.25E+10	0.00E+00	0.00E+00	0.00E+00	0.00E+00	0.00E+00	0.00E+00
19327	60 W	5.46E+18	1.26E+10	0.00E+00	0.00E+00	0.00E+00	0.00E+00	0.00E+00	0.00E+00
19327	50 W	6.11E+18	1.32E+10	0.00E+00	0.00E+00	0.00E+00	0.00E+00	0.00E+00	0.00E+00
19327	40 W	7.26E+18	1.42E+10	0.00E+00	0.00E+00	0.00E+00	0.00E+00	0.00E+00	0.00E+00
19327	30 W	9.03E+18	1.57E+10	0.00E+00	0.00E+00	0.00E+00	0.00E+00	0.00E+00	0.00E+00
19327	20 W	1.13E+19	1.75E+10	0.00E+00	0.00E+00	0.00E+00	0.00E+00	0.00E+00	0.00E+00
19327	10 W	1.43E+19	1.95E+10	0.00E+00	0.00E+00	0.00E+00	0.00E+00	0.00E+00	0.00E+00
19327	0	1.74E+19	2.14E+10	0.00E+00	0.00E+00	0.00E+00	0.00E+00	0.00E+00	0.00E+00
19327	10 E	1.96E+19	2.28E+10	0.00E+00	0.00E+00	0.00E+00	0.00E+00	0.00E+00	0.00E+00
19327	20 E	1.98E+19	2.33E+10	0.00E+00	0.00E+00	0.00E+00	0.00E+00	0.00E+00	0.00E+00
19327	30 E	1.93E+19	2.30E+10	0.00E+00	0.00E+00	0.00E+00	0.00E+00	0.00E+00	0.00E+00
19327	40 E	1.64E+19	2.16E+10	0.00E+00	0.00E+00	0.00E+00	0.00E+00	0.00E+00	0.00E+00
19327	50 E	1.26E+19	1.94E+10	0.00E+00	0.00E+00	0.00E+00	0.00E+00	0.00E+00	0.00E+00
19327	60 E	9.02E+18	1.69E+10	0.00E+00	0.00E+00	0.00E+00	0.00E+00	0.00E+00	0.00E+00
19327	70 E	6.20E+18	1.45E+10	0.00E+00	0.00E+00	0.00E+00	0.00E+00	0.00E+00	0.00E+00
19327	80 E	4.26E+18	1.24E+10	0.00E+00	0.00E+00	0.00E+00	0.00E+00	0.00E+00	0.00E+00
19327	90 E	3.08E+18	1.10E+10	0.00E+00	0.00E+00	0.00E+00	0.00E+00	0.00E+00	0.00E+00
19327	100 E	2.44E+18	1.01E+10	0.00E+00	0.00E+00	0.00E+00	0.00E+00	0.00E+00	0.00E+00
19327	110 E	2.20E+18	9.80E+09	0.00E+00	0.00E+00	0.00E+00	0.00E+00	0.00E+00	0.00E+00
19327	120 E	2.29E+18	1.00E+10	0.00E+00	0.00E+00	0.00E+00	0.00E+00	0.00E+00	0.00E+00
19327	130 E	2.71E+18	1.08E+10	0.00E+00	0.00E+00	0.00E+00	0.00E+00	0.00E+00	0.00E+00
19327	140 E	3.53E+18	1.21E+10	0.00E+00	0.00E+00	0.00E+00	0.00E+00	0.00E+00	0.00E+00
19327	150 E	4.82E+18	1.39E+10	0.00E+00	0.00E+00	0.00E+00	0.00E+00	0.00E+00	0.00E+00
19327	160 E	6.60E+18	1.60E+10	0.00E+00	0.00E+00	0.00E+00	0.00E+00	0.00E+00	0.00E+00
19327	170 E	8.71E+18	1.80E+10	0.00E+00	0.00E+00	0.00E+00	0.00E+00	0.00E+00	0.00E+00

References for Chapter 9

9.1 H.Y. Tada, J.R. Carter, Jr., B.E. Anspaugh, and R.G. Downing, "Solar Cell Radiation Handbook," 3rd Edition, JPL Publication 82-69, Jet Propulsion Laboratory, California Institute of Technology, Pasadena, Calif., November 1982.

9.2 B.E. Anspaugh, "Solar Cell Radiation Handbook, Addendum 1: 1982-1988," JPL Publication 82-69, Addendum 1, Jet Propulsion Laboratory, California Institute of Technology, Pasadena, Calif., February 1989.

9.3 D.W. Sawyer and J.I. Vette, "AP-8 Trapped Proton Environment for Solar Maximum and Solar Minimum," National Space Science Data Center, NSSDC/WDC-A-R&S 76-06, December 1976.

9.4 J.I. Vette, "The AE-8 Trapped Electron Model Environment," National Space Science Data Center, NSSDC/WDC-A-R&S 91-24, November 1991.

9.5 J.A. Barton, B.W. Mar, et al., "Computer Codes for Space Radiation Environment and Shielding," Air Force Weapons Laboratory, Technical Document Report No. WL TDR-64-71, (AD-444602), Vol. I, 1964.

9.6 A. Hassitt and C.E. McIlwain, "Computer Programs for the Computation of B and L (May 1966)," National Space Science Data Center, NSDC 7\67-27, May 1967.

9.7 E.G. Stassinopoulos and G.D. Mead, "ALLMAG, GDALMG, LINTRA: Computer Programs for Geomagnetic Field and Field-Line Calculations," National Space Science Data Center, NSSDC 72-12, 1972.

9.8 H.B. Garrett and D. Hastings, "The Space Radiation Models," Paper No. 94-0590, *32nd AIAA Aerospace Sciences Meeting & Exhibit*, Reno, Nevada, January 1994.

9.9 E.G. Stassinopoulos, Private Communication, May 1995.

9.10 E.G. Stassinopoulos, J.J. Hebert, E.L. Butler, and J.L. Barth, "SOFIP: A Short Orbital Flux Integration Program," National Space Science Data Center, NSSDC/WDC-A-R&S 79-01, January 1979.

APPENDIX

Table A-1. Some Useful Physical Constants[a]

T_0	Absolute Zero	-273.16	°C
AU	Astronomical Unit	1.4959789×10^8	km
AMU	Atomic Mass Unit	1.66057×10^{-27}	kg
		931.5016	MeV
N_A	Avogadro constant	6.02205×10^{23}	mol^{-1}
a_0	Bohr Radius	5.29177×10^{-11}	m
k	Boltzmann constant	1.38066×10^{-23}	joule/K
		8.61735×10^{-5}	eV/K
q	Electron Charge	1.60219×10^{-19}	coulomb (joule/eV)
		4.80286×10^{-10}	esu ([dyne]$^{1/2}$ - cm)
m_e	Electron rest mass	9.10953×10^{-31}	kg
		1/1822.8874	AMU
		0.51100	MeV
R	Gas Constant	1.98719	cal/(mol-K)
μ_0	Permeability of vacuum	1.25664×10^{-6}	henry/m
ϵ_0	Permittivity of vacuum	8.85419×10^{-12}	farad/m
hc/q	Photon energy per unit wavelength	1.23985	eV/ $\lambda(\mu m)$
h	Planck constant	6.62618×10^{-34}	joule-sec
		4.13570×10^{-15}	eV-sec
m_p	Proton rest mass	1.67265×10^{-27}	kg
		938.2796	MeV
E_R	Rydberg Energy (q^2 [esu]/2a$_0$)	2.17991×10^{-18}	Joule
		13.60580	eV
c	Speed of Light in vacuum	2.99792×10^8	m/sec
σ	Stefan-Boltzmann constant	5.67032×10^{-8}	watts/(m^2-K^4)
kT	Thermal energy at 300 K	0.0259	eV
λ	Wavelength of 1-eV quantum	1.23977	μm

[a] Most of these values are from NBS Special Publication 398, August 1974

Table A-2. Properties of GaAs, Si, and Ge at 300 K[a]

Properties	GaAs	Ge	Si
Atoms/cm^3	4.42×10^{22}	4.42×10^{22}	5.0×10^{22}
Atomic Number (Z)	Avg. = 32	32	14
Ga	31		
As	33		
Atomic Weight (M)	144.63	72.60	28.09
Ga	69.72		
As	74.91		
Bandgap Energy (eV)	1.424	0.66	1.12
Breakdown Field (V/cm)	$\approx 4 \times 10^5$	$\approx 10^5$	$\approx 3 \times 10^5$
Carrier Conc. (Intrinsic, cm^{-3})	1.79×10^6	2.4×10^{13}	1.45×10^{10}
Crystal Structure	Zincblende	Diamond	Diamond
Debye Length (Intrinsic, μm)	2250	0.68	24
Debye Temperature (K)	370 K[b]		
Density (g/cm^3)	5.317[c]	5.3267	2.328
Density of States (Effective)			
Conduction Band, N_C (cm^{-3})	4.7×10^{17}	1.04×10^{19}	2.8×10^{19}
Valence Band, N_V (cm^{-3})	7.0×10^{18}	6.0×10^{18}	1.04×10^{19}
Dielectric Constant (ϵ_s/ϵ_0)	13.1	16.0	11.9
Displacement Energy, E_d (eV)	7 - 11 Avg: 10	13	11 - 12.9
Effective Mass, m^*/m_0			
Electrons	0.067	$m_l^* = 1.64$ $m_t^* = 0.082$	$m_l^* = 0.98$ $m_t^* = 0.19$
Holes	$m_{lh}^* = 0.082$ $m_{hh}^* = 0.45$	$m_{lh}^* = 0.044$ $m_{hh}^* = 0.28$	$m_{lh}^* = 0.16$ $m_{hh}^* = 0.49$

[a] From Sze, *Physics of Semiconductor Devices*, John Wiley, New York, 1981 unless otherwise noted.

[b] S. Adachi, "GaAs, AlAs, and Al$_x$Ga$_{1-x}$As: Material Parameters for use in Research and Device Applications," *Journal of Applied Physics*, 58, R1, 1985.

[c] J.S. Blakemore, "Semiconducting and Other Major Properties of Gallium Arsenide," *Journal of Applied Physics*, Vol. 53, R123, 1982.

Table A-2. Properties of GaAs, Si, and Ge at 300 K (Cont'd)[a]

Properties	GaAs	Ge	Si
Electron Affinity, χ (V)	4.07	4.0	4.05
Ionization Potential, Avg. (eV)[b]	385.1	350.0	173.0
Lattice Constant (Å)	5.6533	5.64613	5.43095
Melting Point (°C)	1238	937	1415
Minority Carrier Lifetime (s)	$\approx 10^{-8}$	10^{-3}	2.5×10^{-3}
Mobility, Drift (cm^2/V-s)			
Electrons (μ_n)	8500	3900	1500
Holes (μ_p)	400	1900	450
Optical Phonon Energy (eV)	0.035	0.063	0.037
Resistivity, Intrinsic (Ohm-cm)	10^8	47	2.3×10^5
Specific Heat (J/g-°C)	0.35	0.31	0.7
Thermal Conductivity (W/cm-°C)	0.46	0.6	1.5
Thermal Diffusivity (cm^2/s)	0.24	0.36	0.9
Thermal Expansion Coefficient ($\Delta L/L\Delta T$, K^{-1})	5.73×10^{-6} [c]	5.8×10^{-6}	2.6×10^{-6}

[a] From Sze, *Physics of Semiconductor Devices*, John Wiley, New York, 1981 unless otherwise noted.

[b] From Seltzer, EPSTAR Program.

[c] J.S. Blakemore, "Semiconducting and Other Major Properties of Gallium Arsenide," *Journal of Applied Physics*, Vol. 53, R123, 1982.

Table A-3.   Properties of GaAs and $Ga_{1-x}Al_xAs$ at 300 $K^{(a)}$

Properties	GaAs	$Al_xGa_{1-x}As$
Bandgap Energy (eV)	1.424	$1.424 + 1.247x$ $(0 \leq x \leq 0.45)$ $1.900 + 0.125x + 0.143x^2$ $(0.45 < x \leq 1.0)$
Debye Temperature (K)	370	$370 + 54x + 22x^2$
Density (g/cm^3)	5.360	$5.36 - 1.6x$
Electron Affinity χ (eV)	4.07	$4.07 - 1.1x$ $(0 \leq x \leq 0.45)$ $3.64 - 0.14x$ $(0.45 < x \leq 1.0)$
Lattice Constant (Å)	5.6533	$5.6533 + 0.0078x$
Melting Point (°C)	1238	$1238 - 58x + 560x_2^{(b)}$ $1238 + 1082x - 560x^{2(c)}$
Thermal Expansion Coefficient ($\Delta L/L\Delta T$, K^{-1})	6.4 x 10^{-6}	$(6.4 - 1.2x)$ x 10^{-6}
Specific Heat (J/g-°C)	0.34	$0.34 + 0.13x$

[a] From S. Adachi, "GaAs, AlAs, and $Al_xGa_{1-x}As$: Material Parameters for use in Research and Device Applications," *Journal of Applied Physics*, 58, R1, 1985, unless otherwise noted.

[b] This gives the solidus-surface curve [M.B. Panish and M. Ilegems, *Progress in Solid State Chemistry*, 7, 39, Pergamon, Oxford, 1972].

[c] This gives the liquidus-surface curve, [M.B. Panish and M. Ilegems, *Progress in Solid State Chemistry*, 7, 39, Pergamon, Oxford, 1972].

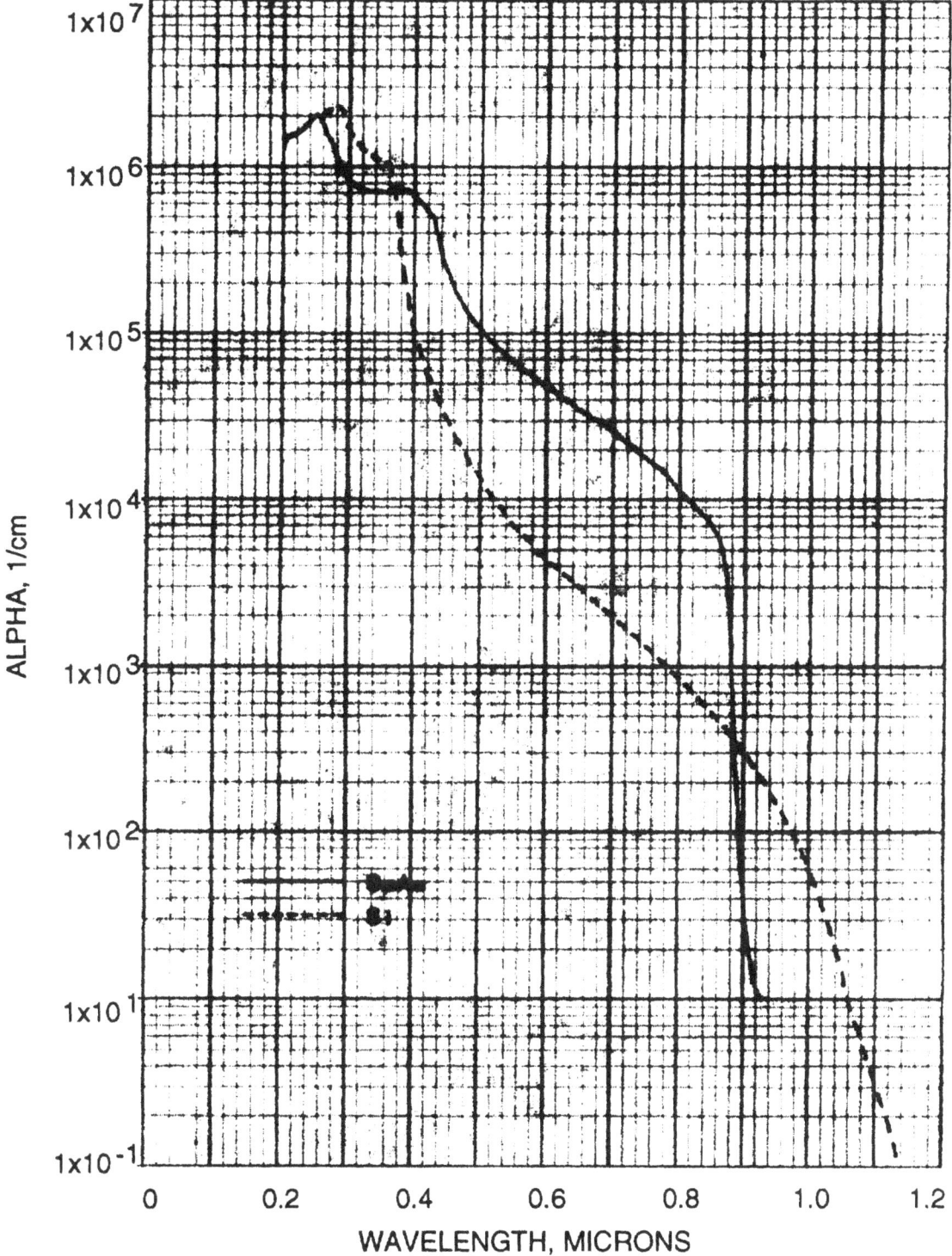

Figure A.1.   Optical Absorption Coefficients for GaAs and Si

Data Sources: GaAs: D.E. Aspnes et al., *J. Appl. Phys.*, 60 (2), 754, 1986.
                       H.C. Casey et al., *J. Appl. Phys.* 46, 250, 1975.
               Si: M.A. Green, *High Efficiency Silicon Solar Cells*, Trans Tech Publ.,
                       Switzerland, 1987.

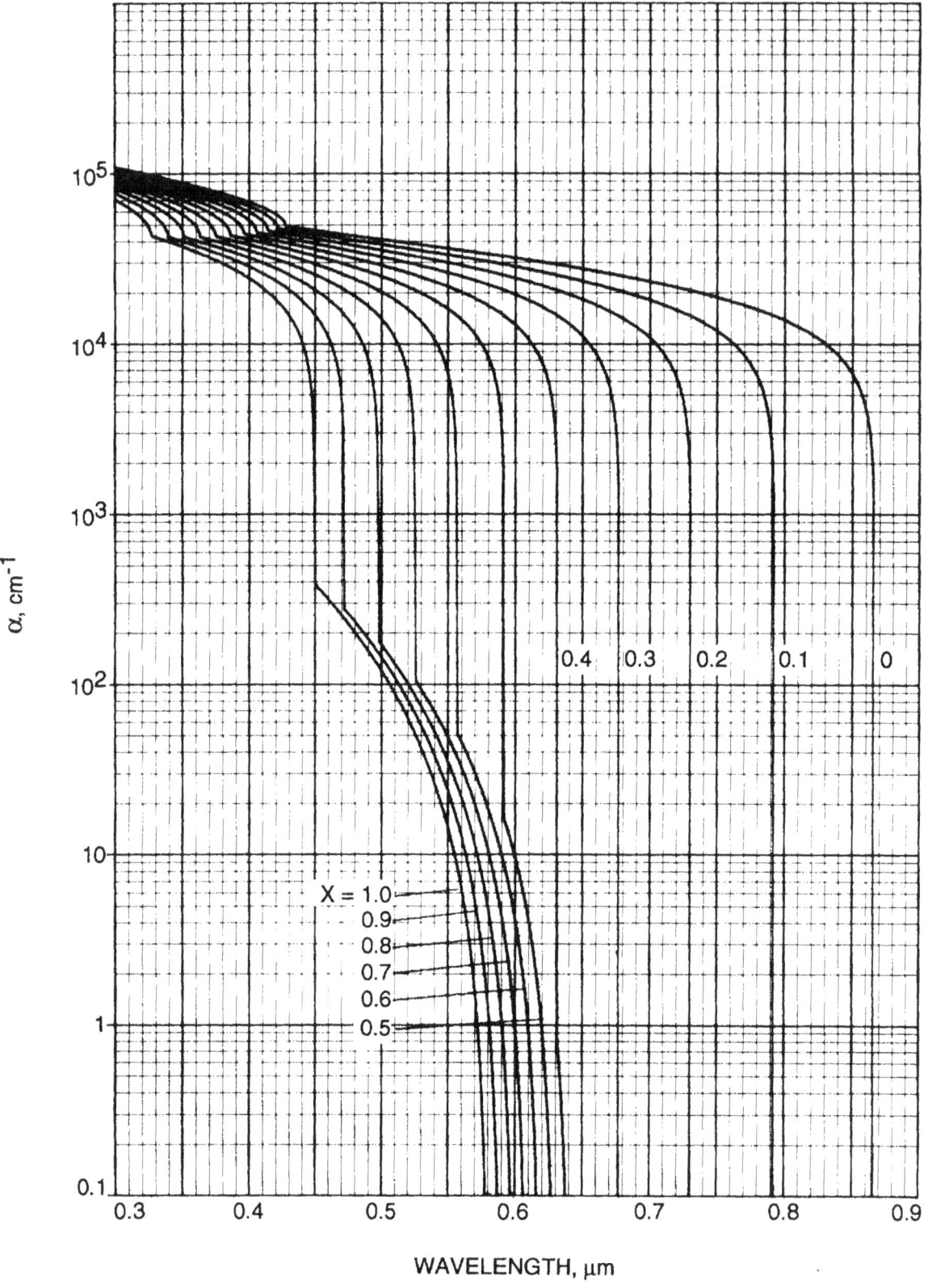

Figure A.2.   Optical Absorption Coefficients for $Ga_{1-x}Al_xAs$ as a Function of Aluminum Concentration, x

Data Sources: J.A. Hutchby and R.L. Fudurich, *J. Appl. Phys.*, 47, No. 11, 3140, 1976.
B.T. Dabney, *J. Appl. Phys.*, 50, No. 11, 7210, 1979.

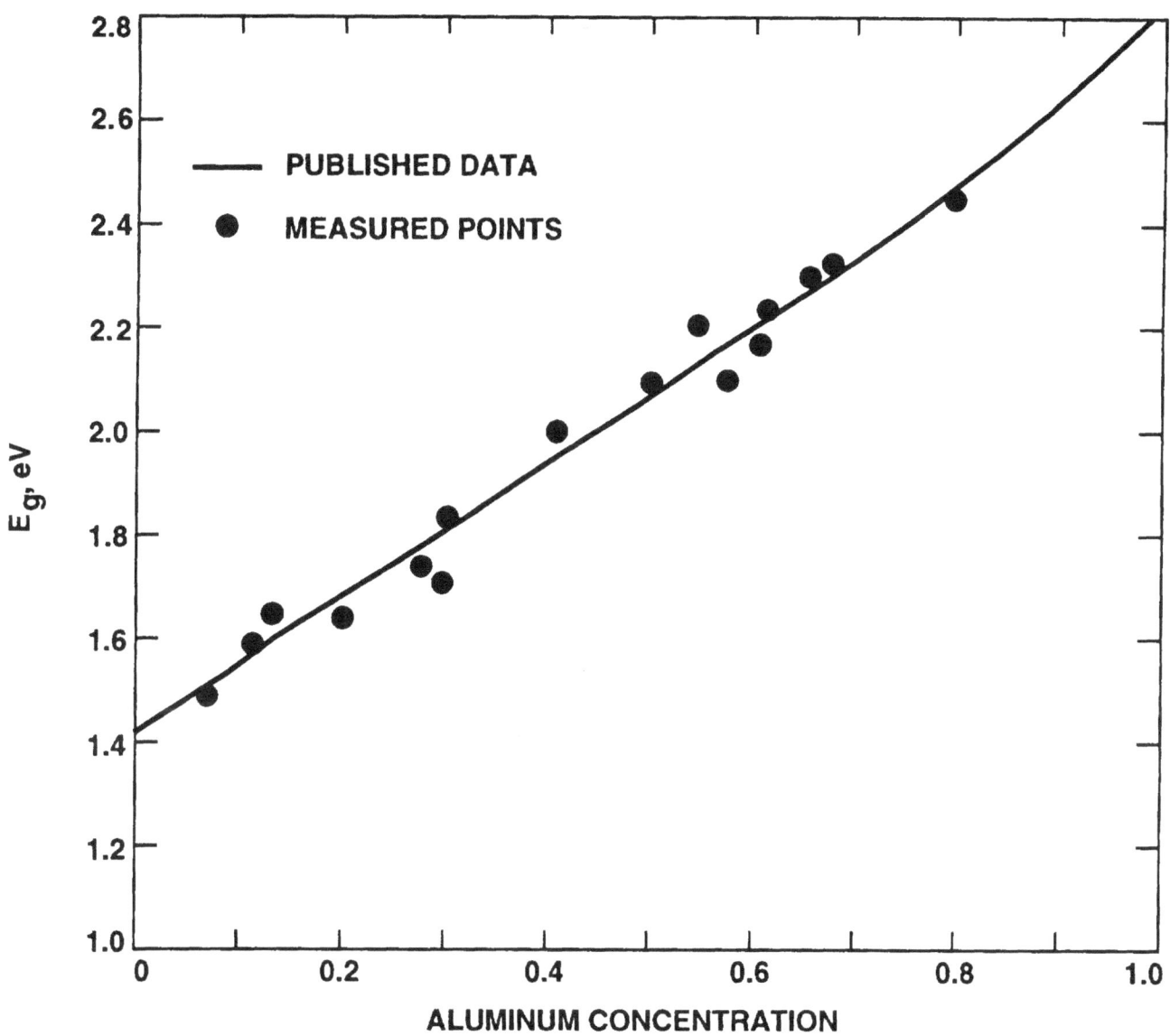

Figure A.3.   Band Gap Energy vs. Aluminum Concentration, x, in $Ga_{1-x}Al_xAs$

Data Source: J. Ewan et al., *Proc. of the 11th IEEE Photovoltaic Specialists Conf.*, 409, 1975.
(© 1975 IEEE, used with permission)

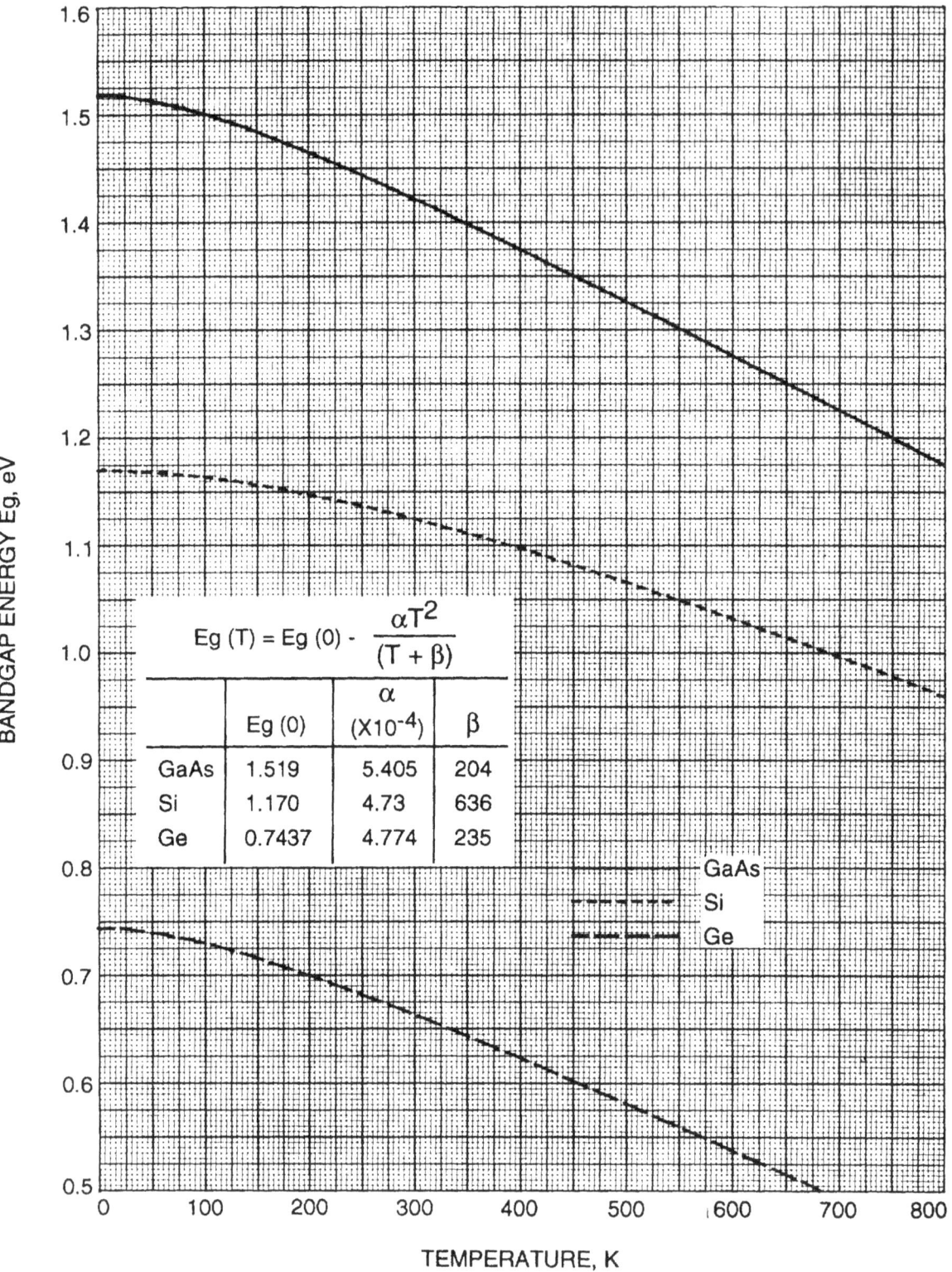

$$Eg\,(T) = Eg\,(0) - \frac{\alpha T^2}{(T + \beta)}$$

	Eg (0)	α $(\times 10^{-4})$	β
GaAs	1.519	5.405	204
Si	1.170	4.73	636
Ge	0.7437	4.774	235

Figure A.4.   Bandgap Energy of GaAs, Si, and Ge as a Function of Temperature

Data Source: C.D. Thurmond, *J. Electrochem. Soc.*, 122, No. 8, 1133, 1975.
(used with permission)

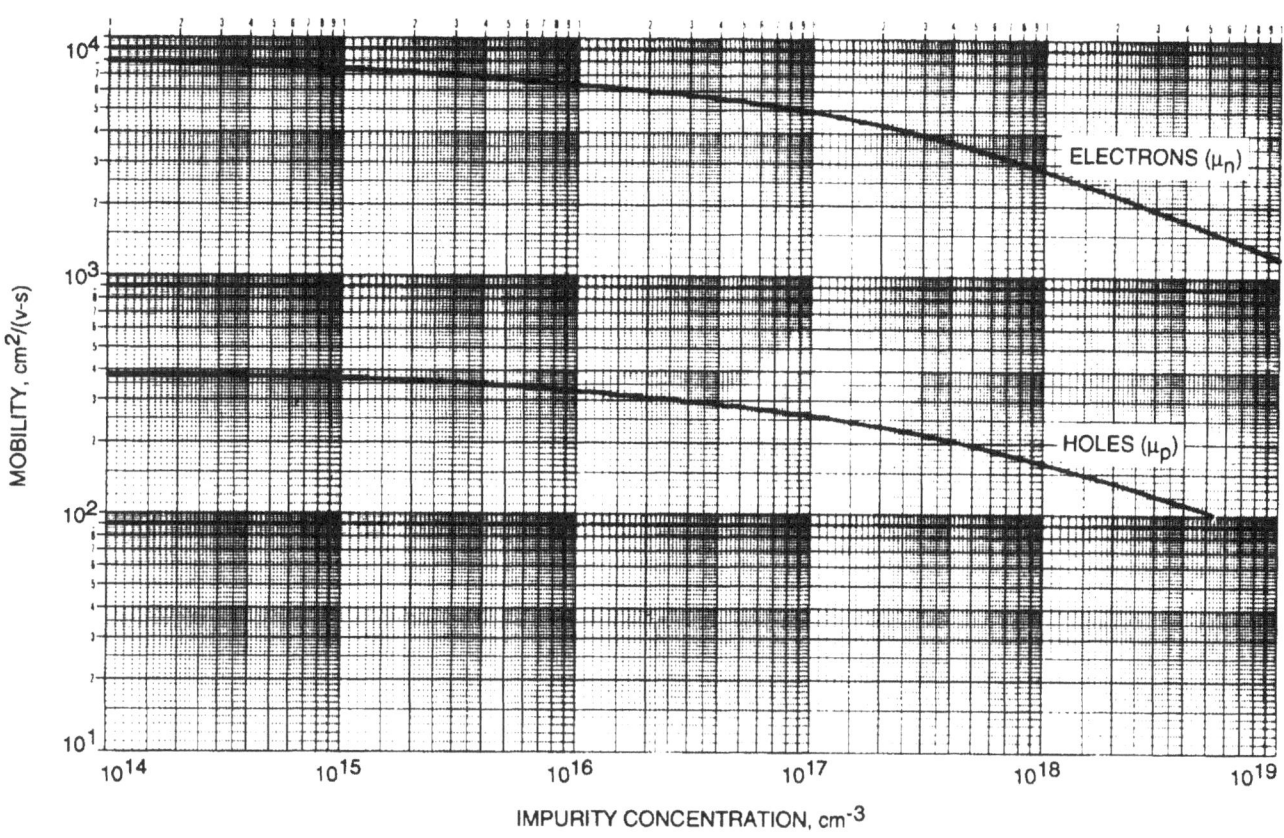

Figure A.5.   Drift Mobility of Electrons and Holes in GaAs at 300 K vs. Impurity Concentration

Data Source: S.M. Sze, *Physics of Semiconductor Devices*, Second Edition, John Wiley & Sons, New York, 1981.

Table A-4.   Indices of Refraction of Some Common Solar Cell Materials at 300 K

1. Si and GaAs[a][b]

Wavelength (nm)	n_d Si	n_d GaAs
0.40	6.0	4.15
0.45	4.75	4.8
0.5	4.25	4.4
0.6	3.9	3.85
0.7	3.75	3.65
0.8	3.65	3.62
0.9	3.6	3.6
1.0	3.5	3.5
1.1	3.5	3.46

2. Indices of Refraction of Miscellaneous Materials at $\lambda = 589$ nm

Material	n
MgF_2	1.38
Organic Film	1.4
SiO_2	1.46
Al_2O_3	1.76
SiO_x	1.8-1.9
Si_3N_4	2.05
Ta_2O_5	2.2
ZnS	2.36
TiO_2	2.62
CMG Coverglass[c]	1.516
CMX Coverglass	1.5265
CMZ Coverglass	1.49
Fused Silica	1.46-1.51

[a]*Encyclopedia of Chemical Technology*, Kirk-Othmer, Eds., Vol. 18, Wiley, New York, 1964.

[b]B.O. Seraphin and H.E. Bennett, *Semiconductors and Semimetals*, Willardson and Beers, eds., Vol. 3, Academic Press, New York, 1967.

[c]Coverglass Data from Product Specification Reports, PS 400, Issue 1, PS 401, Issue 2, and PS 292, Issue 4, Pilkington Space Technology, United Kingdom.

Table A-5.   Physical and Optical Properties of Coverglass Materials

Physical Property	CMX[a][b]	CMZ[c]	CMG[d]	Fused Silica
Density (g/cm^3	2.605	2.385	2.554	2.2
Thermal Expansion Coeff. (1/°C)	5.6×10^{-6}	3.67×10^{-6}	5.6×10^{-6}	0.55×10^{-6}
Youngs Modulus (N/m^2)	7.5×10^{10}	7.05×10^{10}	7.87×10^{10}	7.16×10^{10}
Poisson's Ratio	0.22	0.19	0.175	0.16
Bulk Resistivity, 20-60°C (Ω-m)	2.2×10^{12}	1.0×10^{14}	5.0×10^{11}	
Surface Resistivity, 100°C (Ω/□)	1.9×10^{16}	$>10^{16}$		
50% Cut-on Point, 150-μm thick (nm)	355 ± 5	350 ± 5	345 ± 5	
α, Solar Absorptance, 150-μm thick (λ = 250 - 2500 nm), with AR	0.06	0.053	0.04	0.01
ϵ_n, Normal emittance, 150-μm thick, with AR	0.86	0.845	0.845	
ϵ_h, Hemispherical emittance, 150-μm thick, with AR	0.815			0.78
n_d, Refractive Index	1.5265	1.49	1.516	1.46-1.51

[a]CMX, CMZ, and CMG coverglass materials were developed by Pilkington in the United Kingdom. All three materials are basically borosilicate glasses, doped with cerium dioxide to prevent the formation of color centers by electron and proton irradiation. All three are furnished with antireflection coatings.

[b] "Product Specification for Manufacture and Quality Assurance of CMX Solar Cell Coverglasses with Antireflection Coating," Pilkington Report No. PS292, Issue 4, 1991. CMX glass was developed for use in space for solar cell coverglasses and optical solar reflectors.

[c] "Product Specification for Manufacture and Quality Assurance of CMZ Solar Cell Coverglasses with Antireflection Coating," Pilkington Report No. PS 400, Issue 1, 1991. CMZ glass was developed for use in space as a coverglass for silicon solar cells.

[d] "Product Specification for Manufacture and Quality Assurance of CMG Solar Cell Coverglasses with Antireflection Coating," Pilkington Report No. PS 401, Issue 2, 1992. CMG glass was developed for use in space as a coverglass for GaAs solar cells.

Table A-6.   Ranges, Stopping Powers, and Straggling of Protons Incident on GaAs

Stopping Units =   MeV / (mg/cm2)

Proton Energy	dE/dx Elec.	dE/dx Nuclear	Projected Range	Longitudinal Straggling	Lateral Straggling
10.00 keV	1.369E-01	2.096E-03	930 A	584 A	697 A
12.00 keV	1.486E-01	1.932E-03	1098 A	640 A	776 A
14.00 keV	1.593E-01	1.795E-03	1262 A	689 A	849 A
16.00 keV	1.691E-01	1.679E-03	1422 A	732 A	916 A
18.00 keV	1.783E-01	1.580E-03	1578 A	772 A	977 A
20.00 keV	1.870E-01	1.493E-03	1730 A	807 A	1035 A
22.00 keV	1.952E-01	1.417E-03	1879 A	840 A	1089 A
24.00 keV	2.030E-01	1.350E-03	2024 A	870 A	1139 A
26.00 keV	2.085E-01	1.289E-03	2168 A	897 A	1187 A
28.00 keV	2.109E-01	1.235E-03	2311 A	924 A	1233 A
30.00 keV	2.131E-01	1.185E-03	2454 A	949 A	1278 A
33.00 keV	2.161E-01	1.119E-03	2669 A	984 A	1343 A
36.00 keV	2.188E-01	1.061E-03	2884 A	1018 A	1406 A
40.00 keV	2.221E-01	9.938E-04	3171 A	1060 A	1487 A
45.00 keV	2.255E-01	9.222E-04	3529 A	1109 A	1584 A
50.00 keV	2.284E-01	8.614E-04	3887 A	1155 A	1676 A
55.00 keV	2.309E-01	8.092E-04	4245 A	1199 A	1766 A
60.00 keV	2.329E-01	7.636E-04	4603 A	1240 A	1852 A
65.00 keV	2.346E-01	7.236E-04	4960 A	1279 A	1937 A
70.00 keV	2.359E-01	6.880E-04	5318 A	1316 A	2018 A
80.00 keV	2.377E-01	6.275E-04	6035 A	1388 A	2177 A
90.00 keV	2.386E-01	5.779E-04	6755 A	1456 A	2329 A
100.00 keV	2.387E-01	5.314E-04	7479 A	1519 A	2476 A
110.00 keV	2.381E-01	4.965E-04	8208 A	1580 A	2620 A
120.00 keV	2.371E-01	4.663E-04	8944 A	1639 A	2760 A
130.00 keV	2.356E-01	4.399E-04	9687 A	1695 A	2898 A
140.00 keV	2.338E-01	4.167E-04	1.04 um	1750 A	3034 A
150.00 keV	2.318E-01	3.960E-04	1.12 um	1804 A	3168 A
160.00 keV	2.295E-01	3.775E-04	1.20 um	1857 A	3302 A
170.00 keV	2.270E-01	3.608E-04	1.27 um	1909 A	3435 A
180.00 keV	2.244E-01	3.457E-04	1.35 um	1960 A	3567 A
200.00 keV	2.190E-01	3.193E-04	1.51 um	2066 A	3831 A
220.00 keV	2.135E-01	2.969E-04	1.68 um	2172 A	4096 A
240.00 keV	2.079E-01	2.778E-04	1.85 um	2276 A	4363 A
260.00 keV	2.025E-01	2.612E-04	2.03 um	2381 A	4632 A
280.00 keV	1.972E-01	2.466E-04	2.21 um	2485 A	4904 A
300.00 keV	1.921E-01	2.337E-04	2.39 um	2591 A	5180 A
330.00 keV	1.849E-01	2.169E-04	2.68 um	2759 A	5602 A
360.00 keV	1.782E-01	2.026E-04	2.98 um	2929 A	6033 A
400.00 keV	1.701E-01	1.864E-04	3.40 um	3172 A	6624 A
450.00 keV	1.612E-01	1.697E-04	3.95 um	3497 A	7389 A
500.00 keV	1.534E-01	1.560E-04	4.53 um	3828 A	8183 A
550.00 keV	1.466E-01	1.445E-04	5.14 um	4164 A	9005 A
600.00 keV	1.406E-01	1.347E-04	5.78 um	4506 A	9855 A
650.00 keV	1.352E-01	1.263E-04	6.45 um	4854 A	1.07 um
700.00 keV	1.304E-01	1.189E-04	7.14 um	5207 A	1.16 um
800.00 keV	1.220E-01	1.066E-04	8.59 um	6096 A	1.35 um
900.00 keV	1.150E-01	9.677E-05	10.15 um	6987 A	1.55 um
1.00 MeV	1.090E-01	8.872E-05	11.79 um	7882 A	1.75 um
1.10 MeV	1.037E-01	8.200E-05	13.52 um	8787 A	1.97 um
1.20 MeV	9.911E-02	7.629E-05	15.34 um	9702 A	2.19 um
1.30 MeV	9.498E-02	7.137E-05	17.24 um	1.06 um	2.41 um
1.40 MeV	9.126E-02	6.709E-05	19.23 um	1.16 um	2.65 um
1.50 MeV	8.789E-02	6.333E-05	21.29 um	1.25 um	2.89 um
1.60 MeV	8.480E-02	5.999E-05	23.43 um	1.35 um	3.14 um

Table A-6. Ranges, Stopping Powers, and Straggling of Protons Incident on GaAs (Cont'd)

Proton Energy	dE/dx Elec.	dE/dx Nuclear	Projected Range	Longitudinal Straggling	Lateral Straggling
1.70 MeV	8.198E-02	5.701E-05	25.65 um	1.45 um	3.40 um
1.80 MeV	7.937E-02	5.434E-05	27.94 um	1.55 um	3.66 um
2.00 MeV	7.470E-02	4.972E-05	32.74 um	1.81 um	4.20 um
2.20 MeV	7.065E-02	4.587E-05	37.84 um	2.08 um	4.78 um
2.40 MeV	6.708E-02	4.261E-05	43.22 um	2.34 um	5.37 um
2.60 MeV	6.392E-02	3.980E-05	48.88 um	2.61 um	5.99 um
2.80 MeV	6.108E-02	3.737E-05	54.81 um	2.87 um	6.63 um
3.00 MeV	5.853E-02	3.523E-05	61.01 um	3.15 um	7.30 um
3.30 MeV	5.513E-02	3.248E-05	70.81 um	3.65 um	8.34 um
3.60 MeV	5.216E-02	3.014E-05	81.19 um	4.15 um	9.43 um
4.00 MeV	4.873E-02	2.754E-05	95.91 um	4.93 um	10.97 um
4.50 MeV	4.512E-02	2.488E-05	115.71 um	6.03 um	13.00 um
5.00 MeV	4.207E-02	2.272E-05	137.03 um	7.11 um	15.17 um
5.50 MeV	3.946E-02	2.092E-05	159.84 um	8.20 um	17.46 um
6.00 MeV	3.719E-02	1.940E-05	184.10 um	9.29 um	19.88 um
6.50 MeV	3.521E-02	1.810E-05	209.78 um	10.41 um	22.42 um
7.00 MeV	3.345E-02	1.697E-05	236.87 um	11.54 um	25.08 um
8.00 MeV	3.048E-02	1.511E-05	295.13 um	15.01 um	30.74 um
9.00 MeV	2.805E-02	1.363E-05	358.76 um	18.38 um	36.86 um
10.00 MeV	2.603E-02	1.243E-05	427.63 um	21.73 um	43.41 um
11.00 MeV	2.451E-02	1.143E-05	501.32 um	25.08 um	50.36 um
12.00 MeV	2.307E-02	1.059E-05	579.62 um	28.46 um	57.68 um
13.00 MeV	2.179E-02	9.874E-06	662.69 um	31.90 um	65.39 um
14.00 MeV	2.065E-02	9.251E-06	750.49 um	35.40 um	73.48 um
15.00 MeV	1.963E-02	8.705E-06	843.00 um	38.98 um	81.94 um
16.00 MeV	1.872E-02	8.223E-06	940.17 um	42.64 um	90.79 um
17.00 MeV	1.789E-02	7.795E-06	1.04 mm	46.37 um	100.00 um
18.00 MeV	1.714E-02	7.411E-06	1.15 mm	50.19 um	109.59 um
20.00 MeV	1.583E-02	6.751E-06	1.37 mm	62.25 um	129.87 um
22.00 MeV	1.472E-02	6.204E-06	1.62 mm	74.04 um	151.59 um
24.00 MeV	1.377E-02	5.743E-06	1.88 mm	85.79 um	174.74 um
26.00 MeV	1.295E-02	5.349E-06	2.16 mm	97.62 um	199.28 um
28.00 MeV	1.223E-02	5.008E-06	2.46 mm	109.61 um	225.20 um
30.00 MeV	1.160E-02	4.710E-06	2.77 mm	121.79 um	252.46 um
33.00 MeV	1.077E-02	4.326E-06	3.27 mm	146.20 um	295.85 um
36.00 MeV	1.007E-02	4.003E-06	3.81 mm	170.39 um	342.15 um
40.00 MeV	9.280E-03	3.643E-06	4.58 mm	209.86 um	408.32 um
45.00 MeV	8.473E-03	3.279E-06	5.63 mm	266.67 um	497.91 um
50.00 MeV	7.813E-03	2.984E-06	6.78 mm	322.18 um	594.88 um
55.00 MeV	7.264E-03	2.739E-06	8.02 mm	377.56 um	698.97 um
60.00 MeV	6.799E-03	2.533E-06	9.34 mm	433.35 um	809.91 um
65.00 MeV	6.399E-03	2.358E-06	10.76 mm	489.84 um	927.47 um
70.00 MeV	6.053E-03	2.206E-06	12.26 mm	547.18 um	1.05 mm
80.00 MeV	5.481E-03	1.956E-06	15.50 mm	738.22 um	1.32 mm
90.00 MeV	5.028E-03	1.759E-06	19.06 mm	919.30 um	1.61 mm
100.00 MeV	4.661E-03	1.599E-06	22.92 mm	1.10 mm	1.92 mm

Multiply Stopping by	for Stopping Units
5.3168E+02	MeV / mm
1.0000E+00	keV / (ug/cm^2)
1.0000E+03	keV / (mg/cm^2)
1.0000E+03	MeV / (g/cm^2)
1.0742E+02	L.S.S. reduced units

Table A-6 was computed using the Trim program written by J.P. Biersack, and J.F. Ziegler   (Version 91.14)

A-13

Table A-7.   Ranges, Stopping Powers, and Straggling of Electrons Incident on GaAs

Composition (Constituent Z : Fraction by weight)

31: 0.467247   33: 0.532753

<Z/A> = 0.442411   I = 385.1 eV   DENSITY = 5.320E+00 g/cm3

ENERGY	COLLISION	STOPPING POWER RADIATIVE	TOTAL	CSDA RANGE	RADIATION YIELD	DENSITY EFFECT DELTA
MeV	MeV cm2/g	MeV cm2/g	MeV cm2/g	g/cm2		
0.00100	3.850E+01	4.967E-03	3.850E+01	1.299E-05	6.450E-05	4.118E-04
0.00125	3.703E+01	5.570E-03	3.704E+01	1.959E-05	7.949E-05	5.145E-04
0.00150	3.511E+01	6.098E-03	3.511E+01	2.652E-05	9.323E-05	6.171E-04
0.00175	3.318E+01	6.569E-03	3.319E+01	3.385E-05	1.064E-04	7.196E-04
0.00200	3.138E+01	6.995E-03	3.138E+01	4.159E-05	1.194E-04	8.220E-04
0.00250	2.825E+01	7.737E-03	2.826E+01	5.841E-05	1.452E-04	1.026E-03
0.00300	2.569E+01	8.368E-03	2.570E+01	7.698E-05	1.709E-04	1.231E-03
0.00350	2.359E+01	8.916E-03	2.360E+01	9.731E-05	1.968E-04	1.434E-03
0.00400	2.184E+01	9.399E-03	2.185E+01	1.193E-04	2.227E-04	1.638E-03
0.00450	2.035E+01	9.829E-03	2.036E+01	1.431E-04	2.486E-04	1.840E-03
0.00500	1.908E+01	1.021E-02	1.909E+01	1.684E-04	2.747E-04	2.043E-03
0.00550	1.798E+01	1.056E-02	1.799E+01	1.954E-04	3.007E-04	2.245E-03
0.00600	1.701E+01	1.088E-02	1.702E+01	2.240E-04	3.268E-04	2.447E-03
0.00700	1.540E+01	1.144E-02	1.541E+01	2.859E-04	3.788E-04	2.849E-03
0.00800	1.410E+01	1.192E-02	1.411E+01	3.538E-04	4.307E-04	3.250E-03
0.00900	1.303E+01	1.233E-02	1.304E+01	4.276E-04	4.823E-04	3.650E-03
0.01000	1.213E+01	1.269E-02	1.214E+01	5.071E-04	5.337E-04	4.048E-03
0.01250	1.040E+01	1.343E-02	1.042E+01	7.302E-04	6.606E-04	5.037E-03
0.01500	9.163E+00	1.400E-02	9.177E+00	9.865E-04	7.851E-04	6.018E-03
0.01750	8.224E+00	1.445E-02	8.238E+00	1.275E-03	9.073E-04	6.990E-03
0.02000	7.486E+00	1.482E-02	7.501E+00	1.593E-03	1.027E-03	7.955E-03
0.02500	6.395E+00	1.539E-02	6.411E+00	2.317E-03	1.260E-03	9.863E-03
0.03000	5.625E+00	1.581E-02	5.641E+00	3.151E-03	1.484E-03	1.174E-02
0.03500	5.050E+00	1.615E-02	5.066E+00	4.088E-03	1.700E-03	1.360E-02
0.04000	4.602E+00	1.643E-02	4.619E+00	5.123E-03	1.909E-03	1.543E-02
0.04500	4.244E+00	1.667E-02	4.261E+00	6.251E-03	2.112E-03	1.724E-02
0.05000	3.950E+00	1.689E-02	3.967E+00	7.469E-03	2.310E-03	1.903E-02
0.05500	3.705E+00	1.708E-02	3.722E+00	8.771E-03	2.502E-03	2.080E-02
0.06000	3.496E+00	1.725E-02	3.513E+00	1.015E-02	2.689E-03	2.255E-02
0.07000	3.161E+00	1.757E-02	3.179E+00	1.315E-02	3.051E-03	2.601E-02
0.08000	2.903E+00	1.787E-02	2.921E+00	1.644E-02	3.398E-03	2.941E-02
0.09000	2.699E+00	1.814E-02	2.717E+00	1.999E-02	3.731E-03	3.275E-02
0.10000	2.532E+00	1.839E-02	2.551E+00	2.380E-02	4.053E-03	3.605E-02
0.12500	2.227E+00	1.900E-02	2.246E+00	3.428E-02	4.811E-03	4.414E-02
0.15000	2.019E+00	1.959E-02	2.039E+00	4.599E-02	5.516E-03	5.202E-02
0.17500	1.870E+00	2.019E-02	1.890E+00	5.875E-02	6.179E-03	5.976E-02
0.20000	1.757E+00	2.081E-02	1.778E+00	7.241E-02	6.806E-03	6.737E-02
0.25000	1.600E+00	2.211E-02	1.622E+00	1.019E-01	7.981E-03	8.232E-02
0.30000	1.497E+00	2.349E-02	1.521E+00	1.338E-01	9.075E-03	9.701E-02
0.35000	1.426E+00	2.497E-02	1.451E+00	1.675E-01	1.011E-02	1.115E-01

Table A-7. Ranges, Stopping Powers, and Straggling of Electrons Incident on GaAs (Cont'd)

ENERGY	STOPPING POWER COLLISION	RADIATIVE	TOTAL	CSDA RANGE	RADIATION YIELD	DENSITY EFFECT DELTA
MeV	MeV cm2/g	MeV cm2/g	MeV cm2/g	g/cm2		
0.40000	1.375E+00	2.651E-02	1.402E+00	2.026E-01	1.111E-02	1.260E-01
0.45000	1.337E+00	2.813E-02	1.365E+00	2.388E-01	1.207E-02	1.403E-01
0.50000	1.309E+00	2.980E-02	1.338E+00	2.758E-01	1.300E-02	1.547E-01
0.55000	1.287E+00	3.154E-02	1.318E+00	3.135E-01	1.392E-02	1.691E-01
0.60000	1.270E+00	3.332E-02	1.303E+00	3.516E-01	1.482E-02	1.835E-01
0.70000	1.246E+00	3.705E-02	1.283E+00	4.290E-01	1.660E-02	2.125E-01
0.80000	1.232E+00	4.097E-02	1.273E+00	5.073E-01	1.834E-02	2.417E-01
0.90000	1.223E+00	4.505E-02	1.268E+00	5.861E-01	2.006E-02	2.711E-01
1.00000	1.218E+00	4.929E-02	1.268E+00	6.649E-01	2.178E-02	3.007E-01
1.25000	1.216E+00	6.041E-02	1.277E+00	8.616E-01	2.604E-02	3.747E-01
1.50000	1.221E+00	7.218E-02	1.293E+00	1.056E+00	3.029E-02	4.480E-01
1.75000	1.229E+00	8.449E-02	1.313E+00	1.248E+00	3.455E-02	5.198E-01
2.00000	1.238E+00	9.724E-02	1.335E+00	1.437E+00	3.880E-02	5.896E-01
2.50000	1.257E+00	1.238E-01	1.381E+00	1.805E+00	4.729E-02	7.227E-01
3.00000	1.275E+00	1.516E-01	1.426E+00	2.161E+00	5.575E-02	8.472E-01
3.50000	1.291E+00	1.804E-01	1.472E+00	2.506E+00	6.414E-02	9.635E-01
4.00000	1.306E+00	2.099E-01	1.516E+00	2.841E+00	7.244E-02	1.073E+00
4.50000	1.319E+00	2.400E-01	1.559E+00	3.166E+00	8.064E-02	1.175E+00
5.00000	1.331E+00	2.706E-01	1.602E+00	3.483E+00	8.872E-02	1.272E+00
5.50000	1.343E+00	3.016E-01	1.644E+00	3.791E+00	9.667E-02	1.363E+00
6.00000	1.353E+00	3.331E-01	1.686E+00	4.091E+00	1.045E-01	1.450E+00
7.00000	1.371E+00	3.971E-01	1.768E+00	4.670E+00	1.197E-01	1.611E+00
8.00000	1.386E+00	4.623E-01	1.849E+00	5.223E+00	1.345E-01	1.757E+00
9.00000	1.400E+00	5.286E-01	1.929E+00	5.753E+00	1.487E-01	1.892E+00
10.00000	1.412E+00	5.957E-01	2.008E+00	6.261E+00	1.623E-01	2.016E+00
12.50000	1.438E+00	7.665E-01	2.204E+00	7.449E+00	1.944E-01	2.289E+00
15.00000	1.458E+00	9.407E-01	2.399E+00	8.536E+00	2.238E-01	2.522E+00
17.50000	1.475E+00	1.118E+00	2.593E+00	9.538E+00	2.507E-01	2.726E+00
20.00000	1.490E+00	1.297E+00	2.787E+00	1.047E+01	2.754E-01	2.906E+00
25.00000	1.514E+00	1.661E+00	3.174E+00	1.215E+01	3.194E-01	3.217E+00
30.00000	1.533E+00	2.030E+00	3.562E+00	1.363E+01	3.574E-01	3.481E+00
35.00000	1.548E+00	2.403E+00	3.952E+00	1.497E+01	3.905E-01	3.711E+00
40.00000	1.561E+00	2.780E+00	4.342E+00	1.617E+01	4.198E-01	3.915E+00
45.00000	1.573E+00	3.160E+00	4.733E+00	1.728E+01	4.459E-01	4.100E+00
50.00000	1.583E+00	3.543E+00	5.125E+00	1.829E+01	4.693E-01	4.268E+00
55.00000	1.592E+00	3.927E+00	5.519E+00	1.923E+01	4.904E-01	4.424E+00
60.00000	1.599E+00	4.314E+00	5.913E+00	2.011E+01	5.096E-01	4.568E+00
70.00000	1.613E+00	5.091E+00	6.704E+00	2.169E+01	5.432E-01	4.829E+00
80.00000	1.625E+00	5.874E+00	7.498E+00	2.310E+01	5.718E-01	5.059E+00
90.00000	1.635E+00	6.660E+00	8.295E+00	2.437E+01	5.964E-01	5.266E+00
100.00000	1.643E+00	7.451E+00	9.094E+00	2.552E+01	6.179E-01	5.453E+00
125.00000	1.661E+00	9.438E+00	1.110E+01	2.801E+01	6.615E-01	5.855E+00
150.00000	1.676E+00	1.144E+01	1.311E+01	3.008E+01	6.949E-01	6.189E+00
175.00000	1.688E+00	1.344E+01	1.513E+01	3.185E+01	7.214E-01	6.474E+00
200.00000	1.698E+00	1.546E+01	1.715E+01	3.340E+01	7.431E-01	6.724E+00

To convert ranges in g/cm^2 to cm, divide by 5.32 g/cm^3.

Table A-7 was computed using the EPSTAR program written by S.M. Seltzer.

Table A-8.   Ranges, Stopping Powers, and Straggling of Protons Incident on CMG Glass

Stopping Units =   MeV / (mg/cm²)   CMG Density = 2.554 g/cm³

Proton Energy	dE/dx Elec.	dE/dx Nuclear	Projected Range	Longitudinal Straggling	Lateral Straggling
10.00 keV	2.486E-01	4.091E-03	1458 A	576 A	769 A
12.00 keV	2.697E-01	3.665E-03	1702 A	618 A	844 A
14.00 keV	2.889E-01	3.330E-03	1934 A	653 A	910 A
16.00 keV	3.066E-01	3.058E-03	2157 A	684 A	970 A
18.00 keV	3.232E-01	2.832E-03	2372 A	710 A	1023 A
20.00 keV	3.388E-01	2.642E-03	2579 A	734 A	1072 A
22.00 keV	3.535E-01	2.478E-03	2780 A	755 A	1117 A
24.00 keV	3.675E-01	2.336E-03	2974 A	774 A	1158 A
26.00 keV	3.798E-01	2.211E-03	3163 A	791 A	1197 A
28.00 keV	3.899E-01	2.100E-03	3348 A	807 A	1233 A
30.00 keV	3.993E-01	2.002E-03	3530 A	821 A	1267 A
33.00 keV	4.123E-01	1.871E-03	3797 A	842 A	1314 A
36.00 keV	4.240E-01	1.759E-03	4057 A	860 A	1359 A
40.00 keV	4.378E-01	1.631E-03	4397 A	883 A	1414 A
45.00 keV	4.525E-01	1.497E-03	4811 A	908 A	1476 A
50.00 keV	4.647E-01	1.385E-03	5215 A	931 A	1534 A
55.00 keV	4.746E-01	1.291E-03	5612 A	951 A	1588 A
60.00 keV	4.826E-01	1.210E-03	6003 A	970 A	1638 A
65.00 keV	4.889E-01	1.139E-03	6389 A	988 A	1686 A
70.00 keV	4.936E-01	1.077E-03	6773 A	1004 A	1731 A
80.00 keV	4.992E-01	9.730E-04	7534 A	1038 A	1817 A
90.00 keV	5.005E-01	8.889E-04	8294 A	1068 A	1897 A
100.00 keV	4.987E-01	8.193E-04	9056 A	1096 A	1974 A
110.00 keV	4.945E-01	7.607E-04	9825 A	1124 A	2048 A
120.00 keV	4.885E-01	7.106E-04	1.06 um	1150 A	2120 A
130.00 keV	4.813E-01	6.672E-04	1.14 um	1175 A	2190 A
140.00 keV	4.732E-01	6.293E-04	1.22 um	1200 A	2260 A
150.00 keV	4.646E-01	5.958E-04	1.30 um	1225 A	2330 A
160.00 keV	4.557E-01	5.659E-04	1.39 um	1250 A	2400 A
170.00 keV	4.466E-01	5.392E-04	1.47 um	1275 A	2469 A
180.00 keV	4.375E-01	5.151E-04	1.56 um	1300 A	2539 A
200.00 keV	4.197E-01	4.732E-04	1.74 um	1361 A	2681 A
220.00 keV	4.026E-01	4.382E-04	1.93 um	1424 A	2827 A
240.00 keV	3.865E-01	4.083E-04	2.12 um	1488 A	2976 A
260.00 keV	3.715E-01	3.825E-04	2.33 um	1554 A	3129 A
280.00 keV	3.576E-01	3.601E-04	2.54 um	1622 A	3287 A
300.00 keV	3.447E-01	3.403E-04	2.76 um	1691 A	3450 A
330.00 keV	3.272E-01	3.147E-04	3.10 um	1820 A	3704 A
360.00 keV	3.116E-01	2.929E-04	3.47 um	1953 A	3970 A
400.00 keV	2.934E-01	2.684E-04	3.98 um	2165 A	4343 A
450.00 keV	2.740E-01	2.434E-04	4.67 um	2474 A	4838 A
500.00 keV	2.575E-01	2.229E-04	5.40 um	2787 A	5365 A
550.00 keV	2.434E-01	2.058E-04	6.17 um	3103 A	5922 A
600.00 keV	2.311E-01	1.913E-04	6.99 um	3424 A	6507 A
650.00 keV	2.203E-01	1.789E-04	7.85 um	3749 A	7120 A
700.00 keV	2.107E-01	1.680E-04	8.75 um	4080 A	7759 A
800.00 keV	1.944E-01	1.501E-04	10.67 um	5122 A	9113 A
900.00 keV	1.810E-01	1.358E-04	12.75 um	6128 A	1.06 um
1.00 MeV	1.696E-01	1.241E-04	14.97 um	7118 A	1.21 um
1.10 MeV	1.599E-01	1.144E-04	17.33 um	8105 A	1.37 um
1.20 MeV	1.514E-01	1.062E-04	19.83 um	9096 A	1.54 um
1.30 MeV	1.440E-01	9.918E-05	22.47 um	1.01 um	1.72 um
1.40 MeV	1.373E-01	9.307E-05	25.23 um	1.11 um	1.91 um

Table A-8. Ranges, Stopping Powers, and Straggling of Protons Incident on CMG Glass (Cont'd)

Proton Energy	dE/dx Elec.	dE/dx Nuclear	Projected Range	Longitudinal Straggling	Lateral Straggling
1.50 MeV	1.314E-01	8.770E-05	28.13 um	1.21 um	2.10 um
1.60 MeV	1.260E-01	8.296E-05	31.16 um	1.32 um	2.30 um
1.70 MeV	1.211E-01	7.873E-05	34.31 um	1.42 um	2.51 um
1.80 MeV	1.166E-01	7.493E-05	37.59 um	1.53 um	2.72 um
2.00 MeV	1.087E-01	6.840E-05	44.50 um	1.87 um	3.17 um
2.20 MeV	1.020E-01	6.297E-05	51.90 um	2.21 um	3.65 um
2.40 MeV	9.611E-02	5.839E-05	59.77 um	2.54 um	4.15 um
2.60 MeV	9.096E-02	5.446E-05	68.11 um	2.86 um	4.67 um
2.80 MeV	8.640E-02	5.105E-05	76.89 um	3.19 um	5.23 um
3.00 MeV	8.233E-02	4.806E-05	86.13 um	3.52 um	5.80 um
3.30 MeV	7.698E-02	4.422E-05	100.82 um	4.20 um	6.71 um
3.60 MeV	7.235E-02	4.098E-05	116.49 um	4.86 um	7.67 um
4.00 MeV	6.708E-02	3.736E-05	138.87 um	5.94 um	9.04 um
4.50 MeV	6.158E-02	3.368E-05	169.20 um	7.51 um	10.87 um
5.00 MeV	5.700E-02	3.070E-05	202.11 um	9.03 um	12.84 um
5.50 MeV	5.312E-02	2.822E-05	237.54 um	10.53 um	14.94 um
6.00 MeV	4.978E-02	2.613E-05	275.45 um	12.04 um	17.18 um
6.50 MeV	4.688E-02	2.435E-05	315.81 um	13.55 um	19.54 um
7.00 MeV	4.434E-02	2.280E-05	358.57 um	15.09 um	22.04 um
8.00 MeV	4.007E-02	2.025E-05	451.12 um	20.42 um	27.39 um
9.00 MeV	3.662E-02	1.824E-05	552.96 um	25.46 um	33.24 um
10.00 MeV	3.376E-02	1.661E-05	663.90 um	30.42 um	39.56 um
11.00 MeV	3.164E-02	1.525E-05	783.26 um	35.34 um	46.32 um
12.00 MeV	2.958E-02	1.412E-05	910.79 um	40.27 um	53.49 um
13.00 MeV	2.780E-02	1.314E-05	1.05 mm	45.27 um	61.09 um
14.00 MeV	2.623E-02	1.230E-05	1.19 mm	50.35 um	69.13 um
15.00 MeV	2.485E-02	1.156E-05	1.34 mm	55.52 um	77.58 um
16.00 MeV	2.361E-02	1.091E-05	1.51 mm	60.80 um	86.46 um
17.00 MeV	2.250E-02	1.033E-05	1.67 mm	66.17 um	95.75 um
18.00 MeV	2.151E-02	9.818E-06	1.85 mm	71.65 um	105.45 um
20.00 MeV	1.977E-02	8.932E-06	2.23 mm	91.09 um	126.08 um
22.00 MeV	1.832E-02	8.198E-06	2.64 mm	109.71 um	148.30 um
24.00 MeV	1.709E-02	7.581E-06	3.08 mm	128.05 um	172.09 um
26.00 MeV	1.602E-02	7.053E-06	3.55 mm	146.35 um	197.42 um
28.00 MeV	1.510E-02	6.598E-06	4.06 mm	164.75 um	224.26 um
30.00 MeV	1.429E-02	6.199E-06	4.59 mm	183.33 um	252.57 um
33.00 MeV	1.323E-02	5.688E-06	5.44 mm	222.68 um	297.78 um
36.00 MeV	1.234E-02	5.258E-06	6.35 mm	261.15 um	346.19 um
40.00 MeV	1.135E-02	4.780E-06	7.67 mm	325.64 um	415.58 um
45.00 MeV	1.034E-02	4.296E-06	9.48 mm	419.41 um	509.85 um
50.00 MeV	9.512E-03	3.904E-06	11.44 mm	509.74 um	612.21 um
55.00 MeV	8.827E-03	3.581E-06	13.57 mm	598.97 um	722.35 um
60.00 MeV	8.248E-03	3.309E-06	15.86 mm	688.19 um	840.01 um
65.00 MeV	7.753E-03	3.076E-06	18.30 mm	777.95 um	964.92 um
70.00 MeV	7.324E-03	2.876E-06	20.89 mm	868.53 um	1.10 mm
80.00 MeV	6.618E-03	2.546E-06	26.50 mm	1.19 mm	1.38 mm
90.00 MeV	6.060E-03	2.287E-06	32.67 mm	1.49 mm	1.69 mm
100.00 MeV	5.608E-03	2.078E-06	39.37 mm	1.79 mm	2.02 mm

Multiply Stopping by	for Stopping Units
5.3168E+02	MeV / mm
1.0000E+00	keV / (ug/cm^2)
1.0000E+03	keV / (mg/cm^2)
1.0000E+03	MeV / (g/cm^2)
1.0742E+02	L.S.S. reduced units

Table A-8 was computed using the Trim program written by J.P. Biersack, and J.F. Ziegler   (Version 91.14)

Table A-9.   Ranges, Stopping Powers, and Straggling of Electrons Incident on CMG Glass

$<Z/A>$ = 0.490296   I = 150.4 eV   DENSITY = 2.554 g/cm³

| ENERGY | STOPPING POWER | | | CSDA RANGE | RADIATION YIELD | DENSITY EFFECT DELTA |
| | COLLISION | RADIATIVE | TOTAL | | | |
MeV	MeV cm2/g	MeV cm2/g	MeV cm2/g	g/cm2		
0.01000	1.717E+01	6.357E-03	1.718E+01	3.384E-04	1.943E-04	0.000E+00
0.01250	1.454E+01	6.541E-03	1.454E+01	4.973E-04	2.375E-04	0.000E+00
0.01500	1.268E+01	6.674E-03	1.269E+01	6.818E-04	2.793E-04	0.000E+00
0.01750	1.130E+01	6.774E-03	1.130E+01	8.910E-04	3.198E-04	0.000E+00
0.02000	1.022E+01	6.854E-03	1.023E+01	1.124E-03	3.592E-04	0.000E+00
0.02500	8.649E+00	6.982E-03	8.656E+00	1.658E-03	4.351E-04	0.000E+00
0.03000	7.555E+00	7.082E-03	7.562E+00	2.277E-03	5.079E-04	0.000E+00
0.03500	6.745E+00	7.164E-03	6.752E+00	2.979E-03	5.781E-04	0.000E+00
0.04000	6.120E+00	7.236E-03	6.128E+00	3.757E-03	6.460E-04	0.000E+00
0.04500	5.623E+00	7.300E-03	5.630E+00	4.609E-03	7.119E-04	0.000E+00
0.05000	5.217E+00	7.361E-03	5.225E+00	5.532E-03	7.760E-04	0.000E+00
0.05500	4.879E+00	7.419E-03	4.887E+00	6.523E-03	8.385E-04	0.000E+00
0.06000	4.594E+00	7.476E-03	4.601E+00	7.578E-03	8.996E-04	0.000E+00
0.07000	4.137E+00	7.589E-03	4.144E+00	9.872E-03	1.018E-03	0.000E+00
0.08000	3.787E+00	7.700E-03	3.795E+00	1.240E-02	1.132E-03	0.000E+00
0.09000	3.510E+00	7.810E-03	3.518E+00	1.514E-02	1.242E-03	0.000E+00
0.10000	3.286E+00	7.918E-03	3.294E+00	1.808E-02	1.349E-03	0.000E+00
0.12500	2.877E+00	8.188E-03	2.885E+00	2.622E-02	1.604E-03	0.000E+00
0.15000	2.599E+00	8.462E-03	2.608E+00	3.536E-02	1.844E-03	0.000E+00
0.17500	2.400E+00	8.740E-03	2.409E+00	4.535E-02	2.072E-03	0.000E+00
0.20000	2.250E+00	9.027E-03	2.259E+00	5.608E-02	2.290E-03	0.000E+00
0.25000	2.042E+00	9.637E-03	2.052E+00	7.938E-02	2.701E-03	0.000E+00
0.30000	1.906E+00	1.029E-02	1.917E+00	1.046E-01	3.090E-03	0.000E+00
0.35000	1.812E+00	1.098E-02	1.823E+00	1.314E-01	3.463E-03	4.221E-03
0.40000	1.742E+00	1.171E-02	1.754E+00	1.594E-01	3.824E-03	2.111E-02
0.45000	1.690E+00	1.247E-02	1.702E+00	1.884E-01	4.177E-03	4.041E-02
0.50000	1.650E+00	1.327E-02	1.663E+00	2.181E-01	4.525E-03	6.130E-02
0.55000	1.619E+00	1.409E-02	1.633E+00	2.485E-01	4.868E-03	8.333E-02
0.60000	1.594E+00	1.494E-02	1.609E+00	2.793E-01	5.209E-03	1.062E-01
0.70000	1.559E+00	1.671E-02	1.575E+00	3.422E-01	5.886E-03	1.537E-01
0.80000	1.535E+00	1.858E-02	1.554E+00	4.061E-01	6.560E-03	2.026E-01
0.90000	1.520E+00	2.052E-02	1.541E+00	4.708E-01	7.235E-03	2.519E-01

Table A-9. Ranges, Stopping Powers, and Straggling of Electrons Incident on CMG Glass (Cont'd)

ENERGY	STOPPING POWER			CSDA	RADIATION	DENSITY
	COLLISION	RADIATIVE	TOTAL	RANGE	YIELD	EFFECT
						DELTA
MeV	MeV cm2/g	MeV cm2/g	MeV cm2/g	g/cm2		
1.00000	1.510E+00	2.254E-02	1.533E+00	5.359E-01	7.913E-03	3.012E-01
1.25000	1.499E+00	2.784E-02	1.526E+00	6.995E-01	9.624E-03	4.221E-01
1.50000	1.497E+00	3.347E-02	1.531E+00	8.631E-01	1.136E-02	5.377E-01
1.75000	1.501E+00	3.935E-02	1.540E+00	1.026E+00	1.312E-02	6.472E-01
2.00000	1.507E+00	4.547E-02	1.553E+00	1.188E+00	1.491E-02	7.507E-01
2.50000	1.522E+00	5.826E-02	1.580E+00	1.507E+00	1.854E-02	9.413E-01
3.00000	1.537E+00	7.169E-02	1.609E+00	1.821E+00	2.224E-02	1.113E+00
3.50000	1.551E+00	8.563E-02	1.637E+00	2.129E+00	2.598E-02	1.269E+00
4.00000	1.565E+00	9.999E-02	1.665E+00	2.432E+00	2.976E-02	1.413E+00
4.50000	1.577E+00	1.147E-01	1.692E+00	2.729E+00	3.355E-02	1.546E+00
5.00000	1.588E+00	1.296E-01	1.718E+00	3.023E+00	3.736E-02	1.669E+00
5.50000	1.599E+00	1.449E-01	1.743E+00	3.312E+00	4.117E-02	1.785E+00
6.00000	1.608E+00	1.604E-01	1.768E+00	3.596E+00	4.498E-02	1.893E+00
7.00000	1.625E+00	1.919E-01	1.817E+00	4.154E+00	5.258E-02	2.092E+00
8.00000	1.640E+00	2.242E-01	1.864E+00	4.698E+00	6.013E-02	2.272E+00
9.00000	1.653E+00	2.571E-01	1.910E+00	5.228E+00	6.762E-02	2.435E+00
10.00000	1.664E+00	2.905E-01	1.954E+00	5.745E+00	7.502E-02	2.584E+00
12.50000	1.688E+00	3.757E-01	2.064E+00	6.990E+00	9.312E-02	2.912E+00
15.00000	1.707E+00	4.630E-01	2.170E+00	8.171E+00	1.106E-01	3.191E+00
17.50000	1.723E+00	5.519E-01	2.275E+00	9.296E+00	1.274E-01	3.434E+00
20.00000	1.737E+00	6.420E-01	2.379E+00	1.037E+01	1.435E-01	3.649E+00
25.00000	1.759E+00	8.254E-01	2.584E+00	1.239E+01	1.738E-01	4.019E+00
30.00000	1.776E+00	1.012E+00	2.788E+00	1.425E+01	2.018E-01	4.330E+00
35.00000	1.791E+00	1.201E+00	2.991E+00	1.598E+01	2.276E-01	4.599E+00
40.00000	1.803E+00	1.392E+00	3.195E+00	1.760E+01	2.515E-01	4.835E+00
45.00000	1.813E+00	1.584E+00	3.398E+00	1.911E+01	2.737E-01	5.047E+00
50.00000	1.823E+00	1.778E+00	3.601E+00	2.054E+01	2.944E-01	5.239E+00
55.00000	1.831E+00	1.973E+00	3.804E+00	2.189E+01	3.137E-01	5.414E+00
60.00000	1.838E+00	2.169E+00	4.008E+00	2.317E+01	3.317E-01	5.575E+00
70.00000	1.851E+00	2.564E+00	4.415E+00	2.555E+01	3.646E-01	5.862E+00
80.00000	1.863E+00	2.961E+00	4.824E+00	2.772E+01	3.937E-01	6.114E+00
90.00000	1.872E+00	3.361E+00	5.234E+00	2.971E+01	4.198E-01	6.338E+00
100.00000	1.881E+00	3.763E+00	5.644E+00	3.155E+01	4.433E-01	6.540E+00

Table A-9 was computed using the EPSTAR program written by S.M. Seltzer.

```
      EQGAFLUX
C     ********************************************************************
C     *                                                                 *
C     *   PROGRAM FOR COMPUTING EQUIVALENT FLUENCE FROM SPACE ELECTRON   *
C     *    AND PROTON ENERGY SPECTRA AND RELATIVE DAMAGE COEFFICIENTS    *
C     *  FOR THE PURPOSE OF ESTIMATING GaAs SOLAR CELL DEGRADATION.      *
C     *                                                                 *
C     *                                                                 *
C     * MACHINE / FORTRAN FEATURES NECESSARY                            *
C     *        NAMELIST INPUT/OUTPUT                                    *
C     *        INPUT UNIT (CARD READER) IS FORTRAN UNIT 5               *
C     *        OUTPUT UNIT (PRINTER) IS FORTRAN UNIT 6                  *
C     *        BLOCK DATA SUBPROGRAM                                    *
C     *        PROGRAM WRITTEN FOR UNIVAC 1108 (FORTRAN 4 COMPATIBLE)   *
C     *        ALPHANUMERIC INPUT/OUTPUT ( 'A' FORMAT ) ASSUMES         *
C     *         6-CHARACTER CAPABILITY.  HOLLERITH STRINGS ARE USED AS  *
C     *         CHARACTER COUNT ( 5HABCDE ) AND AS QUOTE STRINGS.       *
C     ********************************************************************
C
      PARAMETER KE=47,KP=77
      COMMON/DAMAGE/EMEV(KE),EDET(KE,8),PMEV(KP),PISC(KP,8),PVOC(KP,8)
C
      DIMENSION TIMIN(2), TIMOUT(2)
      DIMENSION HEADER(14),THICK(8),COND(2)
      DIMENSION ED(KE),PI(KP),PV(KP)
      DIMENSION ESPEC(70,2),PSPEC(70,2)
      DIMENSION EQUIVE(8),EQV10I(8),EQV10V(8),EQV10P(8)
      DIMENSION EMLN(KE),PMLN(KP)
      DIMENSION EPTOTV(8),EPTOTI(8),EPTOTP(8)
      DIMENSION ESPLN(70,2),PSPLN(70,2)
      DIMENSION TTHICK(8),ITHICK(8)
C
      INTEGER PAGE,PCKE(8),PCKP(8)
C
      DATA THICK/0.,5.59E-3,1.68E-2,3.35E-2,6.71E-2,1.12E-1,1.675E-1,
     *3.35E-1/
      DATA NESPEC/0/,NPSPEC/0/,NSTEP/2/,IDIAG/0/
      DATA PEDRI,PEDRV/400.,1400./,PCKE,PCKP/16*0/,INTFLG/0/
      DATA PEDRP/1000./
      DATA TIMIN/12HDAY          /, TIMOUT/12H1 YEAR        /,
     *   TMULT/365.2422/
C
      NAMELIST /MIKE/ NESPEC,ESPEC,NPSPEC,PSPEC,NSTEP,TIMIN,TIMOUT
     *,TMULT,PEDRI,PEDRV,PEDRP,PCKE,PCKP,IDIAG,INTFLG
C
C     ********************************************************************
C     *  INPUT VARIABLES . . .                                          *
C     *                                                                 *
C     *  HEADER    ALPHANUMERIC RECORD (80 CHARACTERS) TO IDENTIFY CASE.*
C     *                                                                 *
C     *  THE FOLLOWING ARE NAMELIST VARIABLES                           *
C     *  PUNCH NAMELIST ITEMS STARTING IN COLUMN 2                      *
C     *                                                                 *
C     *  NESPEC, NPSPEC   NUMBER OF INPUT DATA FOR ELECTRON AND PROTON  *
C     *        ENERGY SPECTRA.                                          *
C     *  ESPEC(I,J),PSPEC(I,J)   INTEGRAL ENERGY SPECTRUM OF SPACE      *
C     *        ELECTRON AND PROTON ENVIRONMENTS.  J=1  ENERGY IN MEV.   *
C     *        J=2  INTEGRAL FLUX IN PARTICLES PER SQUARE CENTIMETER    *
C     *        PER UNIT TIME. INPUT SPECTRAL DATA IN ASCENDING ORDER,   *
C     *        LOWEST ENERGY FIRST, HIGHEST LAST.                       *
C     *  NSTEP   NUMBER OF POINTS BETWEEN ENERGY ENTRIES OF RELATIVE    *
```

Table A-10. Equivalent Fluence Program for GaAs Solar Cells (Cont'd)

```
C    *          DAMAGE COEFFICIENTS FOR INTERPOLATION (DEFAULT VALUE = 2)*
C    *   TIMIN  12 CHARACTER HOLLERITH STRING WHICH DESCRIBES TIME        *
C    *          INTERVAL REPRESENTED BY INPUT SPECTRA.  FOR EXAMPLE IF    *
C    *          INPUT SPECTRA REPRESENT FLUENCES PER DAY                  *
C    *                      TIMIN = 12HDAY                                *
C    *          IF INPUT SPECTRA REPRESENT FLUENCE PER SECOND             *
C    *                      TIMIN = 12HSECOND                             *
C    *   TMULT  NUMBER OF 'TIMIN' UNITS FOR WHICH EQUIVALENT FLUENCE      *
C    *          IS TO BE COMPUTED.  FOR EXAMPLE IF INPUT SPECTRA          *
C    *          REPRESENT FLUENCE PER HOUR AND EQUIVALENT FLUENCE IS      *
C    *          TO BE COMPUTED FOR 24 HOURS                               *
C    *                      TMULT = 24.                                   *
C    *          (TMULT SHOULD BE INPUT SUCH THAT TMULT = TIMOUT/TIMIN)    *
C    *   TIMOUT  12 CHARACTER HOLLERITH STRING WHICH DESCRIBES TOTAL      *
C    *          TIME OF EXPOSURE AND IS THE PRODUCT OF 'TIMIN' UNITS      *
C    *          AND TMULT.  FOR EXAMPLE IF TIMIN = '1 DAY' AND            *
C    *          TMULT = 365.2422                                          *
C    *                      TIMOUT = 12H1 YEAR                            *
C    *          ------------------------------------------------         *
C    *          INCLUDE ALL 12 CHARACTERS IN THE NAMELIST INPUT          *
C    *              INCLUDING TRAILING BLANKS.                           *
C    *          ------------------------------------------------         *
C    *          IF TIMIN = '1 MONTH' AND MISSION DURATION IS 34.2        *
C    *          MONTHS                                                    *
C    *                      TIMOUT = 12H34.2 MONTHS                       *
C    *              OR    TIMOUT = 12H1 MISSION                           *
C    *          ------------------------------------------------         *
C    *          INCLUDE ALL 12 CHARACTERS IN THE NAMELIST INPUT          *
C    *              INCLUDING TRAILING BLANKS.                           *
C    *          ------------------------------------------------         *
C    *   **NOTE**  DEFAULT VALUES ARE:                                    *
C    *                      TIMIN = 12HDAY                                *
C    *                      TMULT = 365.2422                              *
C    *                      TIMOUT = 12H1 YEAR                            *
C    *   PEDRI   DAMAGE RATIO BETWEEN PROTONS AND ELECTRONS FOR ISC.      *
C    *          (DEFAULT VALUE = 400.)                                    *
C    *   PEDRV   DAMAGE RATIO BETWEEN PROTONS AND ELECTRONS FOR VOC       *
C    *          (DEFAULT VALUE = 1400.)                                   *
C    *   PEDRP   DAMAGE RATIO BETWEEN PROTONS AND ELECTRONS FOR PMAX      *
C    *          (DEFAULT VALUE = 1000.)                                   *
C    *   PCKE, PCKP   FLAGS TO CAUSE PRINTING OF DIFFERENTIAL FLUENCE,    *
C    *          DAMAGE COEFFICIENTS, EQUIVALENT FLUENCE, ETC. FOR         *
C    *          ELECTRONS (PCKE) AND/OR PROTONS (PCKP).  8 VALUES FOR     *
C    *          EACH VARIABLE MAY BE INPUT CORRESPONDING TO COVER GLASS   *
C    *          THICKNESSES (DEFAULT VALUE=0 FOR NO PRINT.                *
C    *                      SET=1 FOR PRINT.)                             *
C    *   IDIAG   FLAG TO PRINT NAMELIST INPUT AS A DIAGNOSTIC AID.        *
C    *          (DEFAULT VALUE = 0 FOR NO PRINT.  SET = 1 FOR PRINT.)     *
C    *   INTFLG  FLAG TO ESTABLISH LIMITS OF INTEGRATION                  *
C    *          WHEN INTFLG = 0 INTEGRATION PROCEEDS OVER ALL ENERGIES    *
C    *          FOR WHICH DAMAGE COEFFICIENTS ARE AVAILABLE AND INPUT     *
C    *          SPECTRA ARE EXTRAPOLATED IF NECESSARY.                    *
C    *          WHEN INTFLG = 1 INTEGRATION PROCEEDS ONLY OVER THE INPUT  *
C    *          ENERGY RANGE, NAMELY, ENERGY INTERVALS ESPEC(1,1) TO      *
C    *          ESPEC(NESPEC,1) AND PSPEC(1,1) TO PSPEC(NPSPEC,1)         *
C    *          DEFAULT VALUE = 0                                         *
C    *                                                                    *
C    *                                                                    *
C    **********************************************************************
C
C
```

```
      PAGE=0
C
C     READ HEADER CARD (IDENTIFIER INFORMATION)
C
  100 READ(5,20,END=9999) HEADER
C
C     INITIALIZE TOTAL FLUENCE VECTORS
C
      DO 11 I=1,8
      EPTOTV(I)=0.
      EPTOTP(I)=0.
   11 EPTOTI(I)=0.
C
C     READ INPUT DATA (NAMELIST 'MIKE')
C
      READ(5,MIKE)
      IF(IDIAG .EQ. 0) GO TO 12
      PAGE=PAGE+1
      WRITE(6,25) HEADER,PAGE
      WRITE(6,MIKE)
   12 CONTINUE
C
      IF(NESPEC .EQ. 0 .AND. NPSPEC .EQ. 0) THEN
        GOTO 100
      END IF
      IF(NESPEC .NE. 0) THEN
        PAGE=PAGE+1
        WRITE(6,25) HEADER,PAGE
        WRITE(6,32)TIMIN,((ESPEC(I,J),J=1,2),I=1,NESPEC)
C       TAKE LOGS OF ELECTRON FLUENCES
        DO 101 J=1,NESPEC
  101   ESPLN(J,2) = ALOG(ESPEC(J,2))
      END IF
C
      IF(NPSPEC .NE. 0) THEN
        PAGE=PAGE+1
        WRITE(6,25) HEADER,PAGE
        WRITE(6,33)TIMIN,((PSPEC(I,J),J=1,2),I=1,NPSPEC)
C       TAKE LOGS OF PROTON ENERGIES AND FLUENCES
        DO 106 J=1,NPSPEC
        PSPLN(J,1) = ALOG(PSPEC(J,1))
  106   PSPLN(J,2) = ALOG(PSPEC(J,2))
      END IF
C
  107 DO 9000 L=1,8
C
C     TAKE LOGS OF RELATIVE DAMAGE COEFFICIENTS AND RELATED ENERGIES
C
      IF(NESPEC .EQ. 0) GO TO 190
      DO 187 K=1,KE
      IF(L .GT. 1) GO TO 181
      EMLN(K)=ALOG(EMEV(K))
  181 IF(EDET(K,L))183,183,185
  183 ED(K)=-50.
      GO TO 187
  185 ED(K)=ALOG(EDET(K,L))
  187 CONTINUE
  190 IF(NPSPEC .EQ. 0) GO TO 200
      DO 150 K=1,KP
      IF(L .GT. 1) GO TO 125
      PMLN(K)=ALOG(PMEV(K))
```

```
  125 IF(PISC(K,L))130,130,135
  130 PI(K)=-50.
      GO TO 140
  135 PI(K)=ALOG(PISC(K,L))
  140 IF(PVOC(K,L))145,145,147
  145 PV(K)=-50.
      GO TO 150
  147 PV(K)=ALOG(PVOC(K,L))
  150 CONTINUE
C
C     COMPUTE EQUIVALENT FLUENCE FOR ELECTRON SPECTRUM
C       (BYPASS IF NO ELECTRON SPECTRUM)
C
  200 LINE=1
      IF(NESPEC .EQ. 0) GO TO 400
      EQUIVE(L) = 0.0
      ELLIM = ESPEC(1,1)
      EULIM = ESPEC(NESPEC,1)
C
C     ITERATE OVER ALL ENERGY INCREMENTS
C
      DO 300 K=1,46
      DIFF=EMLN(K+1)-EMLN(K)
      DELTA=DIFF/NSTEP
      DEL2=DELTA/2.
      DO 300 I=1,NSTEP
      SPEC1=EMLN(K)+DELTA*(I-1)
      DSPEC=SPEC1+DEL2
      EK=EXP(SPEC1)
      EK1=EXP(SPEC1+DELTA)
C
C     PERFORM LINEAR INTERPOLATION OF PHI VS. E (SEMI-LOG)
C
      CALL INTP(EK,PHI1,ESPEC(1,1),ESPLN(1,2),NESPEC)
      CALL INTP(EK1,PHI2,ESPEC(1,1),ESPLN(1,2),NESPEC)
      PHI1 = EXP(PHI1)
      PHI2 = EXP(PHI2)
C
C     PERFORM LINEAR INTERPOLATION OF ELECTRON
C     DAMAGE COEFFICIENTS VS. E (LOG-LOG)
C
      CALL INTP(DSPEC,D1,EMLN(1),ED(1),KE)
      D=EXP(D1)
      IF(D .LT. 1.E-4) D=0.0
C
C     USE RESTRICTED INTEGRATION LIMITS IF INTFLG .GT. 0
C
      IF ( INTFLG .EQ. 0 ) GO TO 201
      IF(EK .LT. ELLIM .OR. EK1 .GT. EULIM) GO TO 202
      GO TO 201
  202 PHI1 = 0.0
      PHI2 = 0.0
  201 DPHI = PHI1 - PHI2
      PROD = DPHI * D
C
C     SUM PRODUCTS OVER ALL ENERGY INCREMENTS
C
      EQUIVE(L) = EQUIVE(L) + PROD
      IF(PCKE(L) .EQ. 0) GO TO 300
C
C     PRINT INTERMEDIATE CALCULATIONS OF DIFFERENTIAL FLUX, RELATIVE
```

Table A-10. Equivalent Fluence Program for GaAs Solar Cells (Cont'd)

```
C           DAMAGE COEFFICIENT, AND EQUIVALENT FLUENCE
C
        IF(LINE .NE. 1) GO TO 50
        PAGE=PAGE+1
        WRITE(6,25) HEADER,PAGE
        WRITE(6,26) THICK(L)
        WRITE(6,30)
    50 DSPEC1=EXP(DSPEC)
        WRITE(6,10)EK,EK1,PHI1,PHI2,DPHI,D,DSPEC1,PROD,EQUIVE(L)
        LINE=LINE+1
        IF(LINE .GE. 50) LINE=1
   300 CONTINUE
C
C       COMPUTE EQUIVALENT FLUENCE FOR PROTRON SPECTRUM
C        (BYPASS IF NO PROTON SPECTRUM)
C
   400 IF(NPSPEC .EQ. 0) GO TO 9000
        LINE=1
        EQV10I(L) = 0.0
        EQV10V(L) = 0.0
        EQV10P(L) = 0.0
        PLLIM = ALOG(PSPEC(1,1))
        PULIM = ALOG(PSPEC(NPSPEC,1))
        DO 500 K=1,76
        DIFF=PMLN(K+1)-PMLN(K)
        DELTA=DIFF/NSTEP
        DEL2=DELTA/2.
        DO 500 I=1,NSTEP
        SPEC1=PMLN(K)+DELTA*(I-1)
        SPEC2=SPEC1+DELTA
        DSPEC=SPEC1+DEL2
C
C       PERFORM LINEAR INTERPOLATION OF PHI VS. E (LOG-LOG)
        CALL INTP(SPEC1,PHI1,PSPLN(1,1),PSPLN(1,2),NPSPEC)
        CALL INTP(SPEC2,PHI2,PSPLN(1,1),PSPLN(1,2),NPSPEC)
        PHI1 = EXP(PHI1)
        PHI2 = EXP(PHI2)
C
C
C       PERFORM LINEAR INTERPOLATION OF DAMAGE COEFFICIENT VS. E(LOG-LOG)
        CALL INTP(DSPEC,DCI,PMLN(1),PI(1),KP)
        CALL INTP(DSPEC,DCV,PMLN(1),PV(1),KP)
        DISC=EXP(DCI)
        DVOC=EXP(DCV)
        IF(DISC .LT. 1.E-4) DISC=0.0
        IF(DVOC .LT. 1.E-4) DVOC=0.0
        IF(INTFLG .EQ. 0) GO TO 401
C
C       USE RESTRICTED INTEGRATION LIMITS IF INTFLG .GT. 0
C
        IF(SPEC1 .LT. PLLIM .OR. SPEC2 .GT. PULIM) GO TO 402
        GO TO 401
   402 PHI1 = 0.0
        PHI2 = 0.0
   401 DPHI = PHI1 - PHI2
        PROD1=DPHI*DISC
        EQV10I(L) = EQV10I(L) + PROD1
        EQV10V(L) = EQV10V(L) + DPHI*DVOC
        EQV10P(L) = EQV10P(L) + DPHI*DVOC
        IF(PCKP(L) .EQ. 0) GO TO 500
        IF(LINE .NE. 1) GO TO 60
```

```
       PAGE=PAGE+1
       WRITE(6,25) HEADER,PAGE
       WRITE(6,41) THICK(L)
       WRITE(6,40)
    60 EK=EXP(SPEC1)
       EK1=EXP(SPEC1+DELTA)
       DFXDCV = DPHI*DVOC
       DSPEC1=EXP(DSPEC)
       WRITE(6,10)EK,EK1,PHI1,DPHI,DISC,DVOC,DSPEC1,PROD1,DFXDCV,
      *EQV10I(L),EQV10V(L)
       LINE=LINE+1
       IF(LINE .GE. 50) LINE=1
   500 CONTINUE
  9000 CONTINUE
C
C      PRINT CALCULATION SUMMARY
C
       PAGE=PAGE+1
       WRITE(6,25) HEADER,PAGE
       WRITE(6,2) (THICK(J),J=1,8)
       DO 520 J=1,8
       TTHICK(J)=THICK(J)*178.8908766+.5
       ITHICK(J)=TTHICK(J)
   520 CONTINUE
       WRITE(6,22)(ITHICK(J),J=1,8)
       DO 1000 K=1,8
       EQUIVE(K) = EQUIVE(K) * TMULT
C
C      CONVERT 10 MEV PROTONS TO EQUIVALENT 1 MEV ELECTRONS USING PEDRV
C        AND PEDRI
C
       EQV10I(K) = EQV10I(K) * TMULT * PEDRI
       EQV10V(K) = EQV10V(K) * TMULT * PEDRV
       EQV10P(K) = EQV10P(K) * TMULT * PEDRP
  1000 CONTINUE
       IF(NESPEC .EQ. 0) GO TO 2000
       WRITE(6,3) (EQUIVE(J),J=1,8)
       DO 2001 I=1,8
       EPTOTV(I)=EPTOTV(I)+EQUIVE(I)+EQV10V(I)
       EPTOTP(I)=EPTOTP(I)+EQUIVE(I)+EQV10P(I)
       EPTOTI(I)=EPTOTI(I)+EQUIVE(I)+EQV10I(I)
  2001 CONTINUE
  2000 IF(NPSPEC .EQ. 0) GO TO 3000
       WRITE(6,4) (EQV10V(J),J=1,8)
       WRITE(6,6) (EQV10P(J),J=1,8)
       WRITE(6,5) (EQV10I(J),J=1,8)
       IF(NESPEC .EQ. 0) GO TO 3000
       WRITE(6,28)
       WRITE(6,29) (EPTOTV(J),J=1,8)
       WRITE(6,34) (EPTOTP(J),J=1,8)
       WRITE(6,31) (EPTOTI(J),J=1,8)
  3000 CONTINUE
       WRITE(6,43) TIMOUT,TMULT
       GO TO 100
C
C
     2 FORMAT(1H0,'SHIELD THICKNESS (GM/CM2)',4X8(1PE10.3))
     3 FORMAT(1H0,'ELECTRON FLUENCE'/1H ,2X'EQUIV 1 MEV ELECTRONS/CM2',
      *  2X8(1PE10.3))
     4 FORMAT(1H0,'PROTON FLUENCE'/1H ,2X'EQUIV 1 MEV ELECTRONS/CM2'/
      *  1H ,16X'VOC',10X8(1PE10.3))
```

```
   5 FORMAT(1H ,16X'ISC',10X8(1PE10.3))
   6 FORMAT(1H ,15X,'PMAX',10X,8(1PE10.3))
  10 FORMAT(11E12.4)
  20 FORMAT(13A6,A2)
  22 FORMAT(1H ,17X,'( MILS ) ',8I10)
  25 FORMAT(1H1,14A6,16X4HPAGE,I4/)
  26 FORMAT(1H ,'(ELECTRON SPECTRUM)',10X'COVER SLIDE THICKNESS =',
     *  F10.5,' GM/CM2'/)
  27 FORMAT(1H1,31HSOLAR FLARE PROTON SPECTRUM FOR,I2,1X,A6,A5,
     *  'EVENT(S)'51X,4HPAGE,1X,I3/)
 271 FORMAT (1H1,'SOLAR FLARE PROTON SPECTRUM FOR',A6,A5,'EVENT',
     * 51X,'PAGE',1X,I3/)
 272 FORMAT (1H ,5X,17HMISSION DURATION=, F5.1,8H MONTHS.
     *  /5X,17HCONFIDENCE LEVEL=, I3, 9H PERCENT.
     *  //13X,6HENERGY,10X,13HINTEGRAL FLUX
     *  /14X,5H(MEV),7X,20HPROTONS/CM2-MISSION. /)
  28 FORMAT(1H0,'TOTAL FLUENCE (ELECTRONS + PROTONS)'/
     *  1H ,2X'EQUIV 1 MEV ELECTRONS/CM2')
  29 FORMAT(1H ,16X'VOC',10X8(1PE10.3))
  34 FORMAT(1H ,15X,'PMAX',10X8(1PE10.3))
  30 FORMAT(5X,3HEK ,9X,3HEK1,9X,3HFX1,9X,3HFX2,9X,3HDFX,9X
     *  ,3HDCI,9X,7HEINTERP,5X,7HDFX*DCI,5X,6HEQFLUX  / )
  31 FORMAT(1H ,16X'ISC',10X8(1PE10.3))
  32 FORMAT(1H0,26X,'ELECTRON'/
     *1H ,10X,'ENERGY',10X,'FLUENCE'/
     *1H ,10X, '(MEV)',11X,'(ELECTRONS/CM2-',2A6,')'//
     *(1H ,0PF16.3,1PE18.4))
  33 FORMAT(1H0,26X,'PROTON'/
     *1H ,10X,'ENERGY',10X,'FLUENCE'/
     *1H ,10X, '(MEV)',11X,'(PROTONS/CM2-',2A6,')'//
     *(1H ,0PF16.3,1PE18.4))
  37 FORMAT(0PF20.3,1PE20.4)
  40 FORMAT(5X,2HEK,10X,3HEK1,9X,3HFX1,9X,3HDFX,9X,3HDCI,9X,3HDCV,
     *9X,7HEINTERP,5X,7HDFX*DCI,5X,7HDFX*DCV,5X,4HEQFI,8X,4HEQFV  / )
  41 FORMAT(1H ,'(PROTON SPECTRUM)',10X,'COVER SLIDE THICKNESS =',
     *  F10.5,' GM/CM2'/)
  43 FORMAT(1H0,'TIME OF EXPOSURE:  ',2A6/3X,'(TMULT = ',1PE12.5,')')
  44 FORMAT(1H1)
9999 CONTINUE
     WRITE(6,44)
     STOP
     END




     SUBROUTINE INTP(XT,YT,X,Y,N)
C
C    ******************************************************************
C    *  LINEAR INTERPOLATION SUBROUTINE                               *
C    ******************************************************************
C
     DIMENSION X(1),Y(1)
     DO 10 I=1,N
     II=I
     IF(XT .LE. X(I)) GO TO 12
  10 CONTINUE
  12 IF(II .EQ. 1) II=2
     IM=II-1
     YT=Y(IM)+(XT-X(IM))*(Y(II)-Y(IM))/(X(II)-X(IM))
     RETURN
     END
```

```
      BLOCK DATA
      PARAMETER KE=47,KP=77
      COMMON/DAMAGE/EMEV(KE),EDET(KE,8),PMEV(KP),PISC(KP,8),PVOC(KP,8)
C
C
C
C     EMEV - ELECTRON ENERGIES FOR DAMAGE COEFFICIENT TABLE EDET
C
      DATA (EMEV(I),I=1,47)
     *   /1.500E-01,1.600E-01,1.700E-01,1.800E-01,1.900E-01,2.000E-01,
     *    2.200E-01,2.400E-01,2.600E-01,2.800E-01,3.000E-01,3.200E-01,
     *    3.600E-01,4.000E-01,4.500E-01,5.000E-01,6.000E-01,7.000E-01,
     *    8.000E-01,9.000E-01,1.000E+00,1.200E+00,1.400E+00,1.600E+00,
     *    1.800E+00,2.000E+00,2.250E+00,2.500E+00,2.750E+00,3.000E+00,
     *    3.250E+00,3.500E+00,3.750E+00,4.000E+00,4.500E+00,5.000E+00,
     *    5.500E+00,6.000E+00,7.000E+00,8.000E+00,9.000E+00,1.000E+01,
     *    1.500E+01,2.000E+01,2.500E+01,3.000E+01,4.000E+01/
C
C     0. GM/CM2 COVER GLASS DAMAGE COEFFICIENTS
C
      DATA (EDET(I),I=  1,47)
     *   /2.500E-06,5.000E-06,1.000E-05,2.500E-05,5.000E-05,1.000E-04,
     *    2.500E-04,5.000E-04,5.000E-03,1.250E-02,2.200E-02,3.000E-02,
     *    6.000E-02,8.500E-02,1.250E-01,1.600E-01,2.300E-01,3.000E-01,
     *    3.650E-01,4.300E-01,5.000E-01,6.250E-01,7.500E-01,8.700E-01,
     *    9.950E-01,1.100E+00,1.250E+00,1.300E+00,1.525E+00,1.650E+00,
     *    1.800E+00,1.900E+00,2.050E+00,2.200E+00,2.450E+00,2.700E+00,
     *    2.900E+00,3.150E+00,3.600E+00,4.050E+00,4.500E+00,4.900E+00,
     *    6.900E+00,8.650E+00,1.025E+01,1.185E+01,1.475E+01/
C
C     0.00559 GM/CM2 COVER GLASS DAMAGE COEFFICIENTS
C
      DATA (EDET(I),I= 48,94)
     *   /0.000E+00,0.000E+00,9.582E-07,2.932E-06,6.978E-06,1.679E-05,
     *    6.805E-05,1.818E-04,5.399E-04,3.622E-03,9.504E-03,1.693E-02,
     *    3.612E-02,6.181E-02,9.577E-02,1.318E-01,1.999E-01,2.693E-01,
     *    3.348E-01,3.993E-01,4.677E-01,5.943E-01,7.185E-01,8.384E-01,
     *    9.615E-01,1.069E+00,1.215E+00,1.279E+00,1.478E+00,1.614E+00,
     *    1.760E+00,1.867E+00,2.008E+00,2.156E+00,2.408E+00,2.656E+00,
     *    2.858E+00,3.102E+00,3.550E+00,3.996E+00,4.442E+00,4.840E+00,
     *    6.823E+00,8.560E+00,1.015E+01,1.173E+01,1.461E+01/
C
C     0.0168 GM/CM2 COVER GLASS DAMAGE COEFFICIENTS
C
      DATA (EDET(I),I=95,141)
     *   /0.000E+00,0.000E+00,0.000E+00,0.000E+00,0.000E+00,7.415E-07,
     *    5.762E-06,2.974E-05,9.434E-05,2.535E-04,1.794E-03,5.463E-03,
     *    1.724E-02,3.674E-02,6.445E-02,9.744E-02,1.619E-01,2.286E-01,
     *    2.940E-01,3.574E-01,4.227E-01,5.486E-01,6.715E-01,7.907E-01,
     *    9.112E-01,1.022E+00,1.165E+00,1.249E+00,1.417E+00,1.568E+00,
     *    1.710E+00,1.825E+00,1.959E+00,2.106E+00,2.362E+00,2.610E+00,
     *    2.818E+00,3.056E+00,3.506E+00,3.951E+00,4.396E+00,4.798E+00,
     *    6.777E+00,8.516E+00,1.010E+01,1.169E+01,1.456E+01/
C
C     0.0335 GM/CM2 COVER GLASS DAMAGE COEFFICIENTS
C
      DATA (EDET(I),I=142,188)
     *   /0.000E+00,0.000E+00,0.000E+00,0.000E+00,0.000E+00,0.000E+00,
     *    0.000E+00,6.795E-07,5.553E-06,2.769E-05,8.541E-05,3.417E-04,
     *    5.163E-03,1.533E-02,3.643E-02,6.232E-02,1.215E-01,1.835E-01,
     *    2.467E-01,3.083E-01,3.704E-01,4.947E-01,6.146E-01,7.334E-01,
     *    8.516E-01,9.643E-01,1.103E+00,1.203E+00,1.341E+00,1.501E+00,
```

```
      *      1.637E+00,1.760E+00,1.885E+00,2.030E+00,2.291E+00,2.540E+00,
      *      2.755E+00,2.987E+00,3.440E+00,3.884E+00,4.328E+00,4.733E+00,
      *      6.709E+00,8.450E+00,1.004E+01,1.161E+01,1.449E+01/
C
C     0.0671 GM/CM2 COVER GLASS DAMAGE COEFFICIENTS
C
      DATA (EDET(I),I=189,235)
      *     /0.000E+00,0.000E+00,0.000E+00,0.000E+00,0.000E+00,0.000E+00,
      *      0.000E+00,0.000E+00,0.000E+00,0.000E+00,1.232E-07,1.916E-06,
      *      4.519E-05,1.154E-03,8.242E-03,2.290E-02,6.664E-02,1.190E-01,
      *      1.756E-01,2.332E-01,2.908E-01,4.090E-01,5.246E-01,6.396E-01,
      *      7.539E-01,8.673E-01,9.999E-01,1.119E+00,1.226E+00,1.390E+00,
      *      1.523E+00,1.654E+00,1.770E+00,1.909E+00,2.172E+00,2.416E+00,
      *      2.638E+00,2.859E+00,3.307E+00,3.748E+00,4.191E+00,4.602E+00,
      *      6.574E+00,8.318E+00,9.909E+00,1.147E+01,1.434E+01/
C
C     0.112 GM/CM2 COVER GLASS DAMAGE COEFFICIENTS
C
      DATA (EDET(I),I=236,282)
      *     /0.000E+00,0.000E+00,0.000E+00,0.000E+00,0.000E+00,0.000E+00,
      *      0.000E+00,0.000E+00,0.000E+00,0.000E+00,0.000E+00,0.000E+00,
      *      0.000E+00,1.186E-06,5.982E-05,2.608E-03,2.393E-02,6.257E-02,
      *      1.088E-01,1.595E-01,2.117E-01,3.198E-01,4.295E-01,5.394E-01,
      *      6.482E-01,7.590E-01,8.875E-01,1.017E+00,1.111E+00,1.262E+00,
      *      1.395E+00,1.530E+00,1.644E+00,1.771E+00,2.033E+00,2.276E+00,
      *      2.502E+00,2.716E+00,3.166E+00,3.601E+00,4.033E+00,4.444E+00,
      *      6.393E+00,8.142E+00,9.735E+00,1.129E+01,1.415E+01/
C
C     0.1675 GM/CM2 COVER GLASS DAMAGE COEFFICIENTS
C
      DATA (EDET(I),I=283,329)
      *     /0.000E+00,0.000E+00,0.000E+00,0.000E+00,0.000E+00,0.000E+00,
      *      0.000E+00,0.000E+00,0.000E+00,0.000E+00,0.000E+00,0.000E+00,
      *      0.000E+00,0.000E+00,0.000E+00,1.527E-06,2.558E-03,2.076E-02,
      *      5.335E-02,9.337E-02,1.381E-01,2.330E-01,3.345E-01,4.369E-01,
      *      5.402E-01,6.445E-01,7.711E-01,8.984E-01,1.002E+00,1.123E+00,
      *      1.260E+00,1.389E+00,1.509E+00,1.628E+00,1.882E+00,2.122E+00,
      *      2.352E+00,2.561E+00,3.005E+00,3.433E+00,3.856E+00,4.269E+00,
      *      6.203E+00,7.926E+00,9.514E+00,1.106E+01,1.392E+01/
C
C     0.335 GM/CM2 COVER GLASS DAMAGE COEFFICIENTS
C
      DATA (EDET(I),I=330,376)
      *     /0.000E+00,0.000E+00,0.000E+00,0.000E+00,0.000E+00,0.000E+00,
      *      0.000E+00,0.000E+00,0.000E+00,0.000E+00,0.000E+00,0.000E+00,
      *      0.000E+00,0.000E+00,0.000E+00,0.000E+00,0.000E+00,0.000E+00,
      *      1.348E-05,4.126E-03,1.951E-02,7.296E-02,1.429E-01,2.222E-01,
      *      3.063E-01,3.939E-01,5.040E-01,6.175E-01,7.295E-01,8.334E-01,
      *      9.354E-01,1.058E+00,1.174E+00,1.289E+00,1.514E+00,1.747E+00,
      *      1.970E+00,2.182E+00,2.600E+00,3.011E+00,3.418E+00,3.824E+00,
      *      5.704E+00,7.408E+00,8.976E+00,1.047E+01,1.327E+01/
C
C     PMEV - PROTON ENERGIES FOR DAMAGE COEFFICIENT TABLES PISC AND PVOC
C
      DATA (PMEV(I),I=1,77)
      *     /2.000E-02,2.500E-02,3.000E-02,3.500E-02,4.000E-02,4.500E-02,
      *      5.000E-02,5.500E-02,6.000E-02,7.000E-02,8.000E-02,9.000E-02,
      *      1.000E-01,2.000E-01,3.000E-01,4.000E-01,6.000E-01,8.000E-01,
      *      1.000E+00,1.200E+00,1.300E+00,1.400E+00,1.600E+00,1.800E+00,
      *      2.000E+00,2.200E+00,2.400E+00,2.600E+00,2.800E+00,3.000E+00,
      *      3.200E+00,3.400E+00,3.600E+00,3.800E+00,4.000E+00,4.200E+00,
```

```
      *      4.400E+00,4.600E+00,4.800E+00,5.200E+00,5.600E+00,6.000E+00,
      *      6.400E+00,6.800E+00,7.200E+00,7.600E+00,8.000E+00,9.000E+00,
      *      1.000E+01,1.100E+01,1.200E+01,1.300E+01,1.400E+01,1.500E+01,
      *      1.600E+01,1.800E+01,2.000E+01,2.200E+01,2.400E+01,2.600E+01,
      *      2.800E+01,3.000E+01,3.400E+01,3.800E+01,4.200E+01,4.600E+01,
      *      5.000E+01,5.500E+01,6.000E+01,6.500E+01,7.000E+01,8.000E+01,
      *      9.000E+01,1.000E+02,1.300E+02,1.600E+02,2.000E+02/
C
C     PISC - PROTON DAMAGE COEFFICIENTS  (SHORT-CIRCUIT CURRENT)
C
C     0. GM/CM2 COVER GLASS DAMAGE COEFFICIENTS
C
      DATA (PISC(I),I=  1,77)
      *     /0.000E+00,1.885E-01,2.186E+00,7.022E+00,1.246E+01,1.701E+01,
      *      2.066E+01,2.349E+01,2.560E+01,2.806E+01,2.887E+01,2.877E+01,
      *      2.821E+01,1.997E+01,1.477E+01,1.221E+01,7.742E+00,5.749E+00,
      *      4.610E+00,3.851E+00,3.560E+00,3.315E+00,2.914E+00,2.597E+00,
      *      2.347E+00,2.139E+00,1.961E+00,1.816E+00,1.683E+00,1.596E+00,
      *      1.503E+00,1.413E+00,1.342E+00,1.273E+00,1.233E+00,1.157E+00,
      *      1.110E+00,1.063E+00,1.025E+00,9.450E-01,8.838E-01,8.269E-01,
      *      7.792E-01,7.318E-01,6.928E-01,6.594E-01,6.262E-01,5.586E-01,
      *      4.911E-01,4.613E-01,4.226E-01,3.931E-01,3.686E-01,3.441E-01,
      *      3.246E-01,2.902E-01,2.656E-01,2.459E-01,2.264E-01,2.165E-01,
      *      2.018E-01,1.895E-01,1.747E-01,1.624E-01,1.501E-01,1.403E-01,
      *      1.329E-01,1.230E-01,1.181E-01,1.107E-01,1.073E-01,9.843E-02,
      *      9.203E-02,8.760E-02,7.776E-02,7.087E-02,6.644E-02/
C
C     0.00559 GM/CM2 COVER GLASS DAMAGE COEFFICIENTS
C
      DATA (PISC(I),I= 78,154)
      *     /0.000E+00,0.000E+00,0.000E+00,0.000E+00,0.000E+00,0.000E+00,
      *      0.000E+00,0.000E+00,0.000E+00,0.000E+00,0.000E+00,0.000E+00,
      *      0.000E+00,0.000E+00,0.000E+00,0.000E+00,0.000E+00,0.000E+00,
      *      0.000E+00,0.000E+00,8.283E-01,3.344E+00,4.011E+00,3.695E+00,
      *      3.384E+00,3.070E+00,2.784E+00,2.538E+00,2.325E+00,2.138E+00,
      *      1.975E+00,1.837E+00,1.722E+00,1.620E+00,1.521E+00,1.439E+00,
      *      1.370E+00,1.298E+00,1.227E+00,1.123E+00,1.028E+00,9.618E-01,
      *      8.899E-01,8.381E-01,7.731E-01,7.402E-01,6.977E-01,6.011E-01,
      *      5.594E-01,4.922E-01,4.427E-01,4.188E-01,3.847E-01,3.551E-01,
      *      3.431E-01,3.018E-01,2.736E-01,2.507E-01,2.304E-01,2.186E-01,
      *      2.043E-01,1.916E-01,1.758E-01,1.633E-01,1.509E-01,1.409E-01,
      *      1.333E-01,1.234E-01,1.183E-01,1.110E-01,1.074E-01,9.856E-02,
      *      9.212E-02,8.766E-02,7.779E-02,7.089E-02,6.645E-02/
C
C     0.0168 GM/CM2 COVER GLASS DAMAGE COEFFICIENTS
C
      DATA (PISC(I),I=155,231)
      *     /0.000E+00,0.000E+00,0.000E+00,0.000E+00,0.000E+00,0.000E+00,
      *      0.000E+00,0.000E+00,0.000E+00,0.000E+00,0.000E+00,0.000E+00,
      *      0.000E+00,0.000E+00,0.000E+00,0.000E+00,0.000E+00,0.000E+00,
      *      0.000E+00,0.000E+00,0.000E+00,0.000E+00,0.000E+00,0.000E+00,
      *      0.000E+00,0.000E+00,0.000E+00,8.575E-01,1.604E+00,2.063E+00,
      *      2.039E+00,1.973E+00,1.899E+00,1.811E+00,1.724E+00,1.635E+00,
      *      1.564E+00,1.487E+00,1.414E+00,1.298E+00,1.188E+00,1.095E+00,
      *      1.007E+00,9.438E-01,8.831E-01,8.238E-01,7.624E-01,6.640E-01,
      *      6.148E-01,5.299E-01,4.696E-01,4.347E-01,3.989E-01,3.799E-01,
      *      3.526E-01,3.115E-01,2.851E-01,2.577E-01,2.412E-01,2.242E-01,
      *      2.092E-01,2.006E-01,1.801E-01,1.653E-01,1.527E-01,1.422E-01,
      *      1.342E-01,1.243E-01,1.187E-01,1.115E-01,1.077E-01,9.883E-02,
      *      9.230E-02,8.778E-02,7.785E-02,7.093E-02,6.646E-02/
C
```

Table A-10. Equivalent Fluence Program for GaAs Solar Cells (Cont'd)

```
C      0.0335 GM/CM2 COVER GLASS DAMAGE COEFFICIENTS
C
       DATA (PISC(I),I=232,308)
     *   /0.000E+00,0.000E+00,0.000E+00,0.000E+00,0.000E+00,0.000E+00,
     *    0.000E+00,0.000E+00,0.000E+00,0.000E+00,0.000E+00,0.000E+00,
     *    0.000E+00,0.000E+00,0.000E+00,0.000E+00,0.000E+00,0.000E+00,
     *    0.000E+00,0.000E+00,0.000E+00,0.000E+00,0.000E+00,0.000E+00,
     *    0.000E+00,0.000E+00,0.000E+00,0.000E+00,0.000E+00,0.000E+00,
     *    0.000E+00,0.000E+00,0.000E+00,0.000E+00,8.558E-01,1.134E+00,
     *    1.160E+00,1.153E+00,1.175E+00,1.298E+00,1.230E+00,1.158E+00,
     *    1.086E+00,1.017E+00,9.575E-01,8.971E-01,8.465E-01,7.347E-01,
     *    6.800E-01,5.909E-01,5.055E-01,4.642E-01,4.295E-01,4.067E-01,
     *    3.722E-01,3.263E-01,2.913E-01,2.715E-01,2.439E-01,2.301E-01,
     *    2.133E-01,2.016E-01,1.882E-01,1.683E-01,1.535E-01,1.462E-01,
     *    1.369E-01,1.264E-01,1.196E-01,1.124E-01,1.082E-01,9.926E-02,
     *    9.259E-02,8.796E-02,7.795E-02,7.099E-02,6.649E-02/
C
C      0.0671 GM/CM2 COVER GLASS DAMAGE COEFFICIENTS
C
       DATA (PISC(I),I=309,385)
     *   /0.000E+00,0.000E+00,0.000E+00,0.000E+00,0.000E+00,0.000E+00,
     *    0.000E+00,0.000E+00,0.000E+00,0.000E+00,0.000E+00,0.000E+00,
     *    0.000E+00,0.000E+00,0.000E+00,0.000E+00,0.000E+00,0.000E+00,
     *    0.000E+00,0.000E+00,0.000E+00,0.000E+00,0.000E+00,0.000E+00,
     *    0.000E+00,0.000E+00,0.000E+00,0.000E+00,0.000E+00,0.000E+00,
     *    0.000E+00,0.000E+00,0.000E+00,0.000E+00,0.000E+00,0.000E+00,
     *    0.000E+00,0.000E+00,0.000E+00,0.000E+00,0.000E+00,5.747E-01,
     *    6.661E-01,7.922E-01,8.016E-01,8.778E-01,8.545E-01,7.787E-01,
     *    7.238E-01,6.401E-01,5.652E-01,5.124E-01,4.701E-01,4.206E-01,
     *    3.914E-01,3.531E-01,3.085E-01,2.827E-01,2.538E-01,2.349E-01,
     *    2.258E-01,2.126E-01,1.912E-01,1.736E-01,1.568E-01,1.470E-01,
     *    1.370E-01,1.285E-01,1.209E-01,1.165E-01,1.104E-01,1.008E-01,
     *    9.326E-02,8.839E-02,7.816E-02,7.112E-02,6.654E-02/
C
C      0.112 GM/CM2 COVER GLASS DAMAGE COEFFICIENTS
C
       DATA (PISC(I),I=386,462)
     *   /0.000E+00,0.000E+00,0.000E+00,0.000E+00,0.000E+00,0.000E+00,
     *    0.000E+00,0.000E+00,0.000E+00,0.000E+00,0.000E+00,0.000E+00,
     *    0.000E+00,0.000E+00,0.000E+00,0.000E+00,0.000E+00,0.000E+00,
     *    0.000E+00,0.000E+00,0.000E+00,0.000E+00,0.000E+00,0.000E+00,
     *    0.000E+00,0.000E+00,0.000E+00,0.000E+00,0.000E+00,0.000E+00,
     *    0.000E+00,0.000E+00,0.000E+00,0.000E+00,0.000E+00,0.000E+00,
     *    0.000E+00,0.000E+00,0.000E+00,0.000E+00,0.000E+00,0.000E+00,
     *    0.000E+00,0.000E+00,0.000E+00,0.000E+00,2.305E-01,7.459E-01,
     *    7.284E-01,6.329E-01,5.839E-01,5.359E-01,4.918E-01,4.570E-01,
     *    4.230E-01,3.709E-01,3.242E-01,3.012E-01,2.643E-01,2.415E-01,
     *    2.262E-01,2.162E-01,1.929E-01,1.718E-01,1.625E-01,1.504E-01,
     *    1.382E-01,1.290E-01,1.230E-01,1.149E-01,1.115E-01,1.003E-01,
     *    9.529E-02,8.961E-02,7.846E-02,7.129E-02,6.661E-02/
C
C      0.1675 GM/CM2 COVER GLASS DAMAGE COEFFICIENTS
C
       DATA (PISC(I),I=463,539)
     *   /0.000E+00,0.000E+00,0.000E+00,0.000E+00,0.000E+00,0.000E+00,
     *    0.000E+00,0.000E+00,0.000E+00,0.000E+00,0.000E+00,0.000E+00,
     *    0.000E+00,0.000E+00,0.000E+00,0.000E+00,0.000E+00,0.000E+00,
     *    0.000E+00,0.000E+00,0.000E+00,0.000E+00,0.000E+00,0.000E+00,
     *    0.000E+00,0.000E+00,0.000E+00,0.000E+00,0.000E+00,0.000E+00,
     *    0.000E+00,0.000E+00,0.000E+00,0.000E+00,0.000E+00,0.000E+00,
     *    0.000E+00,0.000E+00,0.000E+00,0.000E+00,0.000E+00,0.000E+00,
```

Table A-10. Equivalent Fluence Program for GaAs Solar Cells (Cont'd)

```
     *    0.000E+00,0.000E+00,0.000E+00,0.000E+00,0.000E+00,0.000E+00,
     *    0.000E+00,3.661E-01,4.471E-01,5.136E-01,4.895E-01,4.636E-01,
     *    4.316E-01,3.827E-01,3.417E-01,3.108E-01,2.753E-01,2.545E-01,
     *    2.356E-01,2.228E-01,1.972E-01,1.777E-01,1.642E-01,1.518E-01,
     *    1.407E-01,1.308E-01,1.232E-01,1.178E-01,1.102E-01,1.017E-01,
     *    9.541E-02,8.880E-02,7.893E-02,7.153E-02,6.671E-02/
C
C     0.335 GM/CM2 COVER GLASS DAMAGE COEFFICIENTS
C
      DATA (PISC(I),I=540,616)
     *    /0.000E+00,0.000E+00,0.000E+00,0.000E+00,0.000E+00,0.000E+00,
     *    0.000E+00,0.000E+00,0.000E+00,0.000E+00,0.000E+00,0.000E+00,
     *    0.000E+00,0.000E+00,0.000E+00,0.000E+00,0.000E+00,0.000E+00,
     *    0.000E+00,0.000E+00,0.000E+00,0.000E+00,0.000E+00,0.000E+00,
     *    0.000E+00,0.000E+00,0.000E+00,0.000E+00,0.000E+00,0.000E+00,
     *    0.000E+00,0.000E+00,0.000E+00,0.000E+00,0.000E+00,0.000E+00,
     *    0.000E+00,0.000E+00,0.000E+00,0.000E+00,0.000E+00,0.000E+00,
     *    0.000E+00,0.000E+00,0.000E+00,0.000E+00,0.000E+00,0.000E+00,
     *    0.000E+00,0.000E+00,0.000E+00,0.000E+00,0.000E+00,8.524E-02,
     *    2.732E-01,3.019E-01,3.367E-01,3.115E-01,2.895E-01,2.683E-01,
     *    2.454E-01,2.294E-01,2.067E-01,1.887E-01,1.685E-01,1.607E-01,
     *    1.440E-01,1.348E-01,1.266E-01,1.207E-01,1.117E-01,1.036E-01,
     *    9.542E-02,9.020E-02,8.036E-02,7.375E-02,6.707E-02/
C
C     PVOC - PROTON DAMAGE COEFFICIENTS (VOC AND PMAX)
C
C     0. GM/CM2 COVER GLASS DAMAGE COEFFICIENTS
C
      DATA (PVOC(I),I=  1,77)
     *    /0.000E+00,3.504E-02,1.900E-01,5.793E-01,1.323E+00,2.376E+00,
     *    3.512E+00,4.631E+00,5.690E+00,7.508E+00,8.966E+00,1.013E+01,
     *    1.104E+01,1.283E+01,1.123E+01,9.781E+00,6.271E+00,4.608E+00,
     *    3.647E+00,3.034E+00,2.807E+00,2.611E+00,2.305E+00,2.070E+00,
     *    1.890E+00,1.735E+00,1.632E+00,1.527E+00,1.442E+00,1.373E+00,
     *    1.294E+00,1.230E+00,1.168E+00,1.123E+00,1.072E+00,1.028E+00,
     *    9.870E-01,9.431E-01,9.120E-01,8.527E-01,7.880E-01,7.209E-01,
     *    6.966E-01,6.512E-01,6.263E-01,5.946E-01,5.764E-01,5.246E-01,
     *    4.856E-01,4.578E-01,4.397E-01,4.249E-01,4.101E-01,4.002E-01,
     *    3.924E-01,3.728E-01,3.629E-01,3.530E-01,3.445E-01,3.396E-01,
     *    3.347E-01,3.297E-01,3.199E-01,3.150E-01,3.100E-01,3.051E-01,
     *    3.002E-01,2.953E-01,2.928E-01,2.904E-01,2.854E-01,2.830E-01,
     *    2.766E-01,2.756E-01,2.667E-01,2.658E-01,2.584E-01/
C
C     0.00559 GM/CM2 COVER GLASS DAMAGE COEFFICIENTS
C
      DATA (PVOC(I),I= 78,154)
     *    /0.000E+00,0.000E+00,0.000E+00,0.000E+00,0.000E+00,0.000E+00,
     *    0.000E+00,0.000E+00,0.000E+00,0.000E+00,0.000E+00,0.000E+00,
     *    0.000E+00,0.000E+00,0.000E+00,0.000E+00,0.000E+00,0.000E+00,
     *    0.000E+00,0.000E+00,1.954E-01,2.227E+00,2.866E+00,2.771E+00,
     *    2.553E+00,2.334E+00,2.139E+00,1.973E+00,1.825E+00,1.705E+00,
     *    1.598E+00,1.505E+00,1.423E+00,1.346E+00,1.273E+00,1.212E+00,
     *    1.159E+00,1.109E+00,1.064E+00,9.783E-01,9.081E-01,8.357E-01,
     *    7.743E-01,7.290E-01,6.838E-01,6.469E-01,6.241E-01,5.557E-01,
     *    5.278E-01,4.808E-01,4.488E-01,4.327E-01,4.160E-01,4.030E-01,
     *    4.051E-01,3.771E-01,3.660E-01,3.555E-01,3.454E-01,3.394E-01,
     *    3.343E-01,3.306E-01,3.206E-01,3.153E-01,3.104E-01,3.054E-01,
     *    3.005E-01,2.955E-01,2.929E-01,2.905E-01,2.856E-01,2.830E-01,
     *    2.767E-01,2.756E-01,2.668E-01,2.658E-01,2.584E-01/
C
C     0.0168 GM/CM2 COVER GLASS DAMAGE COEFFICIENTS
```

Table A-10. Equivalent Fluence Program for GaAs Solar Cells (Cont'd)

```
C
        DATA (PVOC(I),I=155,231)
   *    /0.000E+00,0.000E+00,0.000E+00,0.000E+00,0.000E+00,0.000E+00,
   *     0.000E+00,0.000E+00,0.000E+00,0.000E+00,0.000E+00,0.000E+00,
   *     0.000E+00,0.000E+00,0.000E+00,0.000E+00,0.000E+00,0.000E+00,
   *     0.000E+00,0.000E+00,0.000E+00,0.000E+00,0.000E+00,0.000E+00,
   *     0.000E+00,0.000E+00,0.000E+00,4.618E-01,1.306E+00,1.544E+00,
   *     1.528E+00,1.507E+00,1.468E+00,1.420E+00,1.370E+00,1.321E+00,
   *     1.269E+00,1.223E+00,1.169E+00,1.084E+00,1.006E+00,9.363E-01,
   *     8.714E-01,8.104E-01,7.615E-01,7.141E-01,6.746E-01,5.987E-01,
   *     5.587E-01,5.039E-01,4.608E-01,4.364E-01,4.213E-01,4.091E-01,
   *     4.050E-01,3.778E-01,3.733E-01,3.581E-01,3.461E-01,3.405E-01,
   *     3.348E-01,3.347E-01,3.213E-01,3.150E-01,3.111E-01,3.061E-01,
   *     3.011E-01,2.959E-01,2.931E-01,2.906E-01,2.860E-01,2.831E-01,
   *     2.769E-01,2.756E-01,2.668E-01,2.658E-01,2.584E-01/
C
C       0.0335 GM/CM2 COVER GLASS DAMAGE COEFFICIENTS
C
        DATA (PVOC(I),I=232,308)
   *    /0.000E+00,0.000E+00,0.000E+00,0.000E+00,0.000E+00,0.000E+00,
   *     0.000E+00,0.000E+00,0.000E+00,0.000E+00,0.000E+00,0.000E+00,
   *     0.000E+00,0.000E+00,0.000E+00,0.000E+00,0.000E+00,0.000E+00,
   *     0.000E+00,0.000E+00,0.000E+00,0.000E+00,0.000E+00,0.000E+00,
   *     0.000E+00,0.000E+00,0.000E+00,0.000E+00,0.000E+00,0.000E+00,
   *     0.000E+00,0.000E+00,0.000E+00,0.000E+00,5.910E-01,8.959E-01,
   *     9.438E-01,9.566E-01,9.768E-01,1.034E+00,1.002E+00,9.588E-01,
   *     9.109E-01,8.622E-01,8.161E-01,7.704E-01,7.266E-01,6.423E-01,
   *     5.979E-01,5.371E-01,4.808E-01,4.539E-01,4.350E-01,4.171E-01,
   *     3.995E-01,3.778E-01,3.681E-01,3.517E-01,3.470E-01,3.376E-01,
   *     3.323E-01,3.325E-01,3.223E-01,3.149E-01,3.095E-01,3.064E-01,
   *     3.022E-01,2.967E-01,2.936E-01,2.910E-01,2.865E-01,2.832E-01,
   *     2.771E-01,2.757E-01,2.669E-01,2.658E-01,2.585E-01/
C
C       0.0671 GM/CM2 COVER GLASS DAMAGE COEFFICIENTS
C
        DATA (PVOC(I),I=309,385)
   *    /0.000E+00,0.000E+00,0.000E+00,0.000E+00,0.000E+00,0.000E+00,
   *     0.000E+00,0.000E+00,0.000E+00,0.000E+00,0.000E+00,0.000E+00,
   *     0.000E+00,0.000E+00,0.000E+00,0.000E+00,0.000E+00,0.000E+00,
   *     0.000E+00,0.000E+00,0.000E+00,0.000E+00,0.000E+00,0.000E+00,
   *     0.000E+00,0.000E+00,0.000E+00,0.000E+00,0.000E+00,0.000E+00,
   *     0.000E+00,0.000E+00,0.000E+00,0.000E+00,0.000E+00,0.000E+00,
   *     0.000E+00,0.000E+00,0.000E+00,0.000E+00,0.000E+00,3.840E-01,
   *     5.509E-01,6.679E-01,6.843E-01,7.207E-01,7.110E-01,6.603E-01,
   *     6.216E-01,5.580E-01,5.104E-01,4.698E-01,4.442E-01,4.177E-01,
   *     4.014E-01,3.871E-01,3.598E-01,3.586E-01,3.394E-01,3.366E-01,
   *     3.300E-01,3.262E-01,3.268E-01,3.128E-01,3.064E-01,3.054E-01,
   *     3.014E-01,2.955E-01,2.919E-01,2.914E-01,2.876E-01,2.837E-01,
   *     2.777E-01,2.758E-01,2.671E-01,2.658E-01,2.585E-01/
C
C       0.112 GM/CM2 COVER GLASS DAMAGE COEFFICIENTS
C
        DATA (PVOC(I),I=386,462)
   *    /0.000E+00,0.000E+00,0.000E+00,0.000E+00,0.000E+00,0.000E+00,
   *     0.000E+00,0.000E+00,0.000E+00,0.000E+00,0.000E+00,0.000E+00,
   *     0.000E+00,0.000E+00,0.000E+00,0.000E+00,0.000E+00,0.000E+00,
   *     0.000E+00,0.000E+00,0.000E+00,0.000E+00,0.000E+00,0.000E+00,
   *     0.000E+00,0.000E+00,0.000E+00,0.000E+00,0.000E+00,0.000E+00,
   *     0.000E+00,0.000E+00,0.000E+00,0.000E+00,0.000E+00,0.000E+00,
   *     0.000E+00,0.000E+00,0.000E+00,0.000E+00,0.000E+00,0.000E+00,
   *     0.000E+00,0.000E+00,0.000E+00,0.000E+00,1.844E-01,6.049E-01,
```

Table A-10. Equivalent Fluence Program for GaAs Solar Cells (Cont'd)

```
    *      6.348E-01,5.368E-01,5.037E-01,4.738E-01,4.452E-01,4.214E-01,
    *      4.042E-01,3.834E-01,3.553E-01,3.430E-01,3.332E-01,3.245E-01,
    *      3.201E-01,3.197E-01,3.188E-01,3.070E-01,3.082E-01,3.039E-01,
    *      2.985E-01,2.921E-01,2.943E-01,2.905E-01,2.862E-01,2.817E-01,
    *      2.787E-01,2.762E-01,2.674E-01,2.658E-01,2.587E-01/
C
C    0.1675 GM/CM2 COVER GLASS DAMAGE COEFFICIENTS
C
     DATA (PVOC(I),I=463,539)
    *    /0.000E+00,0.000E+00,0.000E+00,0.000E+00,0.000E+00,0.000E+00,
    *      0.000E+00,0.000E+00,0.000E+00,0.000E+00,0.000E+00,0.000E+00,
    *      0.000E+00,0.000E+00,0.000E+00,0.000E+00,0.000E+00,0.000E+00,
    *      0.000E+00,0.000E+00,0.000E+00,0.000E+00,0.000E+00,0.000E+00,
    *      0.000E+00,0.000E+00,0.000E+00,0.000E+00,0.000E+00,0.000E+00,
    *      0.000E+00,0.000E+00,0.000E+00,0.000E+00,0.000E+00,0.000E+00,
    *      0.000E+00,0.000E+00,0.000E+00,0.000E+00,0.000E+00,0.000E+00,
    *      0.000E+00,0.000E+00,0.000E+00,0.000E+00,0.000E+00,0.000E+00,
    *      0.000E+00,3.042E-01,3.847E-01,4.361E-01,4.244E-01,4.091E-01,
    *      3.941E-01,3.712E-01,3.459E-01,3.378E-01,3.319E-01,3.195E-01,
    *      3.110E-01,3.123E-01,3.126E-01,3.007E-01,2.989E-01,2.989E-01,
    *      2.951E-01,2.891E-01,2.906E-01,2.865E-01,2.825E-01,2.827E-01,
    *      2.769E-01,2.741E-01,2.678E-01,2.659E-01,2.588E-01/
C
C    0.335 GM/CM2 COVER GLASS DAMAGE COEFFICIENTS
C
     DATA (PVOC(I),I=540,616)
    *    /0.000E+00,0.000E+00,0.000E+00,0.000E+00,0.000E+00,0.000E+00,
    *      0.000E+00,0.000E+00,0.000E+00,0.000E+00,0.000E+00,0.000E+00,
    *      0.000E+00,0.000E+00,0.000E+00,0.000E+00,0.000E+00,0.000E+00,
    *      0.000E+00,0.000E+00,0.000E+00,0.000E+00,0.000E+00,0.000E+00,
    *      0.000E+00,0.000E+00,0.000E+00,0.000E+00,0.000E+00,0.000E+00,
    *      0.000E+00,0.000E+00,0.000E+00,0.000E+00,0.000E+00,0.000E+00,
    *      0.000E+00,0.000E+00,0.000E+00,0.000E+00,0.000E+00,0.000E+00,
    *      0.000E+00,0.000E+00,0.000E+00,0.000E+00,0.000E+00,0.000E+00,
    *      0.000E+00,0.000E+00,0.000E+00,0.000E+00,0.000E+00,7.178E-02,
    *      2.328E-01,2.652E-01,3.028E-01,3.005E-01,2.995E-01,2.964E-01,
    *      2.896E-01,2.851E-01,2.827E-01,2.832E-01,2.840E-01,2.830E-01,
    *      2.804E-01,2.827E-01,2.835E-01,2.793E-01,2.762E-01,2.748E-01,
    *      2.748E-01,2.706E-01,2.664E-01,2.654E-01,2.593E-01/
     END
```

www.ingramcontent.com/pod-product-compliance
Lightning Source LLC
Chambersburg PA
CBHW081718170526
45167CB00009B/3623